# Topics in Applied Physics
# Volume 82

## Available Online

Topics in Applied Physics is part of the Springer LINK service. For all customers with standing orders for Topics in Applied Physics we offer the full text in electronic form via LINK free of charge. Please contact your librarian who can receive a password for free access to the full articles by registration at:

http://link.springer.de/orders/index.htm

If you do not have a standing order you can nevertheless browse through the table of contents of the volumes and the abstracts of each article at:

http://link.springer.de/series/tap/

There you will also find more information about the series.

## Springer

*Berlin*
*Heidelberg*
*New York*
*Barcelona*
*Hong Kong*
*London*
*Milan*
*Paris*
*Tokyo*

**Physics and Astronomy**

ONLINE LIBRARY

http://www.springer.de/phys/

# Topics in Applied Physics

Topics in Applied Physics is a well-established series of review books, each of which presents a comprehensive survey of a selected topic within the broad area of applied physics. Edited and written by leading research scientists in the field concerned, each volume contains review contributions covering the various aspects of the topic. Together these provide an overview of the state of the art in the respective field, extending from an introduction to the subject right up to the frontiers of contemporary research.

Topics in Applied Physics is addressed to all scientists at universities and in industry who wish to obtain an overview and to keep abreast of advances in applied physics. The series also provides easy but comprehensive access to the fields for newcomers starting research.

Contributions are specially commissioned. The Managing Editors are open to any suggestions for topics coming from the community of applied physicists no matter what the field and encourage prospective editors to approach them with ideas.

See also: http://www.springer.de/phys/books/tap/

## Managing Editors

### Dr. Claus E. Ascheron

Springer-Verlag Heidelberg
Topics in Applied Physics
Tiergartenstr. 17
69121 Heidelberg
Germany
Email: ascheron@springer.de

### Dr. Hans J. Kölsch

Springer-Verlag Heidelberg
Topics in Applied Physics
Tiergartenstr. 17
69121 Heidelberg
Germany
Email: koelsch@springer.de

## Assistant Editor

### Dr. Werner Skolaut

Springer-Verlag Heidelberg
Topics in Applied Physics
Tiergartenstr. 17
69121 Heidelberg
Germany
Email: skolaut@springer.de

Vladimir M. Shalaev (Ed.)

# Optical Properties of Nanostructured Random Media

With 185 Figures

Springer

Prof. Vladimir M. Shalaev
School of Electronical
and Computer Engineering
Purdue University
West Lafayette, IN 47907-1285
USA

shalaev@purdue.edu

Library of Congress Cataloging-in-Publication Data

Optical properties of nanostructured random media / Vladimir M. Shalaev (ed.).
    p. cm. -- (Topics in applied physics, ISSN 0303-4216 ; v. 82)
    Includes bibliographical references and index.
    ISBN 3540420312 (alk. paper)
    1. Nanostructure materials. 2. Nonlinear optics. I. Shalaev, Vladimir M., 1957- II.
Series.

TA418.9.N35 O395 2002
620.1'1--dc21

                                                                    2001042603

Physics and Astronomy Classification Scheme (PACS):
42.65.-k, 42.70.-a, 73.20.Mf, 78.30.Ly, 78.66.-w, 81.05.-t

ISSN print edition: 0303-4216
ISSN electronic edition: 1437-0859
ISBN 3-540-42031-2 Springer-Verlag Berlin Heidelberg New York

Springer-Verlag Berlin Heidelberg New York
a member of BertelsmannSpringer Science+Business Media GmbH

http://www.springer.de

© Springer-Verlag Berlin Heidelberg 2002
Printed in Germany

Typesetting: DA-TEX Gerd Blumenstein, Leipzig
Cover design: design & production GmbH, Heidelberg

Printed on acid-free paper    SPIN: 10778744    56/3141/mf    5 4 3 2 1 0

# Preface

The search for new materials is one of the defining characteristics of modern science and technology. Novel mechanical, electrical, magnetic, chemical, biological, and optical devices are often the result of the fabrication of new materials. Of specific interest to this book, recent advances in optical science and technology, such as the development of new lasers, detectors, and photonic devices, have relied heavily on advances in materials research.

This book is the result of collective efforts of leading experts in the field of nonlinear optics of nanostructured materials. The authors try to address, in particular, the fundamental problem of the way the symmetry of nanostructured materials affects their physical properties. In other words, given nanometer-size particles, what geometrical structure should be chosen for fabricating a material with desired properties. Typically, different symmetries are best suited for different applications. For some properties a periodic structure is ideal, whereas for others different types of symmetries, often random ones, would result in better performance of a material. Modern nanotechnology allows one to fabricate materials with almost any structure; this opens new avenues in engineering nanomaterials with desired properties.

Much of science is dominated by questions of symmetry. This is especially obvious in condensed matter physics where translational symmetry dominates both concept and language. And with justification; the elegance of symmetry arguments is so appealing that it tends to push aside many other issues. Yet there is a host of phenomena whose symmetries, if they exist at all, are hidden: the dynamics of a pile of sand; the growth through accretion of clusters such as soot particles and algal colonies; thin film growth and surface etching; the structures of cermets, porous media, globular polymers and proteins, randomly branched objects, and so on. These have always been items of fascination. However, their complexity had, in the past, forestalled the same level of deep dynamic and structural understanding as for crystals. All that is changing; interest in "disordered" systems is growing rapidly owing to the advent of powerful and plentiful computational resources which have dragged in their wake the theoretical innovations needed to truly understand these phenomena. Among the resulting insights are, ironically, the discovery of new symmetries: a seemingly unsymmetrical cluster, for example, might, possess dilation symmetry — when portions of it are magnified or reduced they look structurally similar to the whole. The fascination with nonlinear systems has even led to the insight that order is often a parametric accident of chaos.

The symmetry of a nanostructured material plays a key role in its properties. For example, periodically arranged dielectric nanoparticles in photonic band crystals, with certain types of defects, can be used for fabricating waveguides and microcavities with superb properties. On the other hand, in a number of applications, disordered nanomaterials may surpass their geometrically ordered counterparts. For example, random but statistically scale-invariant structures of metal nanoparticles, such as fractal aggregates of colloidal particles and percolation metal–dielectric films, permit achieving the largest local-field enhancements in a broad spectral range. This property is crucial for designing optical materials with the broadband amplification of nonlinear responses and for various types of spectroscopy.

Thus, although in many cases geometrically ordered nanostructured materials have superb performance, in some other cases, their disordered counterparts, or materials combining both ordered and disordered components, may have better properties. The optics of disordered nanomaterials displays a rich variety of effects some of which are hardly intuitive. Field localization of various sorts occur and recur in a wide gamut of disordered systems, most strikingly in those possessing dilation symmetry, leading to the enhancement of many optical phenomena, especially nonlinear processes. Making judicious use of these enhancement effects and of other aspects of the many complex resonances that distinguish these systems can lead to new and unexpected physics and to such applications as very low threshold lasers (whose cavities are self-organizing loops of coherently scattered events in highly unconventional media), superspectroscopy of single molecules and nanocrystals, and new classes of optical amplifiers and switchers. When developed, in the fullness of time, these disordered materials may attain a level of practical importance and versatility surpassing their geometrically ordered counterparts in certain areas of applications.

I need to say a few words as a guide to the contents of the book. The first four chapters (Sipe and Boyd; Bergman and Stroud; Ma et al.; Sibilia et al.) describe recent theoretical and experimental studies on optical nonlinearities of various composite materials. The most important message here is that nonlinear optical responses of a nanocomposite comprising several materials can significantly exceed the individual nonlinearities of the materials. The resultant enhancement of optical nonlinearities depends strongly on the symmetry of nanostructured composites. As mentioned, the dilation symmetry of fractal aggregates (Shalaev; Drachev et al.; Kim et al.) and percolation metal–dielectric composites (Sarychev and Shalaev; Gadenne and Rivoal) may result in particularly large optical nonlinearities in a very broad spectral range, from the near ultraviolet to the midinfrared. The enhanced local responses of random nanocomposites can be used for various types of spectroscopy, including

Raman and hyper-Raman spectroscopy of single molecules (Moskovits et al.; Kneipp et al.). Magneto-optical phenomena in ferromagnetic cermets are discussed in the chapter by Gadenne. Several chapters of this book address the problem of multiple light scattering in random media. The role of magnetic field in light scattering and propagation is considered in the chapter by van Tiggelen and Rikken . Random lasers with coherent feedback provided by multiple scattering are discussed in the Chapter by Cao. The Chapter by Bozhevolnyi describes scattering and localization of surface plasmon polaritons studied by scanning near-field optical microscopy. Multiple scattering phenomena in linear and nonlinear optical processes on rough metal films are reviewed in the Chapter by Leskova et al.. Altogether, the chapters review important recent advances in nonlinear optics of random media.

Finally, I would like to express my deep appreciation for the time and effort that the authors invested in writing their excellent chapters. I thank all the authors for their genuine interest in this publication and for their cooperation. I am also grateful to Viktor Podolskiy at New Mexico State University who helped with the editorial work.

West Lafayette, Indiana, USA
September 2001                                    *Vladimir M. Shalaev*

# Contents

## Linear and Nonlinear Optical Properties of Quasi-Periodic One-Dimensional Structures

Concita Sibilia, Mario Bertolotti, Marco Centini, Giuseppe D'Aguanno,
Michael Scalora, Mark J. Bloemer, and Charles M. Bowden ............. 63

## Optical Nonlinearities of Fractal Composites

Vladimir M. Shalaev .................................................. 93

## Nonlinear Optical Effects and Selective Photomodification of Colloidal Silver Aggregates

Vladimir P. Drachev, Sergey V. Perminov, Sergey G. Rautian,
and Vladimir P. Safonov ............................................. 113

## Fractal-Microcavity Composites: Giant Optical Responses

Won-Tae Kim, Vladimir P. Safonov, Vladimir P. Drachev,
Viktor A. Podolskiy, Vladimir M. Shalaev, and Robert L. Armstrong .. 149

## Theory of Nonlinear Optical Responses in Metal-Dielectric Composites

Andrey K. Sarychev and Vladimir M. Shalaev ......................... 169

## Surface-Plasmon-Enhanced Nonlinearities in Percolating 2-D Metal–Dielectric Films: Calculation of the Localized Giant Field and Their Observation in SNOM

Patrice Gadenne and Jean C. Rivoal ................................. 185

## Localization Phenomena in Elastic Surface Plasmon Polariton Scattering

## Multiple-Scattering Phenomena in the Second-Harmonic Generation of Light Reflected from and Transmitted Through Randomly Rough Metal Surfaces

# Nanocomposite Materials for Nonlinear Optics Based on Local Field Effects

John E. Sipe[1] and Robert W. Boyd[2]

[1] Dept. of Physics, University of Toronto
   Toronto, Ontario M5S 1A7, Canada
[2] Institute of Optics, University of Rochester
   Rochester, NY 14627, USA

**Abstract.** The intent of this article is to give an overview of composite materials for nonlinear optics. Our goal is to avoid the details of both the theorist's formalisms and the experimentalist's techniques and try to convey instead the spirit of the kind of work that has been done in this area. We will begin by identifying the class of composite materials we will discuss, reviewing their linear optical properties, and then discussing their nonlinear optical properties. Finally, we will briefly consider two new developments, the presence of optical bistability in composite materials and the importance of photonic band gap effects.

## 1   Introduction

In this article (see also *Gehr* and *Boyd* [1]), we are interested primarily in composite structures of the sort shown generically in Fig. 1, where any length scales that are present (here $a$ and $b$), are reasonably well-defined and all much less than the wavelength of light, $\lambda$. This condition guarantees that scattering due to the inhomogeneities resulting from the composite nature of the material will be negligible, and one is led to move to an effective medium picture: On length scales larger than $a$ or $b$, but still less than $\lambda$, the real composite material, with host dielectric constant $\epsilon_h$ and inclusion dielectric constant $\epsilon_i$, can for the purposes of calculating the propagation of light be replaced by a uniform effective medium with a dielectric constant $\epsilon_{\text{eff}}$. The goal is to relate

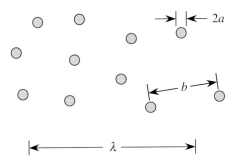

**Fig. 1.** Distance scales relevant to the analysis of nanocomposite optical materials

V. M. Shalaev (Ed.): Optical Properties of Nanostructured Random Media,
Topics Appl. Phys. **82**, 1–18 (2002)
© Springer-Verlag Berlin Heidelberg 2002

$\epsilon_{\text{eff}}$ to $\epsilon_h$ and $\epsilon_i$, and so also for the effective nonlinear response coefficients that will characterize the composite material. Though we will assume that $a$ and $b$ are much less than $\lambda$, we will also assume that they are typically much larger than the underlying interatomic spacing of the materials. Hence, except for perhaps small corrections, the use of expressions for the *macroscopic* dielectric constants $\epsilon_h$ and $\epsilon_i$ appropriate for uniform samples of the host and inclusion media is taken to be accurate. That notwithstanding, the description of the fields and geometry of the material at the level of $\epsilon_h$ and $\epsilon_i$ is typically referred to as the "mesoscopic" level of description, whereas the phrase "macroscopic" is reserved for description at the level of the effective medium. Thus, we can say that our goal is to understand the macroscopic linear and nonlinear optical properties of composite materials in terms of their mesoscopic optical properties. The restriction to (essentially) transparent constituent materials ensures that absorption losses will be negligible. The extension of the work we discuss here to instances where macroscopic physics *cannot* be used to describe the response of the constituent materials takes one into the study of the nonlinear optical properties of superlattices, quantum wells, and quantum dots; this is an interesting area of research that we will pass over in this review.

Maxwell Garnett                    Bruggeman                    Layered

**Fig. 2.** Composite material structures for use in nonlinear optics

One naturally expects that the relation of macroscopic optical properties to the mesoscopic will depend on the topology of the composite material. Three simple models are shown in Fig. 2. The first, with well-defined spherical inclusions in a host background, is called the Maxwell Garnett composite geometry, after the pioneering work by *Maxwell Garnett* [2] on colloidal metallic solutions. The second topology, still disordered but where the constituent materials are more or less interspersed, is referred to as the *Bruggeman* [3] geometry. Finally, one can consider an ordered, layered composite geometry. We stress that in this last geometry, as in all of them, the length scales are all assumed to be much less than the wavelength of light. Hence, interference effects in these multilayered geometry do not arise, and an effective medium description is completely adequate. Only in the last section do we relax this assumption and close with a few comments on phenomena that arise if the layer spacing is comparable to the wavelength of light.

# 2   Linear Optical Properties

Roughly speaking, there are two approaches to deriving the effective medium properties from those of the constituent media properties. The first approach, and in some sense the more straightforward, is to examine, at some level of approximation, the nature of the actual mesoscopic fields in the composite material and perform spatial averages over them to identify the values of the macroscopic fields [4,5]; expressions for the effective medium response coefficients then result, as we will see in examples below. The second approach relies on writing an expression for the internal energy of the composite material and comparing it with an expression for an effective medium [6]. The extraction of effective medium response coefficients can then be done at various levels of approximation.

Of these two methods, the second is the more formal and the more general, and with it the study of the optical properties of composite materials can be set in the context of the study of other ( e. g., acoustic) properties of composite materials. The advantage of the first is that in many instances the underlying physics is more easily grasped and the dependence of the effective medium properties on constituent properties is made clearer. Hence, we will use it in this review.

We begin with Maxwell Garnett geometry, where the topology naturally leads one to think of an analogy to the problem of determining the optical properties of an amorphous collection of atoms in vacuum. We will briefly review this classic problem of early twentieth-century physics, for the benefit of readers who might not have seen it since their graduate student days; it provides an easy introduction to the optical properties of composite materials.

Consider first an ordered lattice of atoms, as illustrated in Fig. 3, each supposed to be characterized by a polarizability $\alpha$ [7]. The atomic polarizability $\alpha$ relates the dipole moment $\mathbf{p}$ induced in a typical atom to the local electric field at the atomic site. But that local field differs from the macroscopic Maxwell electrical field $\mathbf{E}$ for two reasons. The first is that a lattice site is not a typical point in the medium but a very special one; the second is that the field of a given dipole moment $\mathbf{p}$ certainly contributes to the macroscopic Maxwell field but not, except for radiation reaction, to the field to which the dipole responds. These two effects are sometimes referred to as "local field corrections," and we can write schematically

$$\mathbf{p} = \alpha \left( \mathbf{E} + \text{local field corrections} \right) . \tag{1}$$

To estimate the second correction described above, we note that the integral of the microscopic electric field $\mathbf{e}$ over a sphere surrounding a charge distribution with a dipole moment $\mathbf{p}$ is given in the electrostatic limit by [8]

$$\int \mathbf{e} \, dV = -\frac{4\pi}{3} \mathbf{p} . \tag{2}$$

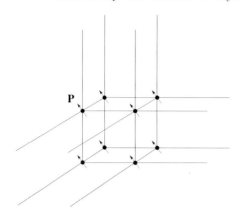

**Fig. 3.** An ordered array of polarizable point particles (adapted from Fig. 4.9 Jackson [7])

If the lattice consists of $N$ atoms in a volume $V$, one would then be led to estimate at least the second local–field correction by writing

$$\mathbf{p} = \alpha \left[ \mathbf{E} - \frac{\left( -\dfrac{4\pi}{3}\mathbf{P} \right)}{(V/N)} \right], \tag{3}$$

since $V/N$ is the volume associated with each atom in the lattice. More rigorous arguments show that, for a cubic lattice or an amorphous material [9], this procedure leads to the correct result without any further modification due to the first correction described above. Hence the dipole moment per unit volume, $\mathbf{P}$, is given by

$$\mathbf{P} = \frac{N}{V}\mathbf{p} = \frac{N}{V}\alpha \left( \mathbf{E} + \frac{4\pi}{3}\mathbf{P} \right). \tag{4}$$

Then writing $\mathbf{D} = \mathbf{E} + 4\pi\mathbf{P} \equiv \epsilon\mathbf{E}$, we find that the dielectric constant $\epsilon$ can be obtained by solving the so-called Clausius–Mossotti equation,

$$\frac{\epsilon - 1}{\epsilon + 2} = \frac{4\pi}{3}\frac{N}{V}\alpha, \tag{5}$$

also referred to as the Lorentz–Lorenz relation when the relation between the index of refraction $n$ and the dielectric constant, $\epsilon = n^2$, is used.

To apply arguments like this to the Maxwell Garnett topology, one first has to identify the effective polarizability of an inclusion sphere. Clearly, the polarization $\mathbf{P}_i$ within the inclusion is given by

$$\mathbf{P}_i = \frac{\epsilon_i - 1}{4\pi}\mathbf{E}_i, \tag{6}$$

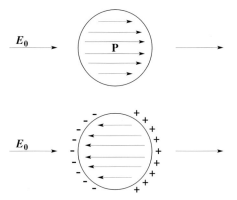

**Fig. 4.** An inclusion sphere in a host dielectric; the field inside the sphere differs from $E_0$ due to the field from the polarization charge

where $\mathbf{E}_i$ is the field within the inclusion. But $\mathbf{E}_i$ is not equal to the field $\mathbf{E}_0$ far from the inclusion (see Fig. 4) because of the depolarization field in the sphere due to $\mathbf{P}_i$ itself. An electrostatic calculation leads to the result [10]

$$\mathbf{P}_i = \frac{3}{4\pi} \frac{\epsilon_i - \epsilon_h}{\epsilon_i + 2\epsilon_h} \mathbf{E}_0 \, . \tag{7}$$

Thus, the dipole moment of the inclusion sphere within the host medium is

$$\mathbf{p}_i = \frac{4\pi}{3} a^3 \mathbf{P}_i = a^3 \frac{\epsilon_i - \epsilon_h}{\epsilon_i + 2\epsilon_h} \mathbf{E}_0 \, , \tag{8}$$

where $a$ is the radius of the sphere, and we can identify an effective polarizability as

$$\alpha = a^3 \frac{\epsilon_i - \epsilon_h}{\epsilon_i + 2\epsilon_h} \, . \tag{9}$$

The actual field lines in the neighborhood of the sphere are shown in Fig. 5, when $\epsilon_i > \epsilon_h$. This result illustrates the general feature that electrostatic fields are concentrated in the host region near a spherical inclusion if $\epsilon_i > \epsilon_h$. We will see later that this has important consequences for nonlinear optical properties.

To apply these results to determine the effective dielectric constant $\epsilon_{\mathrm{eff}}$ in the Maxwell Garnett topology, one would naturally use this expression for the polarizability and take into account the fact that the inclusion spheres are not in vacuum but in a host material with dielectric constant $\epsilon_h$, to write

$$\frac{\epsilon_{\mathrm{eff}} - \epsilon_h}{\epsilon_{\mathrm{eff}} + 2\epsilon_h} = \frac{4\pi}{3} \frac{N}{V} \left( a^3 \frac{\epsilon_i - \epsilon_h}{\epsilon_i + 2\epsilon_h} \right) \tag{10}$$

$$= f \frac{\epsilon_i - \epsilon_h}{\epsilon_i + 2\epsilon_h} ,$$

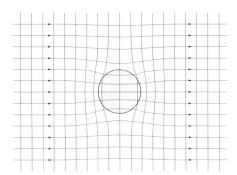

**Fig. 5.** Field lines and equipotential surfaces inside and outside a spherical inclusion particle, plotted for $\varepsilon_i > \varepsilon_h$

[*cf.* (5)], where $f = 4\pi N a^3 / V$ is the fill fraction of the inclusion. More rigorous arguments [4] justify this simple one; the result (10) is the Maxwell Garnett relation.

One can apply these same arguments to consider two (or more) inclusions with fill fractions $f_1$ and $f_2$ and dielectric constants $\epsilon_1$ and $\epsilon_2$, respectively, in a host medium. Not surprisingly, the result is

$$\frac{\epsilon_{\text{eff}} - \epsilon_{\text{h}}}{\epsilon_{\text{eff}} + 2\epsilon_{\text{h}}} = f_1 \frac{\epsilon_1 - \epsilon_{\text{h}}}{\epsilon_1 + 2\epsilon_{\text{h}}} + f_2 \frac{\epsilon_2 - \epsilon_{\text{h}}}{\epsilon_2 + 2\epsilon_{\text{h}}} . \tag{11}$$

This equation provides a convenient bridge to the Bruggeman topology. For there, the two constituents are to be thought of on an equal footing, each interspersed with the other. Thus, one might expect that the "host" medium should here be thought of as the effective medium itself, for it is only within this effective medium that each constituent can be properly thought of as embedded. So one would write eq. (11), with $\epsilon_{\text{h}} = \epsilon_{\text{eff}}$,

$$0 = f_1 \frac{\epsilon_1 - \epsilon_{\text{eff}}}{\epsilon_1 + 2\epsilon_{\text{eff}}} + f_2 \frac{\epsilon_2 - \epsilon_{\text{eff}}}{\epsilon_2 + 2\epsilon_{\text{eff}}} . \tag{12}$$

More sophisticated arguments confirm this result [11], which is the Bruggeman expression for the effective dielectric constant $\epsilon_{\text{eff}}$.

For a given material in the laboratory, it is not always clear whether the Maxwell Garnett or the Bruggeman topology is a better approximation of the mesostructure of the composite. Further, the careful reader will have noted even from the above simple, heuristic arguments leading to the Maxwell Garnett and Bruggeman expressions that these are essentially *mean field* results. Hence, even if the topology is well known, there are still clearly approximations inherent in (10) and (12). It is at least comforting that, in the limit $f_1 \ll f_2$, the Bruggeman result (12) reduces to the Maxwell Garnett expression (10), with $f = f_1$ and $\epsilon_{\text{h}} = \epsilon_2$. But for $f_1 \approx f_2$, typically *neither* the Maxwell Garnett expression (10) *nor* the Bruggeman expression (12) gives a particularly good description of the effective medium dielectric constant $\epsilon_{\text{eff}}$.

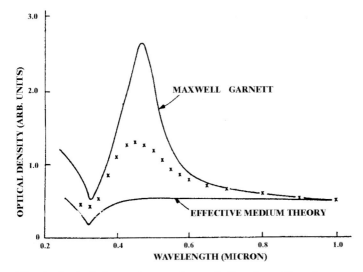

**Fig. 6.** Optical absorption spectrum of silver in a silica glass composite, compared with the predictions of the Maxwell Garnett and Bruggeman (labeled effective medium theory) models. After *Gittleman* and *Abeles* [12]

An example [12] is shown in Fig. 6, where the experimentally observed optical density of silver in a silica glass composite is shown. For metallic inclusions, the Maxwell Garnett expression leads to a surface plasmon resonance when $\epsilon_i + 2\epsilon_h$ is close to zero; no such resonance occurs in the Bruggeman expression. The experimental data shows, perhaps not surprisingly for a 39% fill fraction of silver, that the truth is somewhere in between. *Ping Sheng* [13] and others have developed mean field models that attempt to extrapolate between the Maxwell Garnett and Bruggeman topologies and have achieved some success. Of course, if the topology is well specified, it is possible nowadays to solve the Maxwell equations numerically for the composite material and determine the effective dielectric constant directly.

There is more certainty in the results for the effective dielectric constant of a layered medium than for that of a disordered one. In the layered case [5] (see Fig. 1), the continuity of the tangential component of the mesoscopic electric field leads immediately to an expression for the effective medium dielectric constant $\epsilon_\parallel$ appropriate for an electric field in the plane defined by the layers,

$$\epsilon_\parallel = f_a \epsilon_a + f_b \epsilon_b, \tag{13}$$

where the dielectric constants of the layers are denoted $\epsilon_a$ and $\epsilon_b$ and their fill fractions by $f_a$ and $f_b$, respectively. Similarly, the continuity of the normal component of the mesoscopic electric displacement leads to an expression for the effective medium dielectric constant $\epsilon_\perp$ appropriate for an electric field perpendicular to that plane,

$$\frac{1}{\epsilon_\perp} = \frac{f_a}{\epsilon_a} + \frac{f_b}{\epsilon_b}. \tag{14}$$

Here it is assumed that the constituent media themselves are isotropic at the mesoscopic level (at least with respect to their dielectric constants); the resulting composite material is obviously anisotropic at the macroscopic level. Interestingly, these results illustrate the two extremes for an effective dielectric constant that are identified by the *Wiener limits*; for a constituent with fill fraction $f_i$ and dielectric constant $\epsilon_i$, these rigorous bounds for the effective dielectric constant $\epsilon_{\rm eff}$ are given by [14]

$$\sum_i f_i \epsilon_i \geq \epsilon_{\rm eff} \geq \left[ \sum_i \frac{f_i}{\epsilon_i} \right]^{-1}. \tag{15}$$

## 3    Nonlinear Optical Properties

Now we turn to the nonlinear optical properties of a composite and consider first contributions to the nonlinear index of refraction. For fields

$$\mathbf{E}(\mathbf{r}) e^{-i\omega t} + \text{c.c.}, \tag{16}$$

in the Maxwell Garnett topology, we take the mesoscopic constitutive relation for inclusion particles as

$$\mathbf{D}_i = \epsilon_i \mathbf{E}_i + 4\pi \mathbf{P}_i^{\rm NL}, \tag{17}$$

where the nonlinear polarization $\mathbf{P}_i^{\rm NL}$ in the inclusion is given by [15]

$$\mathbf{P}_i^{\rm NL} = A_i (\mathbf{E}_i \cdot \mathbf{E}_i^*) \mathbf{E}_i + \frac{B_i}{2} (\mathbf{E}_i \cdot \mathbf{E}_i) \mathbf{E}_i^*, \tag{18}$$

where the $i$ denotes the inclusion and $A_i$ and $B_i$ are the usual constants specifying the nonlinear optical response of an isotropic medium. Similar expressions can be written for the host. We then seek an effective medium description of the nonlinear response of the composite; that is, at the macroscopic level we want to write

$$\mathbf{D} = \epsilon_{\rm eff} \mathbf{E} + 4\pi \mathbf{P}_{\rm eff}^{\rm NL}, \tag{19}$$

with

$$\mathbf{P}_{\rm eff}^{\rm NL} = A_{\rm eff} (\mathbf{E} \cdot \mathbf{E}^*) \mathbf{E} + \frac{B_{\rm eff}}{2} (\mathbf{E} \cdot \mathbf{E}) \mathbf{E}^*, \tag{20}$$

where $\epsilon_{\rm eff}$ is the effective dielectric constant discussed in the preceding section, $\mathbf{E}$ is the macrocopic Maxwell field, averaged over the inhomogeneities, and $A_{\text{eff}}$ and $B_{\text{eff}}$ characterize the nonlinear optical response of the composite as an effective medium; the goal is to find $A_{\text{eff}}$ and $B_{\text{eff}}$ in terms of the mesoscopic optical properties.

The simplest case is if nonlinearity need only be considered in the inclusions. Then the linear Maxwell Garnett analysis leads to the result that the linear field in the inclusion is uniform and given by

$$\mathbf{E}_i = \frac{\epsilon_{\text{eff}} + 2\epsilon_{\text{h}}}{\epsilon_i + 2\epsilon_{\text{h}}} \mathbf{E} . \tag{21}$$

If the nonlinearity is weak, this expression can be used as an approximation for the *full* electrical field in the composite in Eq. (18); completing the analysis and working out the expressions for $A_{\text{eff}}$ and $B_{\text{eff}}$, we find that

$$A_{\text{eff}} = f \left| \frac{\epsilon_{\text{eff}} + 2\epsilon_{\text{h}}}{\epsilon_i + 2\epsilon_{\text{h}}} \right|^2 \left( \frac{\epsilon_{\text{eff}} + 2\epsilon_{\text{h}}}{\epsilon_i + 2\epsilon_{\text{h}}} \right)^2 A_i , \tag{22}$$

$$B_{\text{eff}} = f \left| \frac{\epsilon_{\text{eff}} + 2\epsilon_{\text{h}}}{\epsilon_i + 2\epsilon_{\text{h}}} \right|^2 \left( \frac{\epsilon_{\text{eff}} + 2\epsilon_{\text{h}}}{\epsilon_i + 2\epsilon_{\text{h}}} \right)^2 B_i ,$$

where $\epsilon_{\text{eff}}$ is the Maxwell Garnett result (10). A small surprise, perhaps, is that *four* powers of $(\epsilon_{\text{eff}} + 2\epsilon_{\text{h}})/(\epsilon_i + 2\epsilon_{\text{h}})$ appear here. From (18), one might expect only *three* powers to appear. A detailed analysis shows that the fourth arises because the presence of the nonlinear polarization modifies the linear response of the composite itself [4].

The results predicted by (22) are graphed in Fig. 7 as a function of $f$ for various ratios of $\epsilon_i/\epsilon_{\text{h}}$; these Maxwell Garnett results are shown for all $f$, but should be trusted only for reasonably small $f$ where the Maxwell Garnett model has some chance of providing a good description of the optical response. We note that for $\epsilon_i/\epsilon_{\text{h}} < 1$ the effective nonlinearity rises faster as a function

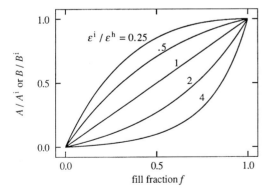

**Fig. 7.** Predicted enhancement in the nonlinear coefficient $A + \frac{1}{2}B$, which is proportional to $\chi^{(3)}$, as a function of the volume fill fraction of nonlinear material for nonlinear inclusion particles embedded in a linear host material. After *Sipe* and *Boyd* [4]

of $f$ than one would expect simply on the basis of increasing fill fraction of nonlinear material; this can be attributed to the effective concentration of electrical fields in regions where the dielectric constant is small, as indicated by Eq. (21). Still, the effective nonlinear optical constants $A_{eff}$, $B_{eff}$ are never predicted to be larger than those of pure inclusion material.

The situation is qualitatively different if the host medium is nonlinear and the index of the inclusion is greater than the host. Then, as is clear from Fig. 4, there will be a "dressing" region around each spherical inclusion where the linear electric field will be large; thus, there will be an enhanced nonlinear polarization induced in the nonlinear host material. The expressions for $A_{eff}$ and $B_{eff}$ are more complicated than (22); we refer the reader to the literature for them and present here only the results in Fig. 8 for the effective $A_{eff}/B_{eff}$. Note that for $\epsilon_i \gg \epsilon_h$, the addition of a small amount of *linear* material to a nonlinear host will lead to an *increase* in the effective nonlinearity of the composite!

Almost as striking is the fact that the *nature* of the nonlinear response can be changed as well. For example, if the nonlinear response in the host material is due to electrostriction ($B_h = 0$) or molecular orientation ($B_h = 6A_h$), a wide range of $B_{eff}/A_{eff}$ ratios can be achieved by adding linear inclusion material at reasonably large values of $\epsilon_i/\epsilon_h$ (see Fig. 9). Only in the limit of electronic nonlinear response in the low frequency limit ($B_h = A_h$) will the qualitative nature of the nonlinear response be independent of the addition of any inclusion material.

A particularly striking example of the ability to modify the qualitative nature of the nonlinear response was demonstrated experimentally by *Smith* et al. [16] who added gold colloids to a solvent containing the nonlinear molecule 1, 1', 3, 3, 3', 3'-hexamethylindotricarbocyanine iodide. Although the

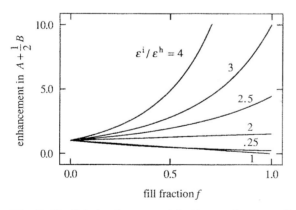

**Fig. 8.** Predicted enhancement in the nonlinear coefficient $A + \frac{1}{2}B$, which is proportional to $\chi^{(3)}$, as a function of the volume fill fraction of nonlinear material for linear inclusion particles embedded in a nonlinear host material. After *Sipe* and *Boyd* [4]

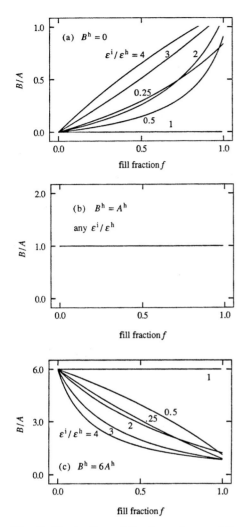

**Fig. 9.** Illustration of the dependence of the ratio $B/A$ on fill fraction of nonlinear material for three different situations. After *Sipe* and *Boyd* [4]

nonlinear absorption is positive for both the pure solvent and the gold colloids, a range of positive and negative nonlinear absorption coefficients could be obtained by varying the gold content. Some of their results are shown in Fig. 10.

In the presence of nonlinearity in both the host and the inclusions, expressions have been worked out for the effective coefficients $A_{\text{eff}}$ and $B_{\text{eff}}$. Extensions to other structures — i. e., Maxwell Garnett-like models with ellipsoids rather than spheres — and other nonlinear optical properties besides the nonlinear index of refraction have been considered by a number of authors

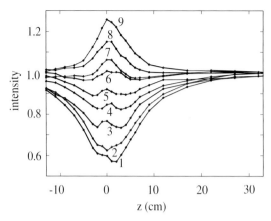

**Fig. 10.** Z-scan results for the imaginary part of $\chi^{(3)}$ for a composite material consisting of gold nanoparticles suspended in a dye solution. The curves are numbered in order of increasing gold concentration. Note that curve 6, which refers to a gold fill fraction of $1.3 \times 10^{-6}$, shows nearly complete cancellation between the two contributions to $\mathrm{Im}\chi^{(3)}$. After *Smith* et al. [16]

within the kind of mean field description outlined here, and numerical studies by *Zhang* and *Stroud* [17] and *Stockman* et al. [18] have considered the exact nonlinear optical response of specified topologies. For transparent materials of the sort that we consider here, these latter authors find good agreement with the Maxwell Garnett results quoted above in the limit that $f$ is small. For larger $f$, there can be significant deviations, particularly if the inclusion particles are metallic. For metallic inclusion particles, the kind of surface plasmon response that enhances the linear absorption (see Fig. 6) can lead to enhanced fluctuations in the local field from inclusion to inclusion even at small $f$, and significant deviations from the Maxwell Garnett results. But this takes us away from our regime of interest of transparent materials.

The very interesting Maxwell Garnett predictions for transparent constituents (Figs. 7–9) still await detailed experimental verification in the laboratory, in large part due to the difficulty of producing well-defined Maxwell Garnett topologies of constituents with significantly different dielectric constants. In the case of the Bruggeman geometry, the indications are that results somewhere between the "nonlinear inclusion" and "nonlinear host" limits of the Maxwell Garnett model should be found. This is as one might expect from an interdispersed structure, where the "dressing" of high electrical fields surrounding the high index element will be perturbed by neighboring high index elements, unlike the situation in Maxwell Garnett geometry with a small fill fraction where the "dressing" can extend well into the host. The nonlinear optical properties of composite materials of the Bruggeman geometry, in the form of nonlinear liquids impregnated into porous glass, have recently been reported by *Gehr* et al. [19].

Materials difficulties are a bit less of a limitation in a layered geometry structure. Here interesting effects arise in the response to an electrical field perpendicular to the layers, where the linear optical response is characterized by (14). The field within medium $a$ is easily found to be $E_a = (\epsilon_\perp/\epsilon_a)E$, where $E$ is the macroscopic field, and if medium $a$ is the nonlinear medium, there will thus be an enhancement factor [5]

$$\mathcal{L}(\omega) = \frac{\epsilon_\perp(\omega)}{\epsilon_a(\omega)} \tag{23}$$

that will enter for each frequency component. For the general four wave mixing susceptibility, one finds

$$\chi_{\text{eff}}^{(3)}(\omega_4 = \omega_1 + \omega_2 + \omega_3)$$
$$= f_a\chi_a^{(3)}(\omega_4 = \omega_1 + \omega_2 + \omega_3)\mathcal{L}(\omega_4)\mathcal{L}(\omega_3)\mathcal{L}(\omega_2)\mathcal{L}(\omega_1).$$

If $\epsilon_b > \epsilon_a$, the electrical field will be concentrated in medium $a$, and in fact the effective composite nonlinearity will be increased by *decreasing* the amount of nonlinear material over a wide range, until of course there is so little nonlinear material that there is essentially no nonlinear response (see Fig. 11). Experiments by *Fischer* et al. [20] confirmed this behavior for the nonlinear index of refraction ($\omega_1 = \omega_2 = -\omega_3 = \omega$), but for materials reasons, a ratio $\epsilon_b/\epsilon_a = 1.77$ was the largest that could be achieved; nonetheless, a 35% enhancement in the nonlinear optical response was observed. More striking results have been achieved by *Nelson* and *Boyd* [21] for the electro-optic effect ($\omega_1 = \omega, \omega_2 = \omega_3 = 0$), where the large static dielectric constants available can be used to achieve a large enhancement factor $\mathcal{L}(0)$. In a layered structure of AF-30/polycarbonate and BaTiO$_3$, an enhancement of the electro-optic effect of about 3 was achieved (see Fig. 13).

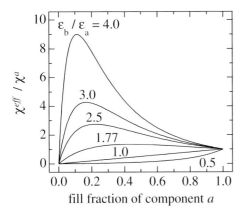

**Fig. 11.** Theoretically predicted enhancement of the third-order susceptibiliity of a layered composite material. After *Boyd* and *Sipe* [5]

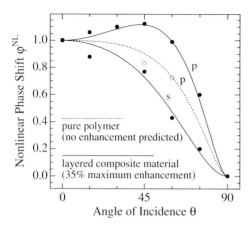

**Fig. 12.** Measurement of the enhanced nonlinear optical response of a layered composite optical materials. After *Fischer* et al. [20]

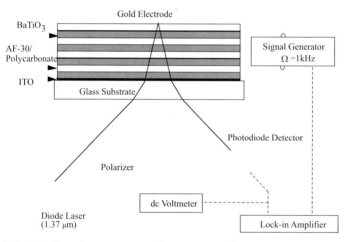

**Fig. 13.** Sample geometry and experimental arrangement used to study an electro-optic composite material. The measured nonlinear response is 3.2 times larger than that of the pure nonlinear material. After *Nelson* and *Boyd* [21]

## 4   Recent Advances

In the analytic expressions derived above, we have assumed throughout that the nonlinearity is sufficiently small that the fields in the nonlinear medium could be taken as those that result from the solution of the linear problem. But now suppose that one goes beyond this approximation. As a simple example, consider an ellipsoid in vacuum subject to an electrical field $E_0$

along one of its principal axes. The full equation that one would need to solve is

$$P = \chi^{(1)}E + 3\chi^{(3)}|E|^2 E \,, \tag{24}$$

where $\chi^{(1)}$ and $\chi^{(3)}$ characterize the nonlinear response of the medium making up the ellipsoid and $E$ is the field in the ellipsoid. It is related to the applied field $E_0$ by a relation of the form

$$E = E_0 + \gamma P \,, \tag{25}$$

where $\gamma$ describes the depolarization fields; for a sphere, for example, $\gamma = -4\pi/3$. Combining (24) and (25) we find an equation for $E$ alone,

$$E_0 = \left[ 1 - \gamma\chi^{(1)} - 3\gamma\chi^{(3)}|E|^2 E \right] \,, \tag{26}$$

which is cubic in $E$; thus it offers the possibility of more than one solution, and hence bistability. Note that the bistability arises through the "feedback effect" of the depolarization field (25), not because saturation or higher order terms in the nonlinear optical response are included.

This kind of scenario was considered theoretically by *Kalyani-walla* et al. [22], and later by *Bergman* et al. [23]. The former actually studied ellipsoids of CuCl coated with silver and predicted that bistability should occur for incident intensities on the order of $50\,\text{kW}/\text{cm}^2$. *Neuendorf* et al. [24] investigated just such a system experimentally and observed a signal indicative of bistability at about the expected incident intensity. Their experiments, however, are subject to thermal effects whose consequences have not been fully investigated.

Very recently, *Yoon* et al. [25] theoretically considered an even simpler geometry, a slab of nonlinear material subject to an incident electrical field with a component $E_0$ perpendicular to the slab. Here, because of the continuity of the normal component of the electric displacement, the particular form of the feedback equation (25) is just

$$E = E_0 - 4\pi P \,. \tag{27}$$

It was found that optical bistability should result if $\chi^{(3)}$ is nearly opposite $\epsilon$ in phase and if the incident field is large enough that $\left| \chi^{(3)} \right| |E_0|^2 \approx 10^{-3}|\epsilon|^3$. There appear to be a number of possibilities for observing optical bistability in this simple geometry.

Finally, we consider new directions involving photonic band gap materials. In a recent communication, *Bennink* et al. [26] proposed using the high $\chi^{(3)}$ of copper, usually inaccessible for optical switching because of the high reflectivity of copper. They showed theoretically that by interspersing thin layers of copper between thin films of silica on the order of the wavelength of light, they could engineer the structure to take advantage of the nonlinear $\chi^{(3)}$ of copper, while letting most of the incident energy proceed through the structure.

(a)

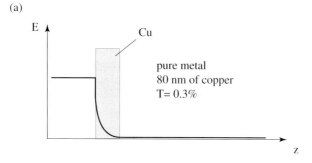

(b)

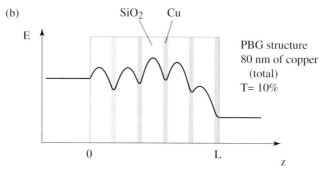

**Fig. 14.** Illustration of the proposal of *Bennink* et al. [26] to utilize metallic/dielectric photonic-band gap structures to access the large optical nonlinearities of metals. Part (**a**) shows that an incident wave is nearly completely attenuated in propagating through 80 nm of pure copper. However, if the same amount of metal is incorporated in a photonic bandgap structure part (**b**), it can allow good transmission and consequently a large nonlinear optical response

This idea is illustrated in Fig. 14. Adopting a value of $\chi^{(3)} = i(1.5 \times 10^{-6})$ esu for copper, they showed that a multilayer slab could be designed to yield an effective nonlinear index of refraction $n_{2,\mathrm{eff}}$, of $(3 + 6i) \times 10^{-9}\,\mathrm{cm}^2/\,\mathrm{W}$. This would lead to a large nonlinear phase change of $\Delta\phi \approx 5\pi/(\mathrm{GW}/\mathrm{cm}^2)$, with little absorption.

Such considerations lead naturally to the general field of nonlinear optical properties of photonic band gap materials. Of course, the periodicity of a photonic band gap (PBG) material is necessarily of a scale of the order of the wavelength of light. Nonetheless, perhaps somewhat surprisingly, an "effective medium" description of the composite material is still possible, though of a rather different nature than in the kinds of composites we have discussed in the earlier sections of this paper. In PBG materials the linear optical properties are characterized by the dispersion relations of the photonic Bloch functions, just as the single-electron properties of metals and semiconductors are described by the dispersion relations of the electronic Bloch functions. A pulse of light propagating through a PBG material near the band gap can be thought of, to first approximation, as a modulated Bloch function. The envelope function of this pulse propagates with the group velocity and suffers the group velocity dispersion of the Bloch function it modulates. If nonlinear effects are taken into account, it also suffers a nonlinear phase change determined by the way the underlying Bloch function samples the nonlinearity of the periodic structure. The group velocity, group velocity dispersion, and effective nonlinearity so derived can be taken to define an "effective medium," whose properties, even for a given PBG structure, can be chosen to some extent by choosing the frequency of the incident pulse and therefore the particular Bloch function that is modulated by the envelope function.

The details of this kind of description have been worked out already for one-dimensional PBG materials [27], and a number of groups are currently looking at higher dimensional geometries. Thus, there should be even more opportunities for constructing interesting structures and bringing geometry to the aid of the designer — or chemical engineer — who wishes to go beyond the kind of nonlinear response coefficients presented by nature in uniform media.

## Acknowledgments

RWB acknowledges support from the U.S. Air Force Office of Scientific Research and from the National Science Foundation. JES acknowledges support from the Natural Sciences and Engineering Research Council of Canada.

# References

1. R. J. Gehr, R. W. Boyd, Chem. Mater. **8**, 1807 (1996)
2. J. C. Maxwell Garnett, Philos. Trans. R. Soc. London **203**, 385 (1904); ibid **205**, 237 (1906)
3. D. A. G. Bruggeman, Ann. Phys. (Leipzig) **24**, 636 (1935)
4. J. E. Sipe, R. W. Boyd, Phys. Rev. A **46**, 1614 (1992)
5. R. W. Boyd, J. E. Sipe, J. Opt. Soc. Am. B **11**, 297 (1994)
6. See, e.g., D. J. Bergman, Phys. Rep. **43**, 377 (1978); X. C. Zeng, D. J. Bergman, P. M. Hui, D. Stroud, Phys. Rev. B **38**, 10970 (1988); O. Levy, D. Stroud, Phys. Rev. B **56**, 8035 (1997); and references therein
7. J. D. Jackson, *Classical Electrodynamics*, 2nd ed. (Wiley, New York 1975)
8. See, for instance, (4.18) of [7]
9. J. Van Kranendonk, J. E. Sipe, *Foundations of the macroscopic electromagnetic theory of dielectric media*, Prog. in Optics **XV**, 247 (1977); J. E. Sipe, Can. J. Phys. **56**, 199 (1978); J. E. Sipe, Can. J. Phys. **58**, 889 (1980) and references therein
10. This equation is a straightforward generalization of (5.57) of [7]
11. See, e.g., D. Stroud, Superlatt. Microstruct. **23**, 567 (1998)
12. J. I. Gittleman, B. Abeles, Phys. Rev. B **15**, 3273 (1977)
13. P. Sheng, Phys. Rev. Lett. **45**, 60 (1980)

14. See, e.g., C. J. F. Böttcher, *Theory of Electric Polarization* (Elsevier, New York 1952) pp. 415–429
15. P. D. Maker, R. W. Terhune, Phys. Rev. A **137**, 801 (1965)
16. D. B. Smith, G. Fischer, R. W. Boyd, D. A. Gregory, J. Opt. Soc. Am. B **14** 1625 (1997)
17. X. Zhang, D. Stroud, Phys. Rev. B **49**, 944 (1994)
18. M. I. Stockman, K. B. Kurlayev, T. F. George, Phys. Rev. B **170**, 71 (2000)
19. R. J. Gehr, G. L. Fischer, R. W. Boyd, J. Opt. Soc. Am. B **14**, 2310 (1997)
20. G. L. Fischer, R. W. Boyd, R. J. Gehr, S. A. Jenekhe, J. A. Osaheni, J. E. Sipe, L. A Weller-Brophy, Phys. Rev. Lett. **74**, 1871 (1995); R. J. Gehr, G. L. Fischer, R. W. Boyd, J. E. Sipe, Phys. Rev. A **53**, 2792 (1996)
21. R. L. Nelson, R. W. Boyd, App. Phys. Lett. **74**, 2417 (1999)
22. N. Kalyaniwalla, J. W. Haus, R.Inguva, M. H. Birnboim, Phys. Rev. A **42**, 5613 (1990)
23. D. J. Bergman, O. Levy, D. Stroud, Phys. Rev. B **49**, 129 (1994)
24. R. Neuendorf, M. Quinten, U. Kreibig, J. Chem. Phys. **104**, 6348 (1996)
25. Y. K. Yoon, R. S. Bennink, R. W. Boyd, J. E. Sipe, Opt. Commun., **179**, 577 (2000)
26. R. S. Bennink, Y.-K. Yoon, R. W. Boyd, J. E. Sipe, Opt. Lett. **24**, 1416 (1999)
27. C. M. De Sterke, D. G. Salinas, J. E. Sipe, Phys. Rev. E **54**, 1969 (1996)

# Response of Composite Media
# Made of Weakly Nonlinear Constituents

David J. Bergman[1] and David G. Stroud[2]

[1] School of Physics and Astronomy
Raymond and Beverly Sackler Faculty of Exact Sciences, Tel Aviv University
Tel Aviv 69978, Israel
bergmann@post.tau.ac.il
[2] Department of Physics, The Ohio State University
Columbus, OH 43210-1106, USA
stroud@mps.ohio-state.edu

**Abstract.** Electrical and optical properties of weakly nonlinear composite media are reviewed. A simple perturbation theory can account for many properties, including enhancement of macroscopic nonlinear response near a quasi-static resonance of a metal/dielectric mixture. However, to describe properties like intrinsic optical bistability, a more elaborate approach is used. Approximations and tools are described for analyzing the properties of composites with generic microstructures, but some solvable microstructures are treated in greater detail, exploiting the existence of closed form expressions for the macroscopic response. The self-consistent effective medium approximation for the weakly nonlinear response of composites is shown to be problematic, but its detailed analysis indicates how an improved self-consistent approximation scheme might be developed in the future.

## 1   Introduction

In standard texts on electrodynamics, the emphasis is placed entirely on *linear* media. In such media, the electric displacement **D** is linearly related to the electric field **E**; the coefficient of proportionality is called the dielectric function $\epsilon$, and at finite frequencies it will typically be a complex, frequency-dependent quantity. There is typically no mention of materials in which **D** and **E** have a nonlinear relationship. But such materials are very common, have numerous technological applications, and are of fundamental interest. A comprehensive review of linear electrical and optical properties of composite media can be found in [1].

This chapter is concerned with certain types of weak nonlinearity in *composite* materials. A *linear* composite is a material in which, as stated above, **D** and **E** are linearly related, but the constant of proportionality $\epsilon$, is complex and frequency-dependent, and also a function of *position*. Thus, a *nonlinear* composite is one in which there is a *nonlinear* and *position-dependent* relation between **D** and **E**. The relationship is taken to be local in space, though not necessarily in time.

V. M. Shalaev (Ed.): Optical Properties of Nanostructured Random Media,
Topics Appl. Phys. **82**, 19–39 (2002)
© Springer-Verlag Berlin Heidelberg 2002

## 2   Perturbation Theory

In this section, we will discuss some cases in which the nonlinearity can be treated as a weak perturbation of the predominant linear relationship between $\mathbf{D}$ and $\mathbf{E}$. We begin by discussing the zero-frequency case, then the generalization to finite frequencies. We will also restrict ourselves, for the most part, to *cubic* nonlinearities and discuss other cases briefly at the end of this section.

In the cubic case at zero frequency, the relation between $\mathbf{D}$ and $\mathbf{E}$ is assumed to be of the form

$$\mathbf{D}(\mathbf{r}) = \epsilon(\mathbf{r})\mathbf{E}(\mathbf{r}) + \chi(\mathbf{r})|\mathbf{E}(\mathbf{r})|^2\mathbf{E}(\mathbf{r})\,. \tag{1}$$

This cubic nonlinearity is often called a Kerr nonlinearity. Both the dielectric constant $\epsilon$ and cubic nonlinear susceptibility $\chi$ are functions of position $\mathbf{r}$. We assume, in this zero-frequency case, that both $\epsilon_e$ and $\chi_e$ are *real*.

Two important quantities in this composite are the *bulk effective* or *macroscopic* dielectric constant $\epsilon_e$ and the *bulk effective* or *macroscopic* Kerr nonlinear susceptibility $\chi_e$. These may be defined in various ways. For example, if we are considering a large system, the relation between the space averages $\langle \mathbf{D}\rangle$ and $\langle \mathbf{E}\rangle$ may be expressed as

$$\langle \mathbf{D}\rangle \equiv \epsilon_e\langle \mathbf{E}\rangle + \chi_e|\langle \mathbf{E}\rangle|^2\langle \mathbf{E}\rangle\,, \tag{2}$$

where $\langle \ldots \rangle$ denotes a space average. It can be considered that this relationship *defines* $\epsilon_e$ and $\chi_e$. In a sufficiently strong applied field, the relationship between $\langle \mathbf{D}\rangle$ and $\langle \mathbf{E}\rangle$ will acquire fifth-order terms, in addition to the first and third-order terms which appear in this equation, but the higher order terms will have no effect on $\epsilon_e$ and $\chi_e$.

A remarkable theorem can be proven about $\chi_e$, namely, it can be evaluated exactly in terms of certain properties of a related *linear* medium. This theorem was first proven by *Stroud* and *Hui* [2], and a similar theorem was also described by *Butenko* et al. [3]. A mathematically analogous result, for the special case where only one of the constituent media has a cubic nonlinearity, was obtained earlier in the context of velocity-dependent corrections to the superfluid density of liquid helium flowing in a superleak [4].

To specify this theorem, we introduce a "related linear medium" in which $\mathbf{D}(\mathbf{r}) = \epsilon(\mathbf{r})\mathbf{E}(\mathbf{r})$, with no third-order nonlinearity. We imagine that this medium has volume $V$ and that an external electric field $\mathbf{E}_0$ is applied to the medium (by fixing the potential on the surface at $-\mathbf{E}_0 \cdot \mathbf{r}$). Then it can be shown that the space average electric field inside the medium is just $\mathbf{E}_0$. The theorem of interest then states that

$$\chi_e E_0^4 = \frac{1}{V}\int_V \chi(\mathbf{r})|\mathbf{E}(\mathbf{r})|^4\mathrm{d}^3r\,, \tag{3}$$

that is, the effective cubic susceptibility is the space average of the susceptibility inside the medium, weighted by the *fourth power of the local electric field.* Thus, those points where the electric field is largest are most emphasized in the average.

Although the proof of this theorem is given in [2], we give a brief sketch of the argument here. The idea is to write the total electrostatic energy stored in the medium as

$$\int \left[ \epsilon(\mathbf{r})|\mathbf{E}|^2 + \chi(\mathbf{r})|\mathbf{E}|^4 \right] \mathrm{d}^3 r \, . \tag{4}$$

This is then evaluated through *first order* in $\chi$. The first-order term consists of two parts. One is the integral of $\chi(\mathbf{r})$ multiplied by the fourth power of $|\mathbf{E}_{\mathrm{lin}}(\mathbf{r})|$, the local electric field that would have existed had the medium been strictly linear. The other part involves $\delta\mathbf{E}(\mathbf{r})$, the first-order change in the local electric field due to $\chi$. The key point is that the integral of this second term is exactly zero (this is proven in [2]). The proof that this second term vanishes is also a special case of a mathematical result known as Tellegen's theorem [5], which is a straightforward consequence of the variational properties of the *linear* dielectric or conductivity problem. Hence, $\chi_{\mathrm{e}}$ is given *exactly* by an integral involving the local electric field in the *linear* medium.

For a random medium, it is not possible, in general, to compute the electric field exactly, even if the medium is linear. Therefore, one usually needs to use approximations. One simple approximation which immediately suggests itself is to decouple the field average according to the rule

$$\langle |\mathbf{E}|^4 \rangle_i \cong \langle |\mathbf{E}|^2 \rangle_i^2 \, , \tag{5}$$

where $\langle \ldots \rangle_i$ denotes a space average of the quantity in brackets within the constituent $i$ (we are now assuming that we have an $n$-constituent composite). Equation (5) is usually called the *nonlinear decoupling approximation* (NDA) [6]. This approximation has the advantage that the quantity $\langle \ldots \rangle_i$ is simply related to the macroscopic *linear* properties of the composite. Specifically, it can be shown that

$$p_i \langle |\mathbf{E}|^2 \rangle_i = \frac{\partial \epsilon_{\mathrm{e}}}{\partial \epsilon_i} E_0^2 \, , \tag{6}$$

where $\partial \epsilon_{\mathrm{e}}/\partial \epsilon_i$ means a derivative of $\epsilon_{\mathrm{e}}$ with respect to $\epsilon_i$, holding the dielectric functions of the other constituents fixed (and keeping the microstructure of the composite also fixed), and $p_i$ is the volume fraction of constituent $i$. Thus, in this approximation,

$$(\chi_{\mathrm{e}})_{\mathrm{NDA}} = \sum_i \frac{\chi_i}{p_i} \left( \frac{\partial \epsilon_{\mathrm{e}}}{\partial \epsilon_i} \right)^2 \, . \tag{7}$$

Another result follows immediately from this decoupling approximation, namely, since $\langle |\mathbf{E}|^4 \rangle_i \geq \langle |\mathbf{E}|^2 \rangle_i^2$, it follows that $\chi_{\mathrm{e}} \geq (\chi_{\mathrm{e}})_{\mathrm{NDA}}$, provided that the derivative $\partial \epsilon_{\mathrm{e}}/\partial \epsilon_i$ is calculated exactly in evaluating the NDA. Thus, the NDA gives a lower bound to the true $\chi_{\mathrm{e}}$.

Finally, we point out that the above formalism is not limited to treating nonlinear dielectrics. Any divergence-free field (not just $\mathbf{D}$) which is related to a curl-free field (not just $\mathbf{E}$) by a relation of the form of (1) has a macroscopic cubic nonlinear susceptibility, which can be defined and discussed similarly to the discussion given above. There are many relations of this kind in many branches of physics: Nonlinear relations between electric current and electric field, between heat current and thermal gradient, and between magnetic induction $\mathbf{B}$ and magnetic field $\mathbf{H}$, to name three well known examples, are in this class of problems and can be treated similarly.

Up to this point, we have discussed only zero-frequency relations between $\mathbf{D}$ and $\mathbf{E}$. In discussing optical properties, one will of course be concerned with finite frequencies. In that case, we must consider the fact that $\epsilon$ and $\chi$ will each be both complex and frequency-dependent, in general. For example, an optical Kerr material will be described by a relation of the form of (1), but with a frequency-dependent, complex dielectric function $\epsilon(\omega)$ and susceptibility $\chi(\omega)$. In a composite, these quantities will be position-dependent as well.

Even in this finite-frequency case, where the bulk effective susceptibility $\chi_e$ is also complex and frequency-dependent, it is still given by a relation involving the fourth power of the local electric field. Specifically, one finds that the finite-frequency generalization of (3) is [2]

$$\chi_e E_0^4 = \langle \chi(\mathbf{E} \cdot \mathbf{E})(\mathbf{E} \cdot \mathbf{E}^*) \rangle, \tag{8}$$

where the angular brackets again denote a space average and $\chi$ and $\mathbf{E}$ depend on position. The somewhat strange looking combination of $\mathbf{E}$ and $\mathbf{E}^*$ comes about because of the particular form of the Kerr relation, which involves the square of the *magnitude* of the electric field. This relation has been discussed further by *Stroud* and *Wood* [7].

*Stroud* and *Wood* also proposed a generalization of the NDA to finite frequencies. The decoupling is then written as

$$\langle (\mathbf{E} \cdot \mathbf{E})(\mathbf{E} \cdot \mathbf{E}^*) \rangle_i \cong \langle \mathbf{E} \cdot \mathbf{E} \rangle_i \langle \mathbf{E} \cdot \mathbf{E}^* \rangle_i. \tag{9}$$

They also expressed each of the quadratic electric field averages as a derivative involving $\epsilon_e$, with the approximate result

$$\chi_e \cong \sum_i \frac{\chi_i}{p_i} \frac{\partial \epsilon_e}{\partial \epsilon_i} \left| \frac{\partial \epsilon_e}{\partial \epsilon_i} \right|. \tag{10}$$

Here the derivative $\partial \epsilon_e / \partial \epsilon_i$ is equal to $\langle \mathbf{E} \cdot \mathbf{E} \rangle_i$, but the absolute value of that derivative is only a lower bound on $\langle \mathbf{E} \cdot \mathbf{E}^* \rangle_i$. Accurate expressions for averages of the second type are in fact obtainable for two-constituent composite media from the spectral representation for $\epsilon_e$ which was first introduced by *Bergman* [1,8], as shown by *Ma* et al. [9]. Those expressions can be written in the form

$$p_1 \langle |\mathbf{E}_{\text{lin}}|^2 \rangle_1 = \sum_n \frac{|s|^2 F_n}{|s - s_n|^2} |\mathbf{E}_0|^2 , \tag{11}$$

$$p_2 \langle |\mathbf{E}_{\text{lin}}|^2 \rangle_2 = \left\{ 1 - \sum_n \frac{(|s|^2 - s_n) F_n}{|s - s_n|^2} \right\} |\mathbf{E}_0|^2 , \tag{12}$$

where $p_1$ and $p_2 \equiv 1 - p_1$ are the volume fractions of the two constituents and $s_n$ and $F_n$ are the positions and weights of the poles in the spectral representation of $\epsilon_e$. That representation takes the form

$$F(s) \equiv 1 - \frac{\epsilon_e}{\epsilon_2} = \sum_n \frac{F_n}{s - s_n} , \qquad s \equiv \frac{\epsilon_2}{\epsilon_2 - \epsilon_1} , \tag{13}$$

$$0 \leq s_n < 1 , \qquad 0 < F_n , \qquad \sum_n F_n = p_1 , \tag{14}$$

which achieves a useful separation of the dependence of $\epsilon_e$ on microstructural details from its dependence on the physical properties of the constituents, namely, $\epsilon_1, \epsilon_2$: The poles and residues depend only on the microstructure, and the physical properties of the two constituents enter only through the parameter $s$. Using this representation, similar expressions for $\langle \mathbf{E} \cdot \mathbf{E} \rangle_i$ can also be written: One has only to replace all absolute value symbols in (11) and (12) by simple parentheses.

Although it is not directly relevant to the topic of optical response, we conclude this section by mentioning the connection between the cubic nonlinear response and the problem of flicker noise [1,2,10]. Flicker noise, as it is known, arises from conductance fluctuations in a *linear* composite material. Such fluctuations can occur for many reasons, including thermal noise. We will characterize flicker noise in terms of fluctuations in the local dielectric function $\epsilon(\mathbf{r})$. Suppose we have a material in which $\mathbf{D}$ and $\mathbf{E}$ are connected by the local relation

$$\mathbf{D}(\mathbf{r}) = \epsilon'(\mathbf{r}) \mathbf{E}(\mathbf{r}) , \tag{15}$$

where $\epsilon'(\mathbf{r})$ is taken from an ensemble with mean value $\epsilon(\mathbf{r})$ and fluctuation $\delta\epsilon(\mathbf{r})$. The mean value $\epsilon(\mathbf{r})$ is the same as that which appears in the nonlinear problem discussed above. The fluctuating part $\delta\epsilon(\mathbf{r})$, it is assumed, has a vanishing ensemble average at each point $\mathbf{r}$ and satisfies the relationship $\langle \delta\epsilon(\mathbf{r}) \delta\epsilon(\mathbf{r}') \rangle_{\text{av}} = \lambda \chi(\mathbf{r}) \delta(\mathbf{r} - \mathbf{r}')$, where $\langle \ldots \rangle_{\text{av}}$ denotes an ensemble average and $\lambda$ is some constant. Then the ensemble average of the *fluctuating part of the bulk effective dielectric function* $\delta\epsilon_e$ also vanishes, and its mean-square fluctuation is given by [2]

$$\langle (\delta\epsilon_e)^2 \rangle_{\text{av}} = \frac{\lambda \chi_e}{V} , \tag{16}$$

where $V$ is the system volume. Thus, $\langle(\delta\epsilon_e)^2\rangle_{\mathrm{av}}$ for this *linear* problem is given, to within a constant of proportionality, by the value of $\chi_e$ for the analogous *weakly nonlinear* problem.

Although (16) was originally derived for the static case, it will probably continue to be valid even at finite frequencies, provided that the fluctuations in $\epsilon$ occur at a frequency scale much lower than $\omega$, that is, if $\epsilon(\mathbf{r}, \omega)$ has very slow fluctuations, then (16) will continue to relate $\langle[\delta\epsilon_e(\omega)]^2\rangle_{\mathrm{av}}$ to $\chi_e(\omega)$. However, this relationship at finite frequencies does not appear to have been discussed in the literature.

Before concluding this section, we mention that all of the results described here for cubic nonlinearity can readily be extended to other powers of weak nonlinearity, i.e., to cases where $\mathbf{D} = \epsilon\mathbf{E} + \chi|\mathbf{E}|^\beta\mathbf{E}$, where $\beta(> -1)$ is a (not necessarily integral) constant. The case of cubic nonlinearity ($\beta = 2$) is, however, usually the most relevant to experiments. The reason is that the seemingly more important case ($\beta = 1$) does not appear in media with a center of inversion. Hence, the first nonlinear term that actually appears in a Taylor series expansion of $\mathbf{D}$ in terms of $\mathbf{E}$ is often the cubic term. We should also note that we have omitted discussion of the tensorial nature of the susceptibility $\chi$. Even the cubic case ($\beta = 2$) may have other terms in the expansion beyond simply a term involving the square of the absolute value of the electric field. In most cases in which the constituents have reasonably high point group symmetry, most of these terms do not contribute; so our discussion is adequate for most cases of interest.

# 3   Limiting Cases and Exactly Solvable Microstructures

In this section, we discuss several limiting cases in which the linear response of a composite material can be calculated in closed form, and we proceed to find their weakly nonlinear response, too.

## 3.1   Parallel Cylinders and Parallel Slabs

The simplest kind of solvable microstructure is parallel cylinders, where all of the interfaces are parallel to a fixed direction and the external or average field $\mathbf{E}_0$ is applied in that direction (see Fig. 1a). For simplicity, we consider such a composite made of two constituents, with volume fractions $p_1$ and $p_2 = 1 - p_1$ and with dielectric constants and Kerr susceptibilities $\epsilon_1$, $\chi_1$ and $\epsilon_2$, $\chi_2$. The electric field is uniform and equal to $\mathbf{E}_0$ everywhere; therefore the macroscopic dielectric constant and Kerr susceptibility are given by

$$\epsilon_{\mathrm{cyls}}^{(e)} = p_1\epsilon_1 + p_2\epsilon_2, \quad \chi_{\mathrm{cyls}}^{(e)} = p_1\chi_1 + p_2\chi_2. \tag{17}$$

Another kind of solvable microstructure is parallel slabs, where all of the interfaces are flat planes perpendicular to a fixed direction and the external or average field $\mathbf{E}_0$ is applied in that direction (see Fig. 1b). Again we consider,

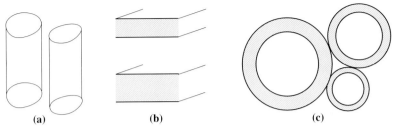

**Fig. 1.** Simple solvable microstructures: (**a**) Parallel cylinders microstructure. The local fields **E** (which has the same value everywhere) and **D** lie along the cylindrical axis of symmetry. (**b**) Parallel slabs microstructure. The applied field, as well as the local fields, are all perpendicular to the slabs. (**c**) Coated sphere assemblage [12]: The entire system is made up of coated particles, each comprised of a spherical core and a concentric spherical shell. The coated spheres must come in an infinite hierarchy of decreasing sizes to fill up the entire volume, but they all have the same core-to-shell volume ratio, equal to $p_1/p_2$

for simplicity, the two-constituent case. Now, the displacement field **D** is uniform everywhere, and the electric field is uniform in each constituent. The macroscopic dielectric constant and Kerr susceptibility are therefore given by

$$\epsilon_{\text{slabs}}^{(e)} = \left( \frac{p_1}{\epsilon_1} + \frac{p_2}{\epsilon_2} \right)^{-1}, \tag{18}$$

$$\chi_{\text{slabs}}^{(e)} = \left( \frac{\chi_1 p_1}{\epsilon_1^2 |\epsilon_1|^2} + \frac{\chi_2 p_2}{\epsilon_2^2 |\epsilon_2|^2} \right) \left( \frac{p_1}{\epsilon_1} + \frac{p_2}{\epsilon_2} \right)^{-2} \left| \frac{p_1}{\epsilon_1} + \frac{p_2}{\epsilon_2} \right|^{-2}. \tag{19}$$

It is easy to check that, in the two cases discussed here, NDA leads to the same results. This is due to the fact that the local electric field has a constant value in each constituent. Thus (5) and (9) cease to be approximations and become exact.

## 3.2   Dilute Regime and Clausius–Mossotti Approximation

An important special case is the *dilute limit*, when a low density of particles is embedded in a host medium. We assume that the particles are spherical, and have dielectric function and Kerr susceptibility $\epsilon_1$ and $\chi_1$, and the host is characterized by $\epsilon_2$, $\chi_2$.

For such a system, a well known approximation for $\epsilon_e$ that is exact to first order in the volume fraction of the inclusions $p_1$, and to second order in the difference of dielectric constants $\epsilon_1 - \epsilon_2$, is the Maxwell-Garnett or Clausius–Mossotti approximation (CMA). The strict "dilute limit" expressions, which are exact to first order in $p_1$, can always be obtained by expanding the CMA results up to that same order. When the microstructure is isotropic, then a convenient representation for $\epsilon_e$ in that approximation is the following simple relation, which is easy to remember and also to transform into an explicit expression for $\epsilon_e$ [1]

$$\frac{\epsilon_e - \epsilon_2}{\epsilon_e + 2\epsilon_2} = p_1 \frac{\epsilon_1 - \epsilon_2}{\epsilon_1 + 2\epsilon_2} . \tag{20}$$

The spectral representation for $\epsilon_e$ in this case is even simpler, namely,

$$F(s) \equiv 1 - \frac{\epsilon_e}{\epsilon_2} = \frac{p_1}{s - p_2/3} . \tag{21}$$

Using (11) and (12) from the previous section, the following result is obtained for the macroscopic nonlinear susceptibility in the NDA (this result was obtained in [6] for the special case where all of the constituent moduli, and hence also $s$, are real)

$$(\chi_e)_{\text{NDA}} = \chi_1 p_1 \left( \frac{s}{s - p_2/3} \right)^2 \left| \frac{s}{s - p_2/3} \right|^2$$
$$+ \chi_2 p_2 \frac{(s - 1/3)^2 + 2p_1/9}{(s - p_2/3)^2} \frac{|s - 1/3|^2 + 2p_1/9}{|s - p_2/3|^2} . \tag{22}$$

In this case, it is also possible to calculate the local field $\mathbf{E}(\mathbf{r})$ both inside and outside of the particles, if they are assumed to be spherical: Interactions between dipole moments induced in such particles are taken into account, but all higher order induced moments are neglected in this CMA. That field can be used for a direct calculation of the fourth moment of $\mathbf{E}(\mathbf{r})$ which appears in (3) or (8). The expression obtained in this way for $\chi_e$ is more exact, since it involves no further approximations beyond those made to obtain the CMA ([11], where the analogous result is obtained for the *static* case, when all the fields and moduli are *real and positive*):

$$(\chi_e)_{\text{CMA}} = [p_2\epsilon_1 + (2 + p_1)\epsilon_2]^{-2} |p_2\epsilon_1 + (2 + p_1)\epsilon_2|^{-2}$$
$$\times \left\{ p_1 \chi_1 (3\epsilon_2)^2 |3\epsilon_2|^2 + p_2 \chi_2 \left[ (\epsilon_1 + 2\epsilon_2)^2 |\epsilon_1 + 2\epsilon_2|^2 \right. \right.$$
$$+ 2p_1 (\epsilon_1 - \epsilon_2)^2 |\epsilon_1 + 2\epsilon_2|^2 + 2p_1 |\epsilon_1 - \epsilon_2|^2 (\epsilon_1 + 2\epsilon_2)^2$$
$$+ \frac{16}{5} p_1 (\epsilon_1 + 2\epsilon_2) |\epsilon_1 + 2\epsilon_2|^2 (\epsilon_1 - \epsilon_2) \text{Re} \left( \frac{\epsilon_1 - \epsilon_2}{\epsilon_1 + 2\epsilon_2} \right)$$
$$+ \frac{4}{5} p_1 (1 + p_1)(\epsilon_1 + 2\epsilon_2)(\epsilon_1 - \epsilon_2) |\epsilon_1 - \epsilon_2|^2$$
$$+ \frac{4}{5} p_1 (1 + p_1)(\epsilon_1 - \epsilon_2)^2 |\epsilon_1 + 2\epsilon_2|^2 \text{Re} \left( \frac{\epsilon_1 - \epsilon_2}{\epsilon_1 + 2\epsilon_2} \right)$$
$$+ \left. \left. \frac{8}{5} p_1 (1 + p_1 + p_1^2)(\epsilon_1 - \epsilon_2)^2 |\epsilon_1 - \epsilon_2|^2 \right] \right\} . \tag{23}$$

Comparison of (22) and (23) shows that the term involving $\chi_1$ is identical in the two expressions, whereas the term involving $\chi_2$ is different. The reason is that, in CMA, the local field is uniform and has the same value inside all of the inclusions; therefore (9), the basic approximation made in NDA, is exact there. By contrast, outside the inclusions, $\mathbf{E}(\mathbf{r})$ is nonuniform, therefore use of NDA there involves a further approximation compared with CMA.

Another fact worth noting is that the CMA results for both $\epsilon_e$ and $\chi_e$ are actually exact in the case of the so-called "coated spheres assemblage" or "Hashin–Shtrikman microstructure" [6], where the system is made up entirely of coated spheres, with a spherical core made of the #1 constituent and a concentric spherical shell made of the #2 constituent. Those spheres must come in an infinitely decreasing hierarchy of sizes to fill up the entire volume, and they all must have the *same* core-to-shell volume ratio, equal to $p_1/p_2$ — see Fig. 1c and [12].

Equations (19), (22), and (23) all exhibit an enhancement of $\chi_e$ when the system is near a sharp quasi-static resonance, due to a vanishing denominator. In the parallel slabs microstructure, there is one such resonance when $p_1\epsilon_2 + p_2\epsilon_1 = 0$, whereas in the CMA there is one such resonance when $p_2\epsilon_1 + (2 + p_1)\epsilon_2 = 0$, which is the same as $s = p_2/3$. More generally, any pole $s_n$ of $F(s)$ represents a quasi-static resonance, in whose vicinity $\chi_e$ will be enhanced by a factor proportional to the inverse fourth power of $s - s_n$. This can be compared with the behavior of $\epsilon_e$ or $F(s)$ itself, which is only proportional to the inverse first power of $s - s_n$. Since all of the poles $s_n$ lie on the semi-closed real segment $(0,1)$ [see (14)], such resonances can occur only when $0 \le s < 1$, or $\epsilon_1/\epsilon_2 < 0$. In practice, they can most easily be approached in a metal/dielectric composite at finite frequencies, when the metal permittivity has a sizable negative real part and a very small imaginary part. They can also be approached in composites of two dielectrics, provided that one of the dielectrics has an intrinsic pole or resonance with large enough weight that the real part of its dielectric constant becomes negative at slightly lower frequencies.

Probably the most important consequence of these resonances is that $\chi_e$ is strongly enhanced near such resonances in dilute collections of coated and multicoated spherical particles. The very large enhancement is due to the fourth-power effect mentioned above. Extensive discussions of this behavior, with numerical examples, can be found, for example, in [9,13,14,15,16,17], [18,19,20].

The resonances also have other effects on nonlinear response at finite frequencies. These are discussed in Sect. 5 below.

# 4   Self-Consistent Effective Medium Approximation and Its Breakdown

A very important and useful class of approximations in the study of macroscopic response of composite media, especially when the microstructure is disordered, is the "self-consistent effective medium approximations (SEMA)".

The first approximation of this type was put forward by *Bruggeman* in 1935, for simple linear, scalar conductivity [21], and then later but independently by *Landauer* in 1952 [22]. This was later extended successfully to elastic stiffness moduli [23,24], to nonscalar electrical conductivity [25,26,27], and to systems with a columnar microstructure [28]. Though these studies were carried out in the language of dc electrical conductivity, they also apply almost without any change to time-dependent electrical phenomena, where one can focus on the response to a monochromatic time-dependent field at a single frequency $\omega$, if that frequency is not too high.

A particularly appealing feature of SEMA is that it exhibits percolation thresholds, in whose neighborhoods dramatic effects appear in the macroscopic response. In disordered metal/dielectric composites, an optical absorption band appears that is due to surface plasmon excitations. This band is absent in either of the pure bulk constituents, and its appearance and details depend on the microstructure of the composite medium. As the relative proportion of metal and dielectric constituents is varied near the percolation threshold of either constituent, this microstructure-dependent absorption band undergoes characteristic changes, which can be quite dramatic—see the review article [1] for details and original references.

The success of SEMA in accounting for various aspects of the linear response of disordered composite media, despite its simplicity, made it natural to try and extend it also to the description of nonlinear phenomena in such media. This was done, successfully, for strongly nonlinear, pure power law conductivity, where it was used to study critical behavior near the percolation threshold of a metal/insulator random composite [29,30]. It was also done for weakly nonlinear, power law conductivity [11]. However, in the latter case, the SEMA breaks down dramatically as the percolation threshold is approached. Understanding the reason for this breakdown has provided insight as to why SEMA works so well in other cases, as well as an indication of what changes need to be made to develop a better approximation for the macroscopic weakly nonlinear response.

Instead of viewing SEMA as an approximation for a general class of microstructures, we consider a microstructure for which it provides an exact description [31]. That is a very artificial microstructure, characterized by an infinite hierarchy of length scales of heterogeneity (see Fig. 2). Starting from a two-constituent composite with an arbitrary microstructure, where only the volume fractions $p_1$, $p_2 = 1 - p_1$ have definite values, we embed in it a small quantity of spherical inclusions of the same two constituents, with radii that are much greater than the heterogeneity length scales of the initial microstructure and with a volume ratio equal to $p_1/p_2$. The small-grain-composite host, inside which these large inclusions are embedded, is treated as homogeneous. The altered composite medium thus retains the original values of $p_1$, $p_2$, but has slightly changed values of its macroscopic linear and nonlinear conductivities $\sigma_e$ and $b_e$, respectively. Those small changes, $d\sigma_e$ and

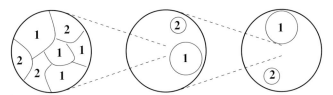

**Fig. 2.** Schematic representation of the hierarchical microstructure for which SEMA is exact [31]

$db_e$, can be calculated exactly to leading order in the total volume fraction $dv$ displaced by the large inclusions that were introduced. This procedure is now repeated, with the altered composite used as a quasi-homogeneous host material, inside of which we again embed a small quantity of spherical inclusions, with radii that are much greater than those used in the previous step and a volume ratio that is again equal to $p_1/p_2$. Such an iterated embedding process changes $\sigma_e$ and $b_e$ in a manner described by the following differential equations [11] (we have assumed, for simplicity, that $\sigma_a$, $b_a$ are all real):

$$\frac{d\sigma_e}{dv} = 3\sigma_e \left( p_1 \frac{\sigma_1 - \sigma_e}{\sigma_1 + 2\sigma_e} + p_2 \frac{\sigma_2 - \sigma_e}{\sigma_2 + 2\sigma_e} \right) , \tag{24}$$

$$\frac{db_e}{dv} = \sum_{a=1}^{2} p_a \left[ (b_a - b_e) \left( \frac{3\sigma_e}{\sigma_a + 2\sigma_e} \right)^4 + b_e \left( \frac{66}{5} A_a^2 - \frac{12}{5} A_a^3 + \frac{13}{5} A_a^4 \right) \right] , \tag{25}$$

where

$$A_a \equiv \frac{\sigma_a - \sigma_e}{\sigma_a + 2\sigma_e} , \tag{26}$$

and where we assumed conductivity with a weak quadratic dependence on the local electric field $\mathbf{E}$. This means that the current density $\mathbf{J}$ depends on $\mathbf{E}$ as follows:

$$\mathbf{J} = \sigma\mathbf{E} + b|\mathbf{E}|^2\mathbf{E} , \quad b|\mathbf{E}|^2 \ll \sigma . \tag{27}$$

Equation (24) has a *fixed point* whenever the right-hand side (r.h.s.) vanishes. This occurs whenever $\sigma_e$ is equal to a solution of the usual SEMA equations. A careful examination of the r.h.s. near the two solutions shows that only one of them is a *stable fixed point*. Thus, whatever the initial value of $\sigma_e$, if the embedding procedure described above is iterated, $\sigma_e$ will eventually tend to that stable fixed point. That is the *physical solution* of the SEMA equations, which is thus also the exact result for the hierarchical microstructure produced by this iterated embedding procedure.

Equation (25) also has some fixed points, which can be studied after substitution of the stable fixed point value of $\sigma_e$ in the r.h.s. of that equation.

In particular, in a metal/insulator mixture, that equation simplifies to (the subscript M denotes the metallic constituent)

$$\frac{db_e}{dv} = b_M \frac{81}{16} \left(p_M - \frac{1}{3}\right)^4 - \frac{b_e}{10} \left(27p_M^3 - 18p_M^2 + 2p_M - 1\right). \tag{28}$$

This has a fixed point whenever the r.h.s. vanishes, which leads to a unique value for $b_e/b_M$. However, though this value is finite and positive when $p_M$ is not much less than 1, it diverges when $p_M$ tends to the only real zero of the cubic polynomial which multiplies $b_e$, namely, $p_M \cong 0.6412$, and then takes on *negative values* for yet smaller values of $p_M$. Such values are clearly unacceptable for a quantity like $\langle \mathbf{E}^4 \rangle$ [see (3)] which must be positive. Moreover, this catastrophe occurs when $p_M$ is still far above the SEMA percolation threshold of $1/3$. A careful examination of (28) near that fixed point reveals that it is a stable fixed point only when $b_e/b_M > 0$. When $p_M < 0.6412$, the negative fixed point value of $b_e/b_M$ constitutes an unstable fixed point, and any solution of that equation which starts out from a positive value of $b_e/b_M$ then tends to $+\infty$, which is the only stable fixed point. But how can an infinite value be the correct physical result for $b_e$?

The physical explanation is as follows: On the largest length scales in the hierarchical microstructure, there exists a region in the quasi-homogeneous host, near the equator of each of the largest insulating inclusions, where the local electric field is enhanced by at least some factor that lies between 1 and 1.5. That region is of order of the inclusion size. Confining our attention to that region, which is homogeneous on the low-resolution scale wherein only the largest inclusions are resolved, we now increase the resolution so as to resolve the inclusions at the next lower length scale. Near the equators of those inclusions that are insulating, the field is again enhanced by the same factor in a volume similar to that of the inclusion. Iteration of those increases in resolution leads to the conclusion that in some exponentially small regions of the composite medium, the local electric field $\mathbf{E}$ attains values that are exponentially enhanced compared with its imposed volume average value $\langle \mathbf{E} \rangle = \mathbf{E}_0$. That these exponentially enhanced values of $\mathbf{E}$ never have a dramatic effect upon $\langle \mathbf{E}^2 \rangle$ is due to the variational properties of the linear conductivity problem: Those dictate that $\langle \sigma \mathbf{E}^2 \rangle / \langle \mathbf{E} \rangle^2$ must always lie between $\sigma_1$ and $\sigma_2$ [1]. Such bounds do not exist for other moments of $\mathbf{E}$. Furthermore, the local enhancements of $\mathbf{E}$ have a greater effect on averages of $\mathbf{E}^n$, the larger the exponent $n$. Apparently, that is why $\langle \mathbf{E}^4 \rangle$ diverges in metal/insulator mixtures with the hierarchical microstructure when $p_M < 0.6412$. We can also expect that higher moments of $\mathbf{E}$ will exhibit even stronger divergences in that microstructure [11].

Obviously, real materials never have the infinite hierarchy of heterogeneity length scales which appears in the microstructure described above, and the resulting $b_e$ will always be finite and positive in real composites. From the above discussion, we can conclude that to develop a sensible approximation for $b_e$ or for the Kerr susceptibility $\chi_e$ in a dielectric composite, we must have a scheme that excludes the possibility of such an infinite hierarchy of length scales. More information about the microstructure must be used than just the volume fractions $p_1$ and $p_2$.

# 5  Quasi-static Resonances: Enhancement of Nonlinearity and Intrinsic Optical Bistability

An isolated metallic inclusion in an otherwise uniform dielectric host exhibits a number of surface plasmon resonances in the frequency region where the real part of the metal permittivity is negative and the imaginary part is small:

$$\epsilon_M = \epsilon'_M + i\epsilon''_M , \tag{29}$$

where $\epsilon'_M < 0$ and $\epsilon''_M \ll |\epsilon'_M|$.

A good example is the surface plasmon resonance of a metal sphere of radius $R$, which can be seen by considering the induced dipole moment $\mathbf{m}$, when such a sphere is embedded in an otherwise uniform dielectric host and a uniform electric field $\mathbf{E}_0$ is applied:

$$\mathbf{m} = R^3 \epsilon_d \frac{\epsilon_M - \epsilon_d}{\epsilon_M + 2\epsilon_d} \mathbf{E}_0 . \tag{30}$$

Here $\epsilon_d$ is the electrical permittivity of the dielectric host, assumed to be real. The induced moment $\mathbf{m}$ would diverge if $\epsilon_M + 2\epsilon_d = 0$. This cannot happen because $\epsilon''_M \neq 0$, but the denominator in (30) can get very small if $\epsilon''_M$ is small and if the frequency is such that $\epsilon'_M + 2\epsilon_d = 0$. When that happens, the induced dipole moment is very large. Consequently, the local electric field in the neighborhood of the inclusion also becomes very large and can exceed $\mathbf{E}_0$ by a large factor.

If the metal has a Drude–Lorentz conductivity $\sigma_M(\omega)$, then $\epsilon_M$ can be written as

$$\epsilon_M = \epsilon_0 + \frac{4\pi i \sigma_M(\omega)}{\omega} = \epsilon_0 + \frac{\omega_p^2}{\omega^2} \frac{i\omega\tau}{1 - i\omega\tau} , \tag{31}$$

where $\omega_p$ is the bulk plasma frequency of the charge carriers, $\tau$ is their conductivity relaxation time, and $\epsilon_0$ is the electrical permittivity of the ionic background, assumed to be real and positive. If $\omega\tau \gg 1$, then the surface plasmon resonance frequency of the embedded metal sphere is

$$\omega_{sp} = \frac{\omega_p}{\sqrt{\epsilon_0 + 2\epsilon_d}} , \tag{32}$$

which reduces to $\omega_{sp} = \omega_p/\sqrt{3}$ if $\epsilon_0 = \epsilon_d = 1$.

For inclusions of other shapes, the resonance occurs at a different frequency, and there is usually more than one resonance frequency. In a nondilute collection of inclusions, the induced polarizations in different inclusions also interact. This changes the isolated inclusion resonances into resonances of the entire system, and those depend on details of the microstructure. The mathematical expression of these resonances are the poles that appear in the spectral expansion of $F(s)$ or $\epsilon_e$, as exhibited in (13). In general, a disordered microstructure results in a broad, quasi-continuous spectrum of poles or resonances, and hence a broad absorption band appears, similar to that found in SEMA [1]. Sharp resonances will appear in two cases: (a) a dilute collection of similarly shaped inclusions, e.g., spherical inclusions; (b) a periodic microstructure.

When $\omega$ is near a sharp resonance, the local field can be greatly enhanced, as we saw for surface plasmon resonance of an isolated spherical inclusion. This enhancement leads to a corresponding enhancement of nonlinear behavior. Such enhancements were already mentioned earlier in Sect. 3.2. An especially dramatic aspect of this is the possibility of achieving "intrinsic optical bistability". That is a situation where, with some values of the externally applied field $\mathbf{E}_0$, the system can be found in two or more different internal states, with different values of the internal local field $\mathbf{E}$. It is interesting to note that, in the quasi-static regime, resonances and bistability can occur only if the electrical permittivity is not real and positive everywhere. In fact, it is possible to prove that, if $\mathbf{E} \cdot \mathbf{D}(\mathbf{E})$ is a real, monotonically increasing function of $|\mathbf{E}|$ or if $\mathbf{D}$ is linear in $\mathbf{E}$, then the local field $\mathbf{E}$ has a uniquely determined value everywhere in the system [1]. To allow for the possibility of bistability, we must therefore have a negative real part of the electrical permittivity in parts of the system, i.e., there must be a metallic constituent and the frequency $\omega$ must satisfy

$$\frac{1}{\tau} < \omega < \omega_{\mathrm{p}} , \tag{33}$$

as well as a nonlinear dependence of $\mathbf{D}$ upon $\mathbf{E}$ in some parts of the system.

## 5.1   Solvable Microstructures

Bistable behavior is most easily demonstrated in the parallel slabs microstructure, shown in Fig. 1b, where an average electric field $\langle \mathbf{E} \rangle = \mathbf{E}_0$ is applied perpendicular to the slabs. We assume there are two types of slabs: metallic slabs with a complex, frequency-dependent but field-independent permittivity $\epsilon_M$, as in (31), and dielectric slabs, with a real but field-dependent permittivity $\epsilon_d(\mathbf{E})$, which we take as quadratic in $|\mathbf{E}|$,

$$\epsilon_d(\mathbf{E}) = \epsilon_{d0} + \chi |\mathbf{E}|^2 . \tag{34}$$

This is an exactly solvable system, even in the nonlinear case, if we assume that $\mathbf{E}$ has the same constant value in all slabs of a certain type, $\mathbf{E}_M$ and $\mathbf{E}_d$ respectively, since the electric displacement $\mathbf{D}$ must then have the same value everywhere:

$$\mathbf{D} \equiv \mathbf{D}_0 = \epsilon_M \mathbf{E}_M = \epsilon_d(\mathbf{E}_d)\mathbf{E}_d . \tag{35}$$

Considering the volume average of $\mathbf{E}$, namely ($p_M$, $p_d = 1 - p_M$ are the volume fractions of the two types of slabs),

$$\langle \mathbf{E} \rangle = \mathbf{E}_0 = p_M \mathbf{E}_M + p_d \mathbf{E}_d , \tag{36}$$

and using (35) to express $\mathbf{E}_M$ in terms of $\mathbf{E}_d$, we get a *complex cubic equation* for $\mathbf{E}_d$, given $\mathbf{E}_0$

$$\mathbf{E}_0 = \left( p_d + p_M \frac{\epsilon_{d0} + \chi |\mathbf{E}_d|^2}{\epsilon_M} \right) \mathbf{E}_d . \tag{37}$$

A solution for $\mathbf{E}_d$ can easily be found once $|\mathbf{E}_d|^2$ is known, and the latter quantity satisfies a *real cubic equation*, obtained by taking the absolute square of (37)

$$\mathbf{E}_0^2 = \left| p_d + p_M \frac{\epsilon_{d0} + \chi |\mathbf{E}_d|^2}{\epsilon_M} \right|^2 |\mathbf{E}_d|^2 . \tag{38}$$

The last equation can be brought to the "canonical form"

$$f(t) \equiv t^3 - 2\mu t^2 + t = \alpha \tag{39}$$

by adopting the following definitions:

$$t \equiv \frac{\chi |\mathbf{E}_d|^2}{\left| \epsilon_{d0} + \frac{p_d}{p_M} \epsilon_M \right|} > 0 , \tag{40}$$

$$\alpha \equiv \chi \mathbf{E}_0^2 \frac{|\epsilon_M|^2 / p_M^2}{\left| \epsilon_{d0} + \frac{p_d}{p_M} \epsilon_M \right|^3} > 0 , \tag{41}$$

$$\mu \equiv - \frac{\mathrm{Re} \left( \epsilon_{d0} + \frac{p_d}{p_M} \epsilon_M \right)}{\left| \epsilon_{d0} + \frac{p_d}{p_M} \epsilon_M \right|} , \quad |\mu| < 1 . \tag{42}$$

Any solution of the canonical equation (39) leads, via (37), to a unique value for $\mathbf{E}_d$ and hence, also for $\mathbf{E}_M$. But (39) will have three real solutions for $\alpha$ within a certain range of values, if $\mu$ satisfies the following inequality:

$$\mu > \frac{\sqrt{3}}{2} . \tag{43}$$

In that case, we get three different solutions for the internal state of the system, as expressed by the values of the complex fields $\mathbf{E}_M$, $\mathbf{E}_d$.

Of course, nature will always choose just one of these states, using physical factors that were ignored in this discussion. In particular, the history should play a significant role, since this is not an equilibrium state. Nevertheless, we do not expect that all three states are equally accessible. We conjecture that, as in the case of thermodynamic equilibrium, only the two extreme values

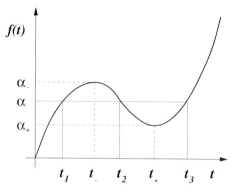

**Fig. 3.** Qualitative plot of the cubic function $f(t)$ for a value of $\mu$ between $\sqrt{3}/2$ and 1. The cubic equation (39) has three real solutions whenever $\alpha$ lies between $\alpha_+$ and $\alpha_-$, which depend on $\mu$ but are always $\mathcal{O}(1)$. The same is true for the three solutions $t_1, t_2, t_3$

of $\mathbf{E}_d$ will be sustainable for any length of time, but not the intermediate value. That is why this phenomenon is called bistability.

An important aspect of the intrinsic optical bistability that we have described is that, the externally applied field $\mathbf{E}_0$ does not have to be very large, and also even the local fields $\mathbf{E}_d$ and $\mathbf{E}_M$ can be quite small. To see this, we note that $\alpha = \mathcal{O}(1)$ over the entire range of values for which (39) has three solutions and that those solutions will also usually be of order 1—see Fig. 3 and [32]. Therefore, the numerators on the r.h.s. of (40) and (41) must have magnitudes similar to those of the denominators; consequently,

$$\chi\mathbf{E}_0^2 = \mathcal{O}\left(\left|\epsilon_{d0} + \frac{p_d}{p_M}\epsilon_M\right|^3\right) , \quad \chi|\mathbf{E}_d|^2 = \mathcal{O}\left(\left|\epsilon_{d0} + \frac{p_d}{p_M}\epsilon_M\right|\right) . \tag{44}$$

Clearly, the field-dependent part of $\epsilon_d$ will be small if $|\epsilon_{d0} + \epsilon_M p_d/p_M|$ is small. The factor $\epsilon_{d0} + \epsilon_M p_d/p_M$ is a "detuning parameter": If the field-dependent part of $\epsilon_d$ were absent, then the system would exhibit a resonance when this detuning parameter vanished—both $\mathbf{E}_d$ and $\mathbf{E}_M$ would become infinite. It is evident from (37) that the field-dependent part of $\epsilon_d$, namely, $\chi|\mathbf{E}_d|^2$, is not competing against the field-independent part $\epsilon_{d0}$, but rather against the detuning parameter. Therefore, when that parameter is small, even a weakly nonlinear permittivity, i.e., $\chi|\mathbf{E}_d|^2 \ll \epsilon_{d0}$, can still have a major effect on the system response, including the appearance of optical bistability. It is important to emphasize the fact that application of (8), which is the result of a simple perturbation treatment of weak nonlinearity, would totally miss the important phenomenon of intrinsic optical bistability.

The requirement of (43) means that $\omega$ should not be too close to the center of the resonance, so that $2\epsilon_M'' < |\epsilon_{d0} + \epsilon_M' p_d/p_M|$.

Similar considerations can be made for a dilute suspension of nonlinear dielectric spherical inclusions, where each sphere is coated by a thin metallic shell. A system with a single inclusion of that type is also exactly solvable and was examined carefully in [33]. Such particles have two sharp resonances, which are zeros of the denominator in the following expression for the ratio of the (uniform) field in the core $\mathbf{E}_{\text{in}}$ to the externally applied uniform field $\mathbf{E}_0$ ($\epsilon_d$, $\epsilon_h$ are the electrical permittivities of the core and host; $\epsilon_M$ is that of the metallic shell; $p_M$ is the volume fraction taken up by that shell as part of the coated sphere):

$$\frac{|\mathbf{E}_{\text{in}}|}{|\mathbf{E}_0|} = \frac{9\epsilon_M\epsilon_h}{(\epsilon_d + 2\epsilon_M)(\epsilon_M + 2\epsilon_h) + 2(1 - p_M)(\epsilon_d - \epsilon_M)(\epsilon_M - \epsilon_h)}. \tag{45}$$

If $p_M \ll 1$, then one of these resonances is much stronger than the other. This is most easily seen if one assumes $\epsilon_h = \epsilon_d$, in which case we can write

$$\frac{|\mathbf{E}_{\text{in}}|}{|\mathbf{E}_0|} \cong -\frac{\frac{9}{2}\frac{\epsilon_d}{p_M}}{\epsilon_M + \frac{9}{2}\frac{\epsilon_d}{p_M}} - \frac{\frac{2}{9}p_M\epsilon_d}{\epsilon_M + \frac{2}{9}p_M\epsilon_d}. \tag{46}$$

Clearly, the first pole or resonance has a much larger residue or weight, and it also corresponds to a much lower frequency, if $\epsilon_M$ is given by (31). Near that resonance, the internal field satisfies a cubic equation that can also be brought to the canonical form of (39). Again, there is a range of values of $\mathbf{E}_0$ for which three different solutions exist for the internal field $\mathbf{E}_{\text{in}}$.

An experiment was performed on a dilute suspension of such spheres in water, where each sphere had a nanometer-size core of CdS, coated by a thin layer of metallic silver [34]. Optical bistability was observed in that system, in agreement with the predictions of [33].

Actually, in such a dilute system of inclusions, with similar shapes and compositions, one could contemplate a situation where different particles (inclusions) are found in states corresponding to different solutions of (39). This would result in a large multiplicity of possible internal states of the system. However, electromagnetic interactions between the different inclusions, largely ignored in this treatment, would have to be taken into account, and they could result in forcing different particles to be in the same internal state. In the parallel slabs microstructure considered above, those interactions are responsible for the fact that the resonance depends on the *total volume fractions* $p_M, p_d$, not just on properties of an individual slab. Nevertheless, it is not presently known whether multiple solutions can be found wherein the local fields $\mathbf{E}_M$, $\mathbf{E}_d$ do not have the same values in all slabs of a given constituent. Clearly, the intriguing possibility of multiple solutions, i.e., of "intrinsic optical multistability" near a sharp resonance of a metal/dielectric composite, deserves further study.

## 5.2   The Principle of "Zero Virtual Work" and Its Application

For most microstructures, the local field is nowhere uniform, and even in the linear case, it cannot be calculated in closed form, not to mention the nonlinear case. However, good approximations and numerical methods exist for the linear problem [1], and it is also possible to exploit such a solution to get a good approximation for nonlinear problems, as long as the field-dependent part of $\epsilon$ is small everywhere. In doing this, we must be careful not to try and use the naive perturbation theory of Sect. 2, since we would thereby miss the possibility of finding nonunique solutions. A tool which avoids that pitfall is based upon the principle of "zero virtual work" [32]. That principle asserts that the solution of the nonlinear partial differential equation for the electric potential $\phi(\mathbf{r})$

$$\nabla \cdot [\epsilon(\mathbf{r}, \nabla\phi)\nabla\phi] = 0 \,, \tag{47}$$

given boundary conditions for $\phi$ on the system surface, also satisfies the following equation:

$$\int dV [\epsilon(\mathbf{r}, \nabla\phi)\nabla\phi]^* \cdot \nabla\delta\phi = 0 \,, \tag{48}$$

where $\delta\phi(\mathbf{r})$ is any scalar function that vanishes on that surface. To use this variational principle, we need to restrict the functions $\phi$ and $\delta\phi$ to a manageable subset of all functions that satisfy the required boundary conditions and includes functions that are sufficiently close to the exact solution. Such a subset is constructed by starting from the solution $\phi_{\mathrm{lin}}$ of the linear problem, where the $\nabla\phi$ dependent part of $\epsilon$ is ignored. Then, we choose our trial function to have the form

$$\phi(\mathbf{r}) = \phi_0(\mathbf{r}) + A[\phi_{\mathrm{lin}}(\mathbf{r}) - \phi_0(\mathbf{r})] \,, \quad \delta\phi(\mathbf{r}) = \delta A[\phi_{\mathrm{lin}}(\mathbf{r}) - \phi_0(\mathbf{r})] \,, \tag{49}$$

where $A$ is a complex parameter, determined by requiring that Eq. (48) be satisfied. The function $\phi_0(\mathbf{r})$ must satisfy the same boundary conditions as $\phi(\mathbf{r})$, so that $\delta\phi(\mathbf{r})$ vanishes at the boundary. The precise form of $\phi_0(\mathbf{r})$ is usually chosen to simplify computation of the integral in (48). For example, if the microstructure is macroscopically homogeneous and the boundary conditions would result in a uniform field $\mathbf{E}_0$ for a uniform medium, then we can choose $\phi_0(\mathbf{r}) = -\mathbf{E}_0 \cdot \mathbf{r}$, so that $-\nabla\phi_0 \equiv \mathbf{E}_0$.

Following this prescription, we arrive at a nonlinear algebraic equation for $A$. When the field dependence of $\epsilon$ on $\mathbf{E}$ is quadratic, as in (34), then the equation for $A$ is cubic, and it can be reduced to the canonical form of (39) by suitable definitions of $t$, $\alpha$, $\mu$. This approach has been applied successfully to a number of situations, including a dilute suspension of metal spheres in a nonlinear dielectric host [32], a "coated sphere assemblage" with a nonlinear dielectric core and a metal shell [32] (see Fig. 1c), and a periodic array of metal inclusions in a nonlinear dielectric host [35]. Experiments on such systems remain to be done. All of these systems have sharp resonances in their linear response. If the frequency is close enough to one of those resonances, then bistable behavior is always possible for some range of $\mathbf{E}_0$, but the field-dependent part of $\epsilon$ is much smaller everywhere than the field-independent part. For that reason, the approximation described here is, in some sense, a lowest order perturbation theory. However, this perturbation approximation is carried out in a fashion that allows nonunique solutions to occur.

## 5.3   Harmonic Generation and Induced Nonlinearity

In general, the nonlinear dependence of $\mathbf{D}$ upon a monochromatic $\mathbf{E}$ introduces higher harmonics into the time dependence of $\mathbf{D}$. These effects are enhanced near a sharp resonance of the linear response of a metal/dielectric composite medium, because of enhancement of the ratio between local field $\mathbf{E}$ and applied field $\mathbf{E}_0$. Another result of this local field enhancement is the appearance of enhanced induced nonlinear behavior at the fundamental frequency $\omega$ of the applied field [36].

To demonstrate this, we focus on a two-constituent, metal/dielectric parallel slabs microstructure, where the basic local response of the metal constituent is linear, and that of the dielectric constituent is limited to linear and quadratic dependence of $\mathbf{D}$ upon $\mathbf{E}$:

$$D_{\mathrm{M},m\omega} = \epsilon_{\mathrm{M},m\omega}E_{\mathrm{M},m\omega} \,, \tag{50}$$

$$D_{\mathrm{d},m\omega} = \epsilon_{\mathrm{d},m\omega}E_{\mathrm{d},m\omega} + \sum_{n=-\infty}^{\infty} d_{(m+n)\omega,-n\omega}E_{\mathrm{d},(m+n)\omega}E_{\mathrm{d},-n\omega} \,. \tag{51}$$

Here $E_{\mathrm{M},m\omega}$, $D_{\mathrm{M},m\omega}$ and $E_{\mathrm{d},m\omega}$, $D_{\mathrm{d},m\omega}$ denote complex amplitudes of monochromatic local fields in the appropriate constituent at the $m$th multiple of the fundamental frequency $\omega$, and $\epsilon_{\mathrm{M},m\omega}$, $\epsilon_{\mathrm{d},m\omega}$, $d_{(m+n)\omega,-n\omega}$ are material coefficients at the appropriate frequencies. The fields are all assumed to be perpendicular to the slabs, hence the use of a scalar notation for the fields. They are also assumed to have the same values in all of the slabs of a particular constituent. The applied or average time-dependent field $\mathbf{E}_0 e^{-i\omega t}$ induces polarization at the second-harmonic frequency, and that beats with the fundamental frequency to produce a contribution to the local field at the fundamental frequency, with an amplitude that is proportional to $|\mathbf{E}_0|^2\mathbf{E}_0$. An equation that describes this effect in the parallel slabs microstructure is

$$\left(p_{\mathrm{d}}\epsilon_{\mathrm{M},\omega} + p_{\mathrm{M}}\epsilon_{\mathrm{d},\omega} - p_{\mathrm{M}}\chi|E_{\mathrm{d},\omega}|^2\right)E_{\mathrm{d},\omega} = \epsilon_{\mathrm{M},\omega}E_{0,\omega} \,, \tag{52}$$

where

$$\chi \equiv \frac{2p_{\mathrm{M}}d_{\omega,\omega}d_{2\omega,-\omega}}{p_{\mathrm{d}}\epsilon_{\mathrm{M},2\omega} + p_{\mathrm{M}}\epsilon_{\mathrm{d},2\omega}} \,. \tag{53}$$

This is similar to (37), but with $\chi$ replaced by $-\chi$. Thus, there are similar consequences, including intrinsic optical bistability near the fundamental frequency resonance of the microstructure, which occurs when $p_{\mathrm{d}}\epsilon_{\mathrm{M},\omega} + p_{\mathrm{M}}\epsilon_{\mathrm{d},\omega} = 0$. We note that the cubic nonlinear susceptibility or nonlinearity coefficient $\chi$ is of second order in the quadratic nonlinearity coefficients $d_{\omega_1,\omega_2}$, and therefore can be expected to be small. However, if the system is also near resonance at the *second-harmonic frequency* $2\omega$, then the denominator in (53) will be small, and $\chi$ will be enhanced. Consequently, intrinsic optical bistability will appear at low applied and local fields if the system is near resonance at *both the fundamental frequency $\omega$ and at its second harmonic.* Though this may be difficult to achieve in the two-constituent parallel slabs system, where only one microstructural parameter is at our disposal, it should be achievable when we allow for more complicated microstructures.

## Acknowledgments

DJB acknowledges the hospitality of Michigan State University, where some of his work on this contribution was performed. That work was also supported, in part, by research grants from the Israel Science Foundation and the US-Israel Binational Science Foundation. The work of DGS was supported by NSF Grants No. DMR 97-31511 and DMR 01-04987.

## References

1. D. J. Bergman, D. Stroud, Solid. State Phys. **46**, 147 (1992)
2. D. Stroud, P. M. Hui, Phys. Rev. B **37**, 8719 (1988)
3. A. V. Butenko, V. M. Shalaev, M. I. Stockman, Z. Phys. D **10**, 81 (1988)
4. D. J. Bergman, B. I. Halperin, P. C. Hohenberg, Phys. Rev. B **11**, 4253 (1975), App. B
5. R. Rammal, C. Tannous, A.-M. S. Tremblay, Phys. Rev. A **31**, 2662 (1985)
6. X. C. Zeng, P. M. Hui, D. J. Bergman, D. Stroud, Phys. Rev. B **38**, 10970 (1988)
7. D. Stroud, V. E. Wood, J. Opt. Soc. Am. **6**, 778 (1989)
8. D. J. Bergman, Phys. Rep. **43**, 377 (1978); also published in *Willis E. Lamb, Jr., a festschrift on the occasion of his 65th birthday*, D. ter Haar, M. O. Scully (Eds.) (North Holland, Amsterdam 1978) pp. 377–407
9. H. Ma, R. Xiao, P. Sheng, J. Opt. Soc. Am. B **15**, 1022 (1998)
10. D. C. Wright, D. J. Bergman, Y. Kantor, Phys. Rev. B **33**, 396 (1986)
11. D. J. Bergman, Phys. Rev. B **39**, 4598 (1989)
12. Z. Hashin, S. Shtrikman, J. Appl. Phys. **33**, 3125 (1962)
13. K. W. Yu, P. M. Hui, D. Stroud, Phys. Rev. B **47**, 14150 (1993)
14. X. Zhang, D. Stroud, Phys. Rev. B **49**, 944 (1994)
15. A. E. Neeves, M. H. Birnboim, J. Opt. Soc. Am. B **6**, 787 (1989)
16. H. B. Liao, R. F. Xiao, J. S. Fu, P. Yu, G. K. L. Wong, P. Sheng, Appl. Phys. Lett. **70**, 1 (1997)
17. K. P. Yuen, M. F. Law, K. W. Yu, P. Sheng, Opt. Commun. **148**, 197 (1998)
18. K. P. Yuen, M. F. Law, K. W. Yu, P. Sheng, Phys. Rev. E **56**, R1322 (1997)
19. J. W. Haus, N. Kalyaniwalla, R. Inguva, M. Bloemer, C. M. Bowden, J. Opt. Soc. Am. B **6**, 797 (1989)
20. M. J. Bloemer, P. R. Ashley, J. W. Haus, N. Kalyaniwalla, C. R. Christensen, IEEE J. Quant. Electron. **26**, 1075 (1990)
21. D. A. G. Bruggeman, Ann. Phys. (Leipzig) **24**, 636 (1935)

22. R. Landauer, J. Appl. Phys. **23**, 779 (1952)
23. B. Budiansky, J. Mech. Phys. Solids **13**, 223 (1965)
24. R. Hill, J. Mech. Phys. Solids **13**, 213 (1965)
25. H. Stachowiak, Physica **45**, 481 (1970)
26. M. H. Cohen, J. Jortner, Phys. Rev. Lett. **30**, 696 (1973)
27. D. Stroud, Phys. Rev. B **12**, 3368 (1975)
28. D. J. Bergman, Y. M. Strelniker, Phys. Rev. B **60**, 13016 (1999)
29. D. J. Bergman, in *Composite Media and Homogenization Theory*, G. Dal Maso, G. F. Dell'Antonio (Eds.) (Birkhäuser, Berlin 1991) pp. 67–79
30. L. Sali, D. J. Bergman, J. Stat. Phys. **86**, 455 (1997)
31. G. Milton, in *Physics and Chemistry of Porous Media*, AIP Conf. Proc. **107**, D. L. Johnson, P. N. Sen (Eds.) (AIP, New York, 1983), pp. 66–67
32. D. J. Bergman, O. Levy, D. Stroud, Phys. Rev. B **49**, 129 (1994)
33. N. Kalyaniwalla, J. W. Haus, R. Inguva, M. H. Birnboim, Phys. Rev. A **42**, 5613 (1990)
34. R. Neuendorf, M. Quinten, U. Kreibig, J. Chem. Phys. **104**, 6348 (1996)
35. R. Levy-Nathansohn, D. J. Bergman, J. Appl. Phys. **77**, 4263 (1995)
36. O. Levy, D. J. Bergman, D. Stroud, Phys. Rev. B **52**, 3184 (1995)

# Third-Order Nonlinear Properties
# of Au Clusters Containing Dielectric
# Thin Films

Hongru Ma[1], Ping Sheng[2], and George K. L. Wong[2]

[1] Department of Physics, Shanghai Jiao Tong University
   Shanghai, China
[2] Department of Physics, The Hong Kong University of Science and Technology
   Clear Water Bay, Kowloon, Hong Kong, China
   phgkwong@ust.hk

**Abstract.** We show experimentally and theoretically that very large enhancement of $\chi^{(3)}$ can be obtained in gold-cluster-containing dielectric thin films. At a high volume fraction of gold clusters, the concentration dependence of $\chi^{(3)}$ cannot be accounted for by the Maxwell Garnett theory. To go beyond the MG theory, we have derived the spectral densities of different microgeometries with related formulae for calculating the effective dielectric constant and the enhancement factors for third-order nonlinearities. It is shown that the largest enhancement source is the anomalous dispersion, realized by the MG and the Sheng theories. Substantially improved agreement between theory and experiment is obtained. Even larger enhancement is predicted as possible by the theory.

## 1 Introduction

Scientific interest in the optical properties of metal-doped glasses has a long history. But study of their nonlinear optical properties only began in the 1980s, when *Flytzanis* and co-workers [1] measured the third-order nonlinear susceptibilities, $\chi^{(3)}$, of Au-doped glasses as well as Ag colloids dispersed in water. It was observed that $\chi^{(3)}$s of the composites were enhanced by several orders over that of pure glass even when only a very small amount of gold clusters (volume fractions of gold $p \sim 10^{-5} - 10^{-6}$) was incorporated. Through a series of extensive studies by the same group [2], the fundamental processes involved in low gold concentration composites are now fairly well understood. In the limit of $p \ll 1$, so that each gold nanoparticle is entirely surrounded by glass or other transparent dielectric hosts and the interparticle spacing is large with respect to the size of the nanoparticles and much smaller than the wavelength of laser, the Maxwell–Garnett theory (MG) [3] works quite well. MG gives the following expression for the effective nonlinear susceptibility of the composites:

$$\chi_{\text{eff}}^{(3)} = p f_1^2(\omega)|f_1(\omega)|^2 \cdot \chi_{\text{m}}^{(3)}, \tag{1}$$

where the local field factor, $f_1(\omega)$, is given by $f_1(\omega) = 3\epsilon_{\text{d}}/[\epsilon_{\text{m}}(\omega)+2\epsilon_{\text{d}}]$, where $\epsilon_{\text{m}}$ and $\epsilon_{\text{d}}$ are the dielectric constant of the gold particles and the surrounding

V. M. Shalaev (Ed.): Optical Properties of Nanostructured Random Media,
Topics Appl. Phys. **82**, 41–61 (2002)
© Springer-Verlag Berlin Heidelberg 2002

dielectric, respectively, and $\chi_m^{(3)}$ is the third-order nonlinear susceptibility of the gold nanoparticle itself. $\chi_m^{(3)}$ can be separated into three contributions:

$$\chi_m^{(3)} \cong \chi_{intra}^{(3)} + \chi_{inter}^{(3)} + \chi_{he}^{(3)}. \tag{2}$$

The first and second terms result from the familiar intraband and interband transitions that also play important roles in the linear susceptibility. The third contribution is due to the modification of the populations of the electrons resulting from the large increase in the electron temperature caused by resonant absorption of photons but before the heat can be transferred to the lattice of the surrounding dielectric. This is the so-called "hot-electron" contribution. After the lattice is heated up, there is an additional thermal contribution, $\chi_{thermal}^{(3)}$, which is a much slower process with a response time of nanoseconds or longer. The thermal contribution could become important when nanosecond laser pulses are used in the measurements. The susceptibilities due to these three electronic contributions are essentially imaginary quantities. Their values have been estimated theoretically as $\mathrm{Im}\chi_{intra}^{(3)} \sim -10^{-10}$ esu, $\mathrm{Im}\chi_{inter}^{(3)} \sim -2 \times 10^{-8}$ esu, $\mathrm{Im}\chi_{he}^{(3)} \sim -10^{-7}$ esu. The above theoretical expectations were largely borne out in the experimental work of *Flytzanis* and co-workers [2]. A prominent enhancement of $\chi^{(3)}$ at the surface plasmon resonance wavelength, as predicted by the fourth power of the local field factor in (1), was also observed in their experiments. The measured value of $\chi_m^{(3)} \sim 5 \times 10^{-8}$ esu is consistent with theoretical estimates.

Despite the large enhancement, the $\chi^{(3)}$ of the composite films studied in those experiments was still in the range of $10^{-12} - 10^{-11}$ esu because of the low concentration of gold ($p \sim 10^{-5} - 10^{-6}$) in those composites. Similar enhancement of $\chi^{(3)}$ has also been observed in Ag cluster colloids [4,5,6]. Since, according to (1), $\chi^{(3)}$ is proportional to the concentration of metal clusters, this suggests the possibility that useful nonlinear optical materials might be obtainable by increasing the gold concentration. Additionally, it is scientifically interesting to investigate to what extent (1) continues to be valid for large $p$ and how the theory should be modified in that limit. In this contribution, we review recent experimental and theoretical work that we undertook in this direction.

## 2 Experimental

Several fabrication methods were used to obtain gold-nanoparticle-containing glasses. These include melt quenching [7,8], sol-gel [9,10], ion implantation [11,12], and sputtering [13,14].

### 2.1 Preparation and Characterization of Samples

Using melt quenching and sol-gel methods, the gold volume fraction that can be incorporated in the glass is usually rather low (typically $p \sim 10^{-5} - 10^{-6}$).

Ion implantation has been used successfully to incorporate a larger fraction [12]. To further increase the metal concentration in the composite film, sputtering is a more effective approach. *Tanahashi* et al. [14] used the sputtering technique to deposit alternative layers of Au and SiO$_2$. After thermal annealing, composite films with a gold volume fraction of $\sim 3\%$ were obtained. In our work [15,16,17,18], we adopted a cosputtering technique to deposit Au and SiO$_2$ simultaneously to obtain composite films first in which gold atoms are homogeneously distributed in the glass matrix. The as-deposited films were then subject to rapid thermal annealing to allow gold atoms to diffuse and nucleate into nanoparticles at nucleation sites. A multitarget magnetron sputtering system, shown schematically in Fig. 1, was used for the cosputtering. An Au target (purity 99.99%) and a fused SiO$_2$ target (purity 99.999%) were used in gun A and gun B, respectively. The two targets are diagonally opposite each other separated by 6 in. The target surfaces are inclined about 45° toward the substrate holder, which is located 5 in. below the targets. Rectangular quartz plates $75 \times 10$ mm were used as a substrate, and the plates were mounted with the long side running gun A to gun B. Codeposition onto a substrate at a temperature of 150° C was carried out in the presence of Ar gas and a small amount of O$_2$. The deposition rates of Au and SiO$_2$ were calibrated in separate runs by sputtering Au or SiO$_2$ to obtain pure Au or SiO$_2$ films. The deposition rate of Au as well as that of SiO$_2$, it was found, varies linearly across the substrate surface, and the highest deposition rate occurs at the edge of the substrate that is closest to the target. Thus by cosputtering both materials, the Au concentration in the deposited films would vary gradually from one edge of the rectangular film to the other. The range of

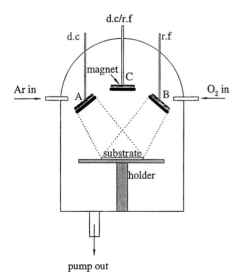

**Fig. 1.** Schematics of the cosputtering geometry

Au concentration in a given composite film could be adjusted by varying the sputtering rate of Au relative to that of SiO$_2$. Due to the slow deposition rate of SiO$_2$, one to two hours were required to grow a film of about 2000 Å thick. Typically, only two deposition runs were needed to cover Au concentrations from 4 to 75% (volume fractions) — one film covered 4 to 40% and the other from 35 to 75%. The composite films prepared this way are ideally suited for composition-dependent studies because uncertainties in other growth conditions are minimized. The as-deposited films were annealed at 850° C for one minute in a rapid annealing furnace (temperature ramped up at a rate up to 200° C/s) in one atmosphere of Ar gas and then cooled down in a flowing Ar atmosphere. We found that rapid thermal annealing was necessary to keep the sizes of nanoparticles small as a result of the high volume fraction of gold in the films. From TEM micrographs, the average radius of gold clusters radius was below 10 nm in films whose volume fraction was lower than 26%. For higher fractions, the size distribution became much broader; average sizes were several tens of nanometers, complicating comparison of experimental results in this range with the theories. After annealing, the color of low Au concentration films changes from light brown to ruby-like. The optical absorption spectra of several annealed samples containing different gold concentrations are shown in Fig. 2. With increasing Au concentration, the resonant peak sharpens and shifts to longer wavelength from 520 nm at 5% Au to 540 nm at 45%. For still higher concentration, the resonant absorption peak decreases, and the off-resonant absorption increases. At 63%, the absorption spectrum becomes completely different. Instead of the appearance of an absorption peak, the absorption spectrum exhibits a transmission min-

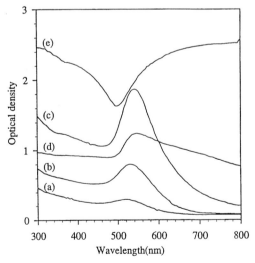

**Fig. 2.** Absorption spectra of SiO$_2$ composite films after thermal annealing (**a**) 5% Au; (**b**) 21% Au; (**c**) 45% Au; (**d**) 59% Au; (**e**) 63% Au

imum near the absorption peak seen at lower concentrations. Conductivity measurements show a clear percolation threshold at 60; the change in the transmission spectrum is associated with the onset of percolation, which is consistent with the theoretical prediction of *Sheng* [19] (Fig. 3).

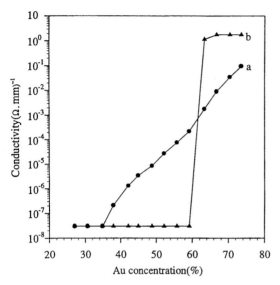

**Fig. 3.** Au composition dependence of electrical conductivity in the Au:SiO$_2$ composite films on a fused substrate. (**a**) Before annealing; (**b**) after annealing at 850° C for 1 min

## 2.2   Nonlinear Optical Measurements

The third-order nonlinear susceptibility $\chi^{(3)}$ of gold-nanoparticle-composite films was studied using degenerated four-wave mixing (DFWM) schemes [15,16,17] using femtosecond as well as picosecond pulse durations. Femtosecond pulses were provided by a UV-pumped $\beta$-barium borate optical parametric amplifier system having a tunable wavelength range from 470 to 780 nm, pulse widths of 200 fs, and a repetition rate of 1 kHz. Two different picosecond lasers were used — a Q-switched CW mode-locked Nd:Yag laser emitting 70 ps pulses at a repetition rate of 500 Hz and a Q-switched active-passive mode-locked laser emitting 35 ps pulses at a repetition rate of 10 Hz. The frequency-doubled outputs (532 nm) of picosecond lasers were used for the third-order nonlinear susceptibility measurements.

A forward DFWM geometry was used for the femtosecond measurements [16], and a backward one for the picosecond cases [15]. The pump and probe beams were polarized in the same direction and perpendicular to the plane containing the pump and probe beams (copolarized beam configuration), so that only the $\chi^{(3)}_{xxxx}$ component was measured. The intensity of the conjugate beam was detected by using a Si photodiode, and the signal was averaged by using a lock-in detection system or a boxcar integrator. The absolute value of $\chi^{(3)}$ was obtained using $CS_2$ as a standard reference. $\chi^{(3)}$ of the films was calculated from that of $CS_2$ by using the equation

$$\chi^{(3)} = \chi^{(3)}_{CS_2} \sqrt{\frac{I_s}{I_{CS_2}}} \frac{n_s^2}{n_{CS_2}^2} \frac{L_{CS_2}}{L_s} \frac{\ln(1/T)}{(1-T)\sqrt{T}}, \tag{3}$$

where $I_s$ and $I_{CS_2}$ are the intensities of the conjugate signals, $n_s$ and $n_{CS_2}$ are the refractive indices, $L_s$ and $L_{CS_2}$ are the thicknesses of the Au:SiO$_2$ films and the $CS_2$, respectively, and $T$ is the transmissivity of the film. The value of $\chi^{(3)}_{xxxx}$ for $CS_2$ is taken as $5 \times 10^{-13}$ esu in the femtosecond regime [20] and $2 \times 10^{-12}$ esu in the picosecond regime [1].

The third-order susceptibility of the annealed films is plotted in Fig. 4 as a function of Au volume fraction. The triangles, squares, and dots represent the results obtained using 200 fs, 35 ps, and 70 ps lasers, respectively, at the

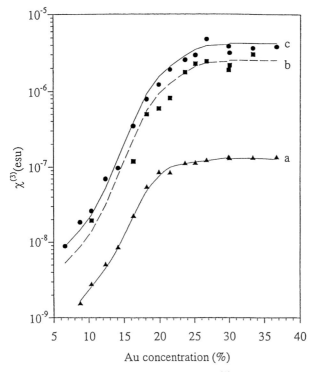

**Fig. 4.** Concentration dependence of $\chi^{(3)}$ in Au:SiO$_2$ composite films measured with three probe lasers (all at 532 nm) with pulse duration of (**a**) 200 fs; (**b**) 36 ps; (**c**) 70 ps

same wavelength of 532 nm. One sees that $\chi^{(3)}$ increases rapidly with increasing Au volume fraction, but the value of $\chi^{(3)}$ depends strongly on the pulse width of the laser used. For the 8% film, the maximum value of $\chi^{(3)}$ obtained using the 70 ps laser is $\sim 4 \times 10^{-6}$ esu, 40% larger than that measured by using the 35-ps laser and 10 times that measured by using the 200-fs laser. This indicates that the measured $\chi^{(3)}$ involves one or more mechanisms that have a response time much longer than 200 fs. To obtain more information about these processes, we used a three-beam forward DFWM scheme, shown in the inset in Fig. 5, to time-resolve the conjugate signal. The measurement was carried out using the 200-fs, 532-nm output of the parametric amplifier system. In this scheme, $b_1$ and $b_2$ are the pump beams that are arranged to arrive at the sample at exactly the same time and form a transient diffraction grating. When the probe beam, $b_3$, enters the sample, a signal beam, $b_s$, is generated as $b_3$ is diffracted off the transient grating. In Fig. 5, the intensity of the signal beam is shown as a function of the arrival time of the probe beam for three samples having different gold volume fractions. For the 12% sample, the nonlinear response has an ultrafast component with a time constant of a few hundred femtoseconds as well as a much slower and smaller component that has a time constant of a few picoseconds. For samples with higher Au

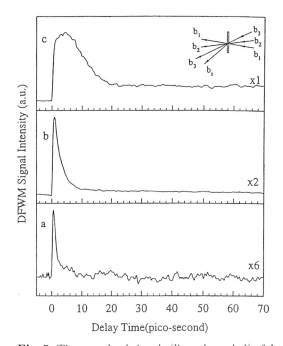

**Fig. 5.** Time-resolved signals (linearly scaled) of Au:SiO$_2$ composite films measured with three-beam forward DFWM and a probe laser with a pulse width of 200 fs at Au concentrations of (**a**) 12%; (**b**) 25% Au; (**c**) 33% Au

concentration, the slower response becomes relatively more important. We attribute the picosecond component to the hot-electron contribution ($\chi_{he}^{(3)}$), which *Flytzanis* et al. [2] also concluded is the dominant contribution in their experiments on low concentration systems using a picosecond laser. The ultrafast component observed in our experiments could contain both intraband and interband contributions ($\chi_{intra}^{(3)}$ and $\chi_{inter}^{(3)}$) as both are ultrafast processes. At higher volume (25 and 33%), the data in Fig. 5 suggest that the hot electron component becomes relatively larger. This could be due to fact that it is expected to decrease rapidly with increasing nanoparticle size. Indeed, from our TEM studies, the average size of the nanoparticles is much larger in our high Au volume fraction films. Beyond 25%, the size of nanoparticles rapidly increases, and average radii become much larger than 10 nm. Our observation that the intraband process also plays an important role in small nanoparticle samples seems to be at variance with the conclusion of *Flytzanis* et al., who did not observe obvious size dependence in their measured nonlinear susceptibility. A possible reason is that when picosecond laser pulses are used, the hot electron contribution would be larger and the size dependence of the nonlinear susceptibility would become less obvious. In addition to the electronic contributions, a very slow component with nanosecond or longer time constant is also evident, especially when the 33-ns laser is used to measure $\chi^{(3)}$ of gold composite films; as was the case in most of the early experiments, the contribution of thermal effect has to be considered.

We also carried out similar studies on Au-nanoparticle-embedded $TiO_2$ and $Al_2O_3$ films [17,18]. These films were fabricated by using the same sputtering system described previously. The experimental results are qualitatively the same as those obtained on $SiO_2$ films and presented above. Readers are referred to the original articles for details. Here, we simply want to mention that the effect of the dielectric host on the nonlinear susceptibility is clearly borne out by these studies. The large nonlinear optical susceptibility of gold-nanoparticle-doped glass stems from the enhancement of the local electrical field near and inside the gold particles at the surface plasmon resonance. Using (1), one can obtain a rough estimate of the effect of the dielectric. Under the resonance condition, i.e., when $|\epsilon_m(\omega) + 2\epsilon_d| \sim 0$, the local-field factor $f_1(\omega)$ is reduced to $3\epsilon_d/i\epsilon_2$, where $\epsilon_2$ is the imaginary part of the dielectric constant of gold ($\epsilon_m = \epsilon_1 + i\epsilon_2$). Hence, the larger the refractive index ($n$) of the host ($\epsilon_d = n^2$ for a nonabsorbing host), the larger the local-field factor $f_1(\omega)$. The refractive index of $TiO_2$ is 2.7 compared to 1.5 of $SiO_2$. Since the enhancement is proportional to the fourth power of $f_1(\omega)$, an order of magnitude larger nonlinear susceptibility should be observed in Au-nanoparticle $TiO_2$ films compared to that of $SiO_2$ films containing a comparable volume fraction of Au. This was indeed borne out in our measurements.

## 3    Theoretical

A key finding of our experiments is that the susceptibility increases linearly with $p$ up to about $p = 0.15$, beyond which $\chi^{(3)}$ increases much faster than a linear dependence. From $p = 0.15$ to $0.28$, the dependence is described reasonably well by a power law $\chi^{(3)} \sim p^t$ with $t$ approximately equal to 3 (see Fig. 6). Beyond $p = 0.28$, the susceptibility levels off and eventually decreases as the percolation threshold is approached. Obviously, the Maxwell Garnett theory breaks down badly for large $p$, and a better theory is needed. The theory of *Shalaev* and co-workers [21,22,23,24] also predicts a linear dependence of $\chi^{(3)}$ on $p$.

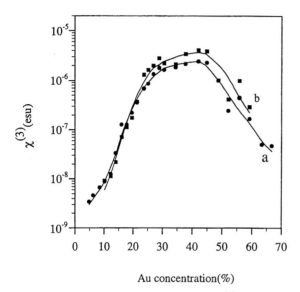

**Fig. 6.** Concentration dependence of $\chi^{(3)}$ measured by forward DFWM (**a**) and backward DFWM (**b**). Both measurements yield essentially the same result. The data between $p = 0.15$ to $0.28$ can be fitted reasonably well by a power law $\chi^{(3)} \sim p^3$

### 3.1    General Theory

The Kerr nonlinear susceptibility of gold-cluster-doped glasses can be enhanced at certain frequencies through the Mie resonance of the constituent materials. The enhancement is expected to depend on the microstructure of the composite. We analyze this effect in the long wavelength limit through the powerful tools of spectral representations [25].

Consider a composite with two constituent materials that form a specific microstructure. The optical properties of the composite are described by the local constitutive relation of the system, given by

$$\mathbf{D} = (\epsilon + \chi^{(3)}|\mathbf{E}|^2)\mathbf{E}\,, \tag{4}$$

where $\epsilon$ is the (spatial-dependent) dielectric constant and $\chi^{(3)}$ is the (spatial-dependent) third-order Kerr nonlinear susceptibility. In the long wavelength limit, the composite acts as a effective medium for the electromagnetic waves; the effective dielectric constant and the effective third-order susceptibility give the refractive index and the Kerr nonlinear effects. These effective quantities are defined through the spatial average of $\mathbf{D}$:

$$\frac{1}{V}\int \mathrm{d}V\mathbf{D} = \frac{1}{V}\int \mathrm{d}V(\epsilon\mathbf{E} + \chi^{(3)}|\mathbf{E}|^2\mathbf{E}) \equiv \bar{\epsilon}\mathbf{E}_0 + \overline{\chi^{(3)}}|\mathbf{E}_0|^2\mathbf{E}_0\,. \tag{5}$$

Here $\mathbf{E}_0 = (\frac{1}{V})\int \mathrm{d}V\mathbf{E}$ is the applied (averaged) electrical field, taken as real. The factor $\mathrm{e}^{-i\omega t}$ is dropped from the expression for simplicity. It is more convenient to calculate the effective quantities through the energy relation instead of the polarization relation (5). Following the general approach of *Stroud* and *Hui* [26], in the quasi-static limit(long wavelength limit), we can obtain the following alternative definitions of effective quantities:

$$\frac{1}{V}\int \mathrm{d}V\mathbf{D}\cdot\mathbf{E} = \bar{\epsilon}\mathbf{E}_0^2 + \overline{\chi^{(3)}}|\mathbf{E}_0|^2\mathbf{E}_0^2 \tag{6}$$

and

$$\frac{1}{V}\int \mathrm{d}V\mathbf{D}\cdot\mathbf{E}^* = \frac{1}{V}\int \mathrm{d}V\mathbf{D}\cdot\mathbf{E} \tag{7}$$

$\mathbf{E}^*$ is the complex conjugate of $\mathbf{E}$. Equation (6) is equivalent to (5) in the quasi-static limit, as can be proved through using $\nabla\cdot\mathbf{D} = 0$ and $\nabla\times\mathbf{E} = 0$. It is usually difficult to have an accurate evaluation of an effective nonlinear optical coefficient due to the lack of an analytic solution to the nonlinear differential equation that results from the nonlinear dielectric constant(s) of the components. However, the magnitude of the nonlinear coefficient is usually small, which means the problem may be treated in the perturbation sense, i.e., the electrical field may be evaluated by ignoring the nonlinearity. To the first order in nonlinearity, therefore,

$$\bar{\epsilon}\mathbf{E}_0^2 = \frac{1}{V}\int \mathrm{d}V\epsilon\mathbf{E}_{\mathrm{lin}}^2\,, \tag{8}$$

$$\begin{aligned}
\overline{\chi^{(3)}}|\mathbf{E}_0|^2\mathbf{E}_0^2 &= \frac{1}{V}\int \mathrm{d}V\chi^{(3)}|\mathbf{E}_{\mathrm{lin}}|^2\mathbf{E}_{\mathrm{lin}}^2 \\
&\equiv p_1\chi_1^{(3)}\langle|\mathbf{E}_{\mathrm{lin}}|^2\mathbf{E}_{\mathrm{lin}}^2\rangle_1 + p_2\chi_2^{(3)}\langle|\mathbf{E}_{\mathrm{lin}}|^2\mathbf{E}_{\mathrm{lin}}^2\rangle_2 \\
&\equiv (\beta_1\chi_1^{(3)} + \beta_2\chi_2^{(3)})|\mathbf{E}_0|^2\mathbf{E}_0^2\,,
\end{aligned} \tag{9}$$

where $\langle\cdots\rangle_i$ means volume averaging inside material $i$; $\chi_1^{(3)}, \chi_2^{(3)}, \beta_1$, and $\beta_2$ are defined as the third-order nonlinear susceptibilities and enhancement factors of material 1 and material 2, respectively; $\mathbf{E}_{\mathrm{lin}}$ is the solution of the linear electrostatic problem, i.e., the problem for which $\chi^{(3)}$ is zero. We see that to get the enhancement factors of nonlinear susceptibilities, the only knowledge needed is the spatial distribution of the linear electrical field. A complete solution of the linear problem is in general also very difficult except for some very special microstructures. To avoid this difficulty, one usually decouples the averages in (9) as

$$\langle|\mathbf{E}_{\text{lin}}|^2\mathbf{E}_{\text{lin}}^2\rangle_i \simeq \langle|\mathbf{E}_{\text{lin}}|^2\rangle_i\langle\mathbf{E}_{\text{lin}}^2\rangle_i,\tag{10}$$

which implies that

$$\beta_i \simeq p_i\frac{\langle|\mathbf{E}_{\text{lin}}|^2\rangle_i\langle\mathbf{E}_{\text{lin}}^2\rangle_i}{|\mathbf{E}_0|^2\mathbf{E}_0^2}.\tag{11}$$

This approximation is exact when the field inside material $i$ is a constant and becomes *poor* when the variation of the field inside material $i$ is large. The merit of this decoupling procedure is that the averages $\langle|\mathbf{E}_{\text{lin}}|^2|\rangle_i$ and $\langle\mathbf{E}_{\text{lin}}^2\rangle_i$ can be obtained from the effective dielectric constant and its spectral representation. One can easily show that

$$p_1\langle\mathbf{E}_{\text{lin}}^2\rangle_1 = \frac{\partial\bar{\epsilon}}{\partial\epsilon_1}\mathbf{E}_0^2,$$

$$p_2\langle\mathbf{E}_{\text{lin}}^2\rangle_2 = \frac{\partial\bar{\epsilon}}{\partial\epsilon_2}\mathbf{E}_0^2.\tag{12}$$

Here $p_1$ and $p_2$ are the volume fraction of components 1 and 2, respectively. Equations (12) are widely used in the literature to calculate the averaged squared electrical field. To calculate $\langle|\mathbf{E}_{\text{lin}}|^2|\rangle_i$, we go to the spectral representation [25,27], where the electrostatic potential of the linear electrostatic equation can be written in operator form as

$$\phi = -\frac{s}{s-\Gamma}z = -\sum_n\frac{s\langle n|z\rangle}{s-s_n}\phi_n,\tag{13}$$

where $s_n$ and $\phi_n$ are the $n$th eigenvalue and eigenfunction of the self-adjoint operator $\Gamma$, defined as

$$\Gamma = \int dV'\eta(\mathbf{r}')\nabla'G(\mathbf{r}-\mathbf{r}')\cdot\nabla'\tag{14}$$

where $G(\mathbf{r}-\mathbf{r}') = 1/(4\pi|\mathbf{r}-\mathbf{r}'|)$ denote Green's function for the Laplacian operator. Here, $s = \epsilon_2/(\epsilon_2-\epsilon_1)$ is the only material parameter of the problem, and $\eta(\mathbf{r})$ is an microstructure indicating function whose value is 1 inside grains of component 1 and zero otherwise. The last equality of (13) is valid only when $\mathbf{r}$ is inside the grains of component 1. From this solution,

$$p_1\langle|\mathbf{E}_{\text{lin}}|^2\rangle_1 = \frac{1}{V}\int_1 dV|\mathbf{E}_{\text{lin}}|^2 = \sum_n\frac{|s|^2f_n}{|s-s_n|^2}\mathbf{E}_0^2,\tag{15}$$

$$p_2\langle|\mathbf{E}_{\text{lin}}|^2\rangle_2 = \frac{\bar{\epsilon}}{\epsilon_2}\mathbf{E}_0^2 - \frac{\epsilon_1}{\epsilon_2}p_1\langle|\mathbf{E}_{\text{lin}}|^2\rangle_1 = \left(1 - \sum_n\frac{(|s|^2-s_n)f_n}{|s-s_n|^2}\right)\mathbf{E}_0^2,\tag{16}$$

with $f_n = |\langle n|z|\rangle|^2$. To see how this expression differs from $\langle \mathbf{E}_{\mathrm{lin}}^2 \rangle$, we also wish to write $\langle \mathbf{E}_{\mathrm{lin}}^2 \rangle$ in the spectral representation form. Since the effective dielectric constant can be expressed as

$$\bar{\epsilon} = \epsilon_2 \left( 1 - \sum_n \frac{f_n}{s - s_n} \right) \equiv \epsilon_2 [1 - F(s)] \,, \tag{17}$$

it follows that

$$p_1 \langle \mathbf{E}_{\mathrm{lin}}^2 \rangle_1 = \sum_n \frac{s^2 f_n}{(s - s_n)^2} \mathbf{E}_0^2 \,, \tag{18}$$

$$p_2 \langle \mathbf{E}_{\mathrm{lin}}^2 \rangle_2 = \left( 1 - \sum_n \frac{(s^2 - s_n) f_n}{(s - s_n)^2} \right) \mathbf{E}_0^2 \,. \tag{19}$$

It is clear that when $\epsilon$'s are real, (15) and (16) are the same as (18) and (19), as expected. However, $\langle \mathbf{E}_{\mathrm{lin}}^2 \rangle \neq \langle |\mathbf{E}_{\mathrm{lin}}|^2 \rangle$ when $\epsilon$'s are complex. A common error in the literature is to treat them as the same even when the $\epsilon$'s are complex. When the $\Gamma$ operator has a continuous spectrum, (15) and (16) may be written as

$$p_1 \langle |\mathbf{E}_{\mathrm{lin}}|^2 \rangle_1 = \int \mathrm{d}x \frac{|s|^2 \mu(x)}{|s - x|^2} \mathbf{E}_0^2 \,,$$

$$p_2 \langle |\mathbf{E}_{\mathrm{lin}}|^2 \rangle_2 = \left( 1 - \int \mathrm{d}x \frac{(|s|^2 - x)\mu(x)}{|s - x|^2} \right) \mathbf{E}_0^2 \,,$$

$$p_1 \langle \mathbf{E}_{\mathrm{lin}}^2 \rangle_1 = \int \mathrm{d}x \frac{s^2 \mu(x)}{(s - x)^2} \mathbf{E}_0^2 \,,$$

$$p_2 \langle \mathbf{E}_{\mathrm{lin}}^2 \rangle_2 = \left( 1 - \int \mathrm{d}x \frac{(s^2 - x)\mu(x)}{(s - x)^2} \right) \mathbf{E}_0^2 \,. \tag{20}$$

Here $\mu(x)$ is the spectral density of the operator $\Gamma$. It is related to $F(s)$ by the relation

$$F(s) = \int \frac{\mu(x)}{s - x} \mathrm{d}x \,. \tag{21}$$

By writing $s$ as $s + \mathrm{i}0^+$, the right-hand side of the above equation becomes $\mathrm{P} \int \frac{\mu(x)}{s - x} \mathrm{d}x - \mathrm{i}\pi\mu(s)$, and thus $\mu(x)$ is given by

$$\mu(x) = -\frac{1}{\pi} \mathrm{Im} \left[ F(x + \mathrm{i}0^+) \right] \,. \tag{22}$$

The advantage of using the spectral density to evaluate the dielectric constant and the optical nonlinearities lies in the separation of the geometric contribution from the material contribution. In effect, once a given type of microstructure is known and its associated spectral density calculated, all effective dielectric and optical properties can be simply evaluated from the material parameters of the components. Below we give explicit formulae for the four types of microstructures associated with their respective effective medium theories. It should be noted that with the known analytic expression of effective dielectric constant, the spectral density can be obtained from formula (22), and most effective quantities of the system related to spectral function can be calculated. In this respect, an analytic expression of the effective dielectric constant contains much more information than the effective dielectric constant itself.

## 3.2    Application of the Theory
## to Four Effective Medium Theories

### 3.2.1    The Dispersion Microstructure (Maxwell Garnett Theory)

The dispersion microstructure pertains to the geometry of colloidal systems, for example. It has a dispersed component (1) in a matrix component (2).

Since the dispersed component can never form an infinite connected network, this particular microstructure precludes the existence of a percolation threshold, except the trivial one at $p = 1$. Since the dispersion microstructure may be regarded as formed by a similar "structural unit" consisting of a sphere of component 1 coated by a layer of matrix material (component 2) [28] with different sizes to fill the space, optically there can be a so-called "anomalous dispersion" if component 1 is a metal [29]. The Maxwell Garnett (MG) effective dielectric constant is the solution of the equation [3,28]

$$\frac{\bar{\epsilon} - \epsilon_2}{\bar{\epsilon} + 2\epsilon_2} = p \frac{\epsilon_1 - \epsilon_2}{\epsilon_1 + 2\epsilon_2} . \tag{23}$$

Here $p_1 = p$, and $p_2 = 1 - p$. A representative picture of this microstructure and its spectral function is shown in Box 1. When the spectral function is available, $\chi^{(3)}$ may be directly evaluated from (11).

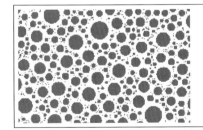

$$F(s) = \frac{p}{s - (1 - p)/3} .$$

$$\mu(x) = p\,\delta(x - \frac{1}{3}(1 - p)).$$

### 3.2.2    The Symmetrical Microstructure
### (Bruggeman's Self-Consistent Theory)

The symmetrical microstructure is so labeled because it satisfies the symmetry of the interchange between $p_1$ and $p_2$, simultaneous with the interchange of the material parameters of the two components. At either $p \longrightarrow 0$ or $p \longrightarrow 1$, the symmetrical microstructure is somewhat similar to the dispersion microstructure. But at intermediate compositions, there can be a matrix inversion and hence a percolation threshold for either component. At $p > 1/3$, component 1 forms an infinite connected network. At $p < 2/3$, component 2 forms an infinite connected network. At $1/3 < p < 2/3$, both components are connected in the symmetrical microstructure. Note that the anomalous dispersion effect is generally absent in the symmetrical microstructure, in contrast to the dispersion microstructure. In Bruggeman's self-consistent theory (EMA), the effective dielectric constant is the solution of [28,30]

$$p\frac{\epsilon_1 - \bar{\epsilon}}{\epsilon_1 + 2\bar{\epsilon}} = (1 - p)\frac{\epsilon_2 - \bar{\epsilon}}{\epsilon_2 + 2\bar{\epsilon}}. \tag{24}$$

Here, $p_1 = p$, $p_2 = 1 - p$. A representative picture of this microstructure and its spectral function is given in Box 2. $\chi^{(3)}$ may be evaluated from (11).

$$F(s) = \frac{1}{4s}\left(-1 + 3p + 3s - 3\sqrt{(s - x_1)(s - x_2)}\right)$$

$$x_1 = \frac{1}{3}\left(1 + p - 2\sqrt{2p(1 - p)}\right),$$

$$x_2 = \frac{1}{3}\left(1 + p + 2\sqrt{2p(1 - p)}\right).$$

$$\mu(x) = \frac{(3p - 1)}{2}\theta(3p - 1)\,\delta(x)$$
$$+ \begin{cases} \frac{3}{4\pi x}\sqrt{(x - x_1)(x_2 - x)} & x_1 < x < x_2 \\ 0 & \text{otherwise.} \end{cases}$$

### 3.2.3   The Granular Metal Microstructure (Sheng Theory)

In granular metals, both the anomalous dispersion effect and the percolation threshold in electrical conductivity were observed. Therefore, the previous two microstructures are inadequate for their description. The granular microstructure proposed by *Sheng* consists of two types of structural units [19]: a sphere of component 1 coated by a layer of component 2, and its reverse. This type of microstructure, it has been shown, gives the best description of the electrical and optical properties of granular metal films [19]. For spherically shaped structural units, the effective dielectric constant is given by the solution of the equation

$$f D_1 + (1 - f)D_2 = 0, \tag{25}$$

where

$$D_1 = \frac{(\bar{\epsilon} - \epsilon_2)(\epsilon_1 + 2\epsilon_2) + (\epsilon_2 - \epsilon_1)(\bar{\epsilon} + 2\epsilon_2)p}{(2\bar{\epsilon} + \epsilon_2)(\epsilon_1 + 2\epsilon_2) + 2(\bar{\epsilon} - \epsilon_2)(\epsilon_2 - \epsilon_1)p},$$

$$D_2 = \frac{(\bar{\epsilon} - \epsilon_1)(2\epsilon_1 + \epsilon_2) + (\epsilon_1 - \epsilon_2)(\bar{\epsilon} + 2\epsilon_1)(1 - p)}{(2\bar{\epsilon} + \epsilon_1)(2\epsilon_1 + \epsilon_2) + 2(\bar{\epsilon} - \epsilon_1)(\epsilon_1 - \epsilon_2)(1 - p)},$$

$$f = \frac{(1 - p^{1/3})^3}{\left(1 - (1 - p)^{1/3}\right)^3 + (1 - p^{1/3})^3}. \tag{26}$$

$p_1 = p$, and $p_2 = 1 - p$. The solution in the $s = \epsilon_2/(\epsilon_2 - \epsilon_1)$ notation can be written as

$$\frac{\bar{\epsilon}}{\epsilon_2} = \frac{-b(s)}{2a(s)} \pm \frac{\sqrt{b(s)^2 - 4a(s)c(s)}}{2a(s)}, \tag{27}$$

where

$$a(s) = 2s(3s - 3 + p)(3s - 1 + p),$$
$$b(s) = 2(2 - 3f)(1 - p)p + s(-3 - 5p + 2p^2 + 12s + 3ps - 9s^2),$$
$$c(s) = (1 - s)(2p + 4p^2 - 3s - 12ps + 9s^2). \tag{28}$$

The $F(s)$ and $\mu(x)$ are shown in Box 3 together with a representative picture of this microstructure. The values of $x_{1\pm}$, $x_{2\pm}$, $x_{3\pm}$ in the formulas in Box 3 are numerical solutions of the equation $b(x)^2 - 4a(x)c(x) = 0$, which is a sixth-order polynomial equation for [25].

$$\mu(x) \quad - \quad \frac{(2 - 3f)\,p}{3 - p}\,\theta(2 - 3f)\,\delta(x)$$

$$+ \frac{(3f - 1)\,p}{2}\,\theta(3f - 1)\,\delta(x - \frac{1 - p}{3})$$

$$+ \frac{(2 - 3f)\,(1 - p)\,p}{2\,(3 - p)}\,\theta(2 - 3f)\,\delta(x - (1 - \frac{p}{3}))$$

$$+ \frac{27}{2|a(x)|\pi}\,\left[\Pi_{i=1}^3 (x - x_{i-})(x_{i+} - x)\right]^{1/2}$$

$$\left[\sum_{i=1}^3 (\theta(x - x_{i-}) - \theta(x - x_{i+}))\right].$$

The region where the spectral function is nonzero is shown in Fig. 7 as a $p$ vs. $x$ plot. Note that the percolation threshold for component 1 is given by the solution to the equation $2 - 3f = 0$, i.e., at $p_l = 0.455070 \cdots$. Similarly, the solution of $3f - 1 = 0$, given by $p_u = 1 - p_l = 0.544929$, denotes the percolation threshold for component 2 in the Sheng theory. Between the two thresholds, the microstructure is characterized by biconnectedness for both components. From the known spectral density $\mu(x)$, $\langle \mathbf{E}_{\text{lin}}^2 \rangle_1$, $\langle \mathbf{E}_{\text{lin}}^2 \rangle_2$, $\langle |\mathbf{E}_{\text{lin}}|^2 \rangle_1$, $\langle |\mathbf{E}_{\text{lin}}|^2 \rangle_2$ can be calculated from (20); $\chi^{(3)}$ can then be evaluated from (11).

### 3.2.4   The Hierarchical Microstructure
### (The Differential Effective Medium Theory)

The hierarchical microstructure is best described as a dispersion of component 1 particles in a matrix which is built up successively through a homogenization process whereby smaller particles of component 1 are dispersed in

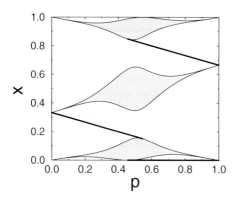

**Fig. 7.** The region where the spectral function is nonzero in Sheng's theory plotted as a function of volume fraction $p$. The *shaded areas* correspond to continuous spectra, and the *thick lines* indicate delta functions

a (previously) homogenized medium. The hierarchical microstructure, it has been shown, gives a very good description of the electrical and elastic properties of sedimentary rocks [31]. The effective dielectric constant of DEM is given by the solution of [30,32]

$$1 - p = \frac{\bar{\epsilon} - \epsilon_1}{\epsilon_2 - \epsilon_1} \left( \frac{\epsilon_2}{\bar{\epsilon}} \right)^{1/3}, \tag{29}$$

where $p_1 = p$ and $p_2 = 1 - p$. The spectral density is

$$\mu(x) = \begin{cases} \dfrac{\sqrt{3}(1-p)(1-x)^{1/3}}{(2x)^{4/3}\pi} \left[ (1+B)^{1/3} - (1-B)^{1/3} \right], & x_l < x < x_u \\ 0, & \text{otherwise,} \end{cases} \tag{30}$$

with

$$B = \sqrt{1 - \frac{4(1-p)^3}{27(1-x)^2 x}}. \tag{31}$$

$x_l$ and $x_u$ are determined by the requirement that $B$ should be real:

$$x_l = \frac{2}{3} \left[ 1 - \cos\left( \frac{\pi - \phi}{3} \right) \right],$$
$$x_u = \frac{2}{3} \left[ 1 - \cos\left( \frac{\pi + \phi}{3} \right) \right], \tag{32}$$

and $\phi = \arccos[2(1-p)^3 - 1]$, $0 < \phi < \pi$. The averaged field intensities and $\chi^{(3)}$ can be obtained from the spectral density as in the other cases.

## 3.3   Illustrations and Comparison of Theory with Experiments

To illustrate the nonlinear enhancement effect, we use the derived formulae for different microgeometries to study the $AuSiO_2$ composite system at the wavelength of 620 nm, where the dielectric constant of Au is $9.97 + 0.822i$ [33] and the dielectric constant of $SiO_2$ is taken as 2.25. Figure 8 gives plots of the spectral densities of different microgeometries at the Au volume fraction of 40%. A vertical line indicates a delta function, with the weight of the delta function given by its height. We see that both the EMA and the DEM have similar spectrum densities, but the EMA has an extra delta function with weight $(3p − 1)/2$ when the volume fraction is larger than $1/3$. Physically, this reflects the fact that the EMA has a percolation threshold at $p = 1/3$ whereas the DEM has none. The Maxwell Garnett theory has the simplest spectrum density, consisting of a delta function at $x = p/3$ with weight $p$. The Sheng theory has a spectrum rich in structures; it contains three branches of continuous spectra with a delta function in between. There is also a percolating threshold at $p = 0.455070 \cdots$.

Plots of the effective dielectric constant as a function of Au volume fraction are shown in Fig. 9. The EMA and the DEM are similar. The real parts, it is noted, vary almost monotonically from $p = 0$ ($SiO_2$) to $p = 1$ (Au). The imaginary parts are relatively flat as a function of $p$. The MG and the Sheng theories exhibit anomalous dispersions in their effective dielectric constant, related physically to the Mie resonance of coated spheres and manifested here as large undulations in the real and imaginary parts of the effective dielectric constant. *This resonance is identified as the cause of large third-order nonlinear susceptibility enhancement*, as seen from Fig. 10, where the enhancement factor $\beta_i$ is plotted as a function of $p$. It is seen that

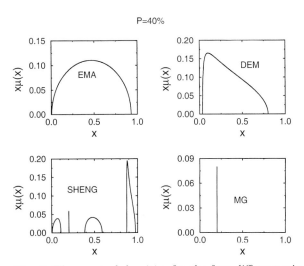

**Fig. 8.** The spectral densities for the four different microstructures at $p = 0.4$

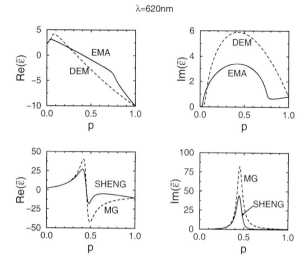

**Fig. 9.** The composite effective dielectric constants for the four different microstructures, calculated at $\lambda = 620$ nm with material parameters given in the text. Note the anomalous dispersions exhibited in the MG theory and the Sheng theory

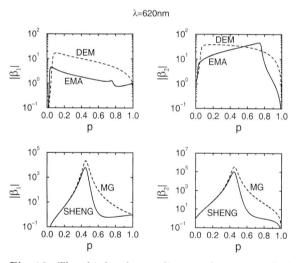

**Fig. 10.** The third-order nonlinear enhancement factor $\beta_i$ plotted as function of volume fraction $p$. *Left side*: The enhancement factor in material 1. *Right side*: The enhancement factor in material 2. The calculation is for $\lambda = 620$ nm and material parameter values given in the text

the enhancements are small for the EMA and the DEM but can be very large in the MG and the Sheng microgeometries. At volume fractions somewhat below that of the percolation threshold, the enhancement can be more then four orders of magnitude!

Our recent experiment [15] on the optical third-order nonlinearity in granular $AuSiO_2$ films has shown very large effective $|\overline{\chi^{(3)}}|$ in this system. In Fig. 11, the calculated $|\overline{\chi^{(3)}}|$ from the Sheng theory is plotted as a function of Au volume fraction $p$, together with the measured data. The parameters used are $\chi_{Au}^{(3)} = 8 \times 10^{-8}$ esu, obtained by fitting the data point at $p = 0.67$, and $\chi_{SiO_2}^{(3)} = 2 \times 10^{-12}$ esu [32]. This value of $\chi_{Au}^{(3)}$ itself should be regarded as an "effective" value since the local fluctuations in the electrical field, neglected in our present calculations, could cause an enhancement relative to its true value. It is seen that the agreement between theory and experiment is reasonably good at $\lambda = 530$ nm, aside from the composition dependence. By using the same parameters, it is predicted that at $\lambda = 620$ nm, the effective $|\overline{\chi^{(3)}}|$ can reach the value of $10^{-3}$ esu.

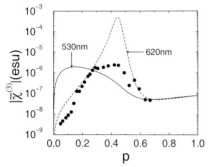

**Fig. 11.** Comparison of the measured third-order nonlinear susceptibility $\overline{\chi^{(3)}}$ in $AuSiO_2$ composites with theoretical values calculated from the Sheng theory at $\lambda = 530$ nm. The dashed line is calculated from the same parameters at $\lambda = 620$ nm. The composition where the maximum occurs is in disagreement. However, the maximum theory and experimental values of $\overline{\chi^{(3)}}$ are very close

## Acknowledgments

This work was supported by a RGC Grant HKUST 6153/00P. The nonlinear optical measurements were carried out in the Joyce M. Kuok Laser and Photonic Laboratory at HKUST. G.K.L. Wong wishes to acknowledge the contributions of Drs. R.F. Xiao and K.S. Wong to the experimental work reported here.

# References

1. F. Hache, D. Ricard, C. Flytzanis, U. Kreibig, Appl. Phys. A **47**, 347 (1988)
2. C. Flytzanis, F. Hache, M. C. Klein, D. Ricard, P. Roussingnol, Prog. Opt. XXIX (North-Holland, Amsterdam 1991) p. 321
3. J. C. M. Garnett, Philos. Trans. R. Soc. London **203**, 385 (1904); ibid **205**, 237 (1906)
4. S. G. Rautian, V. P. Safonov, P. A. Chubakov, V. M. Shalaev, M. I. Shtockman, JEPT Lett. **47**, 243 (1988)
5. Yu. E. Danilova, V. P. Drachev, S. V. Perminov, V. P. Safonov, Bull. Rus. Acad. Sci. Phys., **60**, 342 (1996); ibid. 374; Yu. E. Danilova, N. N. Lepeshkin, S. G. Rautian, V. P. Safonov, Physica A **241**, 231 (1997)
6. A. V. Butenko, P. A. Chubakov, Yu. E. Danilova, S. V. Karpov, A. K. Popov, S. G. Rautian, V. P. Safonov, V. V. Slabko, V. M. Shalaev, M. I. Stockman, Z. Phys. **17**, 283 (1990)
7. S. D. Stookey, J. Am. Ceram. Soc. **32**, 246 (1949)
8. R. D. Maurer, J. Appl. Phys. **29**, 1 (1958)
9. J. Matsuoka, R. Mizutani, S. Kaneko, H. Nasu, K. Kamiya, K. Kadono, T. Sakaguchi, M. Miya, J. Ceram. Soc. Jpn. **101**, 53 (1993)
10. I. Tanahashi, Y. Manabe, T. Tohda, S. Sasaki, A. Nakamura, J. Appl. Phys. **79**, 1224 (1996)
11. K. Fukumi, A. Chayahara, K. Kadono, T. Sakaguchi, Y. Horino, M. Miya, J. Hayakawa, M. Satou, Jpn. Appl. Phys. **30**, L742 (1991)
12. R. H. Magruder III, L. Yang, R. F. Haglund, C. W. White, L. Yang, R. Dorsinville, R. R. Alfano, Appl. Phys. Lett. **62**, 1730 (1993)
13. T. Akai, K. Kadono, H. Yamanaka, T. Sakaguchi, Y. Horino, M. Miya, J. Hayakawa, M. Satou, Jpn. J. Appl. Phys. **30**, 742 (1991)
14. I. Tanahashi, Y. Manabe, T. Tohda, S. Sasaki, A. Nakamura, J. Appl. Phys. **79**, 1224 (1996)
15. H. B. Liao, R. F. Xiao, J. S. Fu, P. Yu, G. K. L. Wong, P. Sheng, Appl. Phys. Lett. **70**, 1 (1997)
16. H. B. Liao, R. F. Xiao, J. S. Fu, H. Wang, K. S. Wong, G. K. L. Wong, Opt. Lett. **23**, 388 (1998)
17. H. B. Liao, R. F. Xiao, H. Wang, K. S. Wong, G. K. L. Wong, Appl. Phys. Lett. **72**, 1817 (1998)
18. H. B. Liao, R. F. Xiao, J. S. Fu, G. K. L. Wong, Appl. Phys. B **65**, 673 (1997)
19. P. Sheng, Phys. Rev. Lett. **45**, 60 (1980)
20. K. Minoshima, M. Taiji, T. Kobayashi, Opt. Lett. **16**, 1683 (1991)
21. V. M. Shalaev, Phys. Rep. **272**, 61 (1966)
22. V. M. Shalaev and M. I. Stockman, Sov. Phys. JEPT **65**, 287 (1987); A. V. Butenko, V. M. Shalaev, M. I. Stockman, Sov. Phys. JEPT **67**, 60 (1988); Z. Phys. D **10**, 71 (1988); Z. Phys. D **10**, 81 (1988)
23. V. A. Markel, V. M. Shalaev, E. B. Stechel, W. Kim, R. L. Armstrong, Phys. Rev. B **53**, 2425 (1996)
24. V. M. Shalaev, E. Y. Poliakov, V. A. Markel, Phys. Rev. B **53**, 2437 (1996)
25. Hongru Ma, Rongfu Xiao, P. Sheng, J. Opt. Soc. Am. B **15**, 1022 (1998)
26. D. Stroud and P. M. Hui, Phys. Rev. B **37**, 8719 (1988)
27. D. J. Bergman, Solid State Phys. **46**, H. Ehrenreich, D. Turnbull (Eds.) (Academic, Boston 1992) p. 147; G. W. Milton, Appl. Phys. A **26**, 1207 (1981); G. W. Milton, J. Appl. Phys. **52**, 5286 (1980)

28. P. Sheng, *Introduction to Wave Scattering, Localization, and Mesoscopic Phenomena* (Academic, Boston 1995) Chap. 3
29. R. W. Cohen, G. D. Cody, M. D. Coutts, B. Abeles, Phys. Rev. B **8**, 3689 (1973)
30. D. A. G. Bruggeman, Ann. Phys. (Leipzig) **24**, 636 (1935)
31. P. Sheng, Phys. Rev. B**41**, 4507 (1990)
32. For a generalized account, see A. N. Norris, A. J. Callegari, P. Sheng, J. Mech. Phys. Solids **33**, 525 (1985)
33. J. H. Weaver, H. P. R. Frederikse, in *CRC Handbook of Chemistry and Physics*, 74th ed., D. R. Lide (Ed.) (CRC, Boca Raton 1993/1994) **12**-109

# Linear and Nonlinear Optical Properties of Quasi-Periodic One-Dimensional Structures

Concita Sibilia[1], Mario Bertolotti[1], Marco Centini[1], Giuseppe D'Aguanno[1], Michael Scalora[2], Mark J. Bloemer[2], and Charles M. Bowden[2]

[1] INFM at Dipartimento di Energetica, Universita' di Roma "La Sapienza", Via Scarpa 16, 00161 Roma, Italy
concita.sibilia@uniroma1.it

[2] AMSAM-RD-WS-ST, U. S. Army Aviation & Missile Research, Development, & Engineering Center
Redstone Arsenal, AL 35898, USA

**Abstract.** The optical properties of self-similar optical multilayer structures are first discussed for low input intensities, thus allowing the neglect of nonlinear effects. The structures under consideration are obtained by alternating two dielectric layers of different refractive indexes following a fractal set. The triadic Cantor and the Fibonacci sets are considered, and some applications of the field localization properties of these structures are discussed. Nonlinear behavior is also discussed, restricted to third-order nonlinear polarization of the dielectric materials constituting the structures.

## 1 Introduction

Quasi-periodic structures with two or more incommensurate periods are intermediate between periodic and random media. The interest in quasi-periodic layered media originated in studies of analogous systems in solid-state physics. The problem of the propagation of electrons in one-dimensional quasi-periodic structures has revealed interesting features, such as the presence of localized, critical, and extended states [1,2,3,4]. Theoretical studies have received great impetus from the experimental discovery of a quasi-crystal phase in metallic alloys [5], which was followed by the realization of quasi-periodic superlattice structures [6].

An analogous problem in optics was addressed by *Kohmoto* et al. [7] who studied a one-dimensional quasi-periodic structure involving a stack of dielectric layers arranged in a Fibonacci sequence. The optical system exhibited several advantages over its solid-state counterpart. In solid-state physics, electron–electron and electron–phonon interactions are inevitable. On the other hand, optical experiments are more "pure," since photons do not interact. Moreover, the polarization of light adds new features to the localization problem which is absent in solid-state physics.

It is of interest to investigate the behavior of quasi-periodic structures mainly because of the peculiar aspects of the localization of light inside fractal

V. M. Shalaev (Ed.): Optical Properties of Nanostructured Random Media,
Topics Appl. Phys. **82**, 63–91 (2002)
© Springer-Verlag Berlin Heidelberg 2002

structures. Recently, many theoretical studies of one-dimensional (1-D) quasi-periodic structures based on a Fibonacci or a Cantor sequence have been performed [7,8,9,10], and interesting experimental work has been done [8,9]. The properties of the structures studied [6,7,8,9,10] are linked to the properties of self-similar spaces also because of the possibility of weak localization of photons. The treatment is so general that it can be applied to any kind of waves propagating in a self-similar medium. Photons that localize in fractal structures have been named "fractons," and the existence of fracton modes [11] has been experimentally proven for acoustic waves in one-dimensional Cantor composites [12]. In optics, scattering and diffraction from fractal objects have also been recently investigated [13,14,15,16,17,18,19].

Generally speaking, layered, quasi-periodic structures can be classified as a type of photonic band gap (PBG) structure or crystal. Periodic PBG structures have been recently intensively studied theoretically as well as experimentally [20,21,22]. An essential property of photonic crystals is the existence of forbidden frequency bands, from which propagating modes, spontaneous emission, and zero-point fluctuations are all absent. On the other hand, it has been observed that the electrical-field intensity strongly increases near the PBG edges in the frequency domain transmission curves [23]. This is related to a spatial field distribution localized inside the structure. Many possible applications have been envisioned, and devices function because of field localization effects [23]. Field localization in periodic structures has been described through the concept of density of modes (DOM) [24], a quantity that can be derived directly from transmission properties. The density of modes increases sharply if defects are introduced inside the structure or if the structure is arranged in a quasi-periodic geometry [25]. In what follows, we will discuss in more detail the linear and nonlinear optical properties of multilayer structures based on a Fibonacci or a Cantor code [26,27,28].

## 2   Something about Fractals

The term "fractal" was introduced by *Mandelbrot* [29] to describe geometric objects with no integral dimension. The definition given by Mandelbrot states that a fractal is a self-similar set whose dimension is different from the topological dimension; a self-similar set is an invariant set with respect to a scale change. Self-similar fractals are generated mathematically by a recursive operation of *generators* and *initiators* [29]. A process is defined on an object, named *initiator*. For example in the Koch fractal, the initiator is a line of unit length. A segment of length 1/3 is erased in the middle of the initiator, and an equilateral triangle, without basis, is built on the segment. This operation can be repeated again at the smallest scale: a line of 1/3 is erased again in the middle of each of the four segments. This fractal has a scale factor of 3 (Fig. 1).

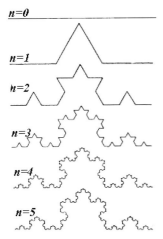

*n=0*

*n=1*

*n=2*

*n=3*

*n=4*

*n=5*

**Fig. 1.** Example of a self-similar fractal: the Koch fractal

The triadic Cantor set has a generation procedure very similar to the Koch fractal; the difference is that one removes $n$ segments of length $l_n = (1/3)^n$ from an unitary line, without adding any more. We start from a straight line segment of unitary length. Then we "wipe away" the central middle third and repeat the process on the remaining two segments of length $1/3$. Repeating the middle-third wiping-out process over and over again, does not leave a single connected segment. In Fig. 2, the first five levels of the Cantor set generation are shown [10]. The Fibonacci set will be considered later.

The optical structures that we will discuss in the following can be suitably constructed by using the criteria used to define the set. For example, a 1-D structure can be realized with a multilayered stack of different materials assembled following the fractal code.

One of the interesting phenomena appearing in fractal structures of this type is wave localization, so that the field (acoustic, electromagnetic, or other) becomes spatially confined in some suitable regions and/or delocalized in some other parts. Many theoretical papers have been written on this

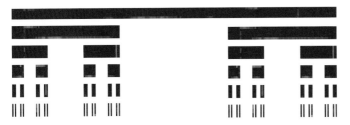

**Fig. 2.** Example of the generation sequence of a Cantor set

subject [13,30]. Although the problem originated in solid-state physics in connection with the theory of electrical conductivity in disordered and quasi-periodic media, the subsequent realization that almost any wave equation with a random (or quasi-periodic) potential may possess localized solutions has made the field quite general. We bring forth the analogy to the electronic problem, where localized, extended, and critical states can be defined. Let us recall that by a localized surface wave, or critical states, we mean that the wave functions $\psi(z)$ for the electron or the electrical field optical waves are described by an envelope function that varies asymptotically as $\exp(-\xi z)$ for exponentially localized states, or as $z^{-\xi}$ for a critical state ($z$ is the propagation coordinate). By drawing the analogy to electronic wave functions, one might observe that forbidden bands of energy correspond to almost null transmission in the optics of multilayers; thus, the electrical field distribution is expected to decay exponentially with increasing distance into the structure, producing a "surface localized state." On the contrary, a band of energies corresponding to almost complete transmission is considered an allowed region where "extended states" are likely to be observed. The transition regions therefore correspond to "critical states." Depending on the number of layers and on the spectral width of the associated transmission function, extended states become "bulk localized states." These states appear when the structure has a transmission spectrum that exhibits isolated peaks in the middle of frequency band gaps, and field localization occurs [31]. All of these properties make fractal structures very attractive from the optical point of view, and even more interesting in the framework of their nonlinear response.

One of the most interesting consequences related to the fractal nature of quasi-periodic structures is self-similarity and scaling behavior. Self-similarity and scaling behavior of the transmission spectra in one-dimensional Fibonacci multilayers consisting of quarter-wave plats of $SiO_2/TiO_2$ were experimentally investigated in [32].

## 3   Optical Properties of Filters Based on a Fractal Code

Let us consider a structure realized by alternating two dielectric layers of different refractive indexes such that the high-index layers belong to a triadic Cantor set, as described in Figs. 2 and 3. This is obtained by alternating two nondispersive, planar, dielectric layers of refractive indexes $n_2$ and $n_1 (n_2 > n_1)$ whose thicknesses are such that their optical paths are the same. Let us take the layer of refractive index $n_2$ as the initiator. If $L$ is the optical thickness of the initiator, the generator is obtained by substituting the central part of the initiator, whose optical thickness is $L/3$, with a layer of refractive index $n_1$ and optical thickness $L/3$. The layered structure is obtained by iterating the operation up and down and stopping the iteration at the $N$th step (Fig. 3). For the sake of simplicity, the incident light is assumed

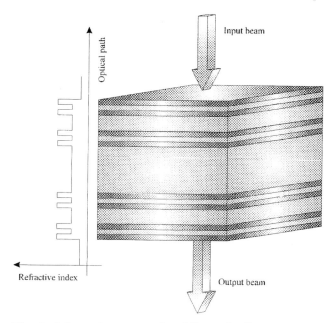

**Fig. 3.** A layered structure that follows the Cantor set

to be a plane wave propagating in a direction at an angle $\theta$ with respect to the normal at the interface planes.

We use the matrix transfer method (see [10] for applications to the Cantor-like multilayer) for calculating the transmission characteristics. The expression of the electrical field transmission (for a linearly polarized wave, parallel to the interface planes of the resonator) at a wavelength $\lambda$ is as follows

$$t(k_0, L) =$$

$$\frac{2}{T_{22}^{(N)}(k_0, 3^N L) - \frac{T_{21}^{(N)}(k_0, 3^N L)}{ik_0} - ik_0 T_{12}^{(N)}(k_0, 3^N L) + T_{11}^{(N)}(k_0, 3^N L)}, \quad (1)$$

where i is the imaginary unit, $k_0$ is the vacuum wave number, and $T_{hk}^{(N)}$ are the elements of the transfer matrix $\boldsymbol{T}^{(N)}$ of the structure, obtained by the $N$th iteration of the recursive relation, which in Cantor layers is [10]

$$\boldsymbol{T}^{(k)}(3^k \varphi) = \boldsymbol{T}^{(k-1)}(3^{k-1}\varphi)\boldsymbol{T}_1(3^{k-1}\varphi)\boldsymbol{T}^{(k-1)}(3^{k-1}\varphi), \quad k = 1, 2, \ldots, N \quad (2)$$

with

$$\boldsymbol{T}^{(0)}(\varphi) = \boldsymbol{T}_2(\varphi) \tag{3}$$

$$\boldsymbol{T}_h(\varphi) = \begin{pmatrix} \cos\left(\frac{\varphi}{3}\right) & \frac{1}{k_0 n_h}\sin\left(\frac{\varphi}{3}\right) \\ -k_0 n_h \sin\left(\frac{\varphi}{3}\right) & \cos\left(\frac{\varphi}{3}\right) \end{pmatrix}, \quad h = 1, 2, \tag{4}$$

where $\varphi = k_0 L$.

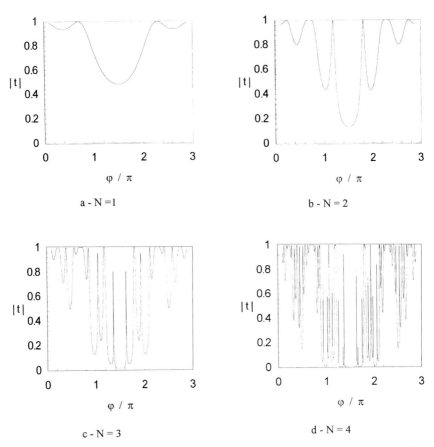

**Fig. 4.** Examples of transmission spectra as a function of the phase $\varphi$ for different levels of the Cantor sequence: **(a)** $N = 1$, **(b)** $N = 2$, **(c)** $N = 3$, **(d)** $N = 4$

Examples of transmission spectra for normal incidence are shown in Fig. 4, where the magnitude of the transmission as a function of $\varphi$ is given for the first four levels of the Cantor sequence.

Let us consider the spectrum. The electrical field in the layers can either fill the whole structure or can be localized in a smaller region. This different behavior depends on the chosen resonance frequency. If the resonance frequency corresponds to an isolated peak of transmission, then the field is stronger in a selected part of the layered structure. If the resonance frequency corresponds to a broad maximum of the spectrum, then the field exists almost everywhere in the layers. This behavior is shown in Fig. 5a and 5c for $\varphi = 0.7\pi$ and $\varphi = 3\pi$, respectively, which correspond to isolated peaks in the transmission spectrum. If the resonance frequency corresponds to a broad maximum of the spectrum, the field exists almost everywhere in the layered structure, as shown in Fig. 5b, d for $\varphi = 0.7\pi$ and $\varphi = 3\pi$, respectively. Note

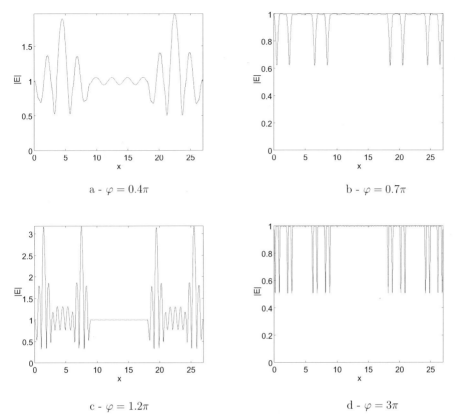

a - $\varphi = 0.4\pi$

b - $\varphi = 0.7\pi$

c - $\varphi = 1.2\pi$

d - $\varphi = 3\pi$

**Fig. 5.** Electrical field spatial distribution inside the layered Cantor sequences $N = 4$, for different values of $\varphi$. ($\varphi_0 = k\lambda$): **(a)** $\varphi = 0.4\pi$; **(b)** $\varphi = 0.7\pi$; **(c)** $\varphi = 1.2\pi$; **(d)** $\varphi = 3\pi$

that the field distribution in Fig. 5d is reminiscent of the structure, as expected. For a Cantor set, the field distribution at the band edge frequency depends on the refractive index of the initiator, in other words, the "central" layer. When the central layer has a low refractive index value, the Cantor set may be considered a resonator with quasi-periodic mirrors, or multiple coupled cavities. For this reason, when we shift the frequency from the low to the high-frequency band edge, the symmetry of the field distribution in the central layer changes.

Similar results can be obtained by using a Fibonacci sequence. An example of a Fibonacci sequence is as follows:

"external medium/ BAABAABAABAABAABABAABABAABAABABAABA/ substrate," where each layer has an optical thickness of one-quarter of a wavelength; the symbol A refers to a layer of refractive index $n_1$, and B refers to a layer of refractive index $n_2$.

In a Fibonacci multilayer $S_j$, there are $F_j$ layers given recursively by $F_{j+1} = F_j + F_{j-1}$ ($j \geq 1$, with $F_0 = A, F_1 = B$).

Transmission properties of such quasi-periodic filters have been discussed in [33] using the method of the "dynamic map", where the $2 \times 2$ matrix map can be reduced to a trace map. All of the spectral properties of the layered structures are contained in the properties of the trace map.

Electrical field localization properties in the spatial domain can be derived by spectral analysis of the transmission function. This can done by introducing the concept of the density of modes. The mode density in a *finite* one-dimensional $N$-period layer stack has been studied analytically in [24], where a correspondence has been found between the group velocity $v = d\omega/dk$ in an *infinite* periodic structure and the density of modes (DOM) $\rho_N = dk_N/d\omega$ of an $N$-period *finite* 1-D lattice.

The magnitude of the density of modes $\rho(\omega)$ is expressed via the real and imaginary parts of the complex transmission coefficient $t \equiv x + iy \equiv \sqrt{T}e^{i\varphi}$ as follows [24]:

$$\rho(\omega) \equiv \frac{dk}{d\omega} = \frac{1}{D} \frac{y'x - x'y}{x^2 + y^2} , \tag{5}$$

where $D$ is the total physical length of the structure; the prime denotes differentiation with respect to $\omega$. The analytical expression for $\rho(\omega)$ for a quarter-wave, $N$-period stack composed of two layer unit cells, it has been shown, is many times larger than that for a homogeneous medium. A finite number of unit cells instead of a periodic, infinite structure allows removing the DOM singularity at the band edge and calculating the value of the DOM for practical devices.

The group velocity and DOM are calculated numerically by using (5). Once the group velocity is known, the group index $n_g = c/v_N$ is also evaluated. To show the main features of fractal structures, we may discuss the following: (a) an $N$-period stack, (b) a Cantor-like multilayer and (c) a Fibonacci multilayer (see Fig. 6). The Cantor-like multilayer is generated by a material having refractive index $n_1 = 2.5$ (initiator). The central part of the initiator is then replaced with a layer of refractive index $n_2 = 1.5$, according to the law of triadic Cantor construction, such that $1/3$ of the generator has the same optical path $L_{opt} = \lambda_0/4$ [25]. Thus, the generator thicknesses are $a = \lambda_0/(4n_1)$ and $b = \lambda_0/(4n_2)$. The total length of the $N$-stage, Cantor-like multilayer is $D = 2^N a + (3^N - 2^N)b$, and its optical path is $L_{opt} = 3^N(\lambda_0/4)$. The Fibonacci multilayer is constructed recursively as the binary, quasi-periodic Fibonacci sequence $S_{j+1} = S_{j-1}S_j$ for $j \geq 1$; with $S_0 = \{B\}$ and $S_1 = \{A\}$. It follows that $S_2 = \{BA\}$, $S_3 = \{ABA\}$, $S_4 = \{BAABA\}$, etc. and in Fig. 6c, the sequence $S_6 = \{BAABAABABAABA\}$ is shown. The layers B and A have the refractive indexes $n_1 = 2.5$ and $n_2 = 1.5$ and thicknesses $a$ and $b$, respectively.

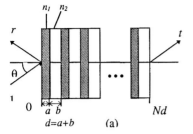

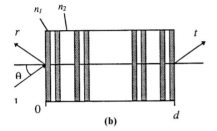

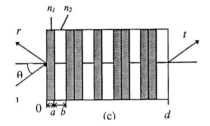

**Fig. 6.** Layered structures with approximately the same optical path (**a**) a periodic stack, (**b**) a Cantor sequence, (**c**) a Fibonacci sequence, $n_1 = 1.5$, $n_2 = 2.5$. $a = \lambda_0/4n_1$, $b = \lambda_0/4n_2$

The total thickness of the multilayer is given by $D_{(j+1)} = D_{(j-2)} + D_{(j-1)}$; the total optical path can be easily calculated if each layer A and B has the same optical path of one-quarter of a wavelength.

The spectral transmission characteristics $T_N$, the DOM $\rho_N$, and the dimensionless group velocity $v_N/c = (c\rho_N)^{-1}$ for the above structures are shown in Figs. 7, 8, 9. Our choice in the numbers of the Cantor prefractal level and Fibonacci sequence is somehow restricted because the total optical path $L_{opt}$ increases as $3^N(\lambda_0/4)$ for the Cantor-like multilayer of level $N$; it also rapidly increases with the number $J$ for the Fibonacci sequence. One should have a sufficiently deep band gap (obtained by using a large number of layers). On the other hand, the total number of layers should not be too large, especially if the structures are to be used for ultrashort pulses propagation, where suitable widths of resonances are required. These considerations should be taken into account considering that, as we will see below, a three-stage Cantor-like multilayer appears to be the most appropriate structure for the materials that we have chosen. Therefore, we consider the three-stage Cantor-like multilayer stack (3CS), having $L_{opt} = 6.75\lambda_0$ and total length $D = 3.79\lambda_0$, as the one for

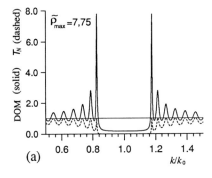

(a)

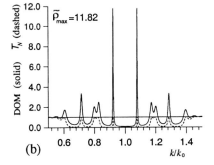

(b)

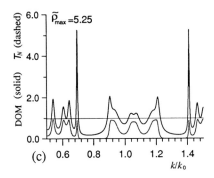

(c)

**Fig. 7.** The spectral transmission characteristics $T_N = |t|^2$ and the dimensionless DOM $\tilde{\rho}_N$ for normal incidence for the structure shown in Fig. 6 (**a**) for the 14-period stack (14PS), (**b**) for the 3CS, and (**c**) for the FMS8

our discussion. We compare it with a 14-periodic stack, having $L_{\text{opt}} = 7\lambda_0$, $D = 3.73\lambda_0$ and with a Fibonacci multilayer $S_8$ (FMS8) with $L_{\text{opt}} = 8.5\lambda_0$, $D = 4.8\lambda_0$ (the use of $S_7$ with $L_{\text{opt}} = 5\lambda_0$ does not allow us to achieve sufficient depth in the band gap).

The spectral transmission characteristics $T_N = |t|^2$ and the dimensionless DOM $\tilde{\rho}_N$ for normal incidence are presented in Fig. 7 for (a) a 14-period stack (14PS), which has approximately the same thickness of the Cantor and Fibonacci layered structures, (b) for the 3CS, and (c) the FMS8. We have normalized the DOM for each structure considered to the dimensionless quantity $\tilde{\rho}_N = v^{\text{bulk}}\rho_N$, as in [24], where $v^{\text{bulk}} = cD/L_{\text{opt}}$ represents the

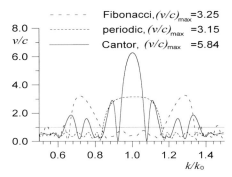

**Fig. 8.** The dimensionless group velocity $v_N/c = (c\rho_N)^{-1}$ for the same structures as in Fig. 6

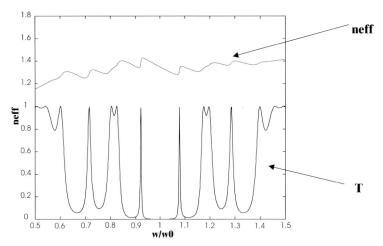

**Fig. 9.** Example of the behavior of the "effective index" (real part) for a Cantor set layered structure with $N = 3$, $n_1 = 1.5$, $n_2 = 2.3$, minimum optical path of $\lambda_0/4$, initiator $n_2$

distance $D$ over the travel time (neglecting reflections), and $c$ is the speed of light; this quantity is different for each structure; it is not a constant for the Cantor sequence but depends on the Cantor set level. However, for the structures that we consider, the value of $v^{\mathrm{bulk}}$ is approximately the same.

As we can see from Fig. 7, the 3CS has the narrowest band gap among all of the structures being considered, and the largest DOM at the band edges: it exceeds the DOM's maxima for both the periodic and Fibonacci $S_8$ multilayers by over 50%.

The FMS8 has the smallest maximum in the DOM; however, the transmission spectrum also exhibits two sufficiently wide band gaps within the same spectral range. This property of the Fibonacci multilayer may be compared with the $\lambda/4 - \lambda/2$ layered structures of [24]; the location of the second-order gap is separated from the first-order gap by approximately a factor of 2, as it is for a mixed half-quarter-wave $N$-period stack [24]. In contrast, a factor of 3 separates the first- and the second-order band edges in both an ordinary quarter-wave $N$-period stack and the Cantor-like multilayer.

We may also show the behavior of the inverse Cantor-like multilayer, which means that the initiator is chosen as the material with smaller refractive index, i.e., $n_1 = 1.5$ and $n_2 = 2.5$. All of the characteristics of the inverse structure are similar to the previous ones; in this case, only a modest decrease of DOM is observed, as it is in all cases in which a weak contrast of refractive indexes exists [25].

The DOM maximum at the band edge increases proportionally to $N^2$ for a periodic structure, for fixed $n_1$ and $n_2$ and for a moderately large value of $N$ [24]. A more complicated scaling law can be found for the Cantor set level: the ratio of the DOM maximum over the optical path increases approximately by a factor 2. The maximum DOM at the band edge corresponds to a "bulk localized state" inside the structure.

The dimensionless group velocity $v_N/c = (c\rho_N)^{-1}$ for the same structures of Fig. 6 is shown in Fig. 8. The group velocity within the band gap corresponds to "superluminal" tunneling velocities of a wave packet through a 1-D photonic band-gap structure [24], which was experimentally measured in [34]. In the middle of the band gap for all structures considered, there are peaks where $v_N/c > 1$. The peak for a Cantor-like structure is the largest.

We find instead a decreased group velocity at the band edge for the Cantor sequence [25] where $v/c_{\min} = 0.045$, *and the group index* $n_g = 1/0.045$, with a different shape and location compared with the periodic structure $(v/c_{\min} = 0.06)$; this makes the Cantor set a good candidate for applications in the pulsed and steady-state regime.

If we increase the level of the multilayer structures, the fourth level of Cantor has optical path $L_{\text{opt}} = 20.25\lambda_0$ and $D = 12.4\lambda_0$. A 41-period stack has $L_{\text{opt}} = 20.5\lambda_0$. The peaks of transmittance are very narrow for both structures, much more so for the Cantor-like sequence. The group velocity peak inside the band gaps of a Cantor-like multilayer exceeds the group velocity for an $N$ periodic structure by a factor of 2. At the band edge, the group velocity also decreases sharply $(v/c_{\min} = 0.0025)$ for the Cantor-like multilayer (for periodic structures, the minimum value of the group velocity is $v/c_{\min} = 0.0093$).

## 4   Dispersive Properties of One-Dimensional Filters

Now we seek an explicit geometric dispersion relation for a multilayer, one-dimensional fractal filter of finite length; this has been proposed in [35] for periodical layered structures. The treatment is general, and it can be applied to any kind of filter, periodic or not. Here, we report the main features of the discussion presented in [35].

We begin by writing the complex transmission coefficient for the structure, obtained via the matrix transfer method:

$$t = x + iy = \sqrt{T}e^{i\phi_t} = e^{i\varphi} , \tag{6}$$

$$\varphi = \phi_t - i \ln \sqrt{T} ,$$

where

$$\varphi_t = \text{tg}^{-1}(y/x) \pm m\pi$$

is the total phase accumulated as light propagates through the medium. The transmission $t = x + iy$ is obtained via the matrix transfer method. $\varphi_t$ contains all information relating to the layered structure, such as refractive indexes, number of layers, and layer thickness. The integer $m$ is uniquely defined by assuming that $\varphi_t(\omega)$ is a monotonically increasing function with the condition that $m = 0$ as $\omega \to 0$. This is important for calculating the effective phase of the field.

Beginning with the analogy of propagation in a homogeneous medium, we can express the total phase associated with the transmitted field as

$$\varphi = k(\omega)D = \frac{\omega}{c}n_{\text{eff}}(\omega)D , \tag{7}$$

where $k(\omega)$ is the effective wave vector; $n_{\text{eff}}$ is therefore the effective refractive index that we attribute to the layered structure whose physical length is $D$. Both $k$ and $n_{\text{eff}}$ are complex numbers.

In particular,

$$\hat{n}_{\text{eff}}(\omega) = (c/\omega D) \left[ \varphi_t - (i/2) \ln(x^2 + y^2) \right] . \tag{8}$$

Equation (8) suggests that at resonance, where $T = x^2 + y^2 = 1$, the imaginary part of the index is identically zero. Inside the gaps, where the transmission is small, scattering losses are expected to be high, leading to evanescent waves ("surface localized states"), and the imaginary part of $n_{\text{eff}}$ is large.

We can also define the effective index as the ratio between the speed of light in vacuum and the effective phase velocity of the wave in the medium:

$$\hat{k}(\omega) = \frac{\omega}{c}\hat{n}_{\text{eff}}(\omega) . \tag{9}$$

Once the effective index has been defined, (9) represents the dispersion relation of the layered structure without any condition of periodicity. It is interesting to note that from the dispersion equation we can also define a "group index" in terms of the effective phase index as follows:

$$n_{\text{g}}(\omega) = \frac{1}{c}\frac{d\hat{k}}{d\omega} = n_{\text{eff}}(\omega) + \omega\frac{dn_{\text{eff}}}{d\omega} . \tag{10}$$

In (7) we have assumed an incident field of unit amplitude, and scattering losses are taken into account by introducing an imaginary component of the effective index of refraction. The effective index defined in (9) thus takes into account the geometric dispersion introduced by the layered structure, including the influence of entry and exit interfaces. An example of the behavior of the real part of the "effective index" for a Cantor ($N = 3$) set layered structure is shown in Fig. 9. The initiator has refractive index $n_2 = 2.3$, $n_1 = 1.4$, and the layer of minimum optical path is $\lambda_0/4$.

## 5    Metal–Dielectric Quasi-Periodic Filters

Recently, 1-D periodical layered structures composed of dielectric/metal layers have also been proposed (MD-PBG) [36,37,38]. In [36], a metallic multilayer arrangement in a PBG geometry has been suggested for enhancing the reflectivity with respect to bulk metal. The theoretical results showed that the reflectivity of 96% for bulk aluminum could be improved to about 98% for a layered structure. Therefore, it may be possible to arrange metals in a layered geometry to have better mirrors. It has been shown in [37,38] that it is also possible to make metals transparent to visible light and opaque to all other wavelengths of electromagnetic radiation up to UV light, even if the total metal thickness is several tens of skin-depths, or hundreds of nanometers.

The phenomenon, discussed at length in [37,38], is due to resonant tunneling of electromagnetic waves through a layered structure that may contain thin metal layers, 30–50 nm or more. By properly spacing the metal layers approximately $\lambda/2$ apart, where $\lambda$ is the desired tunneling wavelength, the structure displays the following unique features: the formation of transmittance passbands, which allow visible light to propagate almost unattenuated; and the presence of a huge stop band that extends on one side to cover the entire electromagnetic spectrum down to static fields, and UV radiation on the other. For this reason, these structures have been referred to as transparent metals. The high reflectivity of the stack at low frequencies is due to the dispersive properties of the metal; the index of refraction, and hence the optical potential, becomes infinitely large, and is accurately described by the Drude model. At high frequencies, near the visible and UV range, the index of refraction of most good conductors can be of the order of unity or less. Therefore, interference may cause the formation of photonic band gap effects, i.e., the formation of passbands and frequency gaps and the effective reduction of absorption losses in the metal. Proposed applications for these structures include sensors, UV blocking films, transparent electrodes for light-emitting polymer stacks, and conductive displays, just to name a few.

In [37,38], the discussion and the examples focused on periodic metallo-dielectric stacks, where individual metal layer thickness varied from 10 to 40 nm. In [39], the discussion was extended to include quasi-periodic structures. It is possible to further increase the transmittance for a given amount of metal by using different geometric arrangements of metallic layers (quasi-periodic), as, for example, in Cantor or Fibonacci sets. We emphasize that the kind of interactions we are discussing can occur in an environment once thought completely inaccessible to light, i.e., through thick metal layers. For comparison, we discuss below the transmission characteristics and the density of modes for normal, TE-polarized, incident waves for periodic and quasi-periodic structures that have approximately the same total metal thickness.

The Cantor-like multilayer is generated by a metallic material having complex refractive index $\hat{n}_1$ (initiator or generator). The central part of the initiator is then replaced with a dielectric layer of refractive index $n_2 = 1.4$, according to the law of triadic Cantor construction, such that $1/3$ of the generator has the same optical path $L_{opt} = \lambda_0/x$ [39], where $\lambda_0$ is the wavelength in vacuum and $x$ is a multiple of 4, to reach a suitable dimension for the metallic layer. Thus, the generator thicknesses are $a = \lambda_0/(x\hat{n}_1)$ and $b = \lambda_0/(xn_2)$. A modified Cantor set can also be considered, in which the geometric thicknesses of the layers (and not the optical path) follow the Cantor law [39].

The Fibonacci multilayer is constructed recursively as the binary, quasi-periodic Fibonacci sequence already discussed with {B} (metallic layer) and $S_1 = \{A\}$ (dielectric layer). The layers B and A have refractive indexes $n_1$ and $n_2 = 1.4$ and thicknesses $a$ and $b$, respectively.

As for the Cantor set, we can recursively generate the Fibonacci code by using geometric thicknesses instead of optical path lengths [39].

Our choice in the numbers of Cantor prefractal level and Fibonacci sequence is again restricted because the total optical path $L_{opt}$ rapidly increases as $3^N(\lambda_0/x)$ for the Cantor-like multilayer of level $N$ and also rapidly increases with the number $J$ of the Fibonacci sequence. For simplicity, we do not change the thickness of all dielectric layers, although that may be considered an additional degree of freedom.

Depending on the desired spectral behavior, it is possible to select a suitable rule for assembling the structures. For example, large transparent spectral regions can be obtained when we use the optical path to construct a Cantor or Fibonacci code [39]. An example is given in Fig. 10, for a three-stage Cantor-like multilayer stack containing eight metallic layers, having $L/3 = \lambda_0/x(\lambda_0 = 1000\,\text{nm}, x = 64)$, total length $D = 700\,\text{nm}$, where the spectral transmission characteristics $T_N = t^2$ and the dimensionless DOM $\tilde{\rho}_N$ for normal incidence are presented. We observe a wide passband in the visible, and a localized weak transmission peak in the IR region ($\sim 1100\,\text{nm}$), where we also calculate a high density of modes. This behavior has been found for other values of metal thickness in the Cantor code layered structure: a strong localized DOM in a region of the spectrum with low transmittivity.

Generally speaking, we remark that if the quasi-periodic sets discussed above are realized by using optical path criteria, then the band structure

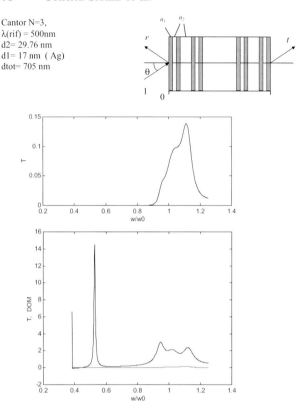

**Fig. 10.** The spectral transmission characteristics $T_N = t^2$ and the dimensionless DOM $\tilde{\rho}_N$ for normal incidence for a three-stage Cantor-set multilayer stack containing eight metallic layers, having $L/3 = \lambda_0/x (\lambda_0 = 1000\,\mathrm{nm}, x = 64)$, total length $D = 700\,\mathrm{nm}$

achieves a different degree of complexity with sharp transmission peaks and a higher density of modes. In all of these filters, a suitable compromise between high transparency and high density of mode can be found, depending on the application. In other cases, wide passbands may be enough to satisfy device performance. So it is possible to get good transparency in the visible range even by increasing the number of metallic layer, thus making the devices practical for many applications.

## 6   Nonlinear Model of the Filter

One of the advantages of field localization in fractal structures is to enhance the nonlinear optical responce. This property has been discussed in [13,40] for Fibonacci quasi-periodic filters, assuming third-order nonlinear polarization in the dielectric layers. Let us discuss here the nonlinear properties of Cantor filters when a third-order nonlinear interaction is taken into account.

To describe nonlinear propagation in a multilayer structure in a steady-state regime, usually a formalism based on the nonlinear transfer matrix method [41,42,43] is applied, with the omission of spatial third harmonics generated inside each layer and of nonlinear terms in the boundary conditions. As extensively discussed in [42], the application of the nonlinear transfer formalism gives reliable results even for a high-finesse structure such as a nonlinear interferential filter, when the central spacer layer thickness exceeds the wavelength inside the material.

To take the nonlinearity into account, the refractive index of the nonlinear layers is taken as the form $n = n_L + n_{2NL}I$, where $n_L$ is the linear index valid for low light intensities, $n_{2NL}$ is the nonlinear coefficient of the refractive index, and $I$ is the light intensity. We adopt the hypothesis that the electrical field in each layer is the sum of two nearly counterpropagating plane waves. Inside a single homogeneous material layer, we assume that the field is constant in amplitude but allow its phase to change nonlinearly, provided the following two conditions are fulfilled: $n_{2NL}I_{CAV} \ll n_L$, and $D > \lambda_0 n_L$, where $I_{CAV}$ is the cavity irradiance level, $D$ is the length of the filter, $\lambda_0$ is the incident wavelength in vacuum (in a stratified structure, the first condition has to be verified for each layer inside the structure). In what follows, we assume isotropic layers, in which the chromatic dispersion of the refractive index is taken into account.

We use the formalism described in [41,43]: the so called "transmission line theory," and we specialize it to our case of a Cantor multilayer structure that can be considered a cascade of nonlinear transmission lines. The field propagating inside the multilayer structure can be modeled by using two functions $V(z)$ and $I(z)$, proportional to the transverse components of the electrical and magnetic fields, respectively, and satisfying the following coupled differential equations:

$$\begin{cases} \frac{dV}{dz} = -ZI\,, \\ \frac{dI}{dz} = -Y_{NL}V\,, \end{cases} \qquad (11)$$

with $Y_{NL} = Y + Y^{(3)}|V|^2$, where $Z$ and $Y_{NL}$ play the role of a linear impedance and nonlinear admittance, respectively, and $Y^{(3)}$ is a third-order nonlinear coefficient; $Y$ is the unperturbed (linear) admittance per unit length. Note that $V(z)$ and $I(z)$ are reminiscent of "voltage" and "current" in a transmission line, respectively. For this reason in the following, we will refer to these quantities in that way. However, it should be understood that there are no ties to the electrical problem, and any such reference should be taken as merely suggestive. Similar arguments are valid for the parameters $Z$ and $Y_{NL}$ as well. It is well known that the solutions of (11) in the linear case ($Y^{(3)} = 0$) are given by

$$V(z) = a(z) + b(z)\,,$$
$$I(z) = \frac{1}{\eta}\left[a(z) - b(z)\right]\,, \qquad (12)$$

where the transmission line parameters $k$ (propagation constant) and $\eta$ (characteristic impedance) are given by the following:

$$k^2 = -ZY,$$
$$\eta = \frac{-Z}{ik}.$$

Using a perturbative approach, in the nonlinear case ($Y^{(3)} \neq 0$) also, the solutions of (11) can be written as in (12), but now the forward and backward waves ($a$ and $b$, respectively) are given by

$$a(z) = \hat{a}(z)e^{i\mathrm{Re}\,(k)z}$$
$$b(z) = \hat{b}(z)e^{-i\mathrm{Re}\,(k)z} \tag{13}$$

where $\hat{a}(z)$ and $\hat{b}(z)$ are not constant, but depend on the $z$ coordinate. After substituting (12) in (11), summing and subtracting the first and the second Eq. (11), and taking into account that

$$\frac{Z}{\eta} = \eta Y = -ik, \tag{14}$$

we obtain the following for the forward and backward waves:

$$\frac{da}{dz} = ika - \frac{Y^{(3)}\eta}{2}|a+b|^2(a+b)$$
$$\frac{db}{dz} = -ikb + \frac{Y^{(3)}\eta}{2}|a+b|^2(a+b). \tag{15}$$

Using (13) in (15), neglecting the spatial third harmonic generated inside each layer and the nonlinear term in the boundary conditions, we have the following for the slowly varying (complex) amplitudes:

$$\frac{d\hat{a}(z)}{dz} = -Im(k)\hat{a} + ik_2\left(|\hat{a}|^2 + 2\left|\hat{b}\right|^2\right)\hat{a}(z)$$
$$\frac{d\hat{b}(z)}{dz} = Im(k)\hat{b} - ik_2\left(2|\hat{a}|^2 + \left|\hat{b}\right|^2\right)\hat{b}(z), \tag{16}$$

where we have introduced the nonlinear parameter

$$k_2 = k\frac{Y^{(3)}}{2Y}. \tag{17}$$

Note that for nonresonant nonlinearity, $Y^{(3)}$ has a pure imaginary value. It is well known that if the transmission line is lossless, $Z$ and $Y$ are pure imaginary parameters; as a consequence [see (13) and (17)] $k$, $k_2$ are also real parameters. It is possible to take into account losses by a complex admittance per unit length $Y = G + iW$, where $G$ is the conductance per unit length, responsible for the traveling wave energy loss in the transmission line. For small losses ($G \ll W$), it is possible to consider $k_2$ approximately real (17):

$$k_2 \cong \frac{1}{2} \mathrm{Im}\left(Y^{(3)}\right) \left|\frac{Z}{Y}\right|^{1/2} . \tag{18}$$

Considering that

$$\hat{a} = |\hat{a}|\, e^{i\varphi_a} ,$$
$$\hat{b} = \left|\hat{b}\right| e^{i\varphi_b} , \tag{19}$$

splitting the two (16) into their real and imaginary parts, we obtain

$$\frac{\mathrm{d}\,|\hat{a}|}{\mathrm{d}z} = -\mathrm{Im}\,(k)\,|\hat{a}| ,$$
$$\frac{\mathrm{d}\left|\hat{b}\right|}{\mathrm{d}z} = \mathrm{Im}\,(k)\left|\hat{b}\right| , \tag{20}$$

for the magnitudes of the amplitudes $\hat{a}$ and $\hat{b}$, and

$$\frac{\mathrm{d}\varphi_a}{\mathrm{d}z} = k_2\left(|\hat{a}|^2 + 2\left|\hat{b}\right|^2\right) ,$$
$$\frac{\mathrm{d}\varphi_b}{\mathrm{d}z} = -k_2\left(2\,|\hat{a}|^2 + \left|\hat{b}\right|^2\right) , \tag{21}$$

for the phases of the amplitudes $\hat{a}$ and $\hat{b}$. It is easy to find a general solution for (20). By subtracting the second part of (21) from the first,

$$\varphi_{\mathrm{NL}} = \varphi_a - \varphi_b = 3k_2 \int_0^z \left(|\hat{a}(\zeta)|^2 + \left|\hat{b}(\zeta)\right|^2\right) \mathrm{d}\zeta . \tag{22}$$

This result can be found in [42] for an electromagnetic plane wave propagating in a third-order nonlinear dielectric medium. From this general formalism, we specialize to the propagation along the $z$ coordinate of a plane wave with linear polarization in the $y$ direction. It is useful to write the complex amplitudes by two new unknown functions $\tilde{a}(z)$ and $\tilde{b}(z)$ as follows:

$$\hat{a}(z) = \tilde{a}(z)e^{-\mathrm{Im}\,(k)z}e^{i\varphi_a} ,$$
$$\hat{b}(z) = \tilde{b}(z)e^{\mathrm{Im}\,(k)z}e^{i\varphi_b} . \tag{23}$$

The multilayer structure can be seen as a cascade of nonlinear transmission lines. In the following, we will label with the index $h$ the parameters of the $h$th transmission line. Imposing the continuity of $V$ and $I$ (i.e., the continuity of electrical and magnetic field tangential components) at the interface between the $h$th and the $(h + 1)$-th transmission lines and using (23), it is possible to describe the $h$th transmission line by the following relation:

$$\begin{pmatrix} \tilde{a}_h \\ \tilde{b}_h \end{pmatrix} = \boldsymbol{S}_h \begin{pmatrix} \tilde{a}_{h+1} \\ \tilde{b}_{h+1} \end{pmatrix} , \tag{24}$$

where the $\boldsymbol{S}_h$ matrix, representing a two-port network, is given by

$$\boldsymbol{S}_h = \frac{e^{-i\varphi_{b_h}}}{t_h} \begin{bmatrix} e^{-i\left(k_h\delta_h+\varphi_{\mathrm{NL}_h}\right)} & r_h e^{-i\left(k_h\delta_h+\varphi_{\mathrm{NL}_h}\right)} \\ r_h e^{ik_h\delta_h} & e^{ik_h\delta_h} \end{bmatrix} \tag{25}$$

with

$$t_h = \frac{2n_h}{n_h + n_{h+1}},$$
$$r_h = \frac{n_h - n_{h+1}}{n_h + n_{h+1}}, \tag{26}$$

and $\delta_h$ is the length of the $h$th transmission line. We also assume that the nonlinearity is weak enough to neglect any reflectivity change at each interface. The input-port parameters $\tilde{a}_h, \tilde{b}_h$ are evaluated to the right of the $h$th interface, and the output-port parameters are evaluated to the left of the $(h+1)$-th interface. Note that the kind of multilayer structure (i.e., periodic, Cantor-like, or other) depends only on the sequence of the transmission line lengths $\delta_h$. In this sense, the approach is applicable to any kind of layered structure.

It is easy to calculate the matrix $\boldsymbol{S}_h$ even if it depends on the port parameters for the presence of $\varphi_{\mathrm{NL}_h}$. To do it, note that by solving (20),

$$|\hat{a}| = |\hat{a}(z_h)| e^{-\mathrm{Im}(k)(z-z_h)},$$
$$\left|\hat{b}\right| = \left|\hat{b}(z_h)\right| e^{\mathrm{Im}(k)(z-z_h)} \tag{27}$$

and taking into account the position (23),

$$|\tilde{a}(z_h)| = |\tilde{a}_h| = |\hat{a}(z_h)|,$$
$$\left|\tilde{b}_h(z_h)\right| = \left|\tilde{b}_h\right| = \left|\hat{b}(z_h)\right|. \tag{28}$$

So, according to (20), (22) becomes

$$\varphi_{\mathrm{NL}_h} = 3k_{2_h} \left[ |\tilde{a}_h|^2 \frac{1 - e^{-2\mathrm{Im}(k_h)\delta_h}}{2\mathrm{Im}(k_h)} + \left|\tilde{b}_h\right|^2 \frac{e^{2\mathrm{Im}(k_h)\delta_h} - 1}{2\mathrm{Im}(k_h)} \right]. \tag{29}$$

It is necessary to know the magnitude of the input-port parameters to calculate the phase shift (29). From (24) and (25),

$$|\tilde{a}_h| = \frac{1}{|t_h|} \left| e^{-ik_h\delta_h}\tilde{a}_{h+1} + r_h e^{-ik_h\delta_h}\tilde{b}_{h+1} \right|,$$
$$\left|\tilde{b}_h\right| = \frac{1}{|t_h|} \left| r_h e^{ik_h\delta_h}\tilde{a}_{h+1} + e^{ik_h\delta_h}\tilde{b}_{h+1} \right|. \tag{30}$$

This means that $|\tilde{a}_h|$ and $\left|\tilde{b}_h\right|$ can be found by using the matrix $\boldsymbol{S}_h$ as if the nonlinearity did not exist at all ($\varphi_{\mathrm{NL}_h} = 0$). Because we consider that the layered structure is modeled by a cascade of $p$ two-port networks, it is possible to calculate the port parameters of the network $p, p-1, \ldots, 1$, once $\tilde{a}_{p+1}$ and $\tilde{b}_{p+1}$ are known. The procedure we use is as follows (the so called "dummy" method [42]). We consider a given transmitted field, hence we determine the incident and reflected amplitudes just inside the output interface, provided that the linear refractive indexes are used in the calculation. Then from the nonlinear propagation, we determine the fields across the final layer. We use these fields to determine the total phase change across the final layer in the presence of the nonlinearity. Hence, we determine the forward and backward fields at the penultimate interface. We iterate this procedure until the incident field is calculated.

When the multilayer structure is made with lossless dielectric layers, so that the propagation constants of the fields are purely real ($\mathrm{Im}\,(k) = 0$), $n_{2\mathrm{nl}}$ is the Kerr coefficient

$$n_{2\mathrm{nl}} = \frac{\chi^{(3)}}{2n_h}$$

$$\left( n_{2\mathrm{nl}} = \frac{1}{2}\varepsilon_0 c n_h n_{2h}^{\mathrm{nl}} \right). \tag{31}$$

It is useful to normalize the intensities to the value

$$I_b = \frac{1}{|n_{2\mathrm{nl}}|}. \tag{32}$$

When losses are taken into account, the following normalization is useful instead:

$$I_b^\alpha = \frac{\alpha}{k_0\,|n_{2\mathrm{nl}}|}. \tag{33}$$

Because the effective intensity in each layer is defined as

$$I_{\mathrm{eff}_h} = \varepsilon_0 c n_h \left[ |\tilde{a}_h|^2 \frac{1 - e^{-2\mathrm{Im}\,(k_h)\delta_h}}{2\mathrm{Im}\,(k_h)} + |\tilde{b}_h|^2 \frac{e^{2\mathrm{Im}\,(k_h)\delta_h} - 1}{3\mathrm{Im}\,(k_h)} \right], \tag{34}$$

the nonlinear phase shift (29) can be written for the lossless case as

$$\varphi_{\mathrm{NL}_h} = 3k_0 \delta_h \frac{n_{2h}^{\mathrm{nl}}}{\varepsilon_0 c n_h} I_{\mathrm{eff}_h}, \tag{35}$$

where $I_{\mathrm{eff}_h}$ is the normalized effective intensity of the $h$th transmission line. When losses are taken into account,

$$\varphi_{\mathrm{NL}_h} = 3k_0 \frac{n_{2h}^{\mathrm{nl}}}{\alpha_h \varepsilon_0 c n_h} I_{\mathrm{eff}_h}. \tag{36}$$

Moreover, we set as incident and transmitted input, respectively,

$$I_{\mathrm{i}} = 1/2\varepsilon_0 c n_0 |\hat{a}_0|^2,$$

$$I_{\mathrm{t}} = 1/2\varepsilon_0 c n_{p+1} |\hat{a}_{p+1}|^2. \tag{37}$$

When the local intensity is very low, the effect of the nonlinear phase change on the forward and backward component of the fields is negligible with respect to the linear phase shift in each layer. To highlight nonlinear effects on the transmission spectrum, let us consider a Cantor filter $N = 3$ realized with AlAs/Al$_{0.3}$ GaAS$_{0.7}$, (the initiator is the AlAs layer), where the minimum optical path is $\lambda/4$, with $\lambda = 1.06\,\mu$m. A negative nonlinear refractive index has been considered for wavelengths around $1.06\,\mu$m of the order of $-4 \times 10^{-13}\,\mathrm{cm}^2/\mathrm{W}$, and no dispersion of the nonlinear coefficient has been taken into account. A nonlinear spectral shift is found for an input intensity of $10^{10}\,\mathrm{W}/\mathrm{cm}^2$ (see Fig. 11). The frequency shift of the nonlinear spectrum is strongly influenced by the refractive index of the initiator (high or low refractive index) and by the refractive index contrast among the constituent layers. The change of the shift, if toward higher or lower frequencies, depends on the sign of the nonlinear contribution [44]. Examples of nonlinear transmission for fixed spectral values are given in Fig. 12. An example of output versus input intensity is shown for the level $N = 2$ (Fig. 12a), and $N = 3$ (Fig. 12c). The corresponding spectral positions are shown in Fig. 12b and Fig. 12d. In this example, a positive nonlinearity of thermal origin has been considered. A more detailed discussion is presented in [26,33] where a comparison has also been made with a traditional layered structure. A reduction of the input threshold intensity for bistability was found. Multistable behavior can also occur, depending on the various system parameters and materials at our disposal.

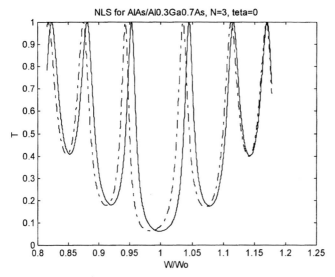

**Fig. 11.** Cantor filter with $N = 3$ realized with AlAs/Al$_{0.3}$ GaAS$_{0.7}$, (the initiator is the AlAs layer); the minimum optical path is $\lambda/4$, with $\lambda = 1.06\,\mu$m: *continuous line*: linear spectrum; *dotted line*: nonlinear spectrum at $10^{10}\,\mathrm{W}/\mathrm{cm}^2$

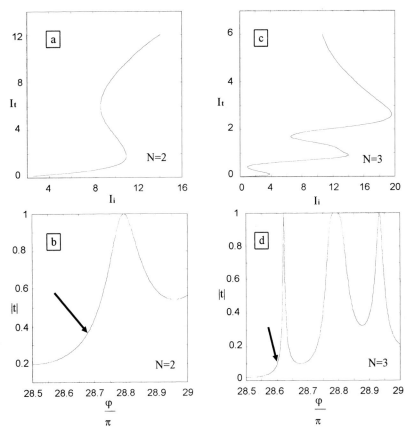

**Fig. 12.** Output $(I_t)$ vs input $(I_i)$ intensity (arbitrary units) for the level $N = 2$ (**a**), and $N = 3$ (**c**) of a Cantor structure. The corresponding spectral positions are shown in (**b**) and (**d**). Refractive indexes are $n_1 = 1.4$, $n_2 = 2.3$. Thermal nonlinearity has been taken into account for the highest refractive index layer, which induces a nonlinear refractive index change of $5 \times 10^{-2}$ when the bistability threshold is reached

The same concepts discussed for multilayers structures can be translated in the form of guided wave geometry, where a quasi-periodic corrugation is "written" on the top of the guide [43]. All of these properties make nonlinear, quasi-periodic structures very interesting for nonlinear filtering properties and applications.

## 7   Mesoscopic Layered Structures

In a stratified structure, the fields inside the structure must be determined by solving a set of transfer-matrix equations. Such an approach is necessary if the layer thicknesses are of the order of the incident wavelength, because interference effects then become important. If individual layer thicknesses are much less than the incident wavelength, the multilayer structure can be considered a uniform effective medium. It is then possible to introduce some simplifying considerations into all calculations of the optical properties of the layers. In fact, in this case the propagation of the light through the structure can be described in terms of effective linear and nonlinear optical susceptibilities.

We assume that the thickness of each layer [28] is much larger than an atomic dimension, but much smaller than the incident wavelength. Results of the analysis depend critically on the polarization of the incident beam. In particular, if the electrical field is TE polarized, then it is spatially uniform within the composite material (because of the boundary condition that states that the tangential component of the electrical field must be continuous at an interface); consequently, the optical constants of the composite [28] become simple averages of those of the constituent materials.

On the other hand, if the incident electrical field is polarized TM, then the electrical field becomes nonuniformly distributed inside the layers of the composite, and, taking advantage of boundary conditions at each layer, the effective linear optical constant is given by [28]

$$\frac{1}{n_{\text{eff}}^2} = \frac{f_a}{n_a^2} + \frac{f_b}{n_b^2}. \tag{38}$$

If we have a periodic layered distribution, the volume fraction $f_a$ and $f_b$ of each material is given by

$$f_a = \frac{d_{\text{tot}}|_a}{d_{\text{tot}}}$$

$$f_b = \frac{d_{\text{tot}}|_b}{d_{\text{tot}}}, \tag{39}$$

where $d_{\text{tot}}$ is the total thickness of the structure and $d_{\text{tot}}|_j (j = a, b)$ is the total thickness of the structure if the initiator is the material $a$ or $b$. If a triadic Cantor structure is considered, then the volume fractions $f_a$ and $f_b$ of each material are given by

$$f_a = \frac{d_{\text{tot}}|_a}{d_{\text{tot}}} = \frac{2^N d_a}{2^N d_a + (3^N - 2^N)d_b},$$

$$f_b = \frac{(3^N - 2^N)d_b}{(3^N - 2^N)d_b + 2^N d_a}. \tag{40}$$

where $d_a$ and $d_b$ are the thicknesses of the smallest layers of the structure. We observe that we have an additional parameter, compared with a periodic structure, that is, the Cantor level $N$.

The layering produces a large enhancement of the effective refractive index. An example is given in Fig. 13a, where the ratio $n_{\text{eff}}/n_a$ is shown for

(a)

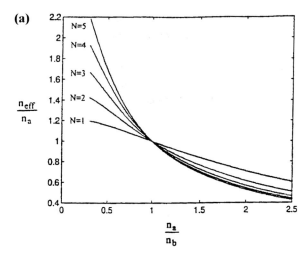

(b)

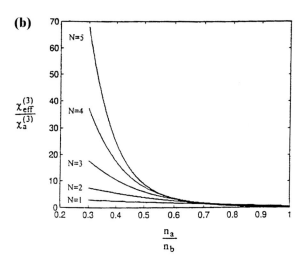

**Fig. 13.** Enhancement of the effective refractive index in the mesoscopic approximation: (**a**) the ratio $n_{\text{eff}}/n_a$ is presented for several values of the Cantor level as a function of $n_a(\omega)/n_b(\omega)$; (**b**) third-order nonlinear susceptibility

several values of the Cantor level as a function of $n_a(\omega)/n_b(\omega)$. In this example, the multilayer is realized with two different materials whose optical path follows the triadic Cantor code. The same can be done with a Fibonacci code, as discussed in [28], where it is shown that an enhancement of the effective index is found when a nonlinear material is taken into account in one of the layers constituting the structure.

The effective index model in the mesoscopic limit, i.e., no interference of the field inside the structure, can show the same value of the "interferometric" effective index, as discussed in Sect. 4, and where it has been evaluated from the transmission properties of the layered structure. This is possible when very low refractive index contrast among the layers is considered in the limit of long wavelengths. An example of the behavior is shown in Fig. 14, where the difference among the two "effective indexes" is reported for several values of the refractive index contrast. Therefore the "interferometric" effective index admits the mesoscopic effective index as a limiting value.

Quasi-periodic layered structures enhance the nonlinear susceptibility of a periodic structure [28]. Again we will consider the fractal structures realized by following a triadic Cantor sequence or a Fibonacci sequence, alternating layers of two different materials $(a, b)$ possessing linear refractive indexes $n_a$ and $n_b$ and nonlinear susceptibilities (we suppose only third-order nonlinearity) $\chi_a^{\mathrm{NL}}$ and $\chi_b^{\mathrm{NL}}$, respectively. The two constituent materials are assumed to be lossless, and the response time of the composite is essentially the same as that of the nonlinear constituent. We assume that each layer is thicker than an atomic dimension but much smaller than the incident wavelength. Consequently, the structural properties of each constituent material are essentially the same as those of a bulk sample, but the propagation of light through the structure can be described in terms of effective linear and nonlinear optical susceptibilities. By using the same notation given in (33) and (34), for TM polarization, the nonlinear effective susceptibility is

$$\chi_{\mathrm{eff}}^{(2)}(\omega = \omega + \omega - \omega) = \frac{\dfrac{f_a \chi_a^{(3)}}{|n_a^2(\omega)|^2 n_a^4(\omega)} + \dfrac{f_b \chi_b^{(3)}}{|n_b^2(\omega)|^2 n_b^4(\omega)}}{\left[\dfrac{f_a}{n_a^2(\omega)} + \dfrac{f_b}{n_b^2(\omega)}\right]^2 \left[\dfrac{f_a}{n_a^2(\omega)} + \dfrac{f_b}{n_b^2(\omega)}\right]^2}. \tag{41}$$

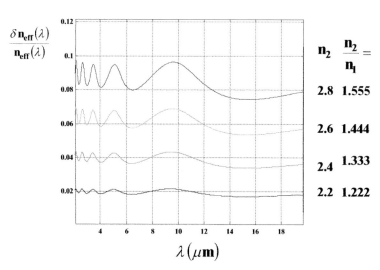

**Fig. 14.** The difference between the "interferometric" and "mesoscopic" effective index reported for several values of the refractive index contrast

Then, for the quasi-periodic Cantor structure, the nonlinear effective refractive index may be much higher than that of the corresponding periodic structure (see Fig. 13b), depending on the material selected as the initiator and, for the Cantor code, from the level of the iteration. Among the quasi-periodic structures studied in [28], a greater enhancement of the Cantor sequence with respect to the Fibonacci sequence can be obtained, on the Cantor level and on the Fibonacci sequence, depending on which material is used as the initiator of the Cantor sequence.

The Fibonacci structure presents an enhancement of the optical constants which is independent of the level of the sequence used [28].

## 8  Conclusions

Fractal layered composites exhibit very interesting and flexible properties. We have discussed here 1-D layered structures following Cantor and Fibonacci fractal codes. The Cantor layered structures offer greater flexibility in handling spectral transmission compared to periodic structures, which may turn out particularly useful for unique filtering properties, mainly when the nonlinear response of the structure is taken into account. The Fibonacci fractal code also has some advantage, even if somehow less interesting than the Cantor code. In the nonlinear case, the enhancement of field localization resulting from the fractal nature may be very usefully exploited in nonlinear optical applications.

Mesoscopic treatment of the structures can be performed when their thickness is much smaller than the incident wavelength, giving rise to interesting enhancements of the linear and nonlinear effective refractive indexes. Finally, we want to point out that the general methods outlined in this review can be very easily applied to other fractal codes.

### Acknowledgments

The author C.S. thanks P. Masciulli and F. Tropea for their contribution to the study of fractal codes.

## References

1. R. Merlin, Structural and electronic properties of nonperiodic superlattices, IEEE J. Quant. Electron. **24**, 1791 (1988)
2. S. A. Gradesluk, V. Freilikher, Localization and wave propagation in randomly layered media, Sov. Phys. Uspekhi **33**, 134 (1990)
3. M. Kohmoto, L. P. Kadanoff, C. Tang, Localization Problem in One Dimension: Mapping and Escape, Phys. Rev. Lett. **50**, 1870 (1983)
4. M. Kohmoto, B. Sutherland, C. Tang, Critical wave functions and a Cantor-set spectrum of a one-dimensional quasicrystal model, Phys. Rev. B **35**, 1020 (1987)

5. D. Shechtman, J. Black, D. Gratias, J. W. Cahn, Metallic Phase with Long-Range Orientational Order and No Translational Symmetry, Phys. Rev. Lett **53**, 1951 (1984)

6. R. Merlin, K. Bajima, R. Clarke, F. T. Juang, P. K. Bhatacharyya, Quasi-periodic GaAs-AlAs Heterostructures, Phys. Rev. Lett. **55**, 1768 (1986)

7. M. Kohmoto, B. Sutherland, K. Iguchi, Localization in Optics, Quasi-periodic media, Phys. Rev. Lett. **58**, 2436 (1987)

8. L. Gourley, C. P. Tiggs, R. P. Schneider, Jr., T. M. Brennan, B. E. Hammons, A. E. McDonald, Optical properties of fractal quantum wells, Appl. Phys. Lett. **62**, 1736 (1993)

9. O. V. Angelsky, P. P. Maksimyak, T. O. Perun, Dimensionality in optical fields and signals, Appl. Opt. **32**, 6066 (1993)

10. M. Bertolotti, P. Masciulli, C. Sibilia, Spectral transmission properties of a self-similar optical Fabry-Perot resonator, Opt. Lett. **19**, 777 (1994)

11. Rammal, G. Toulouse, Random walks on fractal structures and percolation clusters, Phys. Lett. **44**, 412 (1983)

12. F. Craciun, A. Bettucci, E. Molinari, A. Petri, A. Alippi, Direct experimental observation of fracton mode patterns in one dimensional Cantor composite, Phys. Rev. Lett. **68**, 342 (1992)

13. S. Dutta Gupta, D. S. Ray, Localization problem in optics: Nonlinear quasi-periodic media, Phy. Rev. B **41**, 8047 (1990)

14. Y. Sakurada, J. Uozumi, T. Asakura, Fresnel diffraction by one-dimensional regular fractals, Pure Appl. Opt. **1**, 29 (1992)

15. M. V. Berry, Diffractals, J. Phys. A **12**, 781 (1979)

16. J. Uozumi, T. Asakura, Fractal Optics, Curr. Trends Opt. **6**, 83 (1994)

17. C. Allain, M. Cloitre, Optical diffraction on fractals, Phys. Rev. B **33**, 3566 (1986)

18. D. A. Zimnyakov, V. V. Tuchin, Scale properties of the diffraction fields induced by pre-fractal random screens, in *Fractal Frontiers*, M. M. Novak, T. G. Dewey (Eds.) (World Scientific, Singapore 1997) pp. 281–290

19. J. M. Elson, J. M. Bennett, Relation between the angular dependence of scattering and the statistical properties of optical surfaces, J. Opt. Soc. Am. **69**, 31 (1979)

20. J. D. Joannopulos, P. R. Villeneuve, S. Fan, Putting a new twist on light, Nature **386**, 143 (1997)

21. E. Yablonovich, Inhibited spontaneous emission in solid-state physics and electronics, Phys. Rev. Lett. **58**, 2059 (1987)

22. W. Robertson, Measurement of photonic band structures in a two-dimensional periodic dielectric array, Phys. Rev. Lett. **68**, 2023 (1992)

23. M. Scalora, J. P. Dowling, C. M. Bowden, M. J. Bloemer, Optical limiting and switching of ultrashort pulses in Nonlinear PBG Materials, Phys. Rev. Lett. **73**, 136 (1994)

24. J. M. Bendickson, J. P. Dowling, M. Scalora, Analytic expression for electromagnetic mode density of finite 1-D PBG structures, Phys. Rev. E **53**, 4107 (1996)

25. C. Sibilia, I. Nefedov, M. Scalora, M. Bertolotti, Electromagnetic mode density for finite quasi-periodic structures, J. Opt. Soc. Am. B **15**, 1947 (1998)

26. M. Bertolotti, P. Masciulli, P. Ranieri, C. Sibilia, Optical bistability in a nonlinear Cantor corrugated waveguide, J. Opt. Soc. Am. B **13**, 1512 (1996)

27. M. Bertolotti, P. Masciulli, C. Sibilia, F. Wijnands, H. Hoekstra, Transmission properties of A Cantor corrugated waveguide, J. Opt. Soc. Am. B **13**, 628 (1996)

28. C. Sibilia, F. Tropea, M. Bertolotti, Enhanced nonlinear optical response of a Cantor-like and Fibonacci-like quasi-periodic structures, J. Mod. Opt. **45**, 2255 (1998)

29. B. B. Mandelbrot, *The Fractal Geometry of Nature* (Freeman, San Francisco 1992)

30. P. de Vries, H. De Raedt, A. Lagendijk, Localization of waves in fractals: spatial behaviour, Phys. Rev. Lett. **62**, 2515 (1989)

31. Badgasaryan, C. Sibilia, Proc. COST 240 (Springer, Berlin, Heidelberg 1999)

32. W. Gellerman, M. Kohmoto, B. Sutherland, P. C. Taylor, Localization of light waves in Fibonacci dielectric multilayers, Phys. Rev. Lett. **72**, 633 (1994)

33. M. Bertolotti, C. Sibilia, Optical properties of quasiperiodic structures: Linear and nonlinear analysis, Springer Ser. Opt. Sci. **74**, 258 (2000)

34. A. M. Steinberg, R. Y. Chiao, Subfemtosecond determination of transmission delay times for a dielectric mirror (photonic band gap) as a junction of the angle of incidence, Phys. Rev. A **51**, 3525 (1995)

35. M. Centini, C. Sibilia, M. Scalora, G. D'Aguanno, M. Bertolotti, I. Nefedov, M. Bloemer, C. Bowden, Dispersive properties of finite, one-dimensional photonic band gap structures: Applications to nonlinear quadratic interactions, Phys. Rev. E **56**, 4891 (1999)

36. A. J. Ward, J. B. Pendry, W. J. Stewart, Photonic dispersion surfaces, J. Phys. Cond. Matter **7**, 2217 (1995)

37. M. Scalora, M. J. Bloemer, A. S. Pethel, J. P. Dowling, C. M. Bowden, A. S. Manka, Transparent, metallo-dielectric, one-dimensional, photonic band-gap structures, J. Appl. Phys. **83** 2377 (1998)

38. M. J. Bloemer, M. Scalora, Transmissive properties of Ag/MgF2 photonic band gaps, Appl. Phys. Lett. **72**, 1676 (1998)

39. C. Sibilia, M. Scalora, M. Centini, M. Bertolotti, M. J. Bloemer, C. Bowden, Electromagnetic properties of periodic and quasi periodic one-dimensional metallo-dielectric photonic band gap structures, Pure Appl. Optics **1**, 490 (1999)

40. S. Dutta Gupta, Nonlinear Optics of Stratified Media, Prog. Opt. **XXXVIII**, 1 (1998)

41. L. Caleo, C. Sibilia, P. Masciulli, M. Bertolotti, Nonlinear optical filters based on the cascading second order effect, J. Opt. Soc. Am. B **14**, 2315 (1997)

42. J. Danckaert, K. Fobelets, I. Veretennicoff, G. Vitrant, R. Reinish, Dispersive optical bistability in stratified structures, Phys. Rev. B **44**, 8214 (1991)

43. M. Bertolotti, P. Masciulli, P. Ranieri, C. Sibilia, J. Opt. Soc. Am. B**13**, 1517 (1996)

44. C. Sibilia, F. Tropea, M. Bertolotti, V. Rusu, I. Grave, Nonlinear optical properties of Cantor-like quasiperiodic filters, to be published

# Optical Nonlinearities of Fractal Composites

Vladimir M. Shalaev

School of Electrical and Computer Engeneering, Purdue University
West Lafayette, IN 47907-1285, USA
shalaev@purdue.edu

**Abstract.** A theory of optical responses in fractal nanostructured composite materials is outlined. It is shown that the fractal geometry results in localization of plasmon excitations in the "hot" spots, where the local field can exceed the applied field by several orders of magnitude. The high local fields of the localized fractal modes result in dramatic enhancement of optical responses, making surface-enhanced spectroscopy of single molecules and nanocrystals feasible.

## 1  Introduction

Electromagnetic phenomena in random metal–insulator composites, such as rough thin films, cermets, colloidal aggregates and others, have been intensively studied for the last two decades [1,2]. These media typically include small nanometer-scale particles or roughness features. Nanostructured composites possess fascinating electromagnetic properties, which differ greatly from those of ordinary bulk material, and they are likely to become ever more important with the miniaturization of electronic and optoelectronic components.

Giant enhancement of optical responses in a random medium including a metal component, such as metal nanocomposites and rough metal thin films consisting of small nanometer-sized particles or roughness features, is associated with optical excitation of surface plasmons that are collective electromagnetic modes and strongly depend on the geometric structure of the medium. Nanocomposites and rough thin films are often characterized by fractal geometry, where collective optical excitations, such as surface plasmons, tend to be localized in small nanometer-sized areas, namely, hot spots, because the plane running waves are not eigenfunctions of the operator of dilation symmetry that characterizes fractals.

Fractals look similar in different scales; in other words, a part of the object resembles the whole. Regardless of the size, this resemblance persists forever, if the object is mathematically defined. In nature, however, the scale-invariance range is restricted, on the one side, by the size of the structural units (e.g., atoms) and on the other, by the size of the object itself. The emergence of fractal geometry was a significant breakthrough in the description of irregularity [3]. The realization of the fact that the geometry of fractional dimensions is often more successful than Euclidean geometry in describing

V. M. Shalaev (Ed.): Optical Properties of Nanostructured Random Media,
Topics Appl. Phys. **82**, 93–112 (2002)
© Springer-Verlag Berlin Heidelberg 2002

natural shapes and phenomena provided it a major impetus to research and led to better understanding of many processes in physics and other sciences.

Fractal objects do not possess translational invariance and therefore cannot transmit running waves [4]. Accordingly, dynamic excitations, such as, for example, vibrational modes (fractons), tend to be localized in fractals [4]. Formally, this is a consequence of the fact that plane running waves are not eigenfunctions of the operator of dilation symmetry characterizing fractals. The efficiency of fractal structures in damping running waves is probably the key to "self-stabilization" of many of the fractals found in nature [3].

The physical and geometric properties of fractal clusters has attracted growing attention from researchers in the past two decades [3,4]. The reason for this is twofold. First, processes of aggregation of small particles in many cases lead to the formation of fractal clusters rather than regular structures [3]. Examples include the aggregation of colloidal particles in solutions, the formation of fractal soot from little carbon spherules in the process of incomplete combustion of carbohydrates, the growth of self-affine films, and the gelation process and formation of porous media [3]. The second reason is that the properties of fractal clusters are very rich in physics and different from those of either bulk material or isolated particles (monomers) [2,4].

The most simple and extensively used model for fractal clusters is a collection of identical spherically symmetrical particles (monomers) that form a self-supporting geometric structure. It is convenient to think of monomers as identical rigid spheres that form a bond on contact. A cluster is considered self-supporting if each monomer is attached to the rest of the cluster by one or more bonds. Fractal clusters are classified as "geometric" (built as a result of a deterministic iteration process) or "random." Most clusters in nature are random.

The fractal (Hausdorff) dimension of a cluster $D$ is determined through the relation between the number of particles $N$ in a cluster (aggregate) and the cluster's radius of gyration $R_c$:

$$N = (R_c/R_0)^D , \tag{1}$$

where $R_0$ is a constant of the order of the minimum separation distance between monomers. Note that the fractal dimension is, in general, fractional and less than the dimension of the embedding space $d$, i.e., $D < d$. Such a power-law dependence of $N$ on $R_c$ implies a spatial scale-invariance (self-similarity) for the system. For the sake of brevity, we refer to fractal aggregates, or clusters, as fractals.

Another definition of the fractal dimension uses the pair density–density correlation function $\langle \rho(\mathbf{r})\rho(\mathbf{r} + \mathbf{R}) \rangle$:

$$\langle \rho(\mathbf{r})\rho(\mathbf{r} + \mathbf{R}) \rangle \propto R^{D-d}, \quad \text{if } R_0 \ll r \ll R_c . \tag{2}$$

This correlation makes fractals different from truly random systems, such as salt scattered on the top of a desk. Note that the correlation becomes constant, $g(r) = \text{const}$, when $D = d$; this corresponds to conventional media, such as crystals, gases, and liquids. The unusual morphology of fractional dimensions results in unique physical properties of fractals, including the localization of dynamic excitations.

Optical excitations in fractal composites are substantially different from those in other media. For example, there is only one dipolar eigenstate that can be excited by a homogeneous field in a dielectric sphere (for a spheroid, there are three resonances with nonzero total dipole moment); the total dipole moment of all other eigenstates is zero and, therefore, they can be excited only by an inhomogeneous field. In contrast, fractal aggregates possess a variety of dipolar eigenmodes, distributed over a wide spectral range, which can be excited by a homogeneous field.

In continuous media, dipolar eigenstates (polaritons) are running plane waves that are eigenfuctions of the operator of translational symmetry. This also holds for most microscopically disordered media that are homogeneous on average. Dipolar excitations in these cases are typically delocalized over large areas, and all monomers absorb light energy at approximately the same rate in regions that significantly exceed the wavelength. In contrast, fractal composites have optical excitations that are localized in small subwavelength regions. The local fields in these "hot" spots are large, and the absorption by the "hot" monomers is much higher than by other monomers in a fractal composite. This is a consequence of the fact (mentioned above) that fractals do not possess translational symmetry; instead, they are symmetrical with respect to the scale transformation.

In metal fractal composites, the dipolar excitations are represented by plasmon oscillations. Plasmon modes are strongly affected by fractal morphology, leading to the existence of hot spots, which are areas of plasmon localization in fractals. Local enhancements in hot spots can exceed the average surface enhancement by many orders of magnitude because the local peaks of the enhancement are spatially separated by distances much larger than the peak sizes. The spatial distribution of these high-field regions is very sensitive to the frequency and polarization of the applied field [5]. The positions of the hot spots change chaotically but reproducibly with frequency and/or polarization. This is similar to speckles created by laser light scattered from a rough surface; the important difference is that the scale size for fractal plasmons in the hot spots is in the nanometer range rather than in the micrometer range encountered for photons.

The fractal plasmon, as any wave, is scattered from density fluctuations — in other words, fluctuations of polarization. The strongest scattering occurs from inhomogeneities of the same scale as the wavelength. In this case, interference in the process of multiple scattering results in Anderson localization. Anderson localization corresponds typically to uncorrelated disorder. A fractal structure is in some sense disordered, but it is also correlated for all length scales, from the size of constituent particles, in the lower limit, to the total size of the fractal, in the upper limit. Thus, what is unique for fractals is that, because of their scale invariance, there is no characteristic size of inhomogeneity — inhomogeneities of all sizes are present, from the lower to the upper limit. Therefore, whatever the plasmon wavelength, there are always fluctuations in a fractal with similar sizes, so that the plasmon is always strongly scattered and, consequently, can be localized [2,5].

Note that, as shown in [6], a pattern of localization of optical modes in fractals is complicated and can be called inhomogeneous. At any given frequency, individual eigenmodes are dramatically different from each other, and their sizes (their coherent radii) vary in a wide range, from the size of an individual particle to the size of a whole cluster. (In the vicinity of the plasmon resonance of individual particles, even chaotic behavior of the eigenmodes in fractals can be found [6].) However, even delocalized modes typically consist of two or more very sharp peaks that are topologically disconnected, i.e., located at relatively large distances from each other. In any case, the electromagnetic energy is mostly concentrated in these peaks, which can belong to different modes or, sometimes, to the same mode. Thus, despite the complex inhomogeneous pattern of localization in fractals, there are always very sharp peaks, where the local fields are high. These hot spots eventually provide significant enhancement for a number of optical processes, especially the nonlinear ones that are proportional to the local fields to a power greater than one.

Because of the random character of fractal surfaces, the high local fields associated with hot spots look like strong spatial fluctuations. Since a nonlinear optical process is proportional to the local fields raised to a power greater than one, the resulting enhancement associated with the fluctuation area (i.e., with the hot spot) can be extremely large. In a sense, we can say that enhancement of optical nonlinearities is especially large in fractals because of very strong field fluctuations.

Large fluctuations of local electromagnetic fields on the metal surfaces of inhomogeneous metal media result in a number of enhanced optical effects. A well known effect is surface-enhanced Raman scattering (SERS) by molecules adsorbed on a rough metal surface, e.g., in aggregated colloidal particles [7].

In an intense electromagnetic field, a dipole moment induced in a particle can be expanded into a power series: $d = \alpha^{(1)} E(r) + \alpha^{(2)} [E(r)]^2 + \alpha^{(3)} [E(r)]^3 + \ldots$, where $\alpha^{(1)}$ is the linear polarizability of a particle, $\alpha^{(2)}$ and $\alpha^{(3)}$ are the nonlinear polarizabilities, and $E(r)$ is the local field at site $r$. The polarization of a medium (i.e., dipole moment per unit volume), which is a source of the electromagnetic field in a medium, can be represented in an analogous form by the coefficients $\chi^{(n)}$ called "susceptibilities." When the local field exceeds the applied field, $E^{(0)}$, considerably, huge enhancements of nonlinear optical responses occur.

Below we first consider local-field enhancement in nanometer-sized metal particles and then the unique features of the enhancement that are opened up in fractal composites of metal nanoparticles.

## 2  Local-Field Enhancement in Nanospheres and Nanospheroids

Here we consider the local-field enhancement that can be obtained on the surfaces of individual metal nanoparticles, which are much smaller than the wavelength of the incident electromagnetic wave. In such particles, the excitation of free electrons (plasmons) can result in large local fields, much larger than the applied optical field. Plasmon oscillations (also referred to as surface plasmons) aare especially strong at the resonant excitation. This plasmon resonance is associated with collective electron oscillations (surface plasmons). The displacement of free electrons from their equilibrium position in a small particle results in a uncompensated charge on the surface of the particle, leading to its polarization; this polarization, in turn, results in a restoring force that causes electron oscillations. These plasmon oscillations can lead to the large local fields near the surface of the metal particle so that a molecule adsorbed on the metal surface can produce "surface-enhanced" optical signals. Possible shifts of molecular energy levels and creation of new resonances due, for example, to a charge transfer mechanism are not considered in this section; this enhancement (referred sometimes to as chemical enhancement) although important, is not universal and can occur only for special molecules. In contrast, electromagnetic enhancement is universal and takes place regardless of the possible "chemical" renormalization of the molecular cross section.

The metal particles we are concerned with are much smaller than the wavelength and have sizes in the range from 5 to 50 nm. The field enhancement, in this case, can be especially large. In bigger particles, the retardation effects that spoil the quality factor of the plasmon resonance become important; in smaller particles, electron scattering at the metal surface increases the resonance width and thus decreases the quality factor.

We note that since the plasmon resonance is due to the surface charge and thus is of a geometric nature, it depends only on the shape of a particle. For spheres, for example, with radii in the range roughly from 5 to 50 nm, the resonance frequency and its width almost do not depend on the particle's size.

The optical responses of metals can, in many cases, be well approximated with the Drude model. For a Drude metal, the dielectric constant is given by

$$\epsilon = \frac{4\pi i \sigma}{\omega} = \epsilon_0 + \frac{4\pi i \sigma(0)}{\omega[1 - i\omega\tau]}, \tag{3}$$

where the dc conductivity $\sigma(0)$ is related to the plasma frequency $\omega_p$ and relaxation time $\tau$ by $\sigma(0) = \omega_p^2\tau/(4\pi)$ and $\epsilon_0$ is a contribution to $\epsilon$ due to interband electron transitions. (Note that for the relaxation rate of collective plasmon oscillations, $1/\tau$, the following different notations $1/\tau = \omega_\tau = \Gamma$ are interchangeably used in the literature and in this paper.) The Drude model describes well the optical response of free electrons in metals; through the term $\epsilon_0$, it also takes into account the contribution to the dielectric constant due to interband electron transitions. The real and imaginary parts of the Drude dielectric function can also be represented as

$$\epsilon' = \epsilon_0 - \frac{\lambda^2}{\lambda_\mathrm{p}^2} \frac{1}{1 + (\lambda/\lambda_\tau)^2}, \tag{4}$$

and

$$\epsilon'' = \frac{\lambda^3}{\lambda_\mathrm{p}^2 \lambda_\tau} \frac{1}{1 + (\lambda/\lambda_\tau)^2}, \tag{5}$$

where $\lambda/\lambda_\tau \equiv (\omega\tau)^{-1}$ and $\lambda/\lambda_\mathrm{p} \equiv (\omega_\mathrm{p}/\omega)$. For silver and gold, for example, $\lambda_\tau \sim 60\,\mu\mathrm{m}$ and $20\,\mu\mathrm{m}$, respectively, and $\lambda_\mathrm{p} \sim 140\,\mathrm{nm}$.

Below, we consider spheroids, where two out of three semiaxes ($a, b$, and $c$) are equal: $b = c$. In a sphere, all of the semiaxes are the same, $a = b = c$. In prolate and oblate spheroids, $a > b = c$ and $a < b = c$, respectively. As is well known, the largest local fields can be obtained at the tip of sharp structures; therefore, we focus below on cigar-shaped spheroids, where $a \gg b$, and pancake-shaped spheroids, where $a \ll b$. We also assume that the field is polarized along the long axis of a spheroid, so that the largest field enhancement can be obtained. Then the polarizability of a spheroid can be written as

$$\alpha = \frac{V}{4\pi} \frac{\epsilon - 1}{1 + p(\epsilon - 1)}, \tag{6}$$

where $p$ is a depolarization factor, $V$ is the volume of a nanospheroid, and the host medium was chosen, for simplicity, to be a vacuum. For a sphere, $p = 1/3$ and $V = (4\pi/3)R^3$, so that

$$\alpha_0 = R^3 \frac{\epsilon - 1}{\epsilon + 2}, \tag{7}$$

where $R$ is the radius of the sphere. For prolate, cigar-shaped spheroids, $p \approx (b/a)^2[\ln(\sqrt{2}a/b) - 1]$. For all realistic aspect ratios $A = a/b$, the depolarization factor can be estimated as $p \sim (A)^{-2}$. [Really, for $A = 3$, $A = 10$, and $A = 100$, for example, $pA^2 \approx 0.4$, 1.6, and 4, respectively. It is also clear that $A$ cannot be larger than 100 for particles that have a larger axis on a scale of 10 nm; in fact, since the quality factor decreases when any of the spheroid's axes is outside the range of 5–50 nm (see above), the aspect ratio for spheroids with high-quality resonance is limited roughly to 10.] For oblate, pancake-shaped spheroids, $p \approx (\pi/4)A^{-1}$, where the aspect ratio for oblate spheroids is defined as $A = b/a$. It is important to note that at the same aspect ratio $A$, the depolarization factor in the quasi-one-dimensional "cigars" (or needles) is much smaller than in quasi-two-dimensional pancakes (or disks). This fact has important consequences for field enhancement, as discussed below.

Now we consider the field enhancement that can be obtained in nano-sized spheres and spheroids. The largest field enhancement can be obtained at the plasmon resonance when the real part of the denominator in (6) becomes zero, i.e., at

$$\epsilon'_r \equiv \epsilon'(\lambda_r) = 1 - 1/p \,. \tag{8}$$

According to the Drude model (4), the resonant frequency at $\lambda \ll \lambda_\tau$ is given by

$$\lambda_r = \lambda_p \left(1/p + \epsilon_0 - 1\right)^{1/2} \,. \tag{9}$$

For a sphere, $\epsilon'_r = -2$; for spheroids with a large aspect ratio, $A \gg 1$, the depolarization factor is small, $p \ll 1$, so that $\epsilon'_r \approx -1/p$ and $\lambda_r \approx \lambda_p/\sqrt{p}$. The local-field enhancement $E_l/E_0$ at the surface of spheres and at the sharp edges of spheroids is estimated by the resonance quality factor $Q_f$ as

$$\frac{E_l}{E_0} \sim Q_f \sim (4\pi/V)\alpha(\lambda_r) \,, \tag{10}$$

where $\alpha(\lambda_r)$ is the resonant value of the polarizability.

The local-field enhancement for a sphere is given by [see (7) and (10)]

$$Q_f \equiv Q_0 = \frac{|\epsilon'_r - 1|}{\epsilon''_r} = \frac{3}{(\epsilon_0 + 2)^{3/2}} \frac{\lambda_\tau}{\lambda_p} \sim \frac{\lambda_\tau}{\lambda_p} \,. \tag{11}$$

For different noble metals, the magnitudes of $Q_f$ are on the order of 10 to 100 (for silver, it is the largest, about 50 at the resonant frequency $\lambda_r \sim 400\,\text{nm}$).

For a spheroid, according to (6) and (10), the resonant enhancement is estimated as

$$Q_f \sim \frac{1}{p}\frac{|\epsilon'_r|}{\epsilon''_r} \sim \frac{p^{-2}}{(p^{-1} + \epsilon_0 - 1)^{3/2}} \frac{\lambda_\tau}{\lambda_p} \sim \frac{\lambda_\tau}{\sqrt{p}\lambda_p} \,, \tag{12}$$

where for the second estimate, we used (4), (5), and (9), and for the last estimate, we assumed that $1/p \gg \epsilon_0 - 1$. We note that the depolarization factor $p$ is related to the resonance frequency via (9), so that (12) can also be written as

$$Q_f \sim \frac{\left(1 - \epsilon_0 + \lambda_r^2/\lambda_p^2\right)^2}{(\lambda_r/\lambda_p)^3} \frac{\lambda_\tau}{\lambda_p} \sim \frac{\lambda_r\lambda_\tau}{\lambda_p^2} \,, \tag{13}$$

where for the last estimate, we assumed that $(\lambda_r/\lambda_p)^2 \gg \epsilon_0 - 1$.

According to (12), the local-field enhancement is estimated as

$$Q_f \sim A^2 \frac{|\epsilon'_r|}{\epsilon''_r} \sim A\frac{\lambda_\tau}{\lambda_p} \sim AQ_0 \,, \tag{14}$$

for cigar-shaped spheroids, and as

$$Q_{\mathrm{f}} \sim A \frac{|\epsilon_{\mathrm{r}}'|}{\epsilon_{\mathrm{r}}''} \sim \sqrt{A} \frac{\lambda_{\tau}}{\lambda_{\mathrm{p}}} \sim \sqrt{A} Q_0 \, , \tag{15}$$

for pancake-shaped spheroids.

One can see that the local-field enhancement in spheroids can significantly exceed that in spheres, especially, in cigar-shaped nanostructures, where it exceeds the quality factor in spheres roughly by the aspect ratio $A$. These larger values for field-enhancement in spheroids are achieved at the resonant frequencies, $\epsilon'(\lambda_{\mathrm{r}}) = 1 - 1/p$, which are significantly shifted toward the infrared part of the spectrum, where $\epsilon'$ is negative and large in magnitude so that the resonance can occur at very small values of the depolarization factor $p$.

Thus, to obtain strong field enhancement at a particular wavelength $\lambda$, one can use a nanospheroid with the aspect ratio such that $\epsilon'(\lambda) = -1/p+1 \approx -1/p$, for small $p$. However, it is not easy, in general, to fabricate a nanospheroid with a given aspect ratio. Besides, for spectroscopic purposes, we need to have enhancement in a broad spectral range, so that a spectroscopic signal from any optical transition of an arbitrary molecule can be enhanced.

An alternative possibility is to use structures formed by spherical nanoparticles (colloids, for example), which are typically much easier to make. For example, it is clear that a straight chain of $N$ spheres should have roughly the same optical resonances as a prolate spheroid with the aspect ratio $A = N$. By taking different configurations of colloids, one can obtain resonances (and thus enhancement) at different optical frequencies that depend on the geometry of the whole structure. This approach, however, has the same drawback: there are only few frequencies at which the compact structure of nanospheres resonates because the dipole–dipole interactions between particles arranged in "conventional" geometry is long-range, so that a compact system of particles always resonates as a whole at the frequencies depending on the object's external surface. Really, if we integrate the dipole near-field, $\propto 1/r^3$, into the conventional $3d$ space, we obtain a logarithmic divergence; this indicates that the system resonates as a whole. For example, close-packed particles within a spherical volume have roughly the same optical response as the sphere within which the particles are packed, so that all of the resonances are grouped near $\lambda_{\mathrm{r}}$ such that $\epsilon'(\lambda_{\mathrm{r}}) = -2$.

Thus, in both spheroids and compact structures of colloidal particles, large enhancement can be achieved only at few frequencies depending on the geometry of the structure. For spectroscopic studies, however, it is very important to have a *broadband* enhancement. In other words, we would like to have the local-field enhancement as strong as in spheroids, but within a broad spectral range, including the visible and infrared parts of the spectrum, so that an optical signal from any molecular transition could be enhanced and probed spectroscopically.

In that sense, fractal nanostructures considered below seem to be the materials of choice. Because of the scale-invariant geometry of fractals, optical excitations are not spread over the whole structure but rather tend to be localized in small, nanometer-sized areas, which have very different local geometries and thus resonate at different frequencies. As a result, fractals provide enhancement within an unusually large spectral interval, from the near-UV to the far-infrared. Below, we consider two important classes of fractal nanostructures, fractal aggregates of colloidal particles and metal–dielectric films near the percolation threshold.

# 3    Local-Field Enhanced Optical Responses in Fractal Aggregates

Random fractal clusters are complex systems built from simple elementary blocks, e.g., particles, that are called monomers. It is important to emphasize that the rich and complicated properties of fractal clusters are determined by their global geometric structure rather than by the structure of each monomer. In the formulation of a typical problem in the optics of fractal clusters, the properties of monomers and the laws of physics for their interaction with the incident field and with each other are known, whereas the properties of a cluster as a whole must be found.

In Fig. 1, you see a picture of a typical fractal aggregate of colloidal silver particles (obtained via an electron microscope). The fractal dimension of these aggregates is $D \approx 1.78$. Using the well-known model of cluster–cluster aggregation, colloidal aggregate can be readily simulated numerically [2]. Note that voids are present on all scales, from the minimum (about the size of a single particle) to the maximum (about the size of the whole cluster); this is an indication of the statistical self-similarity of a fractal cluster. The size of an individual particle is $\sim 10\,\mathrm{nm}$, whereas the size of the whole cluster is $\sim 1\,\mu\mathrm{m}$.

The process of aggregation, resulting in clusters similar to that shown in Fig. 1, can be described as follows. A large number of initially isolated and randomly distributed nanoparticles executes random walks in the solution. Encounters with other nanoparticles result in their sticking together, first to form small groups, and then, in the course of the random walking, to aggregate into larger formations, and so on. Cluster–cluster aggregates (CCAs), with the fractal dimension $D \approx 1.78$, are thereby eventually formed.

When the constituent particles of a fractal cluster are irradiated by light of amplitude $E^{(0)}$, oscillating dipole moments $d_i$ are induced in them which interact strongly through dipolar forces leading to the formation of collective optical modes. (Note that throughout the text we interchangeably use the

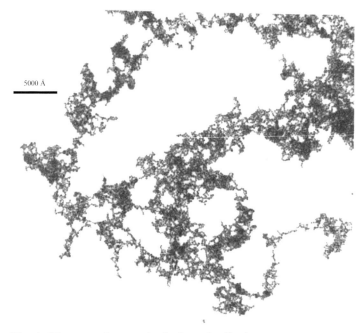

5000 Å

**Fig. 1.** Electron micrograph of a fractal colloid aggregate. Voids corresponding to all length scales are present; the minimum is the size of a single particle, the maximum is the size of the entire cluster. This is a graphic illustration of the statistical self-similarity and hence the fractal nature of the cluster. The radii of the individual particles is $\sim 10$ nm, and the size of the cluster is $\sim 1\,\mu\mathrm{m}$

notations $E_0$ and $E^{(0)}$ for the external field amplitude.) The coupled-dipole equations (CDE) for the induced dipoles acquire the following form [2]:

$$d_{i,\alpha} = \alpha_0 \left( E_\alpha^{(0)} + \sum_{j \neq i} W_{ij,\alpha\beta} d_{j,\beta} \right). \tag{16}$$

In the quasi-static dipole approximation, for example, the interaction operator $W$ between the dipoles has the form

$$W_{ij,\alpha\beta} = \langle i\alpha|W|j\beta \rangle = \left( 3 r_{ij,\alpha} r_{ij,\beta} - \delta_{\alpha\beta} r_{ij}^2 \right) / r_{ij}^5 , \tag{17}$$

where $\mathbf{r}_i$ is the radius vector of the $i$th monomer and $\mathbf{r}_{ij} = \mathbf{r}_i - \mathbf{r}_j$. The Greek indexes denote Cartesian components of vectors and should not be confused with the polarizability $\alpha_0$ of spherical particles forming the aggregate.

Since $W_{ij,\alpha\beta}$ is independent of the frequency $\omega$ in the quasi-static approximation, the spectral dependence of solutions to (16) enters only through $\alpha_0(\omega)$. For convenience, we introduce the variable

$$Z(\omega) \equiv 1/\alpha_0(\omega) = -[X(\omega) + \mathrm{i}\delta(\omega)]. \tag{18}$$

Using (7), we obtain

$$X \equiv -\mathrm{Re}\left[\alpha_0^{-1}\right] = -R^{-3}\left[1 + 3(\epsilon' - 1)/|\epsilon - 1|^2\right], \tag{19}$$

$$\delta \equiv -\mathrm{Im}\left[\alpha_0^{-1}\right] = 3R^{-3}\epsilon''/|\epsilon - 1|^2. \tag{20}$$

The variable $X$ indicates the proximity of $\omega$ to the resonance of an individual particle, that occurs for a spherical particle at $\epsilon' = -2$ (that corresponds to $X \approx 0$), and it plays the role of a frequency parameter; $\delta$ characterizes dielectric losses. The resonance quality factor is proportional to $\delta^{-1}$. At the resonance of a spherical particle when $\epsilon' = -2$, $(R^3\delta)^{-1} = (3/2)|\epsilon'/\epsilon''|$. However, for the collective resonances of an ensemble of particles occurring at $|\epsilon'| \gg 1$ (see below), $(R^3\delta)^{-1} \approx |\epsilon|^2/(3\epsilon'')$, which increases with the wavelength. One can find $X(\lambda)$ and $\delta(\lambda)$ for any material using theoretical or experimental data for $\epsilon(\lambda)$ and formulas (19), (20).

We can write (16) in matrix form. The Cartesian components of three-dimensional vectors $\mathbf{d}_i$ and $\mathbf{E}_{\mathrm{inc}}$ are given by $\langle i\alpha|d\rangle = d_{i,\alpha}$ and $\langle i\alpha|E_{\mathrm{inc}}\rangle = E_\alpha^{(0)}$. The last equality follows from the assumption that the incident field is uniform throughout the sample. The matrix elements of the interaction operator are defined by $\langle i\alpha|\hat{W}|j\beta\rangle = W_{ij,\alpha\beta}$. Then (16) can be written as, $[Z(\omega) - \hat{W}]|d\rangle = |E_{\mathrm{inc}}\rangle$. By diagonalizing the interaction matrix $\hat{W}$ with $\hat{W}|n\rangle = w_n|n\rangle$ and expanding the $3N$-dimensional dipole vectors in terms of the eigenvectors, we obtain the amplitudes of linear dipoles induced by the incident wave and the local fields. The local fields and dipoles are related as [2] $E_{i,\alpha} = \alpha_0^{-1}d_{i,\alpha} = \alpha_0^{-1}\alpha_{i,\alpha\beta}E_\beta^{(0)}$. The local field $E_l$ associated with the $i$th particle, $E_l = E_i$, can be found by solving the CDE as [2]

$$E_{i,\alpha} = \alpha_0^{-1}\sum_{j,n}\frac{\langle i\alpha|n\rangle\langle n|j\beta\rangle}{Z(\omega) - w_n}E_\beta^{(0)}. \tag{21}$$

Equation (21) allows one to express the local fields in terms of the eigenfunctions and eigenfrequencies of the interaction operator. The local fields can then be used to calculate the enhancement of various optical responses. Note that this approach is not limited to a quasi-static approximation. Similar solution can be obtained when the radiative terms in the interaction operator are taken into account [2].

When particles touch each other, the dipole approximation is not adequate. The reason for this is that the dipole field ($\propto r^{-3}$ in the near zone) generated by one monomer is not homogeneous inside the adjacent particle; it is much stronger near the point where the monomers touch than in the center of the neighboring monomer. Effectively, by replacing two touching spheres with two point dipoles located in their centers, we underestimate the actual strength of their interaction. To account for this effect, one typically has to take into account the higher multipolar terms.

Despite the fact that new efficient methods have recently been developed for calculations beyond the dipole approximation [8], they are still computationally applicable only for aggregates with relatively small number of particles or weak interactions between the particles. In aggregates of metal particles, where resonance plasmon oscillations can be excited and the dielectric constant can be very large, $|\epsilon'| \sim 10^3$ to $10^4$, the interparticle interactions are very strong so that any direct computational calculations for aggregates of $10^4$ particles are typically beyond the capability of any computer.

To overcome the inadequacy of the dipole approximation and the overwhelming computational load of the "coupled multipole" methods, a phenomenological procedure (that can be referred to as the cluster renormalization) approach has been suggested by Markel and Shalaev (see, for example, [2]). This method further develops the original idea suggested by Purcell and Pennypacker for odd-shaped objects and applies it to fractals. In this approach, a renormalized cluster is introduced in which neighboring spheres are allowed to intersect geometrically. This allows one effectively to take into account the stronger interaction between the neighboring particles, which, as mentioned, is undervalued by the "conventional" dipole approximation. The radii $R$ of these spheres, as well as the distance $a$ between two neighboring monomers, are chosen to be different from the real experimental ones: $R \neq R_{\exp}$, $a \neq a_{\exp}$, but it is required that the ratio $a/R$ is equal to $(4\pi/3)^{1/3} \approx 1.612$, the same as in the Purcell and Pennypacker model (for details, see [2]). The second equation for $R$ and $a$ can be obtained from the optically important condition that the renormalized cluster has the same fractal dimension, radius of gyration, and total volume as the experimental cluster. It was shown that for fractals, where $D < 3$, all of these conditions are compatible, in contrast to nonfractal clusters of particles. The above model of effective intersecting particles allows one to take into account the stronger depolarization factors for touching particles, remaining within the "renormalized" dipole approximation. The model, it was shown, yields results that are in very good agreement with experimental spectra of fractal clusters [2].

Figure 2 shows the local field $E_i$ distribution excited by light of wavelength $\lambda = 1\,\mu\text{m}$ at the surface of a simulated silver CCA deposited on a plane substrate. This distribution was computed by using solution (21) and the renormalized dipole approximation described above. The largest fields are extremely localized; the local field intensity in the "hot" spots can exceed the applied field by up to $10^5$, and the average enhancement is only $\sim 10^2$ to $10^3$.

The resonance local field is estimated by $|E_{\mathrm{r}}/E^{(0)}| \sim 1/\delta$, as follows from (21), which is the exact solution (for simplicity, hereafter we set $R = 1$). This result can also be obtained from the simple fact that the linear polarizability $\alpha_i = d_i/E^{(0)}$ of the $i$th monomer in a cluster experiences a shift of resonance $w_i$ because of interactions with other particles (where $w_i$ is a real number, in the quasi-static approximation). Therefore, the polarizability of, say, the $i$th particle can always be represented as $\alpha_i = 1/(\alpha_0^{-1} + w_i) = [(w_i - X) - i\delta]^{-1}$, where the shift $w_i$ depends on interactions with all particles. In the limit of noninteracting particles, $w_i = 0$ and $\alpha_i = \alpha_0 = -(X + i\delta)^{-1}$. For resonance particles, $w_i = X(\omega_r) \equiv X_r$, and $\alpha_r = i/\delta$. The local field is related to the local polarizability as $E_i = \alpha_0^{-1}\alpha_i E^{(0)}$, so that for $X \gg \delta$, we obtain $|E_r| \sim (|X_r|/\delta)|E^{(0)}|$ for the resonance field. In the optical spectral range, $|X_r| \sim 1$, and $|E_r| \sim \delta^{-1}|E^{(0)}|$. This estimate agrees qualitatively with the results shown in Fig. 2.

In the long wavelength part of the spectrum, where $|\epsilon'| \gg 1$, we see that $E_r/E^{(0)} \sim \delta^{-1} \sim |\epsilon'|^2/\epsilon''$. To compare with cigar-shaped spheroids, where $|\epsilon'_r| \sim 1/p \sim A^2$, we can write the enhancement in fractals formally as $E_r/E^{(0)} \sim p^{-1}|\epsilon'_r|/\epsilon''_r \sim A^2|\epsilon'_r|/\epsilon''_r \sim A(\lambda_\tau/\lambda_p)$. Since $A \sim |\epsilon'|^{1/2} \sim (\lambda/\lambda_p)$, the local field enhancement in fractals increases with $\lambda$ as $E_r/E^{(0)} \sim (\lambda\lambda_\tau)/\lambda_p^2$. Thus, a fractal, with its variety of local configurations of particles (where optical excitations are localized), can be roughly thought of as a collection of (noninteracting) prolate nanospheroids, with all possible values of the aspect ratio.

It is worth noting again that if a fractal structure is replaced by some $3d$ compact structure of spheroids (or any other particles), then, because of the long-range interactions in nonfractal systems, there would, in general, be no localization of the optical excitations in such a structure. Instead, in most normal modes, every particle contributes significantly to the excitation, so that all the resonances depend on the shape of the whole object and lie typically within a relatively narrow spectral interval.

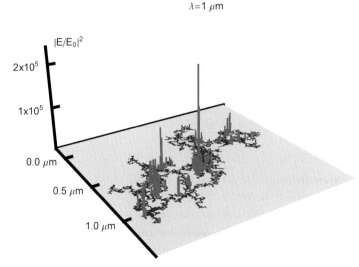

**Fig. 2.** Calculated field distributions on the surface of a silver fractal aggregate deposited on the plane

In fractals, in contrast, the resonance frequency of a localized surface plasmon mode depends on the local configuration of particles at the mode location. A random fractal is composed of a large variety of local geometries, each possessing a different plasmon resonance frequency; as a result, the range of frequencies spanned by the plasmon modes in a fractal cluster is unusually broad, covering the whole of the visible and infrared portions of the spectrum. Additionally, for most metals, the electromagnetic energy and the enhancement of nonlinear optical effects and Raman scattering increase toward longer wavelengths [2] because for most metals, the real part of the dielectric function is negative (that is why metals are such good reflectors) and its magnitude increases strongly toward longer wavelengths, resulting in a parallel increase in the quality factor of the plasmon resonances of fractal clusters composed of metal particles.

Locally intense fields, as shown in Fig. 2, suggest a large number of unusual local optical and photochemical effects, among them, single-molecule spectroscopy. Specifically, because the local enhancement-factor $\propto |E|^4$ [7] for surface-enhanced Raman scattering (SERS), it can reach magnitudes of $10^{12}$, making Raman spectroscopy of single molecules possible [9].

## 4    Enhanced Optical Nonlinearities in Fractals

The local-field enhancement for a nonlinear optical process can, in general, be written as

$$G_n \sim \langle |E_i/E^{(0)}|^k [E_m/E^{(0)}]^m \rangle, \tag{22}$$

where $n = k+m$. In particular, $k = 4$, $m = 0$ describes the field enhancement for SERS, and $k = 2$, $m = 2$ for nonlinear refraction (the optical Kerr effect). Signals associated with coherent nonlinear light scattering, such as four-wave mixing (FWM), are proportional to the average square of the nonlinear polarization; hence the FWM enhancement, for example, is [2] $G_{\mathrm{FWM}} = |G_4|^2$.

For fractals, an estimate for $G_n$ (at $k \neq 0$) can be expressed in terms of the polarizability $\alpha_0 \equiv -[X(\omega) + i\delta(\omega)]^{-1}$ of the individual particles composing the fractal as [2]

$$G_n \sim c_n |X|^n \delta^{1-n} \mathrm{Im}\,[\alpha(X)], \tag{23}$$

where $c_n$ is a frequency-independent constant. In this formula, $(|X|/\delta)^n$ gives the resonant local field (normalized to the applied field) raised to the $n$th power, and $\delta \mathrm{Im}\,[\alpha(X)]$ is the approximate fraction of resonant particles in the fractal for a given light frequency. The factor $\mathrm{Im}\,[\alpha(X)]$ represents the average light extinction, which differs significantly for fractals ($D < d$) and nonfractals ($D = d$).

For the average local-field intensity ($n = 2$), it can be shown that the *exact* result is given by

$$G = G_2 = \frac{(X^2 + \delta^2)}{\delta} \text{Im} \left[ \alpha(X) \right], \tag{24}$$

in agreement with the estimate (23).

It is worth noting here that zero-point field fluctuations, which are responsible for spontaneous emission, are expected to be enhanced in the same manner as their "classical" counterparts, so that formula (24) also characterizes the enhancement of spontaneous emission in fractals.

For nonfractal random systems, the extinction $\text{Im} \left[ \alpha(X) \right]$ typically peaks near the resonance frequency of the individual particles (where $X(\omega) \approx 0$), becoming negligible for $|X| \gg \delta$, so that $G_n$, according to (23), is relatively small [2]. Contrariwise, for fractals, the factor $\text{Im} \left[ \alpha(X) \right]$ remains significant even in the long wavelength part of the spectrum, where $|X(\omega)| \gg \delta$, leading to very large values of $G_n$ [2] ($\text{Im} \, \alpha(X)$ has roughly a box-like distribution in the interval $-4\pi/3 \leq X \leq 4\pi/3$, which in terms of $\lambda$ includes the whole of the visible and infrared parts of the spectrum [2]). Thus, the large enhancement in a broad spectral range, as mentioned, is a direct result of the localization of optical excitations and of the broad variety of resonating local structures.

We also note here that by exciting all (or most of) the fractal modes and matching their phases, one can produce attosecond light pulses because the extremely large spectral range of fractal modes (from the near-UV to the far-IR) in the spectral domain corresponds to attosecond time intervals in time domains.

For example, for Raman scattering with a small Stokes shift, the enhancement is given by $G_{RS} \sim \langle |E|^4 \rangle / |E_0|^4$ [7] so that the above estimate in this case is as follows:

$$G_{RS} \sim C_{RS} X^4 \delta^{-3} \text{Im} \left[ \alpha(X) \right], \tag{25}$$

where $C_{RS}$ is a constant prefactor.

In Fig. 3a, we compare this analytical formula with the results of numerical solution for the CDEs for fractal aggregates with $D = 1.78$ (two different values for parameters $\delta$ are used). One can see that the theoretical formula is in good accord with numerical simulations. As seen in Fig. 3a, the product $G_{RS} \delta^3$ does not depend systematically on $\delta$ in the important region close to the maximum, and its value there is of an order of one. Thus, we conclude that strong enhancement of Raman scattering $G_{RS} \sim \delta^{-3}$, resulting from aggregation of particles into fractal clusters, can be obtained.

In Fig. 3b, surface-enhanced RS data obtained for colloidal silver solutions in experiments [7] is compared with the $G_{RS}$ calculated with the use of (25). [The values of $X$ and $\delta$ at different $\lambda$ were found using formulas (19), (20)]. Note that only the spectral dependence of $G_{RS}$ is informative in this figure since only relative values of $G_{RS}$ were measured in [7]. The experimental data presented in Fig. 3b are normalized by setting $G_{RS} = 28\,250$ at $570\,\text{nm}$, which is a reasonable value. Clearly, the present theory successfully explains the huge enhancement accompanying aggregation of particles into fractals

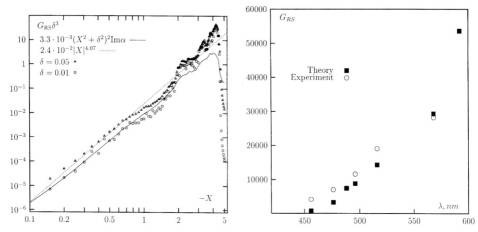

**Fig. 3.** (a) Enhancement of Raman scattering in cluster–cluster aggregates (CCAs) (multiplied by $\delta^3$) for negative $X$, corresponding to the visible and infrared parts of the spectrum. (b) Theoretical and experimental enhancement factors for Raman scattering in colloidal silver aggregates as functions of wavelength $\lambda$

and the observed increase of $G_{RS}$ toward the red part of the spectrum. The strong enhancement toward the red occurs because the local fields associated with collective dipolar modes in CCAs become significantly larger in this part of the spectrum.

We note that the same enhancement characterizes Kerr optical nonlinearity, which is responsible for nonlinear refraction and absorption, $G_K \sim \langle |E|^2 E^2 \rangle / |E_0|^2 E_0^2 \sim G_{RS}$ [2].

For coherent processes, the resultant enhancement $\sim |\langle |E_i/E^{(0)}|^n \rangle|^2$. For quasi-degenerate four wave-mixing (FWM), additional enhancement of the generated nonlinear amplitudes oscillating at almost the same frequency as the applied field should also be included, so that is given by $G_{FWM} \sim |\langle |E|^2 E^2 \rangle|^2 / |E_0|^8$. We can estimate the FWM enhancement factor, $G_{FWM} \sim |G_K|^2$, as [2]

$$G_{FWM} \sim C_{FWM} X^8 \delta^{-6} \left\{ \mathrm{Im} \left[ \alpha(X) \right] \right\}^2. \tag{26}$$

All of these estimates agree with the results of analytical and numerical calculations [2].

Figure 4 shows the results of numerical calculations of the enhancement factor $G_{FWM}$ in silver CCAs as a function (a) of spectral parameter $X$ and as a function (b) of wavelength $\lambda$. The simulations were performed by solving the CDEs, and the simulation results were compared with the above analytical formula. The solid lines in Fig. 4a,b describe the results of calculations based on formula (26), with $C_{FWM}$ found from the relation $G_{FWM} \delta^6 = 1$; its maxima occur at $X \approx \pm 4$. The dashed line in Fig. 4a represents a power-law fit for the range $0.1 \leq |X| \leq 3$, with $\delta = 0.05$. The computed exponent

$(8.31 \pm 1.00)$ is close to 8 (only the negative values of $X$ are shown; the results for positive $X$ are qualitatively similar).

We conclude that formula (26) is in good agreement with numerical simulations in a wide range from the visible to the mid-infrared. Note that for $\lambda > 10\,\mu m$, the resonance condition $\lambda \ll \lambda_\tau$ does not hold, and therefore the theory, strictly speaking, cannot be applied (for silver, $\omega_\tau \equiv \Gamma \approx 0.021\,eV$, i.e., $\lambda_\tau \approx 56\,\mu m$).

As seen in Fig. 4a,b, the enhancement strongly increases toward larger values of $|X|$ (when $X < 0$) or, in other words, toward longer wavelengths, where enhancements for the local fields are stronger.

It also follows from Fig. 4a that the product $G_{FWM}\delta^6$ remains, on average, the same for the two very different values of $\delta$ chosen, namely, 0.01 and 0.05. This indicates that, in accordance with (26), the enhancement is proportional to the sixth power of the resonance quality factor, i.e., $G_{FWM} \propto Q^6$ $(Q \sim \delta^{-1})$ and reaches huge values in the long-wavelength part of the spectrum where $X \approx X_0 = -4\pi/3$.

The nonlinear susceptibility $\bar{\chi}^{(3)}$ of a composite material, consisting of fractal aggregates of colloidal particles in some host medium (e.g., water) is given by $\bar{\chi}^{(3)} = p \cdot G_K \chi_m^{(3)}$ where $G_K \sim G_{FWM}^{1/2}$ is the enhancement of the Kerr optical nonlinearity, $\chi_m^{(3)}$ is the susceptibility of nonaggregated metal particles, and $p$ is the volume fraction filled by metal.

When the initially separated silver particles aggregate and fractal clusters are formed, a huge enhancement of the cubic susceptibility can occur. A millionfold enhancement of degenerate FWM (DFWM) due to the clustering of initially isolated silver particles in colloidal solution was experimentally obtained [10]. The observed enhancement factor for fractal silver composites

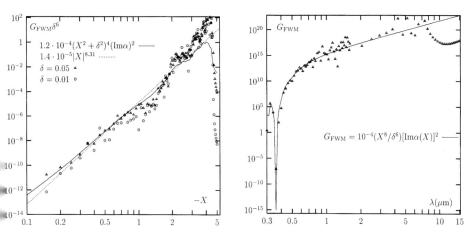

**Fig. 4.** Calculated enhancement for degenerate four-wave mixing, $G_{FWM}$. (a) $G_{FWM}\delta^6$ as a function of $X$ ($X < 0$) for CCAs; (b) $G_{FWM}$ as a function of wavelength for silver CCAs

in these experiments is $G \sim 10^6$ for $\lambda = 532$ nm. Note that in Fig. 4b cal-culations are done with vacuum as a host medium, whereas the experiments were performed in an agueous colloidal solution. The values of $X$ and $\delta$ for silver particles in water at laser wavelength $\lambda = 532$ nm are $X \approx -2.55$ and $\delta \approx 0.05$, respectively. According to Fig. 4a, $G_{\text{FWM}} \sim 10^6$ to $10^7$ for these values of $X$ and $\delta$, which is in reasonable agreement with the experimental observations.

The cubic susceptibility obtained experimentally for an aggregated sam-ple is [10]: $|\bar{\chi}^{(3)}| = 5.7 \times 10^{-10}$ esu with $p \approx 5 \times 10^{-6}$. Note that $p$ is a variable quantity and can be increased. We can assign the value $10^{-4}$ esu to the non-linear susceptibility, $\chi^{(3f)}$, of the fractal clusters, i.e., $\bar{\chi}^{(3)} = p \cdot \chi^{(3f)}$, with $\chi^{(3f)} \sim 10^{-4}$ esu. This is a very large value for a third-order nonlinear sus-ceptibility.

When characterizing potential applications of materials, it is important to have large $\chi^{(3)}$ at a relatively small absorption. As a characteristic of the materials, the figure of merit $F$ can be used that is defined via the ratio of the nonlinear susceptibility $\bar{\chi}^{(3)}$ of a composite material and its linear losses that are given by the imaginary part of the effective (linear) dielectric function, $\bar{\epsilon}'' = 4\pi \text{Im} \left[ \bar{\chi}^{(1)} \right]$. Thus, the figure of merit is defined through the relation $|\bar{\chi}^{(3)}/\bar{\epsilon}''| = F\chi_0^{(3)}$, where $\chi_0^{(3)}$ is a "seed" optical nonlinearity that can be due either to particles forming the composite (then, $\chi_0^{(3)} = \chi_m^{(3)}$) or due to some nonlinear adsorbant molecules (we assume that the former is the case). Results of calculations of $F$ for fractal silver aggregates (CCAs) are shown in Fig. 5, as a function of wavelength (nm). It can be seen that the figure of merit in the fractals is very large, $\sim 10^7$, in the near-infrared part of the spectrum where absorption is relatively small, whereas enhancement of optical nonlinearities is very significant.

A large variety of optical processes can be enhanced and otherwise modi-fied by incorporating fractal clusters in the media or by ensuring that aggrega-tion results in fractals. For example, fractals can be used to improve the per-formance of random lasers, such as powder lasers, and laser paints [11], where lasing emissions can take place as a result of coherent multiple light-scattering in a disordered dielectric (or semiconductor) with appropriate structural el-ements. The notion of creating a "ring" laser cavity in a random medium through a sequence of multiple coherent scattering events along a closed path is itself, a fascinating, and almost counterintuitive prospect. Scattering is nor-mally considered detrimental to lasing since, in a conventional laser cavity, it tends to remove photons from the lasing mode. In a properly constructed random medium, however, strong multiple scattering could return photons to the amplification region resulting in mode amplification [11]. By doping the laser powder or paint medium with fractal aggregates or by imparting a fractal character to the medium as a whole, one could significantly decrease the pump power needed to effect lasing, in other words, one could decrease the lasing threshold.

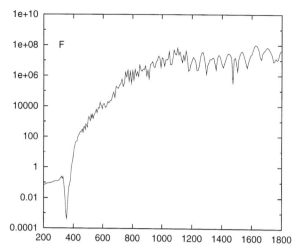

**Fig. 5.** The figure of merit $F$ for fractal silver CCAs; $F$ is the ratio of nonlinear susceptibility $|\bar{\chi}^{(3)}|$ and linear losses $\bar{\epsilon}''$ (see text). The wavelength $\lambda$ is given in nanometers

As discussed in a separate contribution of this book, even more gigantic enhancement can be obtained by combining (multiplicatively) the local-field enhancement in fractals with the enhancement occurring in microcavities [2,12]. In these novel composite materials, microcavities doped with fractals, record-high enhancement of optical phenomena can be obtained.

## Acknowledgements

This work was supported in part by National Science Foundation under Grant No. DMR-0121814, Army Research Office under Grant No. DAAD19-01-1-0682, Petroleum Research Fund, and NASA under Grants No. NAG 8-1710 and NCC-1-01049.

## References

1. J. C. Garland, D. B. Tanner (Eds.), *Electron Transport and Optical Properties of Inhomogeneous Media* (AIP, New York 1978); W. L. Mochan, R. G. Barrera (Eds.), *Electron Transport and Optical Properties of Inhomogeneous Media (ETOPIM 3)* (North-Holland, Amsterdam 1994); A. M. Dykhne, A. N. Lagarkov, A. K. Sarychev. (Eds.), Physica A **241**, (1–2), 1 (1997), Proceedings (ETOPIM 4)
2. V. M. Shalaev, *Nonlinear Optics Of Random Media: Fractal Composites and Metal-Dielectric Films*, Springer Tracts in Modern Physics **158** (Springer, Berlin, Heidelberg 2000)
3. B. B. Mandelbrot, *The Fractal Geometry of Nature* (Freeman, San Francisco 1982); B. Sapoval, *Fractals* (Aditech, Paris 1990); A. L. Barbasi, H. E. Stanley,

*Fractal Concepts in Surface Growth* (Cambridge University Press, Cambridge 1995); R. Jullien, R. Botet, *Aggregation and Fractal Aggregates* (World Scientific, Singapore 1987)

4. S. Alexander, R. Orbach, J. Phys. Lett. **43**, 625 (1982); A. Bunde, S. Havlin (Eds.), *Fractals and Disordered Systems* (Springer, Berlin, Heidelberg 1991); L. Pietroniero, E. Tosatti (Eds.), *Fractals in Physics* (North-Holland, Amsterdam 1986); B. Sapoval, Th. Gorbon, A. Margolina, Phys. Rev. Lett. **67**, 2974 (1991)

5. V. M. Shalaev, M. I. Stockman, Sov. Phys. JETP **65**, 287 (1987); A. V. Butenko, V. M. Shalaev, M. I. Stockman, Z. Phys. D **10**, 81 (1988); V. A. Markel, L. S. Muratov, M. I. Stockman, T. F. George, Phys. Rev. B **43**, 8183 (1991); M. I. Stockman, L. N. Daudey, L. S. Muratov, T. F. George, Phys. Rev. Lett. **72**, 2486 (1994); D. P. Tsai, J. Kovacs, Z. Wang, M. Moskovits, V. M. Shalaev, J. S. Suh, R. Botet, Phys. Rev. Lett. **72**, 4149 (1994); V. M. Shalaev et al., Phys. Rev. B **53**, 2437 (1996); Ibid. **53**, 2425 (1996); V. M. Shalaev, Phys. Rep. **272**, 61 (1996); V. P. Safonov et al., Phys. Rev. Lett. **80**, 1102 (1998)

6. M. I. Stockman, L. N. Pandey, T. F. George, Phys. Rev. B **53**, 2183 (1996); M. I. Stockman, Phys. Rev. E **56**, 6494 (1997); Phys. Rev. Lett. **79**, 4562 (1997)

7. M. Moskovits, Rev. Mod. Phys. **57** , 783 (1985); R. K. Chang, T. E. Furtak (Eds.), *Surface Enhance Raman Scattering*, (Plenum, New York 1982); M. I. Stockman, V. M. Shalaev, M. Moskovits, R. Botet, T. F. George, Phys. Rev. B **46**, 2821 (1992)

8. J. B. Pendry, A. MacKinnon, Phys. Rev. Lett. **69**, 2772 (1992); F. J. Garcia-Vidal, J. M. Pitarke, J. B. Pendry, Phys. Rev. Lett. **78**, 4289 (1997); F. J. Garcia de Abajo, Phys. Rev. Lett. **82**, 2776 (1999)

9. K. Kneipp, Y. Wang, H. Kneipp, L. T. Perelman, I. Itzkan, R. D. Dasari, M. Feld, Phys. Rev. Lett. **78**, 1667 (1997); S. Nie, S. R. Emory, Science **275**, 1102 (1997)

10. S. G. Rautian, V. P. Safonov, P. A. Chubakov, V. M. Shalaev, M. I. Stockman, JETP Lett. **47**, 243 (1988) [transl. from Pis'ma Zh. Eksp. Teor. Fiz. **47**, 200 (1988)]; Yu. E. Danilova, V. P. Drachev, S. V. Perminov, V. P. Safonov, Bull. Russian Acad. Sci. Phys. **60**, 342 (1996); ibid., 374; Yu. E. Danilova, N. N. Lepeshkin, S. G. Rautian, V. P. Safonov, Physica A **241**, 231 (1997); F. A. Zhuravlev, N. A. Orlova, V. V. Shelkovnikov, A. I. Plehanov, S. G. Rautian, V. P. Safonov, JETP Lett. **56**, 260 (1992)

11. V. S. Letokhov, Soviet Phys. JETP **26**, 835 (1968); N. M. Lawandy et al., Nature **368**, 436 (1994); D. S. Wiersma, A. Lagendijk, Phys. Rev. E **54**, 4256 (1996); S. John, G. Pang, Phys. Rev. A **54**, 3642 (1996); G. A. Berger, M. Kempe, A. Z. Genack, Phys. Rev. E **56**, 6118 (1997); H. Cao et al. Phys. Rev. Lett. **82**, 2278 (1999); Phys. Rev. B **61**, 1985 (2000)

12. W. Kim, V. P. Safonov, V. M. Shalaev, R. L. Armstrong, Phys. Rev. Lett. **82**, 4811 (1999); V. P. Drachev, W. Kim, V. P. Safonov, V. A. Podolskiy, N. S. Zakovryazhin, V. M. Shalaev, R. A. Armstrong, J. Mod. Opt., in press (2001); V. A. Podoskiy, V. M. Shalaev, Laser Phys. **11**, 26 (2000)

# Nonlinear Optical Effects and Selective Photomodification of Colloidal Silver Aggregates

Vladimir P. Drachev[1], Sergey V. Perminov[1],
Sergey G. Rautian[2], and Vladimir P. Safonov[2]

[1] Institute of Semiconductor Physics, Siberian Branch of the Russian Academy of Sciences, Novosibirsk 630090, Russia
vdrachev@yahoo.com
[2] Institute of Automation and Electrometry, Siberian Branch of the Russian Academy of Sciences, Novosibirsk 630090, Russia

**Abstract.** Colloidal silver aggregates of nanoparticles were studied experimentally using optical spectroscopy, electron microscopy, near-field optics, and nonlinear optics. Changes in absorption spectra, local structure, and near-field optical response after the irradiation of fractal colloidal aggregates with a laser pulse (selective photomodification) were studied. The diameters of the selectively photomodified domains decreased as the laser wavelength increased, in accordance with the theory of the optics of fractal clusters. Giant enhancements of nonlinear optical responses were found for aggregated nanocomposites compared with nonaggregated. The enhancements are due to excitation of the collective plasmon modes in the aggregates. The plasmon modes are anisotropic and chiral. Nonlinear effects governed by local and nonlocal responses (degenerate four-wave mixing, nonlinear absorption, refraction and gyrotropy, inverse Faraday effect, and ellipse self-rotation) were studied.

The optics of aggregated metal–dielectric composites attracted considerable attention during the last two decades. This attention was due to increasing interest in the physics of the cooperative interaction of particles in random structures and the high application potential of metal nanostructures as nonlinear media, as substances, strongly facilitated to microanalysis, and media for optical data storage. Surface-enhanced Raman scattering [1], enhanced optical nonlinearities from local [2,3,4] and nonlocal [5] response, and selective photomodification of colloidal silver aggregates [6,7] have been studied. The effect of giant fluctuations of local electrical fields in colloidal aggregates, predicted in [8], were demonstrated in these experiments. Together with this, the experiments [3,6] facilitated the development of new techniques for investigating optical processes in aggregated metal nanocomposites.

Traditionally, an essential part of the data regarding the optics of new materials is gained by studying spectra of linear light absorption and scattering. An analysis of linear absorption spectra of metal nanoparticles and small aggregates is given in detail in [9]. The optical properties of a single metal nanosphere are governed to a large extent by excitation of the

V. M. Shalaev (Ed.): Optical Properties of Nanostructured Random Media,
Topics Appl. Phys. **82**, 113–147 (2002)
© Springer-Verlag Berlin Heidelberg 2002

surface plasmon oscillation of conductive electrons. The width of a surface plasmon resonance in a single nanoparticle depends on electron interaction with electrons, phonons, impurities, and with the interface of the particle. Recent investigations of electron dynamics in metal films and nanoparticles by the pump-probe technique with femtosecond pulses revealed femto- and picosecond relaxation processes connected with electron–electron and electron–phonon interactions, whose characteristic times depended on the electron, $T_e$, and lattice, $T_i$, temperatures [10,11]. Electron scattering on the particle interface results in an additional broadening of the surface plasmon resonance, $v_F/R_m$, where $v_F$ is the Fermi velocity and $R_m$ is the particle radius. The resulting width of the resonance of a colloidal silver nanoparticle ($R_m \approx 10\,\mathrm{nm}$) is 60–80 nm; this plasmon resonance is near the wavelength $\lambda_m = 400\,\mathrm{nm}$. Interband transitions in silver are responsible for ultraviolet absorption.

New methods should be applied to study the optical properties of aggregated nanocomposites. An absorption spectrum of aggregated metal nanocomposite is wider than the spectrum of a single particle [9]. When the particles (monomers) form an aggregate (cluster), the distances between the centers of the nearest monomers can be as small as a diameter of the nanoparticles. This means that particles with high polarizability (such as silver and gold particles) strongly interact via dipolar forces, or more generally, multipolar coupling. The interaction changes the linear [8,9] and nonlinear [2] optical properties of the medium substantially. In the simplest aggregate, a pair, plasmon oscillations in different particles interact via restoring forces, and the dipole plasmon resonance splits into two peaks. The long-wavelength peak corresponds to an electric vector of a light field parallel to the axis of a pair, and the short-wavelength peak corresponds to the perpendicular electric vector; hence, the pair is an anisotropic unit. A sample composed of an ensemble of pairs with the different distances has a wide absorption spectrum.

Large colloidal aggregates are typically characterized by fractal geometry [12]. The number of monomers $N$ in a fractal cluster with radius $R_c$ is $N = (R_c/R_0)^D$, where $R_0$ is a parameter of the characteristic distance between the nearest monomers and $D$ is the fractal dimension. The value of $D$ depends on a kind of aggregation process. The process of cluster–cluster aggregation results in $D = 1.78$ [13]. The scaling relation between $N$ and $R_c$ is valid for distances larger than a monomer and smaller than a cluster.

Fractal objects are not translationally invariant; therefore vibrational and optical excitations tend to be localized in fractals [2,3,8,14,15]. The scale invariance of a structure of fractal clusters results in the power law of the angular dependence on the intensity of elastic light scattering [16]. The frequency dependence of inelastic light scattering on localized vibrations obeys the power law, and the exponent is determined by $D$ and the fracton spectral dimension [17]. It was shown theoretically that the light-induced coupling of metal spherical monomers in a fractal cluster resulted in broadening of the absorption spectrum [8,15]. The scaling law was predicted for an absorption spectrum shape of fractal clusters in the model of a diluted fractal [15].

Experimentally, it was found that the absorption spectrum of fractal silver clusters covered a wide spectral range–from the UV to the infrared [18,19]. An essential role of short-range interactions leads to the dependence of the absorption spectrum shape on the distance between the nearest monomer boundaries and to the absence of a scaling law for the absorption spectrum [19].

The random structure of a fractal aggregate results in inhomogeneous broadening of the absorption spectrum. It was found by observation of spectral hole photoburning in colloidal silver aggregates [6]. An electron microscopic study showed that spectral hole photoburning is accompanied by selective local changes in aggregate structure [19] (selective photomodification).

The advanced theory of the optical properties of fractal clusters (free from the limitations of the diluted fractal model) takes into account the light-induced dipole interactions between monomers constituting a fractal [15,20,21,22,23]. These dipole couplings result in localized plasmon eigenmodes. Computer simulations [20,21,22,24] show that the areas of large local-field fluctuations may be localized in small parts of a fractal aggregate. The positions and sizes of these hot spots change with the wavelength and polarization of light. The localization of plasmon excitation is inhomogeneous in the sense that, at any wavelength, modes of different coherent radii are possible and the electrical field of the eigenmode is distributed spikewise in a fractal [22]. The simulations are in qualitative agreement with the experimental results obtained with the help of near-field optics [25]. The areas of high local field, "hot spots," were observed in [25]. Locations of these hot spots were found dependent on light wavelength and polarization. Unfortunately, the spatial resolution of near-field optics techniques was approximately 200 nm, and this was more than a monomer size. Therefore near-field optics does not permit us to know what the real size of the hot spot is.

The selective photomodification technique provides information regarding localization of plasmon excitations in metal aggregates where the spatial resolution $\approx$ 10 nm [7]. With the help of the photomodification effect, one can study the dependence of excitation localization on light wavelength.

Another important problem concerned a contribution of the electromagnetic mechanism to surface-enhanced optical responses. Raman scattering of molecules adsorbed on a metal structure may be enhanced by a charge transfer process and by a high local field in the hot spots [26]. Studying of nonlinearities and the photomodification of metal aggregates permits separating the contribution of electromagnetic enhancement.

The ratio of the size of a particle (10 nm) to the wavelength (500 nm) in a colloidal aggregate is much greater than for an ordinary molecular medium. The size of localization of some plasmon eigenmodes in a fractal aggregate may be comparable with the aggregate size [22] and, consequently, with wavelength. This circumstance is favorable for observing effects caused by spatial dispersion in linear and nonlinear optical processes in metal nanocomposites. The structure of a random aggregate has typically neither a center nor a plane of symmetry, and thus this structure may have handedness. A collective mode localized on such structure exhibits chirality.

It is known [27] that the relation between different components of local nonlinear susceptibility depends on the mechanism of the nonlinearity. The ratio of the nonlocal polarizability to the local one produces information about the ratio between the effective size, which characterizes a gyrotropic medium, and the wavelength [28]. Therefore, it is interesting to do combined measurements of all components of cubic susceptibility tensors, local and nonlocal.

This contribution is concerned with the experimental study of the selective photomodification of colloidal aggregates, spatial dispersion effects, and nonlinear optical effects in colloidal solutions (degenerate four-wave mixing, nonlinear absorption and refraction, nonlinear gyrotropy, and the inverse Faraday effect). The results obtained used were to reveal essential features of optical interactions in aggregated metal nanocomposites and for comparison with theory.

# 1    Spectral Dependence of Selective Photomodification in Colloidal Silver Aggregates

Aqueous and ethanolic colloidal silver solutions were studied. The diameter of the silver monomer in different colloids changed from 10 to 24 nm. Figure 1 shows the typical absorption spectra of a silver colloid with different degrees of aggregation. Increase of the degree of aggregation (increase of a number of particles included in clusters and growth of the clusters sizes) leads to broadening of the absorption spectrum. Curve 4 in Fig. 1 corresponds to the strongly aggregated colloid consisting of fractal clusters. Computer simulations [15,18,21] show that dipole eigenmodes formed by the interactions between monomers in random clusters with a fractal structure span a spectral range much broader than the monomer absorption bandwidth of the surface plasmon resonance. According to the simulations [18], a large ensemble of silver clusters that contains 30 monomers has an absorption spectrum as wide as the spectrum of clusters of 500 particles [21]. As seen in Fig. 2, the spectrum of a strongly aggregated colloid (curve 4 in Fig. 1) coincides with the simulations [18,21]. To get agreement with the experiment, a model of a fractal cluster with intersecting monomers was introduced [18,21,23]. This model takes into account an inhomogeneity of the dipole field, produced by a monomer inside the neighboring monomer. In some sense, this model is equivalent to a description of multipole interactions in the cluster.

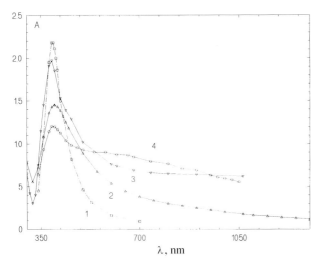

**Fig. 1.** The absorption spectra of the Ag-boronhydride colloid. 1, a fresh quasi-monomeric colloid; 2, 3, and 4 – weakly, medium, and strongly aggregated colloid

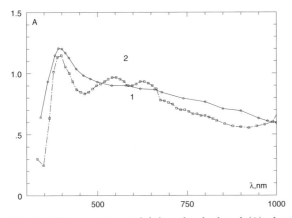

**Fig. 2.** The experimental (**1**) and calculated (**2**) absorption spectra of Ag clusters

Irradiation of aggregated metal nanocomposites by a laser pulse with energy above the threshold leads to "photoburning" of persistent dichroic holes in the absorption spectrum near the laser wavelength [6]. Photomodification of the geometric structure and photoburning of spectral holes have been observed in silver and gold colloids, in polymers doped with metal aggregates, and in films produced by laser evaporation of a silver target [6,29,30]. This phenomenon results from the localization of optical excitations in fractal structures which are prevalent in random nanocomposites. The localization arises because of the scale-invariant distribution of highly polarizable particles (monomers) that interact via light-induced dipole (multipole) fields [15]. As a result, collective plasmon eigenmodes with a very inhomogeneous local-field distribution are formed. Groups of monomers, on which electrical field peaks are localized, may be modified when the energy of incident light pulse $W$ exceeds a certain threshold value $W_{th}$. Let a laser wavelength $\lambda$ be longer than $\lambda_m$. Photomodification of resonance complexes of monomers that contribute to the absorption at a wavelength $\lambda$ results in the appearance of a hole in the absorption spectrum near $\lambda$ for a probe wave polarized parallel to the laser field, and an additional hole at wavelengths shorter than $\lambda_m$ for an orthogonally polarized probe wave [6,29,19].

The example of spectral hole photoburning [31] is shown in Fig. 3. The laser wavelength was 1064 nm. The sample was produced as follows. Silver colloid was prepared by the Creighton technique [32]. After the colloid had aggregated, we added 1% (by weight) of gelatin to the solution after which it was poured onto a glass substrate. Finally, the sample was dried for a week at room temperature. It formed a gelatin film with thickness $l = 1\text{--}25\,\mu m$ doped with fractal Ag clusters.

Let us discuss briefly the results of the experiments on photomodification [6,19,7,33]. (In this review we will pay the main attention to the long-wavelength hole.) First of all we note that the observed dichroism of the spectral holes gives evidence of the anisotropy of collective plasmon eigenmodes. Second, the dependence of the hole depth on laser pulse energy revealed a threshold character. At $W/W_{th} < 3$, the hole depth is proportional to the laser pulse energy, $W$. Third, the width of the spectral hole is maximum near $\lambda_m$, it decreases when the laser wavelength is tuned to the UV, and falls considerably when $\lambda$ is detuned to the infrared range (Fig. 4). This dependence coincides well with the calculated (in arbitrary units) spectral dependence of the width of the eigenmode for a fractal silver cluster [23]. The increase

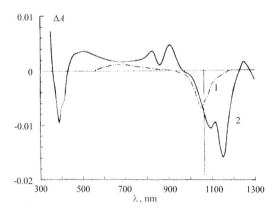

**Fig. 3.** The spectral holes for picosecond (**1**) and nanosecond (**2**) irradiation of Ag-boronhydride aggregates in gelatin. The vertical line corresponds to the laser wavelength

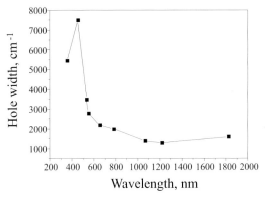

**Fig. 4.** The width of the hole in the absorption spectrum of colloidal Ag aggregates in gelatin film versus the wavelength of the laser radiation. $W/W_{\text{th}} \approx 1.3$–$1.5$

in the quality factor of the eigenmodes in the long-wavelength range (and, hence, the decrease in the hole width) is due to the growth of the absolute value of the real part of the silver dielectric function [23]. The minimal width of the spectral hole measured for one of our samples at $\lambda = 1064\,\text{nm}$ was $600\,\text{cm}^{-1}$. This value is comparable with the width of a resonance curve for a hot spot measured by near-field optics. The minimal width of the local response at $\lambda \approx 750\,\text{nm}$ was $200\,\text{cm}^{-1}$. Fourth, the spectral dependence of the energy absorbed per unit volume, $W_{\text{a}}$, in the layer where photomodification occurs, shows a decrease of $W_{\text{a}}$ when it is detuned from the surface plasmon resonance (Fig. 5). This experiment is worth discussing in more detail.

It is supposed that photomodification occurs when the energy applied to each resonant monomer exceeds a certain value. Therefore, the measured spectral dependence of $W_{\text{a}}(\lambda)$ at a given $\Delta\alpha/\alpha_0$ ($\Delta\alpha$ is a change in absorption coefficient $\alpha_0$ after photoburning of a spectral hole) provides information

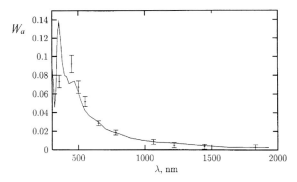

**Fig. 5.** The spectral dependence of the energy absorbed within a unit volume near the photomodification threshold

about the number of photomodified resonance monomers per unit volume as a function of wavelength. In an optically thin sample, $W_a \approx \alpha_0 W_{th}$. The measured value of $W_{th}$ decreases with increasing $\lambda$ ($W_{th} = 28\,\mathrm{mJ/cm^2}$ at $\lambda = 355\,\mathrm{nm}$, and $W_{th} = 7.1\,\mathrm{mJ/cm^2}$ at $\lambda = 1440\,\mathrm{nm}$). The values $W_a$ in Fig. 5 were given for a laser pulse duration of 4–7 ns and for a small hole depth $\Delta\alpha/\alpha_0 = 10^{-2}$). Figure 5 shows that $W_a$, and thus the number of the resonance monomers, decreases as $\lambda$ varies from $\lambda \approx \lambda_m \approx 400\,\mathrm{nm}$ to $\lambda \approx 2000\,\mathrm{nm}$. The results of numerical calculations [7], shown in Fig. 5, support this conclusion.

Theory [15,35] predicts that localization of the optical excitations in a fractal aggregate decreases toward shorter wavelengths, and at a wavelength corresponding to the resonance of an individual monomer, the collective excitations become delocalized. The calculations for silver clusters [7] show that for $\lambda$ near the monomer absorption peak ($\sim 400\,\mathrm{nm}$), the absorption is approximately equally distributed over monomers in a cluster; as $\lambda$ increases toward the infrared region, the absorption becomes more localized, and at $\lambda \approx 2\,\mu\mathrm{m}$, only about 5% of monomers contribute to 50% of the total absorption. This means that there are small highly resonant domains in a fractal that account for a large fraction of the total absorption. Consequently, the threshold photomodification energy tends to decrease with $\lambda$. This can be interpreted as effective "focusing" of the incident light in resonance domains whose size is much smaller than $\lambda$.

The theoretical conclusion regarding excitation localization was also confirmed by electron microscopic study. It was found that spectral photoburning of holes correlates with local changes in the structure of irradiated clusters. Figs. 6a, b, and c are electron micrographs of the samples obtained by the methods of [36,37], and [38], respectively. Figs. 6d, e, and f show the corresponding aggregates after irradiation by laser pulses at three different wavelengths, where, in each case, the pulse energy per unit area exceeds the threshold value, $W_{th}$, that results in local photomodification of the aggregates.

Comparison of the electron micrographs of the cluster before and after irradiation at $\lambda = 1079\,\mathrm{nm}$ (Figs. 6a and d, respectively) shows that the structure of the cluster as a whole remained the same after irradiation, but monomers within small nanometer-sized domains change their size, shape, and local arrangement. The minimal number of monomers in the region of modification is two to three at $\lambda = 1079\,\mathrm{nm}$. Thus, the resonance domain at $\lambda = 1079\,\mathrm{nm}$ may be as small as $\lambda/25$. Although there are fluctuations in both shape and size of the modified domains, Fig. 6d reveals that "hot" zones associated with resonant excitation are highly localized, in accordance with theoretical conclusions [15,23]. Moreover, we found from analysis of the electron micrographs that the total number of monomers in the photomodified domains at $\lambda = 1079\,\mathrm{nm}$ and $W/W_{th} = 1.5$ comprises only $\approx 10\%$ of the total number of monomers in the aggregate.

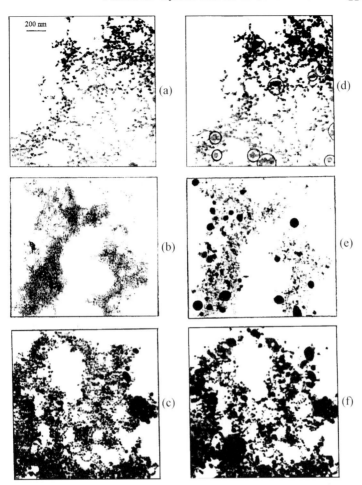

**Fig. 6.** The electron microscopy photographs of colloidal Ag aggregates before (**a**), (**b**), (**c**), and after (**d**), (**e**), (**f**) irradiation by laser pulses at $\lambda = 1079$ nm, $W = 11 \, \mathrm{mJ/cm^2}$ (**d**); $\lambda = 540$ nm, $W = 22.4 \, \mathrm{mJ/cm^2}$ (**e**); $\lambda = 450$ nm, $W = 20 \, \mathrm{mJ/cm^2}$ (**f**)

In a short-wavelength range, localization is weaker: at $\lambda = 532$ nm and $W/W_{\mathrm{th}} = 2$, a larger fraction of monomers, $\approx 30\%$, is in the photomodified domains (Fig. 6e), and at a wavelength close to the monomer absorption peak, $\lambda = 450$ nm, $\approx 70\%$ of the monomers are modified after irradiation (see Figs. 6c and 6f). The sizes of the photomodified domains at $\lambda = 532$ nm are greater than at $\lambda = 1079$ nm. At $\lambda = 532$ nm, the domain may include 10 and more monomers (instead of two to three monomers at $\lambda = 1079$ nm). Thus, the electron microscope studies clearly show that, with increasing $\lambda$, the number of monomers effectively interacting with light and the size of the resonant domains significantly decrease, i.e., localization of the optical modes in fractals becomes stronger.

Now we briefly discuss a possible mechanism for the observed photomodification. An estimate of the laser heating of silver nanospheres in gelatin for an exponential pulse shape was done in [7]. The adiabatic approximation (i.e., the pulse duration was assumed to be much larger than all characteristic times of the system) was used. The temperature dependence of all thermodynamic constants was neglected. From the thermal conductivity equation for an isolated spherical silver nanoparticle in gelatin, with the heat source homogeneously distributed inside the particle, one can estimate the particle temperature as

$$T_P - T_0 \approx 3\sigma W \left\{ 4\pi R^3 c_1 \varrho_1 \left[ 1 + \sqrt{\Gamma t_p} \left( \sqrt{\Gamma t_p} + a \right) \right] \right\}^{-1}, \tag{1}$$

where $a = \sqrt{3c_2\varrho_2/c_1\varrho_1}$, $\Gamma = 3\kappa_2/c_1\varrho_1 R^2$, $c_1, c_2$ and $\varrho_1, \varrho_2$ are the specific heats and mass densities of silver and gelatin, $\kappa_2$ is the thermal conductivity of gelatin, $R$ is the radius of the silver particles, $\sigma$ is the absorption cross-section per monomer, $t_p$ is the pulse duration, and $T_0 = 300\,\mathrm{K}$ is the initial temperature. With the known values of the thermodynamic constants and the measured value of the absorption cross-section, the above formula yields $T_P = 600\,\mathrm{K}$ at the threshold energy $W_{th} = 11.6\,\mathrm{mJ/cm^2}$ for $\lambda = 550\,\mathrm{nm}$.

According to [39], sintering of metal nanoparticles starts when the temperature exceeds half the melting point $T_m$ (for Ag, $T_m = 1234\,\mathrm{K}$ ), and the characteristic time of this process is 100 ps. Thus, the above temperature $T_P$ is sufficient to start the process of sintering in silver colloids at the threshold pulse energy $W_{th}$. Note that enlargement of Ag nanoparticles by approximately the factor of 2 was also observed in island Ag films after thermal heating to $570\,\mathrm{K}$ [40]. The measured time of the photomodification was approximately 100–200 ps [6]. One can conclude that our simple model of photomodification agrees with experimental data.

More detailed treatment of the mechanisms of photomodification should include an account of the heating of the resonant domains and heat transfer along the cluster chains, as well as through the host medium, dispersion, and induced electrostatic forces between monomers. The last problem was discussed in [41].

To discuss the role of laser heating of silver particles, let us consider spectral holes burned by nanosecond and picosecond pulses. Note that shortening of the laser pulse from 5 ns to 5 ps results in a decrease of the $W_{th}$ of less than one order of magnitude. It means that energy is a better parameter for describing photomodification than intensity. Figure 3 shows spectral holes burned by pico- and nanosecond pulses in the same sample. The width of a hole from a nanosecond pulse is less than at a picosecond pulse. Then, the hole from a nanosecond laser pulse is shifted to the long-wavelength side relative to the laser wavelength. The shift grows when a laser wavelength comes to the monomer surface plasmon resonance, $\lambda_m \approx 400\,\mathrm{nm}$. There is no shift at $\lambda \approx 1300$ nm, and the shift changes its sign at $\lambda > 1400$ nm. One of the possible reasons for a shift of a hole from a nanosecond pulse is a change of the silver dielectric function during the laser pulse. Energy transfer from conductive electrons to the crystal lattice in noble metal nanoparticles occurs for several picoseconds [11]. Therefore absorption of a 5-ps light pulse takes place at a fixed lattice temperature and at fixed positions of the monomers in a cluster. In nanosecond irradiation, the crystal lattice has time to be heated during the laser pulse. The heating results in an increase of the electron–phonon relaxation rate, and, hence, in a change of the dielectric function of silver [9]. The change, in turn, results in spatial redistribution of plasmon mode configurations during a laser pulse and leads to broadening of the spectral hole.

The experiments [7] revealed that inhomogeneous broadening of absorption spectrum took place in a wide spectral range – at least up to 2000 nm, which agrees with theory [23].

A wide inhomogeneous absorption band permits recording several spectral holes in the same area of a fractal film. Five spectral holes were recorded with light pulses of the same linear polarization [7]. This experiment demonstrates possible applications of the photomodification effect to dense optical data recording.

Selective local photomodification was also detected with the help of near-field scanning optical microscopy [42,43]. Near-field optical images were taken before and after irradiation of the same fractal cluster by a nanosecond laser pulse at $\lambda = 532$ nm. A probe beam was generated by CW lasers at 543.5 nm or at 632.8 nm. Figure 7 shows that some of the hot spots lost their brightness after irradiation by a pulse whose energy exceeded the threshold value, $W/W_{th} \approx 1.1$. The lower part of Fig. 7 shows the intensity profiles along the marked lines. The irradiation of the sample with the laser pulse energy below the threshold, $W/W_{th} = 0.9$, does not lead to any changes in the near-field images.

Two points are noteworthy: (1) Changes in the local responses arising as a result of photomodification can be two orders of magnitude larger than changes in macroscopic response, far-field optical absorption. Near the threshold after photomodification, the brightness of the hot spots decreases by two times, but the decrease in relative absorption is only 0.01. (2) Together with the decrease in local intensity after photomodification, in Fig. 7 one can see an increase in the brightness of some hot spots. The latter observation may be interpreted as a spatial redistribution of collective plasmon modes because of the photomodification-induced local change of cluster structure.

So, resonant domains can be burnt by a laser pulse. Therefore, one can find the contribution of the resonant domains (hot spots) to enhanced nonlinearities by measuring the nonlinear responses before and after photomodification.

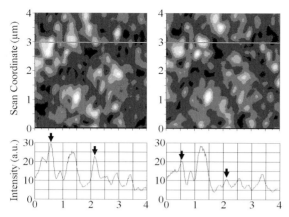

**Fig. 7.** Near-field optical images of a fractal Ag film (**a**) before and (**b**) after photomodification. The pictures were obtained with probe radiation at a wavelength of 543.5 nm. The signal profiles along the lines indicated on the images are shown at the bottom

## 2  Local Optical Nonlinearities in Silver Colloids

Third-order nonlinear polarization can be described by the following relation in the general case:

$$P_i^{(3)} = \chi_{ijkl}^{(3)} E_j E_k E_l + \Gamma_{ijklm}^{(3)} E_j E_k \nabla_m E_l. \tag{2}$$

Here the local nonlinear response is represented by the tensor $\hat{\chi}^{(3)}$, and the tensor $\hat{\Gamma}^{(3)}$ is connected with the nonlocal response that is due to spatial dispersion effects. According to the condition of intrinsic permutation symmetry, the tensor $\hat{\chi}^{(3)}$ for isotropic materials can be represented as [27]

$$\chi_{ijkl}^{(3)}(\omega = \omega + \omega - \omega) = \chi_{1122}(\omega = \omega + \omega - \omega)\,(\delta_{ij}\delta_{kl} + \delta_{ik}\delta_{jl})$$
$$+\chi_{1221}(\omega = \omega + \omega - \omega)\delta_{il}\delta_{jk}. \tag{3}$$

As for the tensor $\hat{\Gamma}^{(3)}$, if the frequency dispersion of the nonlocal response is neglected, $\hat{\Gamma}^{(3)}$ has one nonzero component $g_1$ in an isotropic medium.

If the optical field is of the form

$$\boldsymbol{E}(t, \boldsymbol{r}) = \boldsymbol{A} \exp[-\mathrm{i}(\omega t - \boldsymbol{k}\boldsymbol{r})] + \text{c.c.}, \tag{4}$$

we can obtain the formula for the local part of $\boldsymbol{P}^{(3)}$ as

$$\boldsymbol{P}_{\text{loc}}^{(3)}(\omega) = \chi_1 \boldsymbol{A}(\boldsymbol{A}\boldsymbol{A}^*) + \chi_2 \boldsymbol{A}^*(\boldsymbol{A}\boldsymbol{A}) + \text{c.c.}, \tag{5}$$
$$\chi_1 = 6\chi_{iijj}^{(3)} = 6\chi_{ijij}^{(3)}; \quad \chi_2 = 3\chi_{ijji}^{(3)}; \quad i, j = 1, 2; \quad i \neq j,$$

and that for the nonlocal one as[1]

$$\boldsymbol{P}_{\text{nonloc}}^{(3)}(\omega, \boldsymbol{k}) = -ig_1 \left\{ (\boldsymbol{A}\boldsymbol{A}^*)[\boldsymbol{k}\boldsymbol{A}] + \boldsymbol{A}(\boldsymbol{A}^*[\boldsymbol{k}\boldsymbol{A}]) \right\} + \text{c.c.}. \tag{6}$$

---

[1] The electrical field in [5] was defined with the exponential factor $\exp[\mathrm{i}(\omega t - \boldsymbol{k}\boldsymbol{r})]$, so that the equation for $\boldsymbol{P}_{\text{nonloc}}^{(3)}$ there had the opposite sign.

In this section we will discuss nonlinear optical effects due to the macroscopic cubic susceptibility of a silver nanocomposite $\chi^{(3)}_{1111}$, namely, degenerate four-wave mixing (DFWM), nonlinear absorption, and refraction.

A large nonlinear optical susceptibility, $\hat{\chi}^{(3)}$, was reported in non-aggregated nanocomposites with gold [44,45,46] and silver [44,47] particles at the surface plasmon resonance. High values of the susceptibility are caused by the enhancement of the local fields in metal particles at the surface plasmon resonance and by the high hyperpolarizabilities of metal particles [44]. The third-order polarizabilities of small metal particles can reach values as high as those of resonant atoms and, moreover, metal particles possess a wider spectral resonance than atoms. The mechanisms contributing to the third-order polarizability of the nanoparticles were considered, including nonequilibrium electron heating [46], saturation of the interband transition [46], and the saturation of the absorption of the electron gas confined inside the particle [45,46,48].

The electrical field of a resonance light wave, acting inside the particles in the simplest aggregate, a pair, is stronger than the local field for a single particle. Corresponding enhancement factors $E_i/E_0$ for a local field $E_i$, in comparison with an incident field $E_0$, are $G = \varepsilon_1^2/3\varepsilon_0\varepsilon_2$ for a plasmon resonance in a pair [3] and $f_1 = 3\varepsilon_0/i\varepsilon_2$ for surface plasmon resonance in a single sphere [44]. Here $\varepsilon = \varepsilon_1 + i\varepsilon_2$ and $\varepsilon_0$ are the dielectric constants of a metal particle and a host medium. In metals, $|\varepsilon_1|$ grows in a long-wavelength range, and the value of $G$ increases considerably with the wavelength. A value of $G = 18$ was estimated for a pair of silver particles at $\lambda = 532\,\text{nm}$, and in the near infrared it may be as high as $10^2$. For a single silver particle, the maximum value of $f_1$ is at the surface plasmon resonance, $\lambda = 400\,\text{nm}$, and can be roughly estimated as $|f_1| = 2$.

When the local field exceeds the applied field, $\boldsymbol{E}_0$, huge enhancements of nonlinear optical responses occur. The enhancement factor $G^{(n)}$ for a nonlinear optical process of the $n$th order in an aggregated colloidal solution is defined as the intensity ratio of the radiation generated by a nonlinear process on a monomer incorporated into a metal aggregated on an isolated monomer. The value of $G^{(n)}$ averaged over various fractal cluster realizations can be estimated in the binary approximation [2,23] as $G^{(n)} \approx G^{2(n-1)}$ [2,23]. When the generated frequency occurs in the absorption band of the aggregates, its amplitude can be enhanced as well, compared to the "initial" field. Then the enhancement factor reaches its maximum $G^{(n)} \approx G^{2n}$. Taking into account the estimate for the $G$ given above, one can conclude that for degenerate four-wave mixing (DFWM) ($n = 3$) in silver aggregates, it is possible to obtain an average enhancement $G^{(3)} \approx 10^6$. This estimate coincides by an order of a magnitude with numerical calculations of the enhancement factor arising from collective plasmon mode excitation in fractal clusters [35]. It means that the main advantage of a strongly aggregated nanocomposite (fractal) in comparison with a small aggregate (pair) is not the magnitude of the enhancement factor, but the number of the resonant frequencies that cover a wide spectral range.

It is noteworthy that the local enhancement of nonlinear optical responses in a hot spot may be much stronger than in a macroscopic sample. This was demonstrated by the near-field optics technique for the SERS [49] and for the photomodification process [42].

So, an aggregation of silver particles results in a significant growth of nonlinear responses. Below, we describe the experimental results for a silver colloid which confirm this conclusion [3,50]. Measurements of the DFWM efficiency versus the degree of hydrosol aggregation and laser beam intensity were performed with the help of a Q-switched nanosecond laser [3,50]. All interacting light waves had the same linear polarization. The results were obtained with two silver hydrosols, a boron hydride colloid [32], and collargol at $\lambda = 532$ nm. This wavelength is off the surface plasmon resonance for silver particles. Figure 8 shows the conversion efficiency measured versus the intensity of the strong pump radiation $I_0$ and the degree of hydrosol aggregation. The main result is that in colloids containing clusters, the same conversion efficiency is attained as in nonaggregated monomers at $10^3$ times lower input intensities. This conclusion follows from comparison of the groups of points 1 and 3 in Fig. 8. The intensity dependence of the signal $I_s$ is close to cubic, as is typical of four-photon scattering. Taking into account this cubic dependence, the enhancement factor for DFWM may be estimated as $\sim 10^6$, which agrees with the estimate of $G^{(3)}$ given above. Note that the DFWM in a nonaggregated colloid was measured in focused laser beams. In unfocused beams with diameters of 2 mm, we observed the DFWM signal in a weakly aggregated colloid at a pump intensity of $10 \, \mathrm{MW/cm^2}$ (curve 2 in Fig. 8).

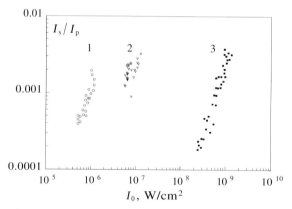

**Fig. 8.** DFWM conversion efficiency in Ag colloids with different aggregation degrees. $I_0$ is the pump radiation intensity at the wavelength 532 nm. 1, nearly nonaggregated colloid; 2, weakly aggregated, and 3, strongly aggregated colloid. The cell is 3 mm thick

Possibly this signal arose from a small number of aggregates in the irradiated volume.

At an increasing degree of hydrosol aggregation, the DFWM signal grows as follows. At $I_0 = 1\,\mathrm{MW/cm^2}$ in a freshly prepared hydrosol, the conversion efficiency $\eta$ was below the detection level ($\eta \approx 2 \times 10^{-5}$), in an hour it increased up to $4 \times 10^{-5}$, and in 7 h it reached $\eta \approx 10^{-3}$. In the most aggregated hydrosols doped by dye Rhodamine 6G, $\eta \approx 0.04$.

The efficiency of the DFWM is governed by the $|\chi^{(3)}|$. We obtained $|\chi^{(3)}_{1111}| - 2 \times 10^{-10}\,\mathrm{esu}$ for the most aggregated silver colloid with a volume fraction of metal of $p = 5 \times 10^{-6}$, at $\lambda = 1064\,\mathrm{nm}$ for a laser intensity of $1\,\mathrm{MW/cm^2}$. This means that the susceptibility per unit volume of the metal is more than $10^{-4}\,\mathrm{esu}$. The measured $|\chi^{(3)}_{1111}|$ at $\lambda = 532\,\mathrm{nm}$ is two times less. The figure of merit (FOM), $|\chi^{(3)}_{1111}|/\alpha$, is about $10^{-9}\,\mathrm{esu \times cm}$. The values of $|\chi^{(3)}_{1111}|/p$ and FOM for the aggregated colloid exceeds substantially (more than one to two orders of magnitude for nanosecond pulses) the corresponding values for nonaggregated composites at the plasmon resonance $\lambda_{\mathrm{m}}$.

Increasing the laser intensity results in the photomodification of strongly aggregated colloids. Well beyond the threshold, photoburning leads to a substantial decrease in nonlinear susceptibility. The efficiency of DFWM at $\lambda = 532\,\mathrm{nm}$ was measured in Ag clusters in a 3-mm cell before photoburning and with a delay of 60 μs after the photoburning pulse with energy ten times more than the threshold energy. The result was that the signal decreased by a factor of 30 in comparison with the nonirradiated sample [51]. This observation means that the resonant modes are responsible to a large extent for the enhancement of the nonlinear response.

To measure the nonlinear refractive index and the nonlinear absorption coefficient, we used the $Z$-scan technique [52]. This method allowed us to determine the nonlinear correction to the refractive index, $n_2$, and to the absorption coefficient, $\alpha_2$, ($n = n_0 + n_2 I$, $\alpha = \alpha_0 + \alpha_2 I$), from experimental data and to calculate $\mathrm{Re}\,\chi^{(3)}$ and $\mathrm{Im}\,\chi^{(3)}$. For aggregated aqueous colloids, the values obtained for $\chi^{(3)}$ were $\mathrm{Re}\,\chi^{(3)} = 10 \times 10^{-11}\,\mathrm{esu}$, $\mathrm{Im}\,\chi^{(3)} = -8.3 \times 10^{-11}\,\mathrm{esu}$ for $\lambda = 540\,\mathrm{nm}$, and $\mathrm{Re}\,\chi^{(3)} = -3.5 \times 10^{-11}\,\mathrm{esu}$, $\mathrm{Im}\,\chi^{(3)} = -2.7 \times 10^{-11}\,\mathrm{esu}$ for $\lambda = 1079\,\mathrm{nm}$. This suggests that nonlinear absorption and nonlinear refraction provide nearly equal contributions to the nonlinearity in the green and near-infrared ranges. This may be regarded as an additional advantage of a aggregated nanocomposite as a nonlinear medium with respect to nonaggregated. The latter is characterized by predominantly imaginary susceptibility at the surface plasmon resonance [46].

At low intensity, the imaginary part of the cubic susceptibility is negative in the blue, green, and near-infrared ranges, and positive in the red region [53]. When the intensity exceeds $3\,\mathrm{MW/cm^2}$, the results of the $Z$-scan fitting according to procedure [52] become unsatisfactory. This implies that the nonlinearity departs from the simple cubic law. In the investigation [53] of a colloidal silver solution at high intensities, significantly exceeding the photomodification threshold, a colloidal solution prepared by the method [38] was used. The concentration of the colloidal solution was adjusted so that $\alpha_0 = 1.6\,\mathrm{cm}^{-1}$ at all wavelengths studied. The solution was placed in a cell 10 mm thick and moved along a focused laser beam to vary the intensity of the incident 4-ns pulse from approximately 1 to $318\,\mathrm{MW/cm^2}$. The absorption coefficient at various times belonging to the incident pulse envelope was measured. Figure 9 presents the dependence of the nonlinear correction to the absorption coefficient, $\Delta\alpha$, on the peak intensity of the incident pulse for $\lambda = 440$, 532 and 650 nm. The colloid displayed three different types of absorption behavior. At $\lambda = 440\,\mathrm{nm}$, the absorption of silver colloid dropped with intensity and demonstrated the phenomenon of so-called nonlinear "bleaching". At $\lambda = 650\,\mathrm{nm}$, a phenomenon of increasing absorption with intensity was observed. At $\lambda = 532\,\mathrm{nm}$, the dependence of absorption on intensity was more complicated; it demonstrated both nonlinear "bleaching" and darkening within certain intensity intervals. Analogous (qualitatively) intensity dependence was observed at $\lambda = 1064\,\mathrm{nm}$.

The dependence $\Delta\alpha(I)$ at $\lambda = 440\,\mathrm{nm}$ is similar to the curve for saturated absorption under the conditions of photoburning of the resonant modes. When the intensity is low, $I \leq I_{\mathrm{th}} = 4\,\mathrm{MW/cm^2}$, the nonlinear absorption obeys the law

$$\frac{\Delta\alpha}{\alpha_0} = \frac{I/I_\mathrm{s}}{(1 + I/I_\mathrm{s})} = \frac{\alpha_2}{\alpha_0} I \tag{7}$$

where $\alpha_2 = -6.6 \times 10^{-8}$ cm/W; however, when the intensity exceeds the threshold of photomodification $I_{\mathrm{th}}$, photoburning of the resonance domains in colloid aggregates occurs, and the dependence $\Delta\alpha(I)$ departs from this law.

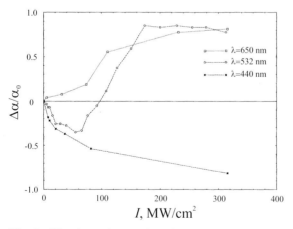

**Fig. 9.** The dependence of nonlinear addition to the absorption coefficient for an aggregated Ag colloid versus the incident peak intensity

Analogous behavior of nonlinear absorption was observed [51] at $\lambda = 540\,\mathrm{nm}$ for $I \leq 3.5\,\mathrm{MW/cm^2}$.

The intensity dependence of a nonlinear correction to the absorption at $\lambda = 532\,\mathrm{nm}$ in Fig. 9 shows that in the range 60–170 MW/cm$^2$, there is a linear increase of absorption versus intensity. This suggests that, in this range, the main nonlinear absorption mechanism is two-photon absorption. The measured coefficient for two-photon absorption is $\alpha_2' = 1.7 \times 10^{-8}\,\mathrm{cm/W}$ at the peak of the incident pulse ($\Delta\alpha = \alpha_2' I$). Approximately the same value of $\alpha_2'$ was obtained at $\lambda = 650\,\mathrm{nm}$ for intensity $I \approx 100\,\mathrm{MW/cm^2}$.

The temporal shape of the nanosecond laser pulses at $\lambda = 532\,\mathrm{nm}$ transmitted through the solution is given in Fig. 10 for different intensities. One can see from this figure that the contribution of two-(or multiphoton-) absorption increases during the laser pulse. The observed phenomenon may be interpreted as follows. Nonlinear optical "bleaching" is caused by the presence of aggregates of nanoparticles in solution and high local fields within fractal structures. Colloid aggregates are modified by high-intensity light within $t \leq 10^{-9}\,\mathrm{s}$, and optical nonlinearity related to fractal structure of the medium decreases. After that, the weaker nonlinearity of isolated particles and bulk metal becomes essential. Therefore, the increasing nonlinear absorption, observed 1 ns after the leading front of the pulse, related probably to two-photon absorption in bulk silver. The above data show that the coefficient of two-photon absorption is 30 times smaller than that of nonlinear "bleaching" for low intensities. According to [54], the probability of two-photon absorption near the L-symmetry point increases with the wavelength as $\lambda^8$. That is why the effect of two-photon absorption is larger in the red and green region of the spectra than that in the blue region.

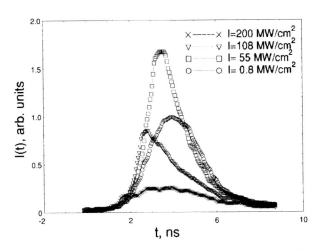

**Fig. 10.** The oscillogram of laser pulses with $\lambda = 532\,\mathrm{nm}$ passed through a Ag colloid at different intensities

Strictly speaking, the observed decrease in the transmittance of a colloid at high intensity may occur not only because of nonlinear absorption but also because of nonlinear scattering. Light scattering in colloids at high intensities was studied. The measured loss of laser energy due to light scattering was at least 10 times smaller than the absorbed energy. It was found that for a nanosecond pulse at $I > 10^8\,\mathrm{W/cm^2}$, an intense conical scattering at the angle to the beam axis of approximately $25\pm20\,\mathrm{mrad}$ arose at the leading front of the incident pulse. The energy conversion to conical scattering was less than 10%. Note that the aperture of our detector received the conical emission when the transmittance was measured.

Nonlinear backscattering in aqueous and ethanolic colloidal silver solutions occurred when the 10-ns laser pulse intensity at $\lambda = 532\,\mathrm{nm}$ exceeded $500\,\mathrm{MW/cm^2}$. At small absorption, $\alpha_0 < 0.35\,\mathrm{cm^{-1}}$, a nonlinear backscattering spectrum consisted of a nonshifted Rayleigh component and a Stokes Mandelstam–Brillouin component. By observation of the spontaneous scattering, it was found that when the absorption coefficient grew, the width of the Mandelstam–Brillouin component increased. At $\alpha_0 > 0.35\,\mathrm{cm^{-1}}$, the nonlinear backscattering consisted only of a stimulated Rayleigh component. The backscattering pulse envelope followed to the incident pulse. Polarization of the Rayleigh backscattering coincided with laser backscattering. The measured backscattering conversion efficiency at $\alpha_0 = 1.6\,\mathrm{cm^{-1}}$ and $I \approx 600\,\mathrm{MW/cm^2}$ was 1%.

Nonlinear darkening was also observed with a 10-ps laser pulse[2] at $\lambda = 532\,\mathrm{nm}$. The dependence $\Delta\alpha = \alpha_2' I$ with $\alpha_2' = 3.4 \times 10^{-9}\,\mathrm{cm/W}$ was found for the intensities $I \sim 100\,\mathrm{MW/cm^2}$. One can see some decrease in the nonlinear coefficient from picosecond excitation in comparison with nanosecond excitation. This decrease is due to a change of the mechanism of nonlinearity. An analogous conclusion was drawn by the authors of [55], who observed high cubic susceptibility, $\chi^{(3)} \approx 4 \times 10^{-6}\,\mathrm{esu}$, of Au:SiO$_2$ films at a pulse duration of 70 ps. Shortening of the pulse duration (up to 200 fs) resulted in a decrease of the susceptibility by a factor of 30. It was supposed that a generation of hot electrons was the main nonlinearity in picosecond excitation, and a saturation of interband transition was the main mechanism in femtosecond excitation. Very large nonlinearity at concentrations below the Au percolation threshold was explained by local-field enhancement in the Au nanocomposite [56].

At nanosecond pulses, both the absorptive and the refractive nonlinearities of an aggregated colloidal silver solution change their signs with wavelength. These changes may be related, on the one hand, to variations of the signs of the enhancement factors [57]. On the other hand, and this is more likely, the sign changes may be due to the band structure of silver, laser heating of the electron gas and the crystal lattice of nanoparticles, and heating

---

[2] The authors are indebted to A.S. Kuch'yanov for providing the picosecond laser for experiments.

of the host medium around the particle by thermal diffusivity. Photoburning of resonance modes results in a decrease of nonlinear responses.

# 3   Nonlocal Optical Nonlinearities in Silver Colloids

It was noted above that large sizes of collective plasmon modes facilitated observation of the effects of spatial dispersion in silver nanocomposites. Optical activity (gyrotropy) is one of such nonlocal effects.

Optical activity is characterized by different responses to right- and left-circularly polarized light and is associated with the nonlocal part of polarizability. Optical activity phenomena include differences in attenuation (circular dichroism), refraction (circular birefringence), and scattering for right- and left-circularly polarized light, which are intensity-dependent in the nonlinear case. As already mentioned above, in aggregates composed of metal nanoparticles, optical excitations are associated with the plasmon modes which can have high resonance quality factors and thus provide particularly strong enhancement in the hot spots [23] which results in large nonlinearities.

This section is concerned with experiments on nonlinear gyrotropy and accompanying polarization effects. But it is worth beginning this section with an examination of the nonlocal properties of plasmon modes by photon scanning tunneling microscopy (PSTM).

## 3.1   Chirality of Plasmon Modes

The localized optical modes of fractals have been studied using PSTM with subwavelength spatial resolution. It has been shown that hot spot locations are very sensitive even to small changes in the linear polarization plane and frequency of the applied field ($\lesssim 1\%$ in the latter case) [25,34,58]. The observed linear polarization dependence, in particular, illustrates strong local anisotropy of plasmon modes in fractals. However, an important question, whether the fractal optical modes exhibit handedness, has not been addressed so far. It is known that two coupled anisotropic oscillators lying in different planes present a simple model of a chiral system. Fractal aggregates are formed by chains of particles with almost all possible local configurations. These local structures have typically neither a center nor a plane of symmetry and thus may have handedness so that a fractal mode localized on such a structure exhibits chirality. Thus, one may anticipate that the fractal modes can be "chiral-active" and manifest the property of optical activity *locally*.

Let us briefly consider the interaction of circularly polarized light with nanoparticle aggregates. The system is described by the coupled-dipole equations (CDE), whose solution permits the determination of the dipoles $d_i$ located at $r_i$ and local field distribution $E(r)$. Elementary dipoles associated with non-gyrotropic particles in a fractal interact with the incident field and with each other via dipolar fields, which are established self-consistently.

Higher multipole contributions could be taken into account by introducing effective intersecting spheres [23].

CDE solutions can be represented in terms of the expansion over the eigenvectors $|n)$ of the dipolar operator: $V|n) = v_n|n)$ [15,22,23]; the operator $V$ includes the near-, intermediate-, and far-zone terms for the dipole field. The local field $\boldsymbol{E}(\boldsymbol{r})$ from all dipoles of the cluster induced by the circularly polarized incident field

$$\boldsymbol{E}_{\mathrm{R,L}} = E_0\hat{\boldsymbol{\sigma}}_{\mathrm{R,L}} = E_0\frac{\hat{\boldsymbol{y}}\cos(k_x x - \omega t) \pm \hat{\boldsymbol{z}}\sin(k_x x - \omega t)}{\sqrt{2}} \tag{8}$$

is expanded over the eigenmodes as

$$E_{\alpha\mathrm{R,L}}(\mathbf{r}) = \sum_n f_{n\alpha}(\mathbf{r})(\bar{n}|\sigma_{\mathrm{R,L}})E_0, \tag{9}$$

where $\alpha$ denotes the Cartesian components [23]. In this expansion,

$$(\bar{n}|\sigma_{\mathrm{R,L}}) = \sum_j [(\bar{n}|jy)\cos(k_x x_j - \omega t) \pm (\bar{n}|jz)\sin(k_x x_j - \omega t)]/\sqrt{2} \tag{10}$$

represents the projection of the mode vector on the circular-polarization unit vector of the incident field and an orthogonal basis $|i\alpha)$ is defined in $3N$-dimensional complex vector space. The mode "weight" distribution $f_{n\alpha}(\mathbf{r})$ is given by

$$f_{n\alpha}(\mathbf{r}) = \sum_i G_{\alpha\beta}(\boldsymbol{r} - \boldsymbol{r}_i)(i\beta|n)[(\bar{n}|n)]^{-1}(\alpha_0^{-1} + v_n)^{-1}, \tag{11}$$

where $G_{\alpha\beta}(\boldsymbol{r} - \boldsymbol{r}_i)$ is the regular part of the free-space dyadic Green function and $\alpha_0$ is the polarizability of the individual particles. (Vector $(\bar{n}|$ is a row-vector with the same entries as a column vector $|n)$, as opposed to $(n|$ which is a row-vector with complex conjugated elements [23].) Note that for a given mode, the distribution function $f_{n\alpha}(\boldsymbol{r})$ does not depend on the incident wave polarization; the polarization dependence is contained in the projection of the mode on the incident field $(\bar{n}|\sigma_{\mathrm{R,L}})$. The difference in optical response to different circular polarizations occurs because the fractal modes may have different projections $(\bar{n}|\sigma_{\mathrm{R,L}})$ on the helical configuration of the incident field with opposite signs of helicity. In other words, the spatial distribution of a fractal eigenmode can have certain handedness and in that sense it possesses chirality. As mentioned, this mode of chirality is possible because local structures in a fractal typically have neither a center nor a plane of symmetry. This results in different distributions of dipole moments:

$$\begin{aligned} d_{i\alpha} &= \sum_n n_{i\alpha}C_n(\bar{n}|\sigma_{\mathrm{R,L}})E_0 \\ &= \sum_n n_{i\alpha}C_n\sum_j[n_{jy}\cos(k_x x_j - \omega t) \\ &\quad \pm n_{jz}\sin(k_x x_j - \omega t)](E_0/\sqrt{2})\,, \end{aligned} \tag{12}$$

where, for simplicity, the mode amplitudes $(i\alpha|n)$ are denoted as $n_{i\alpha}$, and $C_n = [(\bar{n}|n)]^{-1}(\alpha_0^{-1} + v_n)^{-1}$. Consequently, one can obtain different optical responses for right and left polarization. For example, the extinction cross-section which is given from the optical theorem,

$$\sigma_e = \frac{4\pi k}{|E_0|^2} \text{Im} \sum_i d_i E_{R,L}^*(x_i), \tag{13}$$

can be found as

$$\sigma_{e,R,L} = \pi k \text{Im} \sum_n C_n \sum_{i,j} [(n_{jy}n_{iy} + n_{jz}n_{iz}) \cos k_x(x_j - x_i)$$
$$\pm (n_{jz}n_{iy} - n_{jy}n_{iz}) \sin k_x(x_j - x_i)]. \tag{14}$$

One can see that differential optical response,

$$\Delta \sigma_{R,L} = 2\pi k \text{Im} \sum_n C_n \sum_{i,j} (n_{jz}n_{iy} - n_{jy}n_{iz}) \sin k_x(x_j - x_i) \tag{15}$$

increases with an increase of the anisotropic factor $N_{yz}^{(ij)} = (n_{jz}n_{iy} - n_{jy}n_{iz})$ and the cluster size through a geometric factor $\sin k_x(x_j - x_i)$.

Scattering optical activity is denoted as circular intensity differential scattering (CIDS) [59]. The CIDS is characterized by the dimensionless parameter

$$\text{CIDS} = (I_R - I_L)/(I_R + I_L) = S_{14}/S_{11}, \tag{16}$$

where $I_R$ and $I_L$ are the scattered intensities from right-and left-circularly polarized light and $S_{11}$, $S_{14}$ are the elements of the Muller scattering matrix [59], characterizing the average intensity and optical activity, respectively.

The differential scattered intensity is given by $\Delta I_{R,L} = |E_R(\boldsymbol{r})|^2 - |E_L(\boldsymbol{r})|^2$. It can be shown that

$$I_{R,L} = (|E|^2/4) \sum_{m,n} \left\{ f_{n\alpha} f_{m\alpha}^* \right.$$
$$\times \sum_{i,j} [(n_{jy}m_{iy}^* + n_{jz}m_{iz}^*) \cos k_x(x_j - x_i)$$
$$\left. \pm (n_{jz}m_{iy}^* - n_{jy}m_{iz}^*) \sin k_x(x_j - x_i)] \right\}. \tag{17}$$

Because $k_x = k \sin\theta$, the CIDS depends on the angle of incidence $\theta$. Thus, the difference $I_R - I_L$ is proportional to $\sin k_x(x_j - x_i)$, so that the main contributions to CIDS come from the mode amplitudes separated by distances comparable with the wavelength. It means that local CIDS is sensitive to long-range distribution of resonant anisotropic [in the sense nonzero and even large anisotropic factor $(n_{jz}m_{iy}^* - n_{jy}m_{iz}^*)$] modes.

In previous studies, only *macroscopic* optical activity was considered [59]. However, macroscopic effects could be strongly decreased in a racemic set of chiral elements (approximately equal amount of elements with opposite handedness). Therefore, the local optical activity detected using PSTM may be much stronger than the typically observed macroscopic effect and provide data on the chirality of local configurations and their statistical properties.

Numerical simulations performed by using a cluster model, which realistically describes a fractal aggregate deposited on a plane [58,34] show that right–left differences in optical response with respect to their sum could be as much as 0.8 [60].

Experiments on local CIDS were carried out at New Mexico State University [61,60]. Preparation of the fractal aggregates and details of PSTM apparatus are published elsewhere [23,62]. Here, we outline only the most important steps in sample preparation. Solutions of colloid particles $\sim 10\,\text{nm}$ in size were prepared by a citrate reduction method [38]. The addition of organic adsorbates promoted aggregation of colloidal particles into fractal clusters. About $2\,\mu\text{L}$ of the fractal aggregate solution was deposited onto a glass substrate resulting in a thin layer of fractal material; atomic force microscope measurements showed that this layer was about $100\,\text{nm}$ thick. The density of the deposited fractals was roughly $5 \times 10^6$ per $\text{cm}^2$. Transmission electron microscopy images of the samples prepared in this way clearly indicated the fractal structure of the aggregates, which have fractal dimension $D \approx 1.75$. The fractals, typically consisting of several thousand monomers, have a size of 1 to $3\,\mu\text{m}$. The resonance spectra of individual silver particles and their aggregates and the deposition-caused transformation of self-similar aggregates into self-affine structures, with similar optical properties, are described in detail elsewhere [23,34,58].

Fractal colloidal aggregates were placed, using index matching fluid, on the hypotenuse face of a $90°$ prism and illuminated by an evanescent field in total internal reflection geometry. The illumination source was either a helium–neon laser operating at a wavelength of $633\,\text{nm}$ or a tunable diode laser (Newport 2010) operating between 790 and $820\,\text{nm}$. The polarization of the beam was controlled using a Glan–Thompson polarizer (New Focus 5524) and a variable wave plate (New Focus 5540). The local optical signal was collected through an uncoated optical fiber, sharpened to approximately 50-nm radius at its tip using a fiber puller (Sutter Instruments P-2000). The separation between the probe tip and the sample was regulated using nonoptical shear-force feedback when the PSTM was operated in the constant-height mode. Alternatively, the sample was scanned in a constant plane above the sample without active feedback of the probe height.

The scanning plane in the apparatus was tilted by approximately $14°$ with respect to the sample surface, and consequently the tip height changed from about $100\,\text{nm}$ to about $2400\,\text{nm}$ during a $10 \times 10\,\mu\text{m}$ scan in the constant-plane mode. The experimental images shown in Fig. 11 exhibit quite a uniform lateral size of hot spots (i.e., spatial regions of very high intensity shown in white) and contrast over the whole imaging area, although the images were recorded with varying tip heights. In addition, the images obtained in the constant-plane mode show approximately the same spot size and contrast as in the constant-height mode. The average size of hot spots determined from an autocorrelation function is of the order of 400 nm. It gives an upper limit for lateral resolution of the measurements.

Figures 11a and 11b show the local optical images of a fractal sample for right- and left-circularly polarized incident light, respectively. Both images exhibit large variations of light intensity and many hot spots. Visual examination of the two images reveals different spatial distributions of hot spots for the two polarizations, indicating circular differential response in certain regions. The weak correlation of the two images is confirmed by comparison of correlation functions calculated from these images.

Figure 11c shows the distribution for the local CIDS parameter $(I_R - I_L)/(I_R + I_L)$ computed from the optical images shown in Figs. 11a and 11b that were obtained with right- and left-circular polarizations of the evanescent wave. The white and black spots in Fig. 11c represent areas of preferential

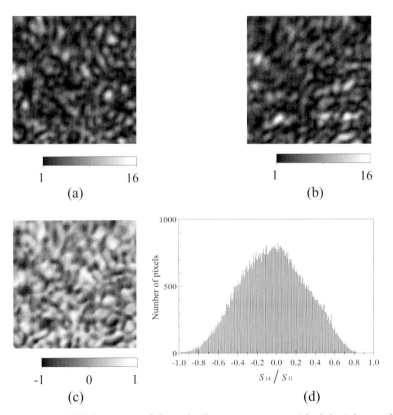

**Fig. 11.** PSTM images of fractal silver aggregates with (**a**) right- and (**b**) left-circularly polarized incident light at 633 nm. (**c**) Local CIDS signal computed from the two images shown in (**a**) and (**b**). (**d**) Histogram of CIDS signal

scattering for right- and left-circularly polarized incident light, respectively. The variations in the parameter CIDS demonstrate that fractal aggregates have the random distribution of chiral systems. Randomness is characterized by a histogram of a number of picture elements with a certain value and sign of CIDS, see Fig. 11d. Note that this distribution is rather wide and has a half width at half height of about 0.45. Despite weak asymmetry the histogram shows an approximately equal number of chiral areas with opposite handedness.

The local optical activity is large $(|(I_R - I_L)/(I_R + I_L)|$ up to 0.6) and varies strongly with wavelength. (Note that the macroscopic absorption spectra of the samples have a flat wavelength dependence in this range, with no spectral structures.) The sign of $(I_R - I_L)/(I_R + I_L)$ can change, even for very small changes of wavelength, $\sim 10\,\text{nm}$. In sharp contrast, the value of the CIDS parameter averaged over the whole $10 \times 10\,\mu\text{m}$ sample, shown in Fig. 11b, is much smaller: $|(I_R - I_L)/(I_R + I_L)| \approx 0.012 \pm 0.04$. These results clearly show that, though locally there exist areas of strong preference for scattering of one circular polarization or the other, this preference is effectively absent when averaged over a macroscopic sample. The observed difference between $I_R(\boldsymbol{r})$ and $I_L(\boldsymbol{r})$ means that different (and spatially separated) modes with opposite signs of handedness may be in resonance with the field of the same frequency. The set of resonating modes is defined by the condition $\text{Re}\,[1/\alpha_0 + v_n] \leq \gamma_n$, where $\gamma_n$ is the resonance width. A change in the wavelength leads to another set of resonant modes excited by the incident field. If the new set contains eigenmodes with different projections $(\bar{n}|E)$, it may result in change in the magnitude (and even sign) of the differential optical response, in accord with experimental data.

## 3.2   Experiments on Nonlinear Gyrotropy in a Macroscopic Sample

Let us first proceed to the self-action of a strong wave (elliptically polarized, in the general case) propagating through a medium with susceptibilities determined by (5) and (6). The interaction of a light wave with such a medium results in rotation of the polarization ellipse of the wave, so that the azimuth of the ellipse is given by the following equation [28] (the $z$ axis is directed along the wave vector):

$$\alpha \equiv \frac{1}{2} \arg\left(A_+ A_-^*\right),$$

$$\frac{d\alpha}{dz} = \rho_0' + \rho_1' \left(|A_+|^2 + |A_-|^2\right) + \sigma_2' \left(|A_-|^2 - |A_+|^2\right). \tag{18}$$

Here $A_+$ and $A_-$ are the amplitudes of the right-hand and left-hand circular components of the wave, $A_\pm = (A_x \pm iA_y)/\sqrt{2}$; $\rho_0 = \rho_0' + i\rho_0''$ is the linear gyration constant, $\rho_1 = 2\pi g_1 \omega^2/c^2 = \rho_1' + i\rho_1''$; $\sigma_{1,2} = 2\pi\omega^2 \chi_{1,2}/kc^2 = \sigma_{1,2}' + i\sigma_{1,2}''$; and $k = (\omega/c)(\text{Re}\,\varepsilon_0)^{1/2}$.

The first term on the right-hand side of (18) corresponds to the natural gyrotropy of the medium. We note that this effect has not been found in our experiments. Study of optical activity in scattering locally shows that plasmon modes possess chirality, but on average, represent a racemic set. The measurements of linear angle rotation in a macroscopic sample using a Perkin Elmer polarimeter do not show any effect, at least at wavelengths 436, 546, 578, 589 nm. The other terms describe the nonlinear rotation of the polarization plane. The one containing $\rho_1'$ is due solely to the nonlocal nature of the nonlinear response and takes place even if the wave is linearly polarized ($|A_+|^2 = |A_-|^2$). This effect is usually referred to as nonlinear gyrotropy, or nonlinear optical activity. This is unlike the term with $\sigma_2'$, connected with the local nonlinearity, which vanishes in purely linear polarization, although it gives an additional rotation for the elliptical case [63]. In principle, a method of separating local and nonlocal contributions in a single measurement has been developed [5]. Here we consider another way, namely, we measure $\sigma_2'$ before to be sure that the local contribution is small at an ellipticity of $5 \times 10^{-5}$ in our case.

For the experiments, we used an ethanolic colloidal solution stabilized with PVP (polyvinylpyrrolidone), prepared as described in [64]. Adding a small quantity of the alkali NaOH ($5 \times 10^{-5}$ parts by weight), we were able to change the degree of aggregation significantly. In Fig. 12, the linear absorption spectra are shown for two basic samples–strongly and weakly aggregated solutions.

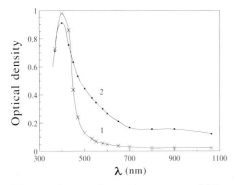

**Fig. 12.** Linear absorption spectra of (1) weakly and (2) strongly aggregated colloid samples

### 3.2.1   Rotation of Ellipse

It follows from (18) and (19) that the technique we just have described is quite appropriate as well for measuring $\chi_2$, provided that the incoming radiation has sufficient ellipticity. For this purpose, we used the same phase element

(Fig. 13), but rotated by 45°. The phase element introduced the phase shift $\approx \pm\pi/4$, which produced either a clockwise or counterclockwise polarization ellipse with a semiaxis ratio $|A_y|^2/|A_x|^2$ of about $1:6$.

According to our experiments, the angles $\alpha_{SR}$ of self-induced rotation of the polarization ellipse were close in value and opposite in sign for left- and right-hand circular polarization of the wave. (We use the convention that left-hand circular polarization corresponds to counterclockwise rotation looking into the beam.)

The value of the rotation in the strongly aggregated Ag solution, as the incident intensity increases from $\approx 0.7\,\mathrm{MW/cm^2}$ to $\approx 5\,\mathrm{MW/cm^2}$, reaches

$$\alpha_{SR} \approx (-1.8 \pm 0.2)\,\mathrm{mrad}, \qquad \text{for a right-hand polarized pump},$$
$$\alpha_{SR} \approx (2.1 \pm 0.2)\,\mathrm{mrad}, \qquad \text{for a left-hand polarized pump}.$$

Taking the average angle for both polarizations of the pump, one can eliminate the influence of nonlinear gyrotropy, so we obtain $\mathrm{Re}\,\chi_2 \approx 1.3 \times 10^{-11}$ esu. Here we also take into account the concentration of the silver particles in this sample; the concentration was 0.7 of that of the aggregated colloid whose spectrum is given in Fig. 12, curve 2.

Let us consider then the experimental observation of *nonlinear gyrotropy* in the colloidal silver solution [5]. As can be seen from (18), for elliptical polarization (which always takes place in real experimental conditions), both local and nonlocal nonlinear responses contribute to the nonlinear rotation of the polarization plane.

The scheme of the polarization measurement is displayed in Fig. 13a. We used the second harmonics of a YAG:Nd pulsed laser. The pulse shape and the transverse distribution in the beam are shown in Fig. 13b.

The radiation passing through the polarizer had an ellipticity of $|A_y|^2/|A_x|^2 \approx 5 \times 10^{-5}$, and using measured value $\chi_2$, one can estimate

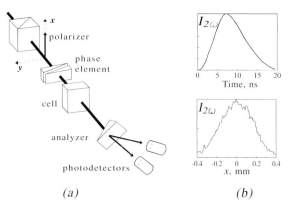

(a)    (b)

**Fig. 13.** Scheme of the polarization measurement. The polarizer is a Glan prism, and the analyzer is a calcite wedge. The cell thickness is 3 mm. In part (**b**) the temporal and spatial properties of the incident radiation are also given

that the local contribution could be omitted. At the exit from the nonlinear medium, the azimuth of the ellipse $\alpha_g(l)$ is given by

$$\alpha_g(l) = \rho_1' \int_0^l |A_x|^2 \, \mathrm{d}z. \tag{19}$$

In Fig. 14, the dependencies of the nonlinear rotational angle on the intensity at the entrance are plotted both for weakly and strongly aggregated colloids. Nonlinear optical activity is strongly dependent on the degree of aggregation of the colloid. For an input intensity $I \lesssim 2\,\mathrm{MW/\,cm^2}$, the measurement data, scaled to the same silver density in both samples, give the values of the nonlinear gyrotropy tensor as

$$\mathrm{Re}\,\Gamma_s^{(3)} \approx 0.9 \times 10^{-16}\,\mathrm{esu}\,,$$
$$\mathrm{Re}\,\Gamma_w^{(3)} \approx 1.1 \times 10^{-18}\,\mathrm{esu}\,, \tag{20}$$

for strongly- and weakly aggregated colloids, correspondingly. In both cases the medium is levorotatory, i.e., causes counterclockwise rotation, looking into the beam.

The strong amplification of the nonlinear gyration constant due to change in the degree of aggregation, as observed in the experiments, is a combination of two factors: enhancement of a local field $G$ and increased average effective size of clusters, $\langle a_{\mathrm{eff}} \rangle / \lambda$, which could be estimated as

$$\frac{\langle a_{\mathrm{eff}} \rangle}{\lambda} \sim \frac{\left\langle \sum_n C_n \sum_{i,j} N_{y,z}^{(i,j)} \sin k_x \, (x_i - x_j) \right\rangle}{\left\langle \sum_n C_n \sum_{i,j} (n_{yi} n_{yj} + n_{iz} n_{jz}) \cos k_x \, (x_i - x_j) \right\rangle} \tag{21}$$

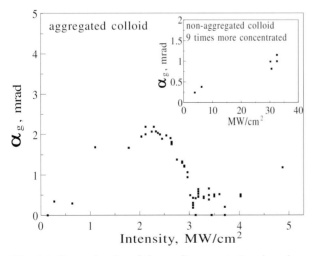

**Fig. 14.** Dependencies of the nonlinear rotational angle on the incident intensity for the strongly aggregated Ag(PVP) colloid (*main chart*) and the weakly aggregated one (*inset*)

and includes an anisotropic factor and a racemic factor due to averaging over the cluster. Here the same notation as in formula (15) is used. Note that local field factor should be the same for nonlinear local response and can be measured separately under the same conditions.

Now we describe the experiments devoted to the determination of the components of $\chi^{(3)}_{ijkl}$ and their changes under aggregation in the same colloid, as $\Gamma^{(3)}$ was measured.

One of the goals of our experiments was an investigation of the dependence of the local nonlinear response versus incoming intensity, both for weakly and strongly aggregated colloids. For this purpose we used a technique of two-beam coupling. The inverse Faraday effect (IFE) was observed.

### 3.2.2   Inverse Faraday Effect

Let there be a strong wave (pump)

$$E_i^{\mathrm{str}} = F_i \exp\left[-\mathrm{i}(\omega t - \boldsymbol{K r})\right] + \text{c.c.} \tag{22}$$

and a weak one (probe)

$$E_i^{\mathrm{weak}} = S_i \exp\left[-\mathrm{i}(\omega t - \boldsymbol{k r})\right] + \text{c.c.}, \tag{23}$$

copropagating within the nonlinear medium. Using (5) and (6), we find the component of the nonlinear polarization that comes along with the probe wave (i.e., has the wave vector $\boldsymbol{k}$). If the angle between $\boldsymbol{K}$ and $\boldsymbol{k}$ is small enough, we can obtain the following expression for an amplitude of this component (for a right-hand circular pump):

$$\begin{aligned} P_+ &= |F_+|^2 [2\chi_1 + 2g_1(k_3 + K_3)]\, S_+ \\ P_- &= |F_+|^2 (\chi_1 + 2\chi_2 + 2g_1 K_3)\, S_-, \end{aligned} \tag{24}$$

that leads to circular birefringence of the probe wave

$$\Delta n_+ - \Delta n_- = \frac{2\pi}{n_0} |F_+|^2 \mathrm{Re}\left(\chi_1 - 2\chi_2 + 2g_1 k_3\right). \tag{25}$$

This circular birefringence generally results in rotation of the probe wave polarization plane. The rotational angle, while the wave is propagating through the medium, is given by

$$\frac{\mathrm{d}\alpha_{\mathrm{IFE}}}{\mathrm{d}z} = \frac{\pi\omega}{cn_0} |F_+|^2 \mathrm{Re}\left(\chi_1 - 2\chi_2 + 2g_1 k_3\right). \tag{26}$$

The experimental setup for the IFE is shown in Fig. 15. The probe wave was linearly polarized in the $x$ direction and the pump wave had right-hand circular polarization. The measured value was the angle $\alpha_{\mathrm{IFE}}$ of rotation of the probe wave polarization plane in the presence of the pump wave.

In Fig. 16, the value of induced circular birefringence (averaged over the pulse duration and the spatial coordinates) is plotted versus the incident

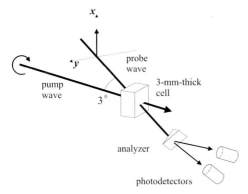

**Fig. 15.** Experimental setup for IFE observation. A noncollinear pump-probe configuration is used, where the probe wave is linearly polarized and the pump wave has a circular polarization. The analyzer (calcite wedge) is turned through 45° relative to the probe wave polarization direction

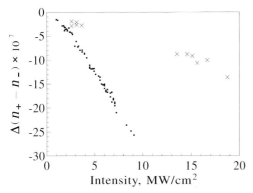

**Fig. 16.** Results of IFE experiments. The *solid points* correspond to strongly aggregated and *cross signs* represent weakly aggregated Ag(PVP) Ag(PVP) colloid. The incident radiation parameters: $\lambda = 532\,\text{nm}$, $\tau \approx 10\,\text{ns}$

intensity, both for weakly and strongly aggregated colloids. The linear dependencies exhibited are in good agreement with those predicted by (25).

If we combine the data for nonlinear gyrotropy, $\chi_2$, and IFE measurements, we obtain the value of $\text{Re}\,\chi_1$:

$$\text{Re}\,\chi_1\,[\text{esu}] = 2\text{Re}\,\chi_2 - 2k_3\text{Re}\,g_1$$
$$+2\sqrt{2}\frac{(\Delta n_+ - \Delta n_-)}{8\pi^2 10^{13} I_{\text{eff}}\left[\text{MW/cm}^2\right]}n_0^2 c\,[\text{esu}]. \tag{27}$$

Here $I_{\text{eff}} = \left(\int_0^l I(z)\,\mathrm{d}z\right)\Big/l$ is the effective value of the intensity inside the cell with the colloid, taking into account both linear and nonlinear absorption. The factor $2\sqrt{2}$ originates from the temporal and spatial averaging that took place in the experiments because we registered the entire beam by slow photodetectors. We assume here a Gaussian distribution of the radiation both in time and space domains (Fig. 13b). For a strongly-aggregated colloid, one can obtain $\mathrm{Re}\,\chi_1 \approx -1.8 \times 10^{-10}$ esu from our measurements, taking $I_{\mathrm{eff}} \approx 0.72 I(0)$.

From measurements of the local constant, $\chi_1$, and the nonlocal constant, $\Gamma^{(3)}$, for both weakly and strongly aggregated colloids, one can conclude that the ratio $\Gamma^{(3)}/\chi_1$ grows for a more aggregated colloid. It means that main enhancement of nonlinear gyrotropy results from an increase of the effective size of aggregates.

# 4    Conclusion

A basis for a theoretical approach in the optics of aggregated metal nanocomposites is the idea of localization of excitation and large fluctuations of local fields in fractal clusters [2,8,15]. Light-induced dipole (multipole) coupling of monomers, that constitute an aggregate, results in collective plasmon modes that cover a wide spectral range.

Experimental investigation of the selective photomodification and optical nonlinearities of colloidal silver solutions have provided essential information regarding the optical properties of aggregated metal nanocomposites. Observation of selective photomodification gave evidence that a broadening of the absorption spectrum of colloidal metal aggregates was inhomogeneous. Collective plasmon modes with different frequencies and polarizations are localized on different collections of monomers. Photomodification occurs when the local intensity at a given monomer group exceeds a certain threshold value. Study of the spectral dependence of selective photomodification showed that the localization of optical excitations became stronger with wavelength. The number of monomers undergoing photomodification in a metal aggregate decreases when a laser pulse wavelength tunes from the monomer absorption peak to the long-wavelength range. In the infrared range, the smallest photomodified domain (hot spot) includes only two monomers, and a fraction of the modified monomers in a cluster is 10%. In the blue range, the number of the photomodified domains and the number of monomers in the domain grow.

The enhancement of optical nonlinearities of colloidal silver aggregates compared to isolated monomers was studied for different processes (four-wave mixing, nonlinear absorption, nonlinear gyrotropy, inverse Faraday effect). The enhancement was associated with excitation of collective plasmon modes in colloidal aggregates and corresponding large fluctuations of local fields. One of the arguments supporting this conclusion is that photoburning of the resonant plasmon modes in the aggregates leads to a significant decrease in nonlinear coefficients for four-wave mixing, absorption, and gyrotropy.

Photomodification and near-field experiments showed an anisotropy of the collective plasmon modes in colloidal aggregates. This conclusion is valid for large fractal aggregates and for small aggregates. Near-field studies [61] revealed chirality of the collective plasmon modes in fractal aggregates. Effects of nonlocal interactions become very essential in large aggregates.

The measured values of the local (IFE) and nonlocal (gyrotropy) nonlinear susceptibilities in macroscopic samples show that the effective size of the domain in an aggregate, responsible for nonlinear optical activity, grows for large clusters. But the effective size of the "chiral" domain in a large cluster, estimated from macroscopic measurements, appears to be smaller than the monomer size. Therefore, one can conclude that the violation of the racemism of silver aggregates in colloidal solution is weak.

## Acknowledgments

This work was supported in part by the Russian Foundation for Basic Research, Grant No 96-02-19331, 99-02-16670, the Program of Governmental Support of the Leading Scientific Schools, Grant No 00-15-96808, and the Civilian Research and Development Foundation, Grant No RE1-2229.

# References

1. R. K. Chang, T. E. Furtak (Eds.), *Surface-Enhanced Raman Scattering* (Plenum Press, New York 1982)

2. A. V. Butenko, V. M. Shalaev, M. I. Stockman, Giant impurity nonlinearities in optics of fractal clusters, Sov. Phys. JETP **67**, 60 (1988) [In Russian: Zh. Eksp. Teor. Fiz., **94**, 107 (1988)]; Z. Phys. D **10**, 81 (1988)

3. S. G. Rautian, V. P. Safonov, P. A. Chubakov, V. M. Shalaev, M. I. Stockman, Surface-enhanced parametric scattering of light by silver clusters, JETP Lett. **47**, 243 (1988) [In Russian: Pis'ma v Zh. Eksp. Teor. Fiz., **47**, 200 (1988)]

4. M. P. Andrews, M. G. Kuzyk, F. Ghbremichael, Local field enhancement of the cubic optical nonlinearity in fractal silver nanosphere/poly(methylmethacrylate) composites, Nonlinear Opt. **6**, 103 (1993)

5. V. P. Drachev, S. V. Perminov, S. G. Rautian, V. P. Safonov, Giant nonlinear optical activity in an aggregated silver nanocomposite, JETP Lett. **68**, 651–656 (1998) [In Russian: Pis'ma v ZhETF, **68**, 618 (1998)]

6. A. V. Karpov, A. K. Popov, S. G. Rautian, V. P. Safonov, V. V. Slabko, V. M. Shalaev, M. I. Stockman, Revealing of wavelength- and polarization-selective photomodification of silver clusters, Sov. JETP Lett. **48**, 571 (1988) [In Russian: Pis'ma v Zh. Eksp. Teor. Fiz., **48**, 528 (1988)]

7. V. P. Safonov, V. M. Shalaev, V. A. Markel, Yu.E. Danilova, N. N. Lepeshkin, W. Kim, S. G. Rautian, R. A. Armstrong, Spectral dependence of selective photomodification in fractal aggregates of colloidal particles, Phys. Rev. Lett. **80**, 1102–1105 (1998)

8. V. M. Shalaev, M. I. Stockman, Optical properties of fractal clusters (susceptibility, giant combination scattering by impurities), Sov. Phys. JETP **65**, 287

(1987) [In Russian: Zh. Eksp. Teor. Fiz., **92**, 509 (1987)]; Z. Phys. D **10**, 71 (1988)

9. U. Kreibig, M. Vollmer, *Optical Properties of Metal Clusters* (Springer, Berlin, Heidelberg 1995)

10. R. H. M. Groeneveld, R. Sprik, A. Lagendijk, Femtosecond spectroscopy of electron-electron and electron-phonon energy relaxation in Ag and Au, Phys. Rev. B **51**, 11433–11445 (1995)

11. M. Perner, P. Post, U. Lemmer, G. von Plessen, J. Feldmann, U. Becker, M. Mennig, M. Schmitt, H. Schmidt: Optically induced damping of the surface plasmon resonance in gold colloids, Phys. Rev. Lett. **78**, 2192–2195 (1997)

12. D. Weitz, M. Oliveria, Fractal structures formed by kinetic aggregation of aqueous gold colloids, Phys. Rev. Lett. **52**, 1433–1436 (1984)

13. B. M. Smirnov, Fractal Clusters, Sov. Adv. Phys. **149**, 177–219 (1986)

14. S. Alexander, R. Orbach, Density of states on fractals: fractons, J. Phys. Lett. (Paris) **43**, L625–L631 (1982)

15. V. A. Markel, L. A. Muratov, M. A. Stockman, T. F. George, Theory and numerical simulation of optical properties of fractal clusters, Phys. Rev. B **43**, 8183 (1991)

16. D. W. Schaefer, J. E. Martin, P. Wiltzius, S. Cannell, Fractal geometry of colloidal aggregates., Phys. Rev. Lett. **52**, 2371 (1984)

17. A. Boukenter, B. Champagnon, E. Duval, Low-frequency Raman scattering from fractal vibrational modes in silica gel., Phys. Rev. Lett. **57**, 2391–2395 (1986)

18. Yu.E. Danilova, V. A. Markel, V. P. Safonov, The Absorption of Light by the Random Silver Clusters [In Russian: Optika atmosfery i okeana **6**, 1436 (1993)]

19. Yu.E. Danilova, V. P. Safonov, Absorption spectra and photomodification of silver fractal clusters, in *Fractal Reviews in the Natural and Applied Sciences*, M. M. Novak (Ed.) (Chapman, London 1995) pp.101–111

20. M. I. Stockman, L. N. Pandey, L. S. Muratov, T. F. George, Optical absorption and localization of eigenmodes in disordered clusters, Phys. Rev. B **51**, 185–195 (1995)

21. V. A. Markel, V. M. Shalaev, E. B. Stechel, W. Kim, R. A. Armstrong, Small-particle composites. I. Linear optical properties, Phys. Rev. B **53**, 2425–2436 (1996)

22. M. I. Stockman, Inhomogeneous eigenmode localization, chaos, and correlations in large disordered clusters, Phys. Rev. E **56**, 6494–6507 (1997)

23. V. M. Shalaev, *Nonlinear Optics of Random Media: Fractal Composites and Metal-Dielectric Films* (Springer, Berlin, Heidelberg 1999)

24. V. M. Shalaev, R. Botet, A. V. Butenko, Localization of collective dipole excitations on fractals, Phys. Rev. B **48**, 6662 (1993)

25. D. P. Tsai, J. Kovacs, Z. Wang, M. Moskovits, V. M. Shalaev, J. S. Suh, R. Botet, Photon scanning tunneling microscopy images of optical excitations of fractal metal colloid clusters, Phys. Rev. Lett. **72**, 4149 (1994)

26. A. Campion, P. Kambhampati, Surface-enhanced Raman scattering, Chem. Soc. Rev. **27**, 241–250 (1998)

27. R. W. Boyd, *Nonlinear Optics* (Academic, San Diego 1992)

28. S. A. Akhmanov, G. A. Lyakhov, V. A. Makarov, V. I. Zharikov, Theory of nonlinear optical activity in isotropic media and liquid crystals, Optica Acta **29**, 1359–1369 (1982)

29. A. I. Plekhanov, G. L. Plotnikov, V. P. Safonov, Production and spectroscopic study of silver fractal clusters by laser evaporation of target, [In Russian: Optika spektroskopiya **71**, 775 (1991)]

30. Yu. E. Danilova, A. I. Plekhanov, V. P. Safonov, Experimental study of polarization-selective holes, burning in absorption spectra of metal fractal clusters., Physica A **185**, 61–65 (1992)

31. Yu. E. Danilova, S. G. Rautian, V. P. Safonov, Interaction of intense light with fractal clusters: absorption, optical phase conjugation, photomodification, Bull. Russian Acad. Sci., Phys. **60**, 56 (1996) [In Russian: Izvestiya RAN ser. fizich., **60**, 56 (1996)]

32. J. A. Creighton, C. G. Blatchford, M. G. Albrecht, Plasma-resonance enhancement of Raman scattering by pyridine adsorbed on silver or gold sol particles of size comparable to the excitation wavelength, J. Chem. Soc. Faraday Trans. II **75**, 790 (1979)

33. V. P. Safonov, Yu. E. Danilova, V. P. Drachev, S. V. Perminov, *Optical nonlinearities in metal colloidal solutions*, in *Optics of Nanostructured Materials*, T. F. George, V. A. Markel (Eds.) (Wiley, New York, 2001) pp. 283–312

34. V. A. Markel, V. M. Shalaev, P. Zhang, W. Huynh, L. Tay, T. L. Haslett, M. Moskovits, Near-field optical spectroscopy of individual surface-plasmon modes in colloid clusters, Phys. Rev. B **59**, 10903–10909 (1999)

35. V. M. Shalaev, Electromagnetic properties of small-particle composites, Phys. Rep. **272**, 61–137 (1996)

36. S. M. F. Griezer, F. Heard, C. G. Barraclough, J. V. Sanders, The characterization of Ag Sols by Electron Microscopy, Optical Absorption and Electrophoresis, J. Colloid. Interfacial Sci. **93**, 545–555 (1983)

37. G. Frens, Th. G. Overbeek, Carey Lea's colloidal silver, Kolloid Z. Polym. **233**, 922 (1969)

38. P. C. Lee, D. Meisel, Adsorption and surface-enhanced Raman scattering of dyes on silver and gold soles, J. Phys. Chem. **86**, 3391 (1982)

39. H. Zhu, R. S. Averback, Sintering process of two nanoparticles:a study by molecular dynamics simulations, Philos. Mag. Lett. **73**, 27–33 (1996)

40. K. Baba, K. Yamaki, M. Miyagi, Metal island films for write-once optical data storage media., In Chemistry and Physics of small-scale Structures, Opt. Soc. Am. Tech. Dig. Ser. **2**, P. 52–54 (1997)

41. F. Claro, The effect of laser irradiation on the formation and destruction of clusters and cluster arrays., Physica A **241**, 223–225 (1997)

42. W. D. Bragg, V. P. Safonov, W. Kim, K. Banerjee, M. R. Young, J. G. Zhu, Z. C. Ying, R. L. Armstrong, V. M. Shalaev, Near-field optical studies of local photomodification in nanostructured materials., J. Microsc. **194**, 574–577 (1999)

43. W. D. Bragg, K. Banerjee, V. A. Podolskiy, V. P. Safonov, J. G. Zhu, V. M. Shalaev, Z. C. Ying, Study of local photomodification of nanomaterials using near-field optics, SPIE Proc. **3791**, 85–92 (1999)

44. D. Ricard, P. Roussignol, C. Flytzanis, Surface-mediated enhancement of optical phase conjugation in metal colloids, Opt. Lett. **10**, 511–513 (1985)

45. F. Hache, D. Ricard, C. Flytzanis, Optical nonlineariries of small metal particles: surface-mediated resonance and quantum size effects, J. Opt. Soc. Am. B **3**, 1647–1655 (1986)

46. F. Hache, D. Ricard, C. Flytzanis, U. Kreibig, The optical Kerr effect in small metal particles and metal colloids — the case of gold, Appl. Phys. A **47**, 347–357 (1988)

47. K. Uchida, S. Kaneko, S. Omi, C. Hate, H. Tanji, Y. Asahara, A. J. Ikushima, T. Tokizaki, A. Nakamura: Optical nonlinearities of a high-concentration of small metal particles dispersed in glass-copper and silver particles, J. Opt. Soc. Am. B **11**, 1236–1243 (1994)

48. S. G. Rautian, Nonlinear saturation spectroscopy of the degenerate electron gas in spherical metallic particles, Sov. JETP **85**, 451–461 (1997)

49. S. M. Nie, S. R. Emory, Probing single molecules and single nanoparticles by surface-enhanced Raman scattering, Science **275**, 1102 (1997)

50. A. V. Butenko, Yu.E. Danilova, S. V. Karpov, A. K. Popov, S. G. Rautian, V. P. Safonov, V. V. Slabko, P. A. Chubakov, V. M. Shalaev, M. I. Stockman, Nonlinear optics of metal fractal clusters, Z. Phys. D **17**, 283–289 (1990)

51. Yu. E. Danilova, N. N. Lepeshkin, S. G. Rautian, V. P. Safonov, Excitation localization and nonlinear optical processes in colloidal silver aggregates, Physica A **241**, 231–235 (1997)

52. M. Sheik-Bahae, A. A. Said, T. H. Wei, D. J. Hagan, E. W. Van Stryland, Sensitive measurement of optical nonlinearities using a single beam, IEEE J. Quant. Electron **26**, 760–769 (1990)

53. R. A. Armstrong, V. P. Safonov, N. N. Lepeshkin, W. Kim, V. M. Shalaev, Giant optical nonlinearities of fractal colloid aggregates, In *Nonlinear Optical Liquids and Power Limiters*, SPIE Proc. **3146** 107–115 (1997)

54. V. M. Shalaev, C. Douketis, T. Haslet, T. Stuckless, M. Moskovits, Two-photon electron emission from smooth and rough metal film in the threshold region, Phys. Rev. B **53**, 11193–11206 (1996)

55. H. B. Liao, R. F. Xiao, J. S. Fu, H. Wang, K. S. Wong, G. K. L. Wong, Origin of third-order optical nonlinearity in $Au:SiO_2$ composite films on femtosecond and picosecond time scales, Opt. Lett. **23**, 388–390 (1998)

56. H. B. Liao, R. F. Xiao, J. S. Fu, P.Yu, G. K. L. Wong, P. Sheng, Large third-order optical nonlinearity in $Au:SiO_2$ composite films near the percolation threshold, Appl. Phys. Lett. **70**, 1–3 (1997)

57. V. M. Shalaev, E. Y. Poliakov, V. A. Markel, Small-particle composites. II. Nonlinear optical properties, Phys. Rev. B **53**, 2437–2449 (1996)

58. S. I. Bozhevolnyi, V. A. Markel, V. Coello, W. Kim, S. Berntsen, V. M. Shalaev: Direct observation of localized dipolar excitations on rough nanostructured surfaces, Phys. Rev. B **58**, 11441 (1998)

59. L. D. Barron, *Molecular Light Scattering and Optical Activity* (Cambridge University Press, Cambridge 1982)

60. V. P. Drachev, W. D. Bragg, V. P. Safonov, V. A. Podolskiy, W. Kim, Z. C. Ying, R. L. Armstrong, V. M. Shalaev, Large Local Optical Activity in Fractal Aggregates of Nanoparticles, J. Opt. Soc. Am. B **18**, issue 12 (2001) (to be published)

61. S. Ducourtieux, S. Gresillon, A. Boccara, J. Rivoal, X. Quelin, P. Gadenne, V. P. Drachev, W. D. Bragg, V. P. Safonov, V. A. Podolskiy, Z. C. Ying, R. A. Armstrong, V. M. Shalaev, Percolation and fractal composites: Optical studies, J. Nonlinear Opt. Phys. Mater. **9**, 105–116 (2000)

62. W. D. Bragg, V. P. Safonov, V. M. Shalaev, W. Kim, Z. C. Ying, K. Banerijee, R. L. Armstrong, V. A. Markel: Near-field observation of selective photomodification in fractal aggregates, Conf. Lasers and Electro-Optics, Baltimore, 1999. Tech. Dig., CWO7 (Optical Society of America, Washington 1999)
63. P. Maker, R. Terhune, C. Savage, Intensity-dependent changes in the refractive index of liquids, Phys. Rev. Lett. **12**, 507–509 (1964)
64. H. Hirai, Formation and catalytic functionality of synthetic polymer-noble metal colloid, J. Macromol. Sci. Chem. A **13**, 633–649 (1979)

# Fractal-Microcavity Composites: Giant Optical Responses

Won-Tae Kim[1], Vladimir P. Safonov[2], Vladimir P. Drachev[3],
Viktor A. Podolskiy[1], Vladimir M. Shalaev[4], and Robert L. Armstrong[1]

[1] Department of Physics, New Mexico State University,
   Las Cruces, NM 88003, USA
[2] Institute of Automation and Electrometry,
   Novosibirsk 630090, Russia
[3] Institute of Semiconductor Physics,
   Novosibirsk 630090, Russia
[4] School of Electrical and Computer Engineering, Purdue University,
   West Lafayette, IN 47907-1285, USA
   shalaev@purdue.edu

**Abstract.** A novel class of materials, fractal-microcavity composites, that provides unprecented enhancement of optical responses is developed. These composites combine the energy-concentrating effects due to localization of optical excitations in fractals with the strong morphology-dependent resonances (MDRs) of dielectric microcavities. Optical excitation of fractal-microcavity composites may result in truly gargantuan local fields. By coupling the localized plasmon modes in fractals with MDRs, strong lasing and nonlinear Raman scattering can be obtained at low light intensities. This allows developing efficient microlasers and makes Raman and hyper-Raman spectroscopy a powerful analytical tool for detecting and characterizing a small number of, or even single molecules.

Advances in science and technology often result from the development of new or improved materials. For example, the arrival of advanced optical media has heralded companion advances in photonics and in laser and detector technology. In this contribution, we demonstrate that nanostructured materials fabricated in our laboratory exhibit dramatically superior optical performance characteristics, foreshadowing great promise for both fundamental and applied studies.

## 1 Introduction

Since the advent of the laser nearly 40 years ago, its unique characteristics have been used to generate a variety of linear and, especially, nonlinear optical emissions such as stimulated scattering and harmonic generation. The initiation of nonlinear emissions typically requires intense pumping sources so that, in practice, emissions are observed only for powerful laser sources whose intensity exceeds some threshold value. The threshold depends on many factors, notably, the quality ($Q$) factor of the optical resonator containing the emitting medium and the strength of the nonlinear interaction, expressed in terms of a nonlinear susceptibility.

V. M. Shalaev (Ed.): Optical Properties of Nanostructured Random Media,
Topics Appl. Phys. **82**, 149–167 (2002)
© Springer-Verlag Berlin Heidelberg 2002

Given the dependence of the threshold pump intensity on medium parameters, it is natural to inquire whether media exist for which initiation of nonlinear emissions occurs for threshold pump intensities dramatically lower than for conventional media. We suggest that a defining characteristic of such media is the existence of greatly enhanced local fields; here, we quantify this concept in terms of an "enhancement factor" given by the magnitude of the ratio of the local to the incident electric field amplitudes. If the enhancement factor is sufficiently large, even extremely weak pump beams will drive local field amplitudes above the threshold required for a particular nonlinear process with the result that emission generated by this process is excited in the medium.

The media of interest here are composites consisting of colloidal silver fractal aggregates and dielectric microcavities. The colloidal silver particles have an average diameter of the order of 10 nm. Using well-established chemical or other fabrication methods, fractal aggregation of the particles may be achieved, and the resulting aggregates, typically comprising $10^2$–$10^3$ silver particles, have dimensions of approximately 1 μm. Composites are formed when aggregates are seeded into a dielectric microcavity; in our experiments, this consisted of a hollow quartz tube of outer diameter 1 mm and inner diameter 0.7 mm. As we shall see, since each component possesses a large enhancement factor and the enhancement factor of the composite depends multiplicatively on the enhancement factors of the individual components, the enhancement factor of the composite can become enormous. Because of this gigantic enhancement, the lasing threshold can decrease significantly, and the efficiency of various optical processes can dramatically increase. In particular, we show here that record-high enhancement for Raman and hyper-Raman scattering can be achieved in the fractal-microcavity composites making possible nonlinear Raman spectroscopy at very low light intensities.

In the remainder of this contribution, we first review the important properties of fractal aggregates and dielectric microcavities and discuss fabrication of the composites. We follow this by a discussion of selected experiments performed in our laboratory that exhibit the remarkable optical characteristics of the composites.

## 2    Optical Properties of the Composites

The giant enhancement of optical responses in metal nanocomposites and thin metallic films containing nanoscale surface features has been intensively studied in recent years [1,2]. This enhancement is associated with the excitation of surface plasmons, collective electromagnetic modes whose characteristics depend strongly on the geometric structure of the metallic component of the medium.

## 2.1   Fractal Silver Aggregates

Nanocomposites often are scale-invariant, and their structure is characterized by fractal geometry [3]. Since fractal objects do not possess translational invariance, they cannot transmit the running waves that are characteristic of homogeneous media. Rather, collective optical excitations in fractals result from near-field multiple scattering within subwavelength regions of the fractal medium; hence, optical excitations, such as surface plasmons, are often spatially localized in fractals [2,4]. The localization is inhomogeneous in the sense that, at any wavelength, modes of different coherence radii are excited [5]. This localization leads to the presence of nanometer-scale spatial regions of high local electrical fields, "hot" spots, and accordingly, to significant enhancement for a variety of optical processes, such as Raman scattering, four-wave mixing, the quadratic electro-optical effect, and nonlinear absorption and refraction [2,6].

The symmetry of fractals is distinctly different from that of continuous media, for example, crystals or gases. Continuous media possess translational invariance so that the medium has the same appearance (at least statistically) at different spatial locations. However, fractal media possess the distinct symmetry of scale invariance so that a fractal has the same appearance (again, at least statistically) when viewed on different spatial scales. These fundamentally different symmetries strongly influence the nature of their resonant excitations. Resonance modes of continuous media are typically delocalized; moreover, medium constituents within a given region absorb light at approximately equal rates and possess approximately equal excitation amplitudes. In contrast, resonance modes of fractal media are spatially localized within subwavelength regions, "hot spots", and the excitation of constituents within these hot spots can greatly exceed those of constituents at other spatial locations [2].

The localization of optical excitations in fractals within small, subwavelength, spatial regions results in extremely large enhancement factors and the corresponding enhancement of many optical effects such as Rayleigh, Raman, and nonlinear light scattering [2]. The enhancement is especially large for nonlinear processes since, in this case, the intensity depends on the power of the enhanced local field. For example, consider degenerate four-wave mixing (DFWM), where both the driving field and the nonlinear amplitude are enhanced. Theory predicts [2] a large enhancement, $G \propto Q^6$, confirmed experimentally [4], where an enhancement factor of the order of $10^6$ occurs upon fractal aggregation of colloidal silver nanoparticles.

## 2.2   Microcavities

Microcavities, such as liquid or solid microspheres and microcylinders, are a second essential component of a composite. Microcavities exhibit a rich spectrum of electromagnetic resonances [7], termed morphology-dependent resonances or MDRs. These resonances, which may have extremely high quality factors ($Q = 10^3$–$10^9$), result from confinement of the radiation within a microcavity by total internal reflection. Light emitted or scattered in a microcavity may couple to the high-$Q$ MDRs lying within its spectral bandwidth, leading to enhancement of both spontaneous and stimulated optical emissions. For example, enhanced fluorescence emission from an organic dye-doped cylindrical or spherical microcavity occurs when either the laser pump or the fluorescence (or both) couple to microcavity MDRs [8]. Moreover, the increased feedback produced by MDRs is sufficient to obtain laser emission from a dye-doped microdroplet under both CW [9] and pulsed [10] laser excitation, with the threshold CW pump intensity three orders of magnitude lower than that of a conventional dye laser in an external cavity. Lasing emission from a microcavity doped with two fluorescent species (a dye-doped sol and a liquid dye) occurs via enhanced radiative (MDR) and nonradiative (Forster) energy transfer [11]. The existence of high-$Q$ microcavity modes is also responsible for numerous stimulated nonlinear effects, including stimulated Raman and Rayleigh-wing scattering and four-wave parametric oscillation under moderate intensity CW excitation [12]. The microcavities used to obtain the results discussed in this contribution consist of hollow quartz tubes (approximately 1 mm o.d., 0.7 mm i.d.) in which we place silver colloids or colloidal silver fractal aggregates for spectroscopic study. These microcavities possess resonance modes having $Q$s of the order of $10^3$–$10^5$.

With increasing pump power, even in the absence of fractal silver or colloidal silver material in the microcavity, resonance enhancement provided by the MDRs may drive the emission into the nonlinear regime. However, with the introduction of such material, SERS- and fractal aggregate-related enhancement effects may be many orders of magnitude greater; hence, numerous nonlinear processes may be observed, even at extremely low concentrations of the emitting species and extremely low pump power.

## 2.3  Composites

To perform optical experiments with fractal-microcavity composites, the initial problem is to be able to fabricate them reliably. This process, in turn, requires fabrication of fractal aggregates, microcavities, and subsequent coupling of the two. Fractal aggregates have been fabricated by other groups [1], and we too developed fabrication capabilities in our laboratory. Normally, the fractal fabrication process consists of two distinct phases: the growth of colloidal particles and the formation of fractal aggregates of these particles. We have used two chemical methods to grow colloidal silver particles, the Lee–Meisel method [13] and Creighton's method [1]. Although both methods result in the formation of colloidal solutions of silver particles with diameters of the order of 10 nm, our experiments show that the Lee–Meisel method results in colloidal solutions possessing a somewhat higher silver concentration; this, in turn, results in more intense optical emissions during spectroscopic investigations. Hence, in the remainder of this contribution, most of our results for chemically produced fractals were obtained using the Lee–Meisel method.

The formation of the colloidal silver solution is the most delicate step in the chemical fractal fabrication process. Careful control of both the constituent concentrations and the ambient temperature is needed to obtain uniform size distribution of silver particles; this, in turn, strongly influences the optical properties of aggregates formed from the particles. Once the colloidal solution is obtained, fractal aggregation may readily be induced by the addition of a suitable organic acid; we used fumaric acid in our experiments. The acid pH is a convenient parameter that may be used to control the aggregation rate; the lower the pH, the faster the aggregation rate. Fractal-microcavity composites are formed by the simple expedient of dipping the open end of the quartz microcavity into the fractal solution where capillary action readily introduces the fractal solution into the microcavity.

Two nonchemical methods were also used to fabricate fractal aggregates: the laser ablation technique and the photoaggregation method. In the laser ablation method, a vacuum chamber containing a silver target is first evacuated and subsequently backfilled to a selected pressure with an inert gas such as helium. Irradiating the target by a high-power laser (entering through a window in the vacuum chamber) causes ablation of silver; in our experiments, we used a Nd:YAG laser to accomplish this. As a function of gas pressure and pump power, first silver monomers, and subsequently silver fractal aggregates, are formed in the ambient gas above the target; this material is subsequently deposited on a substrate inside the chamber. This method provided us with fractal aggregate colloidal silver samples primarily used in experiments designed to increase our understanding of fractal media.

The photoaggregation method, first suggested in [14], was further developed during a series of our experiments discussed later in this contribution. These experiments clearly indicated that irradiation of a colloidal solution by an external pump laser results in the formation of fractal aggregates. Subsequent experiments employing scanning electron microscopy revealed that the aggregation rate as well as the maximum size of the resulting fractal aggregates exhibits a strong dependence on both the wavelength and power of the pump laser.

We conclude this section by noting that there are several reasons for the giant enhancement factors exhibited by fractal-microcavity composites. First, local-field amplitudes are increased in the neighborhood of individual silver monomers via coupling to the localized surface plasmon resonances of fractal aggregates. Second, seeding the aggregates into microcavities further increases the local fields because of light trapping by microcavity resonance modes [7]. Finally, evidence will be presented here that chemical modification of adsorbed molecules may occur resulting in the presence of a sizable chemical enhancement factor [1].

For the spectral emissions reported in this contribution, our results indicate that the enhancement factor may be enormous, exceeding $10^{26}$ in one of our experiments. Such gigantic enhancement factors foreshadow many other results, including the detection of emissions from single molecules and nanoparticles.

## 3   Lasing in Fractal-Microcavity Composites

The initial optical study of fractal-microcavity media was an investigation of lasing [15]. Fractal aggregate optical excitations may be concentrated in regions smaller than the diffraction limit of conventional optics, resulting in large local fields. Seeding the aggregates into microcavities further increases the local fields because of light trapping by microcavity resonance modes. These composites possess unique optical characteristics, including extremely large probabilities of radiative processes. In our experiments, we observed lasing at extremely low threshold pump intensities, well below 1 mW (for some experimental trials, below 50 μW), and dramatically enhanced Raman scattering.

Strong existing evidence suggests that fractal nanocomposites and microcavity resonators individually result in large enhancements of optical emissions. We demonstrate that huge, multiplicative enhancement factors are obtained under the simultaneous, combined action of these two resonant processes when the emitting species is adsorbed onto fractal aggregates contained within high-$Q$ microcavities. We found that doping a dye solution inside a microcavity containing fractal silver aggregates results in a giant enhancement of the efficiency of lasing and nonlinear Raman scattering. Note that, although colloidal Ag aggregates introduce absorption and, hence, linear losses inside the microresonator, at the same time they increase the efficiency of dye excitation and emission. We believe that results discussed in this contribution demonstrate the unique potential of such devices in the development of ultra low threshold microlasers, nonlinear optical devices for photonics, as well as new opportunities for micro analysis, including spectroscopy of single molecules.

Colloidal silver solutions, prepared using the Lee–Meisel method [13], result in the formation of silver nanoparticles (monomers) with an average diameter of 25 nm. Addition of an organic acid (e.g., 0.03 M saturated fumaric acid) to the monomer solution promotes the aggregation of colloidal nanoparticles into fractal clusters, containing, typically, $10^3$ monomers. Electron microscopic analysis of the aggregates (see inset in Fig. 1a) reveals that they possess a fractal structure with the fractal dimension D ≈ 1.8 characteristic of the cluster–cluster aggregation of monomer particles.

Extinction spectra of nonaggregated silver colloids in the visible region of the spectrum exhibit a single resonance feature centered at 420 nm and width 80 nm; this peak is due to the surface plasmon excitation in single silver

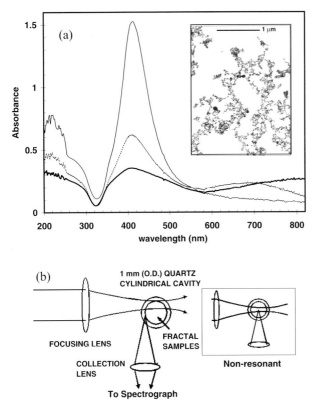

**Fig. 1.** (a) Absorption spectra of silver colloids in different aggregated states (from 1 cm path length cuvettes) and TEM image of fractal aggregates. With an increase of the aggregation degree, the peak at 420 nm decreases and the broad long-wavelength wing becomes stronger. (b) Experimental illumination scheme

nanoparticles. Aggregation leads to the appearance of a broad wing extending toward the long-wavelength part of the spectrum. Enhanced extinction in the long-wavelength region (see Fig. 1a) is a consequence of induced high-$Q$ optical modes in metal fractal aggregates [2].

Lasing experiments were performed using Rhodamine 6G (R6G) dye. A small amount of a parent solution of $10^{-4}$ M R6G in methanol was added to the colloidal silver solution; the resulting dye concentrations ranged from $10^{-8}$ to $10^{-5}$ M. Cylindrical microcavities were fabricated from cylindrical quartz tubes (diameter, 700 μm, outer diameter, 1000 μm), with the dye/colloidal solution placed within the tube.

A 10-mW CW Ar-ion laser ($\lambda = 514.5$ nm) and a 0.75 mW CW green He–Ne laser ($\lambda = 543.5$ nm) were used as pumping sources in these experiments. The pump beam (approximately 2 mm in diameter) was focused into the tube by a 75-mm focal length lens; focal plane beam diameters were 70 μm and 35 μm for Ar- and He–Ne lasers, respectively. Pump beam polarization was vertical (along the axis of the tube), and output radiation was collected at 90° to the incident radiation, as shown in Fig. 1b; different configurations shown in Fig. 1b were used when MDRs were (were not) excited in the cylindrical microcavity. Spectroscopic measurements were performed using a CCD camera mounted to a SpectraPro-300i spectrograph; an 1800/mm grating provided an instrumental spectral resolution of 0.028 nm (full width at half maximum, FWHM).

Elastic scattering of a laser beam passed through the outer edge of the empty cylindrical tube exhibited a well-defined, MDR angular structure; alternatively, when the beam is passed through the inner edge of the empty tube, the intensity of the MDR peaks is significantly reduced. However, filling the tube with a colloidal solution again resulted in strong elastic scattering with a clearly resolved MDR angular structure. Most of our experiments were performed using this illumination geometry. Our observations imply that elastic scattering by fractal aggregates and monomers contributes to output coupling of radiation from microcavity MDRs. Scattering, together with absorption, decreases the quality factor $Q$ of the cavity modes according to $Q^{-1} = Q_a^{-1} + Q_{sv}^{-1} + Q_{ss}^{-1}$, where $Q_a^{-1}$, $Q_{sv}^{-1}$, and $Q_{ss}^{-1}$ are losses due to absorption, volume scattering, and surface scattering, respectively [6]. (Diffraction losses are negligible in our case.) If colloidal aggregates are present in the microcavity, the volume absorption is the most important loss mechanism. The measured absorption coefficient $\alpha = 5\,\mathrm{cm}^{-1}$ at $\lambda = 543.5\,\mathrm{nm}$, so that $Q_a = 2\pi n/\alpha\lambda_L = 3.4 \times 10^4$, where $n = 1.46$ is the refractive index. The measured scattering loss, $Q_{sv}^{-1} + Q_{ss}^{-1}$, is smaller than $Q_a^{-1}$ by, at least, one order of magnitude, implying that scattering may be regarded in our experiments primarily as an output-coupling mechanism for microcavity radiation.

Figure 2 contrasts the luminescent spectrum of a $5 \times 10^{-7}$ M aqueous dye solution in a microcavity with and without the presence of fractal silver aggregates. Without aggregates, a weak, broad luminescent band is observed with a maximum at $\lambda = 560\,\mathrm{nm}$ and a FWHM of 30 nm for $\lambda_L = 514.5\,\mathrm{nm}$ excitation; the lower trace in Fig. 2 shows the central portion of this spectrum, and the inset provides an expanded view. In the inset, representative groupings of small amplitude peaks may be seen corresponding to luminescent emission coupled to microcavity MDRs. The intermode spacing between these peaks is approximately $\Delta\lambda \approx 0.066\,\mathrm{nm}$. This spacing is slightly smaller than the theoretical intermode spacing of $\Delta\lambda = (\lambda^2/2\pi a)[(n^2-1)^{-1/2}\tan^{-1}(n^2-1)^{1/2}] \approx 0.076\,\mathrm{nm}$, calculated for a quartz microcavity of radius $a = 0.5\,\mathrm{mm}$ and refractive index $n \approx 1.46$. The spectral widths of the peaks were limited by our instrumental resolution.

In the presence of fractal aggregates, the luminescent intensity and spectrum are changed dramatically. Figure 2 illustrates the huge increase in MDR peak intensities within a narrow spectral region centered near $\lambda = 561\,\mathrm{nm}$ with a bandwidth of approximately 3 nm for the tube containing the Ag-

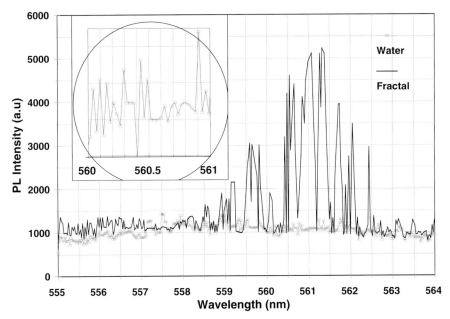

**Fig. 2.** Luminescent spectrum of $5 \times 10^{-7}$ M R6G dye solution in microcavity, with (*heavy line*, neutral density o. d. = 5 filters are used) and without (*circles, thin line*) fractals, for $\lambda_L = 514.5$ nm, CW Ar laser excitation. *Inset* gives detail of spectrum without fractals showing typical mode structure

R6G composite. Closely spaced but spectrally distinct modes in this region have intermode spacings essentially identical to those for the aggregate-free spectra discussed in the preceding paragraph. The measured value of a single peak FWHM, $\delta\lambda = 0.04$ nm (it is close to our instrumental width), allows us to estimate a lower bound for the quality factors, $Q > \lambda/\delta\lambda = 1.5 \times 10^4$. This lower bound value is consistent with the previous estimate of $Q \cong 3.1 \times 10^4$.

Analogous enhanced emission spectra from a dye/fractal/microcavity system are observed under He–Ne laser excitation. Huge MDR peaks are centered near $\lambda = 600$ nm in this case. The narrowing of the emission spectrum (from 30 nm to 3 nm) is characteristic of laser action. To test this point, we studied the emission intensity of different spectral components as a function of the pump intensity. It was found that this dependence is linear for low excitation intensities for all components. However, when the pump intensity exceeds a certain critical value in the range between 20 and 50 W/cm$^2$, some peaks grow dramatically, exhibiting a lasing threshold dependence (Fig. 3). The threshold power for $\lambda_L = 543.5$ nm He–Ne laser excitation is as small as $2 \times 10^{-4}$ W.

The enhanced emission was found confined within an approximately 50-μm region of the tube in a vertical direction, which contained the incident

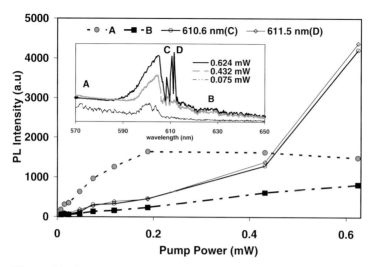

**Fig. 3.** Nonlinear dependence of luminescent peak intensity on pump power. A typical luminescent spectrum near 572 nm is shown as the *dashed line*. Peaks in the presence of fractals are shown as the *solid lines*

pump light; moreover, emission from this region exhibits angular patterns characteristic of microcavity MDRs.

Thus, the spectral, threshold, and spatial dependencies confirm the laser nature of the observed emission. It is noteworthy that the R6G concentration was only $5 \times 10^{-7}$ M in these experiments, three orders of magnitude lower than that for conventional dye lasers with an external cavity and for a microdroplet laser without fractal silver aggregates [12]. In our experiments, the minimum R6G concentration that results in lasing can be as low as $10^{-8}$ M. These findings suggest that the lasing effect is due to dye molecules adsorbed on the surface of silver aggregates. This conclusion is also supported by the fact that increasing the R6G concentration to $10^{-5}$ M does not result in additional growth of the lasing peak intensities compared with the $5 \times 10^{-7}$ M concentration; the additional dye concentration is apparently not adsorbed onto the silver particles, but remains in solution as free molecules, where it does not effectively contribute to the enhanced lasing effect. We conclude that for our composites, the effect of increasing local pump and emitted fields losses is connected with optical absorption by fractals and nonradiative quenching of excitation. Possibly, saturation of fractal absorption facilitates the lasing. The problem of luminescent excitation and amplification (and a possible role of cooperative radiative effects) for molecules adsorbed onto a metal fractal requires further studies.

We note that during the course of the lasing experiments, we also observed pump-induced photoaggregation, reported earlier in [14]. Our experiments carried out to investigate the spectroscopic consequences of the photoaggregation effect had dramatic results. A system consisting of an approximately $10^{-8}$ M R6G dye solution together with unaggregated colloidal silver particles was placed within a microcavity and irradiated by a 543.5-nm, CW, He–Ne pump laser with a power of approximately $50\,\mu$W. For more than 20 min, only very weak luminescent emission was observed from the microcavity. Then, during a very brief period (a minute or two), intense laser emission became visible, coupled to a small number of microcavity MDRs. This emission persisted for a time of the order of 1 h before disappearing. At that time, examination revealed that no fractal material remained in solution within the microcavity; the fractal aggregates presumably grew large enough that they precipitated out. Corroborative experiments using the electron microscope confirmed that, for irradiation with this pump laser, significant aggregation occurred on a timescale of the order of 15–30 min, and sedimentation of the fractals occurred within an hour or so. We interpret these novel findings as direct evidence for the existence of enhancement due to fractal resonance modes. A pump power of $50\,\mu$W corresponds to below-threshold conditions for the dye-colloid-microcavity system; however, if aggregation occurs, thereby increasing the enhancement because of the contribution of fractal resonance modes, the system will be driven above threshold (for the same pump power). Our experiments strongly suggest that photoaggregation has resulted in this increased enhancement. Another important factor in the observed time-dependent effect can be related to the light-induced pulling of fractals into the high-intensity area associated with the whispering gallery modes of the microcavity.

To summarize this section, results reported here promise an advance in the design of micro/nanolasers, operating on a small number of, or even on individual, molecules adsorbed on metal nanostructures within a micro/nanocavity, as well as offering the possibility of combining surface-enhanced radiative processes and high-$Q$ morphology-dependent resonances in microcavities.

## 4    Ultra-Broadband 200–800 nm Light Emission

The Raman effect has long been used in spectroscopic analysis. However, until recently, Raman methods were seldom the method of choice due to the extremely small Raman scattering (RS) cross section, which is of the order $10^{-29}$ cm$^2$. Nonlinear Raman spectroscopy is, in principle, a useful analytical tool, but requires intense laser pumping sources which may introduce unwanted side effects such as photo-induced damage and plasma breakdown. However, even with the use of intense pumping sources, nonlinear Raman processes may still be extremely weak. Consider hyper-Raman scattering (HRS), where the Raman emission is excited by harmonics of the pump laser. The ratio of HRS to ordinary Raman scattering intensities depends on the incident laser pump intensity and is typically $10^{-8} - 10^{-9}$ for pump intensities of the order $10^7$ W/cm$^2$ [16]. A significant increase in the RS intensity results from the mechanism of surface-enhanced RS (SERS), where samples of interest are adsorbed onto nanoscale roughened metal surfaces or nanostructured metal colloids. SERS, which has proven to be a very sensitive spectroscopic method with high molecular specificity, increases the average Raman intensity by a factor of $\sim 10^6$ [1]. Hyper-Raman scattering from adsorbed molecules also benefits from surface enhancement [17,18], which generally includes both enhancements of local electromagnetic fields and the resonant character of nonlinear processes resulting from the charge-transfer band [17]. Under the resonance conditions, hyper-Raman processes could be accompanied by cascade, step-like processes, in which the intermediate and upper energy levels are first populated and then the radiative decay occurs from these states. Since one may not distinguish these processes, we call them nonlinear resonance scattering processes. However, to be observable, it still requires very large intensities, $\sim 10^6$ W/cm$^2$ or larger. Here, we present evidence that far greater enhancement of Raman and may be achieved at low pump intensity in fractal-microcavity composite media. With the aid of these composites, nonlinear Raman spectroscopy can become an efficient analytical tool for detecting and characterizing a small number of, or even single molecules.

We have observed enhanced light emission on the metal surface in the stokes and anti-stokes sides of a spectrum from such media using a CW He–Ne pump laser with a power level as low as 1 mW (the corresponding intensity is as low as 20 W/cm$^2$). To obtain the enhanced light emission spectra, fractal-microcavity composites were irradiated by a Spectra-Physics 632.8 nm laser with a maximum power of approximately 15 mW. The He–Ne laser was focused near the rim of the microcavity in a plane perpendicular to its axis; this geometry insures efficient coupling of the pump light to microcavity MDRs (see Fig. 1b). The He–Ne pump excites Raman and luminescent emissions from Ag-particles/ sodium citrate complex, present in concentrations of $5 \times 10^{-4}$ M as a by-product of fractal preparation. The role of each component of this complex is under study. These emissions, likewise coupled to MDRs, emanate from the microcavity rim where a portion of the light is gathered by a collecting lens and input to an Acton imaging spectrograph fitted with either a 300- or an 1800-groove/mm holographic grating. Spectrally analysed emissions are recorded with the aid of a Princeton Instruments two-dimensional CCD detector.

To avoid possible spectral contamination by laser plasma lines, a laser filter was placed before the sample. A blocking notch filter was inserted after the sample to minimize the amount of stray light at the pump wavelength reaching the CCD. Composite spectra were obtained by superposing eight individual overlapping spectral regions from the 200–800 nm band, using long- or short-pass filters placed after the sample to isolate each region. Individual spectra were combined by matching the relative values of peak intensities in the overlapped portions accounting for the spectral sensitivity of the CCD. This procedure eliminates possible spectral contamination by grating ghosts or overlapping diffraction orders; thus, there is no doubt in the credibility of the measured spectra. However, we should note weak reproducibility of the spectra described here and in Section 3, which could be a result of the short-time stability of the used colloid. Figure 4a shows a low-resolution spectrum obtained using a 300 grooves/mm grating. The spectra obtained are extremely broadband, spanning a range from at least 200 to 800 nm; spectra

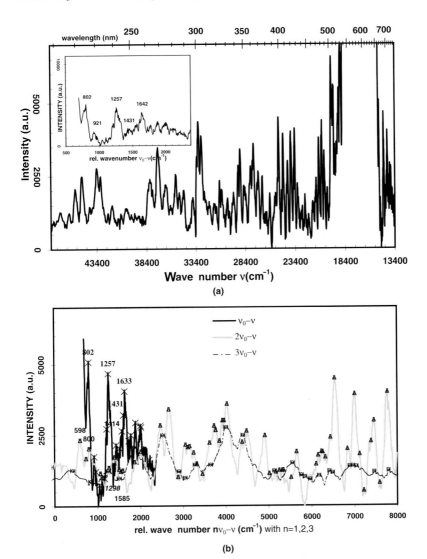

**Fig. 4.** (a) Emission spectrum of the composite and Raman Stokes spectrum of sodium citrate (inset). (b) Stokes sides of the spectra with respect to $\nu_0$, $2\nu_0$, and $3\nu_0$

were not recorded outside of this range because of the degraded response of the spectrometer and the CCD. The spectrum on the anti-Stokes side consists of several groups of well-resolved lines in the $\nu_0$ to $2\nu_0$ region with the spectral resolution from 5 to 20 cm$^{-1}$. In the $2\nu_0$ to $3\nu_0$ region, CCD "pixel" errors are larger, 47 cm$^{-1}$, and spectral structures are poorly resolved or unresolved. The features in this spectrum will be discussed below. The insert displays the Stokes-side spectrum, which contains Raman peaks and a broad emission band.

To estimate the contribution of hyper-Raman scattering from pure molecular transitions to the anti-Stokes spectrum, we should specifically compare the emission spectrum with Raman peak positions, and estimate the hyper-Raman/Raman intensity ratio.

According to previous studies[21,22], SERS spectra from sodium citrate on Carey-Lea citrate-stabilized silver hydrosol show a weak band near 650–660 cm$^{-1}$, no band near 1640–1660 cm$^{-1}$, and a dominant band at 1400 cm$^{-1}$. As described above, we prepared silver sol using the Lee-Meisel method with addition of NaCl. In this case, sodium citrate, a salt of a propanetricarboxylic acid $C_6H_8O_7$, exhibits a modified SERS spectrum (see Fig. 4a insert) characteristic of some carboxylic acids, such as substituted acetic [23] and formic acids [24]. The spectral signature of the modified spectrum includes a reduced intensity of the COO$^-$ vibration near 1410 cm$^{-1}$ (the dominant band in bulk spectra) and increased intensity of Raman bands near 1640 cm$^{-1}$ (C=0 stretch), with the ratio of intensities strongly depending on adsorption details. Previous studies of SERS spectra of different carboxylic acids [21,22,23,24,25,26] exhibit slightly different spectra for the same adsorbate and equivalent SERS spectra for different adsorbates. Explanations for the modified spectrum include geometric orientation effects of the adsorbate with respect to the metal surface [23,26] and chemical transformation during formation of the adsorbate-metal complex [24], so that the spectra depend on sol preparation and age. The presence of Cl$^-$ results in an intense 1636 cm$^{-1}$ band and an unobservable 1400 cm$^{-1}$ band, indicating that COO$^-$ has been desorbed from the silver surface and the molecules are adsorbed mainly via the OH group [25].

Summarizing, the comparison of spectral lines near $2\nu_0$ with Raman peaks near $\nu_0$ (see Fig. 4b) reveal that only a small number can be interpreted as hyper-Raman scattering from adsorbate molecules: specifically, two peaks (at 800 and 933 cm$^{-1}$) have shifts from $2\nu_0$ approximately equal to the SERS shifts 807 and 921 cm$^{-1}$; in addition, a number of other peaks may be regarded as overtones of the Raman fundamentals (e.g., $2 \times 1256$ cm$^{-1}$, $2 \times 1642$ cm$^{-1}$, $3 \times 1256$ cm$^{-1}$). Thus, although we can identify a fraction of peaks as resulting from surface-enhanced Raman and hyper-Raman scattering, many peaks in the broad-band spectrum may not be attributed to either Raman or hyper-Raman scattering from adsorbate molecules and should be treated as nonlinear resonant scattering resulting from multi-photon-pumped luminescence from discrete states of metal particles or adsorbate-molecule/metal-particle complexes [26]. This conclusion is also supported by comparison of the measured ratio of the RS/nonlinear scattering intensities and theoretical estimates for hyper-Raman scattering discussed below.

The measured enhancement factor for Raman scattering, $G^{(\mathrm{RS})}$, is defined as the product of two measured RS intensity ratios. The first ratio comes from comparison of RS in citrate adsorbed on fractal aggregates and in concentrated sodium citrate solution without colloid, $I_f/I_w \sim 10^5$–$10^6$; and the second one comes from comparison of RS from fractal solutions with and without a microcavity, $I_{f-mc}/I_f \sim 10^4$. The resulting factor $G^{(\mathrm{RS})} \sim 10^9$–$10^{10}$, exceeds by up to four orders of magnitude the *average* (macroscopic) RS enhancement on rough metal surfaces and colloidal aggregates.

With the proven fact that the RS enhancement in fractals is concentrated in nanometer-sized hot spots where it exceeds the average RS-enhancement by up to six orders of magnitude [2], we conclude that the *local* RS-enhancement in the hot spots of the fractal-microcavity composites can be as large as $10^{15}$–$10^{16}$ [15]. These enhancement factors exceed the local enhancements for single molecule SERS ($10^{12} \sim 10^{15}$) observed in [19,20]. Therefore, we expect that placing fractal nanostructures in a microcavity will facilitate new possibilities for optical microanalysis and studies of lasing and nonlinear optical effects in single molecules.

Following Reference [17], the ratio of hyper-Raman/Raman intensity may be estimated from the ratio of RS and HRS intensities in bulk solution, which is $I_{\mathrm{HRS}}/I_{\mathrm{RS}} \sim 10^{-14}$–$10^{-15}$ for a pump intensity, $I \sim 50\,\mathrm{W/cm^2}$, used in our experiments. For pure molecular transitions, the cross-sections are assumed to be unchanged; hence, this ratio may increase only as a result of the local field enhancement in fractal-microcavity composites.

In general, the intensity for $n$-photon pumped HRS can be approximated by $I_{n\mathrm{HRS}} = \sigma_{n\mathrm{HRS}} I^n G^{(n)}_{\mathrm{fract}}\, g^{(n)}_{\mathrm{cav}}$, where $\sigma_{n\mathrm{HRS}}$ is the $n$-photon-pumped HRS cross section, $I$ is the intensity of the pump at the fundamental frequency, $G^{(n)}_{\mathrm{fract}}$ is the local-field HRS enhancement factor in fractals, and $g^{(n)}_{\mathrm{cav}}$ is the local-field HRS enhancement of the cavity's MDRs. The case of $n = 1$ corresponds to conventional RS so that $G^{(1)}_{\mathrm{fract}} = G^{(\mathrm{RS})}_{\mathrm{fract}}$; the cases of $n = 2$ and $n = 3$ correspond to two- and three-photon pumped HRS, respectively. MDR and fractal enhancements are decoupled in this approximation.

The fractal-field enhancement $G^{(n)}_{\mathrm{fract}}$ for HRS is defined through the averaging of a ratio (raised to a proper power) of the local and external fields. Following the general approach described in [2], we found an analytical expression for $G^{(n)}_{\mathrm{fract}}$ in terms of $X(\omega)$ and $\delta(\omega)$, which are the known functions of the wavelength and defined via $-[\alpha_0]^{-1} = Z = X + i\delta$, where $\alpha_0$ is the polarizability of individual spherical monomers forming the fractal. The corresponding formulas are as follows. For conventional RS, with a relatively small Stokes shift (so that $X(\omega_s) \approx X(\omega) = X$), $G^{(\mathrm{RS})}_{\mathrm{fract}} \approx |Z/\delta|^4 \delta \mathrm{Im}\,\alpha\,(X)$.

For RS ($n = 1$) with a large shift and for HRS, $G_{\text{fract}}^{(n)} \approx \left[ |Z/\delta|^{2n} \delta \text{Im}\, \alpha\,(X) \right]$ $\times \left[ |Z_s/\delta_s|^2 \delta_s Im\alpha\,(X_s) \right]$, where $X_s = X(\omega_s)$. In these formulas, $\text{Im}\,\alpha\,(X)$ is the average absorption by fractals (Fig. 1a) calculated in [2]. As seen in Fig. 5, the formulas above are in good accord with our simulations based on numerical solution of the coupled-dipole equations in fractals [2]. Both theory and simulations show strong enhancement for RS, especially, for HRS, that increases toward the infrared, where quality factors of fractal plasmon modes are much larger [2].

The microcavity enhancement $g_{\text{cav}}^{(n)}$ depends on whether the fundamental and Stokes waves couple to the MDRs of the cavity. It can range from $g_{\text{cav}}^{(n)} \sim Q_\lambda^n$ (only the fundamental wave is coupled to the MDRs) and $g_{\text{cav}}^{(n)} \sim Q_{\lambda_s}$ (only the Stokes wave is coupled to MDRs) up to $g_{\text{cav}}^{(n)} \sim Q_\lambda^n Q_{\lambda_s}$, when both waves are coupled to MDRs ($Q_\lambda$ and $Q_{\lambda_s}$ are the cavity quality factors at $\lambda$ and $\lambda_s$). However, if none of the waves couples to MDRs, $g_{\text{cav}}^{(n)} \sim 1$. Typical quality factors for our microcavities range from $10^4$ to $10^5$. Thus, even if we assume that the cavity enhancement for the two-photon pumped HRS is close to its maximum, $g_{\text{cav}}^{(2)} \sim Q_\lambda^2 Q_{\lambda_s} \sim 10^{15}$, and multiply it by the fractal enhancement $G_{\text{fract}}^{(2)} \sim 10^6$ at $\lambda = 632$ nm, the resultant factor is still much less than experimentally observed. Since the experimental spectra contain many anti-Stokes emission peaks whose intensities are approximately equal to the Stokes Raman peak intensities, we conclude that they cannot arise from hyper-Raman scattering from pure molecular transitions. We believe that there is an additional source of strong enhancement in our composites, which we assume to be the chemical enhancement. We also note that to

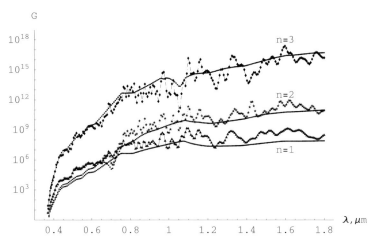

**Fig. 5.** Theoretical and simulated SERS and SEHRS fractal-field enhancement factors, $G_{\text{fract}}^{(n)}$, in silver fractals

describe properly this enhancement one needs to take into consideration the discrete metal-particle states, as was pointed out in[26].

Chemical enhancement results in modification of the cross section $\sigma_{nHRS}$, occurring when adsorbates bind strongly to the SEHRS-active surface so that Raman emission does not arise from the adsorbate alone, but rather from an adsorbate–surface complex (e.g., see [28,29,30]). These new charge-transfer states can contribute resonantly to the Raman and hyper-Raman cross section of the surface complex, greatly increasing its magnitude.

The fractal plasmon modes can play an important role in the chemical enhancement, as they do in electromagnetic (EM) enhancement. For example, energy transfer, which is facilitated by the $Ag_n$ clusters, can occur between fractal plasmon modes and vibronic excitations of molecules adsorbed onto the metal surface. Plasmons can decay into single electron excitations that lead to the resonant enhancement of RS and HRS enhancement [31]. The process can also be mediated by so-called ballistic electrons arising through the plasmon excitations (see [32] and references therein).

We believe that the giant nonlinear emission observed is due to combined contributions from the electromagnetic and chemical enhancements, both strongly benefiting from the fractal plasmon modes.

The dramatic EM enhancement in fractal-microcavity composites and the resonant character of Raman processes resulting from the charge-transfer bands can result in an unusual dependence of the observed emission on pump intensity because of saturation effects. For example, if a two-photon transition is saturated so that the upper level population does not further increase with intensity, then, the intensity dependence for three-photon-pumped HRS is quasi-linear rather than $I^3$ because the hyper-Raman process, in this case, can be roughly thought of as a conventional Raman process from the upper level of the two-photon transition. Similarly, if the one-photon transition is saturated, then two-photon-pumped HRS should have a quasi-linear dependence. Our experimental studies of the intensity dependence (not shown) support this conclusion. The measurements indicate that with an increase in the laser pump power, the intensity dependence (initially nonlinear) quickly becomes quasi-linear for pump power above $2\,mW$. The saturation means that the HRS cross section becomes intensity-dependent. We stress that the lasing effect [15] observed in the same system also provides strong evidence for saturation at low pump power. Finally, we note that at saturation, one may not distinguish between truly multiphoton processes (such as conventional HRS) and cascade, step-like processes, in which the upper energy level is first populated and then a Raman process occurs from that state [33].

Thus, in fractal-microcavity composites, nonlinear Raman spectroscopy becomes possible at low light intensities that makes it a powerful analytical tool for detecting and characterizing molecules.

Because of giant multiplicative enhancement of optical responses in fractal-microcavity composites, we can also expect generation of higher harmonics and cascaded Stokes and anti-Stokes spectral components when the higher vibrational levels are populated. The experiments described above indicate that there is efficient generation of new spectral components covering a very large spectral interval from the near-infrared to the ultraviolet, when the fractal-microcavity composite is irradiated by a low-power He–Ne laser. We speculate that various optical nonlinearities, such as harmonic generation, hyper-Raman and cascaded Stokes and anti-Stokes Raman scattering and their phase-matched counterpart, coherent anti-Stokes Raman (and hyper-Raman) scattering (CARS and CAHRS), can be dramatically enhanced in the composites and generated simultaneously, even at very moderate pump power.

It is interesting to note that the enhancement is expected to increase with the order of the optical nonlinearity because, in this case, the process is proportional to the enhanced local field raised to a greater power; in other words, the larger the nonlinearity, the stronger the enhancement. Under these circumstances, different order optical nonlinearities can coexist and compete with each other, leading to a number of very interesting complex phenomena involving high nonlinearities and feedback, such as optical multistability, optical chaos, and optical self-organization and pattern formation.

### Acknowledgments

This work was supported in part by National Science Foundation (DMR-98101183 and DMR-0071901), Petropleum Research Fund (35028-AC5), Civilian Research and Development Foundation (RE1-2229), Army Research Office (DAAG55-98-1-0425 and DAAD19-01-1-0682), and NASA under Grants NAG 8-1710 and NCC-1-01049.

# References

1. K. Chang, T. E. Furtak (Eds.), *Surface Enhanced Raman Scattering* (Plenum, New York 1982); M. Moskovits, Rev. Mod. Phys. **57**, 783 (1985)
2. V. M. Shalaev, *Nonlinear Optics of Random Media: Fractal Composites and Metal-Dielectric Films* (Springer, Berlin, Heidelberg 2000)
3. B. B. Mandelbrot, *The Fractal Geometry of Nature* (Freeman, San Francisco 1982); A. L. Barbasi, H. E. Stanley, *Fractal Concepts in Surface Growth*, (Cambridge University Press, Cambridge 1995); R. Jullien, R. Botet, *Aggregation and Fractal Aggregates*, (World Scientific, Singapore 1987); F. Family, T. Viscek, *Dynamics of Fractal Surfaces* (World Scientific, Singapore 1990)
4. V. M. Shalaev, M. I. Stockman, Sov. Phys. JETP **65**, 287 (1987); A. V. Butenko, V. M. Shalaev, M. I. Stockman, Sov. Phys. JETP **67**, 60 (1988); S. G. Rautian, V. P. Safonov, P. A. Chubakov, V. M. Shalaev, M. I. Stockman, JETP Lett. **47**, 243 (1988); V. A. Markel, L. S. Muratov, M. I. Stockman, T. F. George, Phys. Rev. B **43**, 8183 (1991)
5. M. I. Stockman, L. N. Pandey, T. F. George, Phys. Rev. B **53**, 2183 (1996); M. I. Stockman, Phys. Rev. E **56**, 6494 (1997); Phys. Rev. Lett. **79**, 4562 (1997)

6. V. M. Shalaev, Phys. Reports **272**, 61 (1996); V. A. Markel, V. M. Shalaev, E. B. Stechel, W. Kim, R. L. Armstrong, Phys. Rev. B **53**, 2425 (1996); V. M. Shalaev, E. Y. Poliakov, V. A. Markel, Phys. Rev. B **53**, 2437 (1996); Yu. E. Danilova, V. P. Drachev, S. V. Perminov, V. P. Safonov, Bull. Russian Acad. Sci. Phys. 60, 342 (1996); Ibid. 374; Yu. E. Danilova, N. N. Lepeshkin, S. G. Rautian, V. P. Safonov, Physica A **241**, 231 (1997)

7. R. K. Chang, A. J. Campillo (Eds.), *Optical Processes in Microcavities* (World Scientific, Singapore 1996)

8. J. F. Owen, P. W. Barber, P. B. Dorain, R. K. Chang, Phys. Rev. Lett. **47**, 1075 (1981)

9. H.-M. Tzeng, K. F. Wall, M. B. Long, R. K. Chang, Opt. Lett. **9**, 499 (1984)

10. A. Biswas, H. Latifi, R. L. Armstrong, R. G. Pinnick, Opt. Lett. **14**, 214 (1988)

11. R. L. Armstrong, J. G. Xie, T. E. Ruekgauer, R. G. Pinnick, Opt. Lett. **17**, 943 (1992)

12. M. B. Lin, A. J. Campillo, Phys. Rev. Lett **73**, 2440 (1994)

13. P. C. Lee, D. Meisel, J. Phys. Chem. **86**, 3391 (1982)

14. S. V. Karpov, A. K. Popov, and V. V. Slabko, JETP Lett. **66**, 106 (1997)

15. W. Kim, V. P. Safonov, V. M. Shalaev, R. L. Armstrong, Phys. Rev. Lett. **82**, 4811–4814 (1999)

16. L. D. Ziegler, J. Raman Spectrosc. **21**,769–779 (1990)

17. J. T. Golab et al., J. Chem. Phys. **81**, 7942 (1988)

18. K. Kneipp et al., Chem. Phys. **247**, 155 (1999)

19. K. Kneipp K. et al., Phys. Rev. Lett. **78**, 1667 (1997)

20. S. Nie, S. R. Emory, Science **275**, 1102 (1997)

21. M. Kerker, O. Siiman, L. A. Bumm, D.-S. Wang, Appl. Opt.,19, 3253–3255 (1980)

22. O. Siiman, L. A. Bumm, R. Callaghan, C. G. Blatchford, M. Kerker, J. Phys. Chem., 87, 1014–1023 (1983)

23. S. Kai, W. Chaozhi, X. Guangzhi, Spectrochimica Acta, 45 A, 1029–1032 (1989)

24. J. L. Castro, J. C. Otero, J. I. Marcos, J. Raman Spectr. 28, 765–769 (1997)

25. Y. Fang, J. Raman Spectr., 30, 85–89 (1999)

26. V. P. Drachev W. Kim, V. P. Safonov, V. A. Podolskiy, N.S. Zakovryashin, E. N. Khaliullin, Vladimir M. Shalaev and R. L. Armstrong, J.of Mod. Opt, 2001 To be published

27. M. Moskovits, J. S. Suh, J. Am. Chem. Soc., 107, 6826–6829 (1985)

28. F. J. Adrian, J. Chem. Phys. **77**, 5302 (1982)

29. A. Campion, P. Kamphampati, Chem. Soc. Rev. **27**, 241 (1998)

30. J. R. Lombardi et al., J. Chem. Phys. **84**, 4174 (1986)

31. J. Lehman et al., J. Chem. Phys. **112**, 5428 (2000)

32. A. M. Michaels et al. J. Am. Chem. Soc. **121**, 9932 (1999)

33. T. Ya. Popova et al., Soviet Phys. JETP **30**, 243–247 (1970)

# Theory of Nonlinear Optical Responses in Metal-Dielectric Composites

Andrey K. Sarychev[1,2] and Vladimir M. Shalaev[3]

[1] Center for Applied Problems of Electrodynamics,
Moscow, 127412, Russia
[2] Department of Physics, New Mexico State University,
Las Cruces, NM 88003, USA
[3] School of Electrical and Computer Engineering, Purdue University,
West Lafayette, IN 47907-1285, USA
shalaev@purdue.edu

**Abstract.** In random metal–dielectric composites near the percolation threshold, surface plasmons are localized in small nanometer-sized areas, hot spots, where the local field can exceed the applied field by several orders of magnitude. The high local fields result in dramatic enhancement of optical responses, especially, nonlinear ones. The local-field distributions and enhanced optical nonlinearities are described using scale renormalization. A theory predicts that the local fields consist of spatially separated clusters of sharp peaks representing localized surface plasmons. Experimental observations are in good accord with theoretical predictions. The localization of plasmons maps the Anderson localization problem described by the random Hamiltonian with both on- and off-diagonal disorder. The feasibility of nonlinear surface-enhanced spectroscopy of single molecules and nanocrystals on percolation films is shown.

## 1 Introduction

Metal–dielectric composites attract much attention because of their unique optical properties, which are significantly different from those of constituents forming the composite [1,2,3]. Semicontinuous metal films can be produced by thermal evaporation or sputtering of metal onto an insulating substrate. In the growing process, first, small metallic grains are formed on the substrate. The typical size $a$ of a metal grain is about 5 to 50 nm. As the film grows, the metal filling factor increases, and coalescence occurs, so that irregularly shaped self-similar clusters (fractals) are formed on the substrate. The concept of scale-invariance (fractality) plays an important role in the description of various properties of percolation systems [2,4]. The sizes of the fractal structures diverge in the vicinity of the percolation threshold, where an "infinite" percolation cluster of metal is eventually formed, representing a continuous conducting path between the ends of a sample. At the percolation threshold, the metal–insulator transition occurs in the system. At higher surface coverage, the film is mostly metallic, with voids of irregular shape. With further coverage increase, the film becomes uniform.

V. M. Shalaev (Ed.): Optical Properties of Nanostructured Random Media,
Topics Appl. Phys. **82**, 169–184 (2002)
© Springer-Verlag Berlin Heidelberg 2002

In random metal–dielectric films, surface plasmon excitations are localized in small nanometer-scale areas referred to as "hot spots" [2,5,6]. As discussed below, the localization can be attributed to the Anderson localization of plasmons in semicontinuous metal films near a percolation threshold (in this case, referred to as percolation films). Electromagnetic energy is accumulated in the hot spots associated with localized plasmons, leading to local fields that can exceed the intensity of the applied field by four to five orders of magnitude. The high local fields in the hot spots also result in dramatically enhanced *nonlinear* optical responses proportional to the local field raised to a power greater than one.

Local electromagnetic field fluctuations and related enhancement of non-linear optical phenomena in metal–dielectric composites near percolation threshold (percolation composites) have recently become an area of active studies because of many fundamental problems involved and high potential for various applications. Percolation systems are very sensitive to the external electrical field since their transport and optical properties are determined by a rather sparse network of conducting channels and the field concentrates in the "weak" points of the channels. Therefore, composite materials can have much larger nonlinear susceptibilities at zero and finite frequencies than those of its constituents. The distinguished feature of percolation composites, to amplify nonlinearities of its components, was recognized very early [7,8,9,10,11,12], and nonlinear conductivities and susceptibilities have been intensively studied during the last decade (see, for example, [1,13,14,15,16,17]).

Here, we consider relatively weak nonlinearities when the conductivity $\sigma(E)$ can be expanded in a power series of the applied electrical field $E$ and the leading term, i.e., the linear conductivity $\sigma^{(1)}$, is much larger than others. This is typical for various nonlinearities in the optical and infrared spectral ranges considered here. Even weak nonlinearities lead to qualitatively new physical effects. For example, generation of higher harmonics can be much enhanced in percolation composites, and bistable behavior of the effective conductivity can occur when the conductivity switches between two stable values, etc. [18]. We note that the "languages" of nonlinear currents/conductivities and nonlinear polarizations/susceptibilities (or dielectric constants) are completely equivalent, and they will be used here interchangeably.

Local-field fluctuations can be strongly enhanced in the optical and infrared spectral ranges for a composite material containing metal particles that are characterized by a dielectric constant with negative real and small imaginary parts. Then, the enhancement is due to the surface plasmon resonance in metallic granules and their clusters [1,14,19,20]. The strong fluctuations of the local electrical field lead to the enhancement of various nonlinear effects. Nonlinear percolation composites are potentially of great practical importance [21] as media with intensity-dependent dielectric functions and, in particular, as nonlinear filters and optical bistable elements. The optical response of nonlinear composites can be tuned by controlling the volume fraction and morphology of the constituents.

The theory of nonlinear optical processes in metal–dielectric composites is based on the fact that the problem of optical excitations in percolation composites mathematically maps the Anderson transition problem. This allowed us to predict the localization of plasmon excitations in percolation metal–dielectric composites and describe in detail the localization pattern. In areas where the resonant plasmons are localized, "hot spots," a high concentration of electromagnetic energy results in very large local fields and dramatic enhancement of optical responses. We show that the plasmon eigenstates are localized on a scale much smaller than the wavelength of the incident light. Plasmon eigenstates with eigenvalues close to zero (resonant modes) are excited most efficiently by an external field. Since the eigenstates are localized and only a small portion of them are excited by the incident beam, the overlapping of the eigenstates can typically be neglected, that significantly simplifies theoretical consideration, and allows one to obtain relatively simple expressions for enhancements of linear and nonlinear optical responses. It is important to stress again that the plasmon localization length is much smaller than the light wavelength; in that sense, the predicted subwavelength localization of plasmons differs quite from the well-known localization of light due to strong scattering in a random homogeneous medium [23].

We also note that a developed scaling theory of optical nonlinearities in percolation composites opens new means to study the classical Anderson problem, taking advantage of the unique characteristics of laser radiation, namely, its coherence and high intensity.

In spite of major efforts, most of the theoretical considerations of local optical fields in percolation composites are restricted to mean-field theories and computer simulations (for references, see [15,16,17]). The effective medium theory [24] that has the virtue of relative mathematical and conceptual simplicity was extended for the nonlinear response of percolating composites [1,13,25,26,27,28,29,30,31] and fractal clusters [28]. For linear problems, predictions of the effective medium theory are usually sensible physically and offer quick insight into problems that are difficult to attack by other means [1]. The effective medium theory, however, has disadvantages typical of all mean-field theories, namely, it diminishes the role of fluctuations in a system. In this approach, it is assumed that local electrical fields are the same in the volume occupied by each component of a composite. For example, the effective medium theory predicts that the local electrical field should be the same in all metal grains regardless of their local arrangement in a metal–dielectric composite. Therefore, the local field is predicted to be almost uniform, in particular, in metal–dielectric composites near percolation. This is, of course, counterintuitive since percolation represents a phase transition, where according to the basic principles, fluctuations play a crucial role and determine a system's physical properties. Moreover, in the optical spectral range, the fluctuations are anticipated to be dramatically enhanced because of the resonance with the sp modes of a composite.

We developed a rather effective numerical method [32] and performed comprehensive simulations of the local-field distribution and various nonlinear effects in two-dimensional percolation composites - that were random metal–dielectric films [15,33,34,35]. The effective medium approach fails to explain the results of the computer simulations performed. It appears that electrical fields in such films consist of strongly localized sharp peaks resulting in very inhomogeneous spatial distributions of the local fields. In peaks ("hot" spots), the local fields exceed the applied field by several orders of magnitudes. These peaks are localized in nanometer-size areas and can be associated with the sp modes of metal clusters in a semicontinuous metal film. The peak distribution is not random but appears to be spatially correlated and organized in some chains. The length of the chains and the average distance between them increase toward the infrared part of the spectrum.

Nonlinear optical effects depend not only on the magnitude of the field but also on its *phase*, so that a nonlinear signal, in general, is proportional to $\langle |E(\mathbf{r})|^k E^m(\mathbf{r}) \rangle$. In this contribution we describe a scaling theory for enhancement of *arbitrary* nonlinear optical process (for both 2d and 3d percolation composites) and show that enhancement differs significantly for nonlinear optical processes that include photon subtraction (annihilation) and for those that do not. Photon subtraction implies that the corresponding field amplitude in the expression for nonlinear polarization (current) $P^{(n)}$ is complex conjugated [22]. For example, the optical process known as coherent anti-Stokes Raman scattering is driven by the nonlinear polarization $P^{(3)} \propto E^2(\omega_1)E^*(\omega_2)$ which results in generation of a wave at the frequency $\omega_g = 2\omega_1 - \omega_2$, i.e., in one elementary act of this process, the $\omega_2$ photon is subtracted (annihilated); the corresponding amplitude $E(\omega_2)$ in the expression for $P^{(3)}$ is complex conjugated.

In this review, we develop a simple scaling approach explaining extremely inhomogeneous field distribution and giant optical nonlinearities of metal–dielectric composites for an arbitrary optical process. We also show the great potential of percolation films for surface-enhanced local spectroscopy of single molecules and nanocrystals.

## 2    Percolation and Anderson Transition Problem

We consider here the general case of a three-dimensional random composite. As mentioned, the typical size $a$ of the metal grains in percolation nanocomposites is of the order of 10 nm, i.e., much smaller than the wavelength $\lambda$ in the visible and infrared spectral ranges, so that we can introduce a potential $\phi(\mathbf{r})$ for the local electrical field. The field distribution problem reduces to the solution of the equation representing the current conservation law:

$$\nabla \cdot \left\{ \sigma(\mathbf{r}) \left[ -\nabla\phi(\mathbf{r}) + \mathbf{E}^{(0)}(\mathbf{r}) \right] \right\} = 0, \tag{1}$$

where $\mathbf{E}^{(0)} \equiv \mathbf{E}_0$ is the applied field and $\sigma(\mathbf{r})$ is the local conductivity that takes $\sigma_m$ and $\sigma_d$ values for the metal and dielectric, respectively. In the discretized form, this relation acquires the form of Kirchhoff's equations defined, for example, on a cubic lattice [1].

We can write Kirchhoff's equations in terms of the local dielectric constant $\epsilon = (4\pi i/\omega)\sigma$, rather than conductivity $\sigma$. We assume that the external electrical field $\mathbf{E}^{(0)}$ is directed, say, along the $z$ axis. Thus, in the discretized form, (1) is equivalent to the following set of equations:

$$\sum_j \epsilon_{ij} \left( \phi_i - \phi_j + E_{ij} \right) = 0, \tag{2}$$

where $\phi_i$ and $\phi_j$ are the electrical potentials determined at the sites of the cubic lattice (or the square lattice, for a two-dimensional system); the summation is over the six nearest neighbors of the site $i$. The electromotive force $E_{ij}$ takes the value $E^{(0)}a_0$ for the bond $\langle ij \rangle$ aligned in the positive $z$ direction, where $a_0$ is the spatial period of the cubic lattice (which can coincide with the size grain $a$). For the bond $\langle ij \rangle$ aligned in the $-z$ direction, the electromotive force $E_{ij}$ takes the value $-E^{(0)}a_0$; for the other four bonds related to site $i$, $E_{kj} = 0$. The permittivities $\epsilon_{ij}$ take values $\epsilon_m$ and $\epsilon_d$, with probabilities $p$ and $1-p$, respectively. Thus a percolation composite is modeled by a random network, including the electromotive forces $E_{ij}$ that represent the external field.

For simplicity, we can assume that the cubic lattice has a very large but finite number of sites $N$ and rewrite (2) in the matrix form with the "interaction matrix" $\widehat{H}$ defined in terms of the local dielectric constants:

$$\widehat{H}\,\Phi = \mathcal{E}, \tag{3}$$

where $\Phi$ is the vector of the local potentials $\Phi = \{\phi_1, \phi_2, \ldots, \phi_N\}$ determined at $N$ sites of the lattice. Vector $\mathcal{E}$ also has $N$ components $\mathcal{E}_i = \sum_j \epsilon_{ij} E_{ij}$, as follows from (2). The $N \times N$ matrix $\widehat{H}$ has off-diagonal elements $H_{ij} = -\epsilon_{ij}$ and diagonal elements $H_{ii} = \sum_j \epsilon_{ij}$ ($j$ refers to the nearest neighbors of the site $i$). The off-diagonal elements $H_{ij}$ randomly take values $\epsilon_d > 0$ and $\epsilon_m = (-1 + i\kappa)\,|\epsilon'_m|$, where the loss factor $\kappa = \epsilon''_m / |\epsilon'_m|$ is small in the optical and infrared spectral ranges, i.e., $\kappa \ll 1$. The diagonal elements $H_{ii}$ are also random numbers distributed between $2d\epsilon_m$ and $2d\epsilon_d$, where $2d$ is the number of nearest neighbors in the lattice for a $d$-dimensional system.

It is important to note that the matrix $\widehat{H}$ is similar to the quantum-mechanical Hamiltonian for Anderson's transition problem with both on- and off-diagonal correlated disorder [36,37]. We will refer hereafter to operator $\widehat{H}$ as Kirchhoff's Hamiltonian (KH). In the approach considered here, the field distribution problem, i.e., the problem of finding a solution to the system of linear equations (2), can be translated into the problem of finding the eigenmodes of KH $\widehat{H}$.

Suppose we have found the eigenvalues $\Lambda_n$ for $\widehat{H}$. In the optical and infrared spectral ranges, the real part $\epsilon'_m$ of the metal dielectric function $\epsilon_m$ is negative ($\epsilon'_m < 0$), whereas the permittivity of a dielectric host is positive ($\epsilon_d > 0$); as mentioned, the loss factor is small, $\kappa = \epsilon''_m / |\epsilon'_m| \ll 1$. Therefore, the manifold of the KH eigenvalues $\Lambda_n$ contains the eigenvalues that have their real parts equal (or close) to zero with very small imaginary parts ($\kappa \ll 1$). Then the eigenstates that correspond to the eigenvalues $|\Lambda_n/\epsilon_m| \ll 1$ are strongly excited by the external field and are seen as giant field fluctuations, representing nonuniform plasmon resonances of a percolation system.

The localized optical excitations can be thought of as field peaks separated, on average, by the distance $\xi_e \propto a\,(N/n)^{1/d}$, where $n$ is the number of the resonant KH eigenmodes excited by the external field and $N$ is the total number of the eigenstates. In the limit $\kappa \ll 1$, only a small part $n \sim \kappa N$ of the eigenstates is effectively excited by the external field. Therefore, the distance $\xi_e$, which we call the field correlation length, is large: $\xi_e/a \propto \kappa^{1/d} \gg 1$.

According to the one-parameter scaling theory, the eigenstates $\Psi_n$ are, it is thought, all localized for the two-dimensional case (see, however, the discussion in [38,39]). On the other hand, it was shown that there is a transition from chaotic eigenstates [40,41] to the strongly localized eigenstates in the two-dimensional Anderson problem [42] with an intermediate crossover region. The KH also has strong off-diagonal disorder, $\langle H'_{ij} \rangle = 0$ ($i \neq j$), which usually favors localization [43,44]. Our conjecture is that the eigenstates $\Psi_n$ are localized, at least those with $\Lambda_n \approx 0$, in a two-dimensional system. (We cannot, however, rule out the possibility of inhomogeneous localization similar to that obtained for fractals [45] or power-law localization [36,46].)

The Anderson transition in a three-dimensional system is less understood and little is known about the eigenfunctions [36,47]. We conjecture that the eigenstates with $\Lambda_n \approx 0$ are also localized in the three-dimensional case.

# 3    Scaling in Local-Field Distribution

For the films concerned, gaps between metal grains are filled by a dielectric substrate, so that a semicontinuous metal film can be thought of as a $2d$ array of metal and dielectric grains randomly distributed over the plane. The dielectric constant of a metal can be approximated by the Drude formula

$$\epsilon_m = \epsilon_b - (\omega_p/\omega)^2/(1 + i\omega_\tau/\omega)\,, \tag{4}$$

where $\epsilon_b$ is the interband contribution, $\omega_p$ is the plasma frequency, and $\omega_\tau$ is the plasmon relaxation rate ($\omega_\tau \ll \omega_p$). In the high-frequency range considered here, losses in metal grains are relatively small, $\omega_\tau \ll \omega$. Therefore, the real part $\epsilon'_m$ of the metal dielectric function $\epsilon_m$ is much larger (in modulus) than the imaginary part $\epsilon''_m$, i.e., the loss parameter $\kappa$ is small,

$\kappa = \epsilon''_m / |\epsilon'_m| \cong \omega_\tau/\omega \ll 1$. We note that $\epsilon'_m$ is negative for frequencies $\omega$ less than the renormalized plasma frequency,

$$\tilde{\omega}_p = \omega_p/\sqrt{\epsilon_b}. \tag{5}$$

It is instructive to consider first the special case of $-\epsilon'_m = \epsilon_d$, where $\epsilon_m \equiv \epsilon'_m + i\epsilon''_m$ and $\epsilon_d$ are the dielectric constants of the metallic and dielectric components, respectively. The condition $-\epsilon'_m = \epsilon_d$ corresponds to the resonance of individual metal particles in a dielectric host in the two-dimensional case. For simplicity, we also set $-\epsilon_m = \epsilon_d = 1$, which can always be done by simply renormalizing the corresponding quantities.

It can be shown that the field distribution on a percolation film at $-\epsilon_m = \epsilon_d = 1$ formally maps the Anderson metal–insulator transition problem [2,5,6]. In accord with this, the field potential representing the plasmon modes of a percolation film must be characterized by the same spatial distribution as the electron wave function in the Anderson transition problem. Such mathematical equivalence of the two physically different problems stems from the fact that the current conservation law for a percolation film acquires (when written in the discretized form) the form of Kirchhoff's equations, which, in turn, (when written in the matrix form) become identical to the equations describing the Anderson transition problem [6]. The corresponding Kirchhoff Hamiltonian for the field distribution problem is given by a matrix with random elements which can be expressed in terms of the dielectric constants for metal and dielectric bonds of the lattice representing the film. In this matrix, the values $\epsilon_m = -1$ and $\epsilon_d = 1$ appear in the matrix elements with probability $p$ and $(1 - p)$, respectively (where $p$ is the metal filling factor given at percolation by $p = p_c$, with $p_c = 1/2$ for a self-dual system). In such a form, the Kirchhoff Hamiltonian is characterized by a random matrix, similar to that in the Anderson transition problem, with both on- and off-diagonal disorder. Based on this mathematical equivalence, it was concluded in [2,5,6] that the plasmons in a percolation film can experience Anderson-type localization within small areas, and the size is given by the Anderson length $\xi_A$. For most localized plasmon modes, $\xi_A$ can be as small as one grain $a$.

Below, we develop a simple scaling approach that explains the nontrivial field distribution predicted and observed in percolation films. This scale-renormalization method supports the main conclusions of a rigorous (but tedious) theory of [2,5,6] and has the virtue of being simple and clear, which is important for understanding and interpreting future experiments.

First, we estimate the field in the hot spots for $-\epsilon'_m = \epsilon_d$. Hereafter, we use the sign * (not to be confused with complex conjugation) to indicate that the quantity concerned is given for $-\epsilon'_m = \epsilon_d$ (with $\epsilon_d \sim 1$); for $\xi_A$, however, we omit this sign since this quantity always refers to $-\epsilon'_m = \epsilon_d$.

Since $\epsilon'_m$ is negative at optical frequencies, metal particles can be roughly thought of as inductor-resistor ($L$–$R$) elements, whereas the dielectric gaps between the particles can be treated as capacitive ($C$) elements. Then, the condition $\epsilon'_m = -\epsilon_d$ means that the conductivities of the $L$–$R$ and $C$ elements are equal in magnitude and opposite in sign, i.e., there is a resonance in the equivalent $L$–$R$–$C$ circuit corresponding to individual particles.

The local field in resonating particles is enhanced by the resonance quality factor $Q$ which is the inverse of the loss factor, $Q = \kappa^{-1}$, so that

$$E^*_m \sim E_0 \kappa^{-1} (a/\xi_A)^d , \tag{6}$$

where the factor $(a/\xi_A)^2$ takes into account that the resonating mode is localized within $\xi_A$. The resonant modes excited by a monochromatic light represent only the fraction $\kappa$ of all modes so that the average distance (referred to as the field correlation length $\xi^*_e$) between the field peaks is given by

$$\xi^*_e \sim a/\kappa^{1/d} \gg \xi_A . \tag{7}$$

Note that the field peaks associated with the resonance plasmon modes represent in fact the normal modes, with the near-zero eigennumbers, of Kirchhoff's Hamiltonian discussed above [2,6]. These modes are strongly excited by the applied field and seen as giant field fluctuations on the surface of the film.

Now we turn to the important case of "high contrast," with $|\epsilon_m| \gg \epsilon_d$, that corresponds to the long-wavelength part of the spectrum where the local-field enhancement can be especially strong. From the basic principles of Anderson localization [2], it is clear that a higher contrast favors localization, so that plasmon modes are expected to be localized in this case as well.

It is clear that at $|\epsilon_m| \gg \epsilon_d$, individual metal particles cannot resonate. We can renormalize, however, the high-contrast system to the case of $-\epsilon'_m = \epsilon_d$ considered above by formally "dividing" the film into square elements of the special resonant size

$$l_r = a\left(|\epsilon_m|/\epsilon_d\right)^{\nu/(t+s)} \tag{8}$$

and considering these squares as new renormalized elements of the film. Really, using the known scaling dependences [1,4] for "metal" and "dielectric" squares of size $l$ (which, respectively, do or do not contain a metal continuous path through the square):

$$\epsilon_m(l) \sim (l/a)^{-t/\nu} \epsilon_m \tag{9}$$

and

$$\epsilon_d(l) \sim (l/a)^{s/\nu} \epsilon_d , \tag{10}$$

we find that the dielectric constants of the renormalized elements with the size $l = l_r$ are equal in magnitude and opposite in sign,

$$- \epsilon_m(l_r) = \epsilon_d(l_r) . \tag{11}$$

Thus, for these renormalized elements of size $l_r$, there is a resonance similar to the resonance in the $R$–$L$–$C$ circuit describing individual metal particles in a dielectric host. In this case, however, some effective (renormalized) $R$–$L$–$C$ circuits represent resonating square elements.

For a two-dimensional percolation film, the critical exponents are given by $t \approx s \approx \nu \approx 4/3$; they represent the percolation critical exponents for conductivity, dielectric constant, and percolation correlation length, respectively [1,4]. Below, for simplicity we consider the two-dimensional case (d = 2), though all results can be easily generalized for arbitrary d.

In the renormalized system, the estimate obtained above for field peaks still holds. Since the electrical field and eigenfunction both scale as $l_r$, we arrive at the conclusion that in the high-contrast system (with $|\epsilon_m| \gg \epsilon_d$), the field maxima can be estimated as

$$E_m \sim (l_r/a)E_m^* \sim E_0 \kappa^{-1}(l_r/a)(a/\xi_A)^2 \sim E_0 \kappa^{-1}(l_r/\xi_A)^2$$
$$\sim E_0(a/\xi_A)^2|\epsilon_m|^{3/2} / \left(\epsilon_d^{1/2}\epsilon_m''\right). \tag{12}$$

The light-induced eigenmodes in the high-contrast system are separated, on average, by the distance $\xi_e$ that exceeds the mode separation $\xi_e^*$ at $\epsilon_m = -\epsilon_d$ by factor $l_r/a$,

$$\xi_e \sim (l_r/a)\xi_e^* \sim l_r/\sqrt{\kappa} \sim a|\epsilon_m|/\sqrt{\epsilon_m''\epsilon_d}. \tag{13}$$

For a Drude metal at $\omega \ll \omega_p$, the local field peaks, according to (4) and (12), are given by

$$E_m/E_0 \sim \epsilon_d^{-1/2}(a/\xi_A)^2(\omega_p/\omega_\tau), \tag{14}$$

and the distance between the excited modes (13) is estimated as

$$\xi_e \sim a\omega_p/\sqrt{\epsilon_d\omega\omega_\tau}. \tag{15}$$

Figure 1 illustrates, as described above, the renormalization of the field peaks and their spatial separations at the transition between the reference (renormalized) system with $-\epsilon_m = \epsilon_d = 1$ and the high-contrast system of $|\epsilon_m/\epsilon_d| \gg 1$.

As follows from the figure, the largest local fields of amplitude $E_m$ result from excitation of the resonant clusters of size $l_r$. At $-\epsilon_m = \epsilon_d = 1$, $l_r = a$ (8), as in the reference system. With increasing wavelength (and thus the contrast $|\epsilon_m|/\epsilon_d$), the resonant size $l_r$ and the distance $\xi_e$ between the resonating modes both increase.

The above results have a clear physical interpretation and can also be obtained from the following complementary considerations. Let us consider two metal clusters, with conductance $\Sigma_m = -i(a/4\pi)\omega\epsilon_m(l)$, separated by a dielectric gap, with conductance $\Sigma_d = -i(a/4\pi)\omega\epsilon_d(l)$, as shown in Fig. 2a. The clusters and the gap are both of size $l$, and $\epsilon_m(l)$ and $\epsilon_d(l)$ are defined in (9) and (10), respectively. The equivalent conductance $\Sigma_e$ for $\Sigma_m$ and $\Sigma_d$

$\varepsilon_d = -\varepsilon_m = 1$

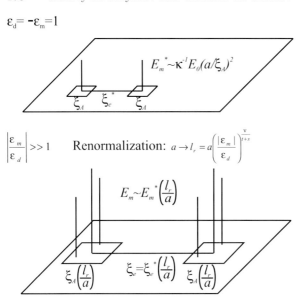

**Fig. 1.** Renormalization of the field distribution at the transition between the reference case with $-\epsilon_m/\epsilon_d = 1$ and the high-contrast case of $-\epsilon_m/\epsilon_d \gg 1$

in series is given by $\Sigma_e = \Sigma_m \Sigma_d / (\Sigma_m + \Sigma_d)$, and the current $j$ through the system is $j = \Sigma_e E_0 l$. The local field, however, is strongly inhomogeneous, and the largest field occurs at the point of the closest approach between the clusters, where the separation between clusters can be as small as $a$; then, the maximum field $E_m$ is estimated as $E_m = (j/\Sigma_d)/a \sim E_0(l/a)/\left[1 + (l/a)^{(t+s)/\nu}\epsilon_d/\epsilon_m\right]$ [where we used (9) and (10)]. For the "resonant" size $l = l_r$, the real part of the denominator in the expression for $E_m$ becomes zero, and the field $E_m$ reaches its maximum, where it is estimated as $E_m/E_0 \sim \kappa^{-1}(l_r/a)$.

In the estimate obtained, we assumed, for simplicity, that $\xi_A \sim a$ and, in this limit, we reproduced the result (12). To obtain the "extra factor" $(a/\xi_A)^2$ of (12), we have to take into account that the localization area for the field is $\xi_A$ rather than $a$, so that the field peak is "spread over" for the distance $\xi_A$. With this correction, we immediately arrive at formula (12).

It is clear that for any frequency of the applied field $\omega$, there are always resonant clusters of the size (8)

$$l = l_r(\omega) \sim a(\omega/\tilde{\omega}_p)^{2\nu/(t+s)}, \tag{16}$$

where the local field reaches its maximum $E_m$. The resonant size $l_r$ increases with the wavelength. It is important that at percolation, the system is scale-invariant so that all possible sizes needed for resonant excitation are present, as schematically illustrated in Fig. 1b. At some large wavelength, only large clusters of appropriate sizes resonate, leading to field peaks at the points of closest approach between the metal clusters; with the decrease of the wavelength of the applied field, the smaller clusters begin to resonate, whereas the larger ones (as well as the smaller ones) are off the resonance, as shown in Fig. 1b.

We can also estimate the number $n(l_r)$ of field peaks within one resonating square of size $l_r$. In the high-contrast system (with $|\epsilon_m/\epsilon_d| \gg 1$), each field maximum of the renormalized system (with $|\epsilon_m/\epsilon_d| = 1$) splits into $n(l_r)$ peaks of $E_m$ amplitude located along a dielectric gap in the "dielectric" square of size $l_r$ (Figs. 1 and 2). The gap "area" scales as the capacitance of the dielectric gap, and so must the number of field peaks in the resonance square. Therefore, we estimate that

$$n(l_r) \propto (l_r/a)^{s/\nu_p}. \tag{17}$$

In accordance with the above considerations, the average (over the film surface) intensity of the local field is enhanced as

$$\left\langle \left| \frac{E}{E_0} \right|^2 \right\rangle \sim (E_m/E_0)^2 n(l_r)(\xi_A/\xi_e)^2 \sim (a/\xi_A)^2 |\epsilon_m|^{3/2} \Big/ (\epsilon_m'' \epsilon_d), \tag{18}$$

where we used (8), (13), and (17) and the critical exponents $t = s = \nu = 4/3$.

In Fig. 3, we also show the simulated field distribution on a silver–glass percolation film at two different wavelengths. In accordance with the consideration above, we see that the local-field distribution consists of clusters of very sharp peaks where the spatial separation increases with the wavelength. A qualitatively similar field distribution was detected in recent experiments [5] using scanning near-field optical microscopy.

Thus, using simple arguments based on the scaling dependences of $\epsilon_m(l)$ and $\epsilon_d(l)$ on $l$ and the resonance condition $-\epsilon_m(l_r) = \epsilon_d(l_r)$, one can define the renormalization procedure that allows one to rescale the "high-contrast" system to the renormalized one with $-\epsilon_m = \epsilon_d = 1$.

Below we show that the enhanced local field in the hot spots results in giant enhancement of *nonlinear* optical responses of semicontinuous films.

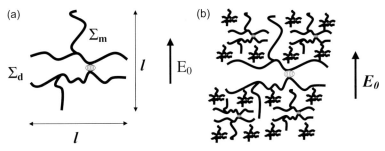

**Fig. 2. (a)** A typical element of a percolation film consisting of two conducting metal clusters with a dielectric gap in between. **(b)** Different resonating elements of a percolation film at different wavelengths

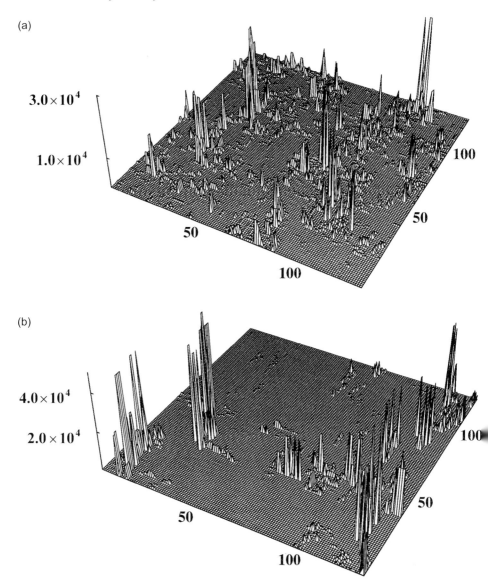

**Fig. 3.** Local-field distribution on a silver-glass percolation film at different wavelengths: (**a**) $\lambda = 1.5\,\mu m$ and (**b**) $\lambda = 10\,\mu m$

## 4   Enhanced Optical Nonlinearities

In general, we can define the high-order field moments as

$$M_{n,m} = \frac{1}{SE_0^m |E_0|^n} \int |E(\mathbf{r})|^n E^m(\mathbf{r}) \, d\mathbf{r}, \tag{19}$$

where, as above, $E_0$ is the amplitude of the external field and $E(\mathbf{r})$ is the local field [note that $E^2(\mathbf{r}) \equiv \mathbf{E}(\mathbf{r}) \cdot \mathbf{E}(\mathbf{r})$]. The integration is over the entire surface $S$ of the film.

The high-order field moment $M_{2k,m} \propto E^{k+m} E^{*k}$ represents a nonlinear optical process in which, in one elementary act, $k+m$ photons are added and $k$ photons are subtracted [22] because the complex conjugated field in the general expression for nonlinear polarization implies photon subtraction, so that the corresponding frequency enters the nonlinear susceptibility with a minus sign [22]. As first shown in [6], enhancement is significantly different for nonlinear processes with photon subtraction in comparison with those where all photons enter the nonlinear susceptibility with a plus sign. The enhancement of Kerr optical nonlinearity $G_K$ (see below) is equal to $M_{2,2}$, second-harmonic generation (SHG) and third-harmonic generation (THG) enhancements are given by $|M_{0,2}|^2$ and $|M_{0,3}|^2$, respectively, and surface-enhanced Raman scattering (SERS) is represented by $M_{4,0}$.

The high-order moments of the local field in $d = 2$ percolation films can be estimated as $M_{n,m} \sim (E_m/E_0)^{n+m} n(l_r)(\xi_A/\xi_e)^2$. Using the scaling formulas (8)–(17) for the field distribution, we obtain the following estimate for the field moments:

$$M_{n,m} \sim \left(\frac{E_m}{E_0}\right)^{n+m} \frac{(l_r/a)^{s/\nu}}{(\xi_e/\xi_A)^2} \sim \left(\frac{|\epsilon_m|^{3/2}}{(\xi_A/a)^2 \epsilon_d^{1/2} \epsilon_m''}\right)^{n+m-1}, \tag{20}$$

for $n+m > 1$ and $n > 0$ (where we took into account that for two-dimensional percolation composites, the critical exponents are given by $t \cong s \cong \nu \cong 4/3$).

Since $|\epsilon_m| \gg \epsilon_d$ and the ratio $|\epsilon_m|/\epsilon_m'' \gg 1$, the moments of the local field are very large, i.e., $M_{n,m} \gg 1$, in the visible and infrared spectral ranges. Note that the first moment, $M_{0,1} \simeq 1$, corresponds to the equation $\langle \mathbf{E}(\mathbf{r}) \rangle = \mathbf{E}_0$.

Now consider the moments $M_{n,m}$ for $n = 0$, i.e., $M_{0,m} = \langle E^m(\mathbf{r}) \rangle / (E_0)^m$. In the renormalized system where $\epsilon_m(l_r)/\epsilon_d(l_r) \cong -1 + i\kappa$, the field distribution coincides with the field distribution in the system with $\epsilon_d \simeq -\epsilon_m' \sim 1$. In that system, field peaks $E_m^*$, differ in phase and cancel each other out, resulting in the moment $M_{0,m}^* \sim O(1)$ [6]. In transition to the original system, the peaks increase by the factor $l_r$, leading to an increase in the moment $M_{0,m}$. Then, using (8), (13), and (17), we obtain the following equation for the moment:

$$M_{0,m} \sim M_{0,m}^* (l_r/a)^m \left[\frac{n(l_r)}{(\xi_e/a)^2}\right] \sim \kappa(l_r/a)^{m-2+s/\nu} \sim \frac{\epsilon_m'' |\epsilon_m|^{(m-3)/2}}{\epsilon_d^{(m-1)/2}}, \tag{21}$$

for $m > 1$ (where again we used the critical exponents $t \cong s \cong \nu \cong 4/3$).

For a Drude metal (4) and $\omega \ll \omega_{\mathrm{p}}$, from (20) and (21), we obtain

$$M_{n,m} \sim \epsilon_{\mathrm{d}}^{(1-n-m)/2} (a/\xi_{\mathrm{A}})^{2(n+m-1)} (\omega_{\mathrm{p}}/\omega_\tau)^{n+m-1} , \qquad (22)$$

for $n + m > 1$ and $n > 0$, and

$$M_{0,m} \sim \epsilon_{\mathrm{d}}^{(1-m)/2} \left( \frac{\omega_{\mathrm{p}}^{m-1} \omega_\tau}{\omega^m} \right) , \qquad (23)$$

for $m > 1$.

Note that for all moments, the maximum in (22) and (23) is approximately the same (if $\xi_{\mathrm{A}} \sim a$), so that

$$M_{n,m}^{(\mathrm{max})} \sim \epsilon_{\mathrm{d}}^{(1-n-m)/2} \left( \frac{\omega_{\mathrm{p}}}{\omega_\tau} \right)^{n+m-1} . \qquad (24)$$

However, in the spectral range $\omega_{\mathrm{p}} \gg \omega \gg \omega_\tau$, moments $M_{0,m}$ gradually increase with wavelength, and the maximum is reached only at $\omega \sim \omega_\tau$, whereas the moments $M_{n,m}$ (with $n > 1$) reach this maximum at much shorter wavelengths (roughly, at $\omega \approx \tilde{\omega}_{\mathrm{p}}/2$) and remain almost constant in the indicated spectral interval. This conclusion is supported by the numerical simulations for the silver–glass percolation films shown in Fig. 4; one can see that the above scaling formulas are in good accord with the simulations.

For silver–glass percolation films, with $\omega_{\mathrm{p}} = 9.1\,\mathrm{eV}$ and $\omega_\tau = 0.021\,\mathrm{eV}$, we find that the average field enhancement can be as large as $G_{\mathrm{RS}} \sim M_{4,0} \sim 10^7$ for Raman scattering (see also Fig. 4), and as large as $G_{\mathrm{FWM}} \sim |M_{2,2}|^2 \sim 10^{14}$ for degenerate four-wave mixing. According to Fig. 3, the local-field intensity in the hot spots can approach the magnitude $10^5$ so that the local enhancement of nonlinear optical responses can be truly gigantic, up to $10^{10}$, for Raman scattering, and up to $10^{20}$, for four-wave mixing signals. With this level

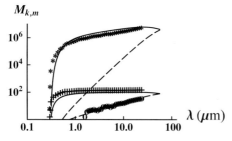

**Fig. 4.** Average enhancement of the high-order field moments $M_{n,m}$ in a percolation silver-glass two-dimensional film as a function of wavelength: $M_{4,0}$ [scaling formula (20) – *upper solid line* and numerical simulations – *]; $M_{0,4}$ [scaling formula (21) – *upper dashed line*]; $M_{2,0}$ [scaling formula (20) – *lower solid line* and numerical simulations – +]; $M_{0,2}$ [scaling formula (21) – *lower dashed line* and numerical simulations – 0]

of enhancement, one can perform *nonlinear* spectroscopy of single molecules and nanocrystals. It is important that the enhancement be obtained in the huge spectral range, from the near-UV to the far-infrared, which is a major virtue for spectroscopic studies of different molecules and nanocrystals. We also note that the field enhancement provided by semicontinuous metal films can be used for various photobiological and photochemical processes.

## Acknowledgments

This work was supported in part by National Science Foundation under Grant DMR-0121 814, Petroleum Research Fund, Army Research Office under Grant DAAD19-01-1-0682, and NASA under Grants NAG 8-1710 and NCC-1-01049.

# References

1. D. J. Bergman, D. Stroud, Solid State Phys. **46**, 147 (1992)
2. V. M. Shalaev, *Nonlinear Optics Of Random Media: Fractal Composites and Metal-Dielectric Films* (Springer, Berlin Heidelberg 2000)
3. A. K. Sarychev, V. M. Shalaev, Phys. Rep. **335**, 275 (2000)
4. D. Stauffer, A. Aharony, *Introduction to Percolation Theory*, 2 ed. (Taylor Francis, Philadelphia 1991)
5. S. Grésillon, L. Aigouy, A. C. Boccara, J. C. Rivoal, X. Quelin, C. Desmarest, P. Gadenne, V. A. Shubin, A. K. Sarychev, V. M. Shalaev, Phys. Rev. Lett. **82**, 4520 (1999)
6. A. K. Sarychev, V. A. Shubin, V. M. Shalaev, Phys. Rev. B **60**, 16389 (1999); A. K. Sarychev, V. M. Shalaev, Physica A **266**, 115 (1999); A. K. Sarychev, V. A. Shubin, V. M. Shalaev, Phys. Rev. E **59**, 7239 (1999)
7. H. J. Herrmann, S. Roux (Eds.) *Statistical Models for the Fracture of Disordered Media* (Elsevier North-Holland, Amsterdam 1990) and references therein
8. A. Aharony, Phys. Rev. Lett. **58**, 2726 (1987)
9. D. Stroud, P. M. Hui, Phys. Rev. B **37**, 8719 (1988)
10. V. M. Shalaev, M. I. Stockman, Sov. Phys. JETP **65**, 287 (1987); A. V. Butenko, V. M. Shalaev, M. I. Shtockman, Sov. Phys. JETP **67**, 60 (1988)
11. A. P. Vinogradov, A. V. Goldenshtein, A. K. Sarychev, Zh. Tekh. Fiz. **59**, 208 (1989) Engl. trans. in Sov. Phys. Techn. Phys. **34**, 125 (1989); V. A. Garanov, A. A. Kalachev, A. M. Karimov, A. N. Lagarkov, S. M. Matytsin, A. B. Pakhomov, B. P. Peregood, A. K. Sarychev, A. P. Vinogradov, A. M. Virnic, J. Phys. **3**, 3367 (1991); V. A. Garanov, A. A. Kalachev, S. M. Matytsin, I. I. Oblakova, A. B. Pakhomov, A. K. Sarychev, Zh. Tekh. Fiz. **62**, 44 (1992) [Engl. trans. Sov. Phys. Techn. Phys. (1992)]
12. D. J. Bergman, Phys. Rev. B **39**, 4598 (1989)
13. P. M. Hui, in *Nonlinearity and Breakdown in Soft Condensed Matter*, K. K. Bardhan, B. K. Chakrabarty, A. Hansen (Eds.), Lecture Notes Phys. **437** (Springer, Berlin, Heidelberg 1996)
14. V. M. Shalaev, Phys. Rep. **272**, 61 (1996)
15. V. M. Shalaev, A. K. Sarychev, Phys. Rev. B **57**, 13265 (1998)

16. Hongru Ma, Rongfu Xiao, Ping Sheng, J. Opt. Soc. Am. B **15**, 1022 (1998)
17. D. Stroud, Superlattice Microstruct. **23**, 567 (1998)
18. D. Bergman, O. Levy, D. Stroud, Phys. Rev. B **49**, 129 (1994); O. Levy, D. Bergman, Physica A **207**, 157 (1994); O. Levy, D. J. Bergman, D. G. Stroud, Phys. Rev. E **52**, 3184 (1995)
19. R. W. Cohen, G. D. Cody, M. D. Coutts, B. Abeles, Phys. Rev. B **8**, 3689 (1973)
20. J. P. Clerc, G. Giraud, J. M. Luck, Adv. Phys. **39**, 191 (1990)
21. C. Flytzanis, Prog. Opt. **29**, 2539 (1992) and references therein
22. R. W. Boyd, *Nonlinear Optics* (Academic, New York 1992); L. D. Landau, E. M. Lifshits, L. P. Pitaevskii, *Electromagnetics of Continuous Media*, 2nd ed. (Pergamon, Oxford 1984)
23. P. Sheng, *Introduction to Wave Scattering, Localization, and Mesoscopic Phenomena* (Academic, San Diego 1995)
24. D. A. G. Bruggeman, Ann. Phys. (Leipzig) **24**, 636 (1935)
25. X. C. Zeng, D. J. Bergman, P. M. Hui, D. Stroud, Phys. Rev. B **38**, 10970 (1988); X. C. Zeng, P. M. Hui, D. J. Bergman, D. Stroud, Physica A **157**, 10970 (1989)
26. D. J. Bergman, in *Composite Media and Homogenization Theory*, G. Dal Maso, G. F. Dell'Antinio (Eds.) (Birkhauser, Boston 1991) p. 67
27. O. Levy, D. J. Bergman, J. Phys. C **5**, 7095 (1993)
28. P. M. Hui, D. Stroud, Phys. Rev. B **49**, 11729 (1994)
29. H. C. Lee, K. P. Yuen, K. W. Yu, Phys. Rev. B **51**, 9317 (1995)
30. W. M. V. Wan, H. C. Lee, P. M. Hui, K. W. Yu, Phys. Rev. B **54**, 3946 (1996)
31. P. M. Hui, P. Cheung, Y. R. Kwong, Physica A **241**, 301 (1997) and references therein
32. F. Brouers et al., in *Fractals in the Natural and Applied Sciences* (Chapman Hall, London 1995) Chap. 24
33. F. Brouers, A. K. Sarychev, S. Blacher, O. Lothaire, Physica A **241**, 146 (1997)
34. F. Brouers, S. Blacher, A. N. Lagarkov, A. K. Sarychev, P. Gadenne, V. M. Shalaev, Phys. Rev. B **55**, 13234, (1997)
35. P. Gadenne, F. Brouers, V. M. Shalaev, A. K. Sarychev, J. Opt. Soc. Am. B **15**, 68 (1998)
36. B. Kramer, A. MacKinnon, Rep. Prog. Phys. **56**, 1469 (1993)
37. D. Belitz, T. R. Kirkpatrick, Rev. Mod. Phys. **66**, 261 (1994); M. V. Sadovskii, Phys. Rep. **282**, 225 (1997)
38. V. I. Fal'ko, K. B. Efetov, Phys. Rev. B **52**, 17413 (1995)
39. K. B. Efetov, *Supersymmetry in Disorder and Chaos* (Cambridge University Press, Cambridge 1997)
40. M. V. Berry, J. Phys. A **10**, 2083 (1977)
41. A. V. Andreev et al., Phys. Rev. Lett. **76**, 3947 (1996)
42. K. Muller et. al., Phys. Rev. Lett. **78**, 215 (1997)
43. J. A. Verges, Phys. Rev. B **57**, 870 (1998)
44. A. Elimes, R. A. Romer, M. Schreiber, Eur. Phys. J. B **1**, 29 (1998)
45. M. I. Stockman, Phys. Rev. E **56**, 6494 (1997); Phys. Rev. Lett. **79**, 4562 (1997); M. I. Stockman, L. N. Pandey, L. S. Muratov, T. F. George, Phys. Rev. B **51**, (1995); M. I. Stockman, L. N. Pandey, T. F. George, ibid. **53**, 2183 (1996)
46. M. Kaveh, N. F. Mott, J. Phys. A **14**, 259 (1981)
47. T. Kawarabayashi, B. Kramer, T. Ohtsuki, Phys. Rev. B **57**, 11842 (1998)

# Surface-Plasmon-Enhanced Nonlinearities in Percolating 2-D Metal–Dielectric Films: Calculation of the Localized Giant Field and Their Observation in SNOM

Patrice Gadenne[1] and Jean C. Rivoal[2]

[1] Université de Versailles Saint Quentin, Laboratoire de Magnétisme et d'Optique, CNRS UMR 8624, F-78035 Versailles, France
gadenne@physique.uvsq.fr

[2] Université Pierre et Marie Curie, Laboratoire d'Optique Physique ESPCI, CNRS UPR 5, F-75231 Paris, France
rivoal@optique.espci.fr

**Abstract.** The surfaces of percolating random 2-D metal–dielectric films consist of several spectral resonances, which have been calculated and afterward observed by near-field optical microscopy. These films show anomalous optical properties which are investigated in the first section. Nonlinear electrical and optical properties of metal–dielectric film percolation composites, though recognized very early, were not well understood. It is only recently that calculation of local fields in semicontinuous films allows us to define the enhancement factors of optical nonlinearities. These calculations are outlined from basic principles in the second section and compared with experimental results. An insightful approach to the same problem is to use a network description to represent the random system and discretize the equations satisfied by the scalar potential of the electrical field. We recall in the third section how such discretization leads to a Hamiltonian which is paradigmatic in the theory of Anderson localization. The imaging and spectroscopy of localized optical excitation in gold-on-glass percolation films was performed using near-field optical microscopy (SNOM), and the fourth section recalls the basic features of the experimental technique and describes the first experimental observation of "hot spots" in a nanometer-scale area.

## 1 Introduction

In a significantly wide range close to the percolation second-order transition, granular metal thin films are known to manifest electromagnetic properties that are absent for both components: bulk metal and dielectric.

Two-dimensional metal–dielectric films consist of a planar distribution of nanometer-sized metal grains randomly distributed on the surface of an insulating substrate; each metallic grain has about the same height (thickness) above the substrate. If the edges do not dominate the shape of the grains, we can assume that the metallic grains are not too far from cylinders, and then the ratio between the metal covered surface and the total surface of the

V. M. Shalaev (Ed.): Optical Properties of Nanostructured Random Media,
Topics Appl. Phys. **82**, 185–213 (2002)
© Springer-Verlag Berlin Heidelberg 2002

substrate can be regarded as representing the metal filling factor $p$, or metallic concentration of the film. When this filling factor increases, coalescence between initially isolated metallic grains occurs, resulting in the formation of very irregularly shaped clusters. When $p$ increases and reaches the value $p = p_c$ at the percolation threshold, one extended metal cluster spans the entire sample, and electrical transition from insulating to dc conducting behavior occurs. The films in this concentration region show very remarkable linear and nonlinear optical properties, like second-harmonic generation and white-light generation [1]. We will focus on the linear localized enhancement of the fields in this review.

The first part is devoted to the general characterization of these kinds of films. The recent calculation of the field distribution allowing local enhancements in semicontinuous films is outlined in the second part and compared with linear optical properties. In the third part, the scaling theory of field spatial distribution in metal–dielectric composites is briefly reviewed to introduce the Anderson localization concept [2]. Basic features of the experimental technique [3], scanning near-field optical microscopy (SNOM), are described in the fourth part, together with the first imaging and spectroscopy of localized optical excitations, so called "hot spots, in gold-on-glass percolation films in nanometer-sized areas.

## 2    Semicontinuous Fractal Metallic Films

Since the 1970s, a large extra absorption has been pointed out [4,5], which was partly but not sufficiently explained by the presence of grain surface plasmons (SP), like the well-known Maxwell-Garnett resonance [6]. Moreover, the mean field models [7,8] never gave a reasonably good account of both reflectance $R$ and transmittance $T$ spectra together in the ultraviolet, visible, and infrared ranges.

Some of the models give an account of the transmittance in a defined wavelength range, some others of the reflectance, sometimes of the sum of both which gives the absorbance $A = 1 - (R + T)$. Most of them give an acceptable account in a small wavelength range, but nothing good out of it. In fact, it has been pointed out several times that the optical properties of such granular metal films would depend mostly on geometric morphology, which was demonstrated as be fractal years ago [9,10,11]. Then, the only optical models which can more or less reproduce linear optical properties are based on percolation theory [5,12], which takes into account the specific fractal morphology of such films. The point that cannot be easily taken into account is the following: the length scale on which the sample can be regarded as homogeneous varies and becomes different for each wavelength and for each frequency. That is the reason that all of the "shape parameters" supposed to be able to describe the system in a homogenization process cannot be taken as constant.

However, even with a poor understanding of the hidden fundamental physics, this kind of thin metallic film has been used for surface-enhanced Raman scattering (SERS) analysis since the 1970s. Mainly silver, but also aluminum and gold are presently used as active layers for SERS [13].

## 2.1  Thin Film Deposition

Granular metal films can easily be obtained by several more or less sophisticated processes.

The simplest is deposition from a heated crucible under classical vacuum conditions. In that case the deposition time has to be short enough (few tens of seconds) to avoid pollution of the sample by residual gas in the deposition chamber. But, the final amount of deposited metal is not easily controlled, and the geometric structure on the nanometric scale is not fully predictable. Ultrahigh vacuum conditions are the price to pay for avoiding high pollution and controlling the deposition process well, and also a long distance between the crucible and substrate is required for good homogeneity of the samples over a few centimeters. The mean free path in the vacuum chamber has then to be significantly longer than the deposition chamber size. These techniques allow easy physical (optical, electrical, and thickness) in situ characterizations during deposition, and as a consequence, good real time control of the deposit.

The Knudsen cell [14] can be used for high repeatability of very slowly deposited films, monolayer by monolayer, to accurately determine the characteristic properties step by step [15,16].

Radio-frequency (RF) sputtering from a metallic target produces thin granular films on silica or glass substrates. This is an easy technique, which is fast, but requires good knowledge of the deposition chamber, because physical in situ characterization during deposition is not easy to implement in the geometric arrangement of target and substrate and because of the perturbations caused by the presence of the argon RF plasma during deposition. However, when using a composite target, this RF sputtering method allows easy production of another class of samples, called cermets, which are made of a 3-D nanoscale mixture of tortuous metallic grains and a solid-state insulator. The optical properties of such cermet samples are close to those of granular metal films, but will not be discussed here. In cermet thin films, the percolation threshold and the optical as well as the electrical properties, depend not only on the respective relative amount of metal and insulator giving the metal filling factor $p$, but also depend on the total thickness $d$ of the 3-D thin sample [17]. This unusual effect occurs because the direction of the thickness is not equivalent to the two others, which are "in the plane" of the 3-D thin film.

## 2.2   Nonoptical Characterizations

We arbitrarily divide the usual characterization techniques into in situ measurements, which give information about the film during deposition but require fast data acquisition, and ex situ measurements, which can be slow and more accurate but give information at one special fixed stage of the growing process.

The mass thickness $d_m$ is continuously measured in situ by using a quartz microbalance of very high sensitivity (of the order of one hundredth of a monolayer). This mass thickness is the thickness that would have an homogeneous bulk metallic film of the same weight. The actual thickness of the grains in the film can then be calculated by taking into account the filling factor $p$, which is determined by transmission electron microscopy (TEM) ex situ measurements and image analysis ($d \approx d_m/p$). As soon as the metallic film becomes continuous, the mass thickness is equal to the actual thickness of the film. The quartz microbalance has to be accurately calibrated by measuring the thickness of a continuous thin film, deposited under the same conditions, by the X-ray Kiessig fringe method [18].

DC electrical conductivity $\sigma(d_m)$ can be followed in situ as soon as the percolation threshold is reached. In fact, very low tunneling or hopping conductivity between metallic grains, through the insulator channels, can already be measured in the "nonmetallic" conduction regime below the percolation threshold. Analysis of the data displayed in Fig. 1 gives the experimental

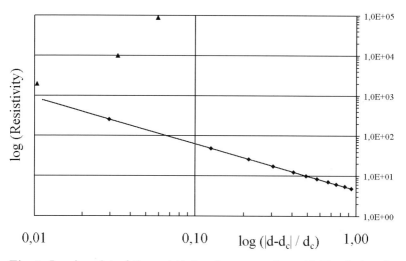

**Fig. 1.** Log–log plot of the resistivity of one granular gold film during deposition: The $x$ axis quantity $\frac{|d_m - d_c|}{d_c}$ is proportional to the usual percolation scaling parameter $p - p_c$. The measurements are diamonds on the "metallic conduction" side of the percolation threshold, and triangles on the "hopping conduction" side. The slope of the straight line is $\approx -1.2$, which is not far from the theoretical value

percolation exponent $\mu$ given by $\sigma \approx (p - p_c)^\mu$, which is in reasonably good agreement $(1.1 \leq \mu \leq 1.4)$ with the results expected from theory [19,20,21].

As the geometric morphologies of the films are mainly responsible for the electrical and optical properties, it is very important to get pictures of them as good as possible. This is achieved by obtaining first ex situ TEM images and afterward some image analysis giving the filling factor $p$ and the characteristics of the fractal morphology of the deposits. Because the films are discontinuous, they do not have their own integrity, and they cannot be easily pulled away from the substrate to set them on the TEM copper grids. The best way to process them, to fully maintain the fine geometry, is first to add a graphite thin film to recover the necessary integrity, and then to dissolve the substrate locally. Evidence is that this is a destructive process. Another possible way, which has been experimentally proved sufficient in some cases, is to deposit the films directly on TEM copper grids. To recover the interaction conditions between the metal and the silica substrate, these grids should be previously covered by a very thin silica film, transparent to the electrons.

It has to be emphasized here that the exact value $p_c$ of the percolation threshold filling factor depends on the experimental conditions and is not characteristic by itself. In granular metal films deposited on amorphous dielectric substrates, it has been found that the value is close to 0.5, but most of the times slightly above 0.5, sometimes reaching values as high as 0.65. This points out some geometric dissymmetry between the insulating channels and the conducting paths. So many different parameters play a role here (the nature of the substrate and the polishing process and quality, the possible presence of underlayers, the nature of the metal, the deposition temperature and deposition process, the deposition rate, the vacuum and residual gas conditions, etc.), that the only way to proceed is the experimental empirical one. In fact, everyone of these parameters partly controls the probability for one atom to arrive at the surface of the substrate to be captured and then to be adsorbed at certain sites whose adhesive efficiency depends on the mobility of the atoms on the surface. This probability can also increase during one deposition process, just because of the presence of the previous metallic atoms and clusters at the surface of the substrate. One then can understand why it is preferable to use TEM for every deposit. In Figs. 2 and 3, the representative samples are classified by their optical spectra (see Sect. 2.3 below). When looking carefully at the micrographs at the left-hand side of the Figs. 2 and 3, it is easy to conclude that the mass thickness is not the only parameter giving the morphology of the film, and then it does not by itself give the distance $(p - p_c)$ from the percolation threshold, which governs the optical properties. Mass thickness, filling factor, and percolation threshold can only be set in direct relation during one single deposition process, because those parameters then remain the same.

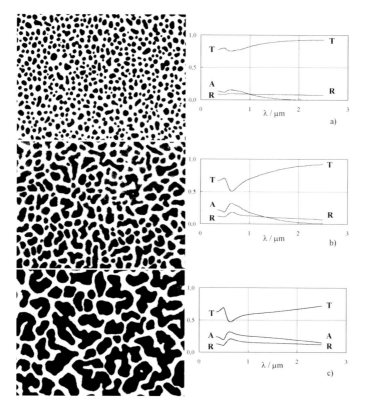

**Fig. 2.** Transmission electron micrographs of gold samples deposited on well polished silica substrates ($456 \times 333$ nm). White is vacuum, gray and black are metal. The three samples are granular gold of increasing mass thickness and filling factors (**a**) isolated quasi-spherical grains, $d_m = 2.0$ nm; (**b**) $p \ll p_c, d_m = 3.5$ nm; (**c**) $p \ll p_c, d_m = 5.8$ nm. Corresponding linear optical properties are plotted facing the electron micrographs. Emphasis should be laid on the respective absorption of each granular metal film

After digitization of the micrographs and thresholding them, one gets an image of square pixels, for example, white for vacuum and black for metal. This kind of image has been extensively analyzed for granular gold-on-glass films [9,10,22]. The point we would emphasize here is that the geometric morphology of such films is fractal in the percolation regime [19,20,21,23]. As evidence, this fractality cannot be mathematically verified over all of the length scales, but at least over one or two orders of magnitude (a factor of 10 to 100). There are several methods for studying the fractality or multifractality of metallic clusters [24].

In 2-D thin films, it is possible to compare the way the perimeter of a cluster increases and the way the surface of the same cluster increases. In Euclidean space, the perimeter increases as the square root of its surface.

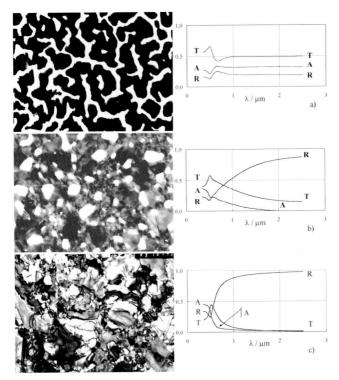

**Fig. 3.** Transmission electron micrographs of gold samples deposited on well polished silica substrates ($456 \times 333$ nm). White is vacuum, gray and black are metal. The three samples are granular gold of increasing mass thickness and filling factors close to or above the percolation threshold $p_c$ (**a**) $p \approx p_c, d_m = 6.8$ nm; (**b**) $p > p_c, d_m = 9.8$ nm; (**c**) ordinary continuous, not annealed, gold thin film made of small randomly oriented crystallites, $d_m = 21.3$ nm. The differences in orientation of the crystallites are visible on the **3b** and **3c** micrographs, this gives the differences in levels of gray. Corresponding linear optical properties are plotted facing the electron micrographs. Emphasis should be laid on the respective absorption of each granular metal film

Figure 4 shows up this behavior for small metal clusters; the points for the low values fit well the straight line of slope $1/2$. But for higher values, the perimeter progressively scales as the surface raised to the power of 0.94, which is smaller but very close to unity. This demonstrates that the perimeter becomes very tortuous, as can be seen on the micrographs, and scales as it should (exponent of 0.95) when following the theoretical percolation model. The 2-D percolating metallic clusters can then be regarded as fractals. The theoretical results of the percolation theory provide a good model to account for the properties of 2-D granular metal films, as has been experimentally checked.

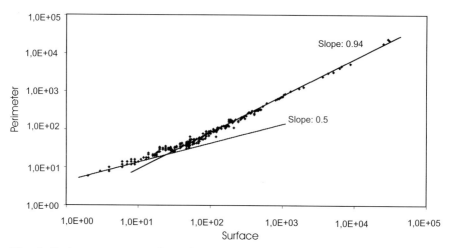

**Fig. 4.** Perimeter versus surface of metallic clusters: Percolation exponent

For 3-D cermet films, composition $p$ and stoichiometry can be determined by Castaing microprobe analysis, Rutherford Back Scattering (RBS), and X-ray photoemission spectroscopy (XPS). TEM graphs can be achieved only on thin samples (10 to 40 nm), which are no longer representative of the optical and electrical properties of thicker samples of the same composition.

## 2.3   Linear Optical Measurements

Reflectance $R$, Transmittance $T$, and therefore absorbance $A = 1 - (R + T)$, which are all linear optical measurements, are performed ex situ by using a spectrophotometer of good accuracy in the UV, visible, and near-infrared ranges. These measurements allow calculating intrinsic optical properties, namely, the $\epsilon = \epsilon' + i\epsilon''$ dielectric function (which is equivalent to the complex refractive index $n = n' + in''$) of the finished samples after the deposition process is stopped. However, these $R$ and $T$ measurements are done at the length scale of the optical beam, which is macroscopic, and therefore can only give a macroscopically averaged $\epsilon$ value. As pointed out in the introduction of Sect. 2, this macroscopic averaging erases the local fluctuations, which will be proved, in the fourth section, to be giant and then dominant in the percolation regime. This remark holds as normally pointed out for the order parameter in every phase transition. The value of the $\epsilon = \epsilon' + i\epsilon''$ dielectric function, which could be deduced from the $R$ and $T$ measurements, is not totally convincing. We should remark that in infrared range out of the interband transitions, when increasing $p$, the real part $\epsilon'$ of $\epsilon$ has to cross zero, from positive values to negative ones, close to the percolation threshold $p_c$. However, the accuracy of such a determination of the real part $\epsilon'$ of $\epsilon$ is very poor close to $p_c$, since the calculation process does not converge well enough. This can be better understood, when knowing that nearly every value in the range $[-10, +10]$ is acceptable at $p_c$ [25]. On the contrary, the imaginary part $\epsilon''$ of $\epsilon$ is given approximately [26] by the ratio

$$\frac{\epsilon''}{\lambda} \approx \frac{n_s}{2\pi d} \frac{1 - (R+T)}{T}, \tag{1}$$

where $\lambda$ is the wavelength, $n_s$ is the refractive index of the substrate, and $d$ is the actual thickness of the film, which is not easily determined. Representative values of $\epsilon = \epsilon' + i\epsilon''$ can be found in the literature [5,8,27], and because they do not bring more to the present task, we will not present them. We will discuss only the experimental data $R$, $T$ and $A$ in Sect. 2.4, after presentation of the usual in situ optical measurement techniques.

In situ measurements of optical reflectance and transmittance and their spectral dependence can be achieved by putting a dispersive stage in front of a CCD camera or an array of photodiodes linked to a computer [5,28]. Even a much simpler setup can give very sensitive information about film growth by following the transmittance at only one accurately chosen wavelength for increasing mass thickness $d_m$, for example (see Fig. 5), using the 632.8-nm ray of an He–Ne laser when depositing granular gold films [25]. In that special case, measurements are done in the most sensitive range of wavelength because it is close enough to the Maxwell-Garnett resonance of single particles and not inside the interband transition region allowing us to follow the presence of high extra absorption all along the percolation regime (see Sect. 2.4).

Other in situ optical characterizations of granular metal film growth have been used by different authors [15,16]. We could mention spectroscopic ellipsometry measurements in a reflection or transmission arrangement. Spectro-

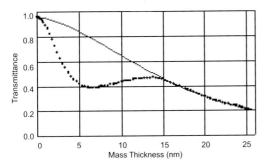

**Fig. 5.** In situ variations of the transmittance $T$ at $l = 632.8\,\mathrm{nm}$ (close to the plasmon resonance of individual grains) versus the deposited mass thickness during deposition. The points are measurements, and the solid line is a theoretical calculation for a continuous film of the same thickness. Note the big difference due to extra absorption for thickness less than 15 nm

scopic differential reflectance measurement

$$\frac{\Delta R}{R} = \frac{R(d_\mathrm{m}) - R(d_\mathrm{m} = 0)}{R(d_\mathrm{m} = 0)} \tag{2}$$

is another powerful technique, which is very sensitive because the starting $R(d_\mathrm{m} = 0)$ values are of the order of few percent. This last technique can point out and characterize the narrow resonance peaks, for increasing metal quantity, starting at the very early stage of growth, namely, for a fraction of one monolayer to several monolayers in mass thickness. Moreover, by using this technique, it is easy to know [16,29] which kind of process is implied in early growth of the thin film for each particular case: full layer after full layer growth, distribution of spheroids, or mixtures of both cases.

## 2.4   Linear Optical Properties in the Percolation Regime

The right-hand parts of Figs. 2 and 3 show the $R$, $T$, and $A$ spectra for different stages of thin granular gold films, when increasing the values of the metal filling factor $p$, starting with the low values, crossing the percolation transition at $p_\mathrm{c}$, and ending with one homogeneous deposited thin gold film. It has to be said again here that both insulator, which is vacuum here, and metal cannot absorb in this infrared range. The interband transitions for gold end close to 510 nm (2.45 eV) [30,31]. There is no interband transition for silver in this range. The intraband transitions, which occur in this spectral range, do not give rise to actual absorption because the metal reflects. This can be clearly seen in Fig. 3c for the continuous gold film, where the interband transition edge is clearly identified below 0.6 μm and absorption is very low and decreasing for increasing wavelength.

When increasing the metal filling factor $p$, the general optical behavior of the films starts from insulating-like (Fig. 2a everywhere high $T$ and low $R$, increasing $T$ and decreasing $R$ toward the long wavelength) and ends with metallic-like (Fig. 3c everywhere low $T$ and high $R$, decreasing $T$ and increasing $R$ toward the long wavelength).

What is important is that in the mid $p$ percolation regime, very large ($\approx 30\,\%$) flat absorption appears, nearly independent of the wavelength in the infrared range. The same point is easily pointed out for a silver percolating film. Figure 6 shows the same spectra for one silver granular film close to the percolation threshold. There is no interband transition for silver in this range, and the total absorption of the percolating film is even higher for every wavelength longer than that of grain plasmon resonance.

Figure 5 shows in situ optical measurements of transmittance at 0.6328 nm for increasing mass thickness $d_\mathrm{m}$. Because reflectance is much less sensitive to absorption than transmittance, the difference between the full line curve, which gives the transmittance of a continuous film of the same thickness, and the cross curve, which gives the actual transmittance of the film, represents approximately the absorbance of the film at each stage of growth. Because the

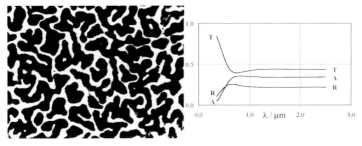

**Fig. 6.** Silver percolating film: TEM micrograph (456 × 333 nm) and optical linear properties

chosen wavelength is close enough to the Maxwell-Garnett-like resonances, some absorbance exists from the very beginning of the deposition process. This absorbance reaches a value as high as 35 % just before the percolation threshold, which occurs for $d_m = 7.0$ nm for this sample.

There is a possible interpretation of the absorption at wavelengths close to 0.6 μm, by mentioning the grain plasmon resonances and their short range interactions, but nothing like this can explain "every wavelength flat giant absorption", which we have shown in the spectra of Figs. 2c, 3a, and 6 all over the near-infrared range. Moreover, this extra absorption of electromagnetic waves has been found much further away in the infrared for granular gold films [27].

# 3   Enhancement of Optical Processes in Semicontinuous Metallic Films

Before analyzing the previous experimental results, let us recall that it was recognized several years ago that local electromagnetic fields in small-particle composites exhibit strong fluctuations that significantly exceed the applied field ([37] and references therein). It was shown that local fields can be found from linear optical response theory.

More recently, the optical properties of metal–dielectric films, such as those described in the previous section, were investigated theoretically [38]. It was shown that the local fields can be very inhomogeneous and consist of strongly localized sharp peaks. In these peaks (called "hot spots"), the local field can exceed the applied field by several orders of magnitude, resulting in giant enhancements of the optical nonlinearities.

## 3.1   Plasmon Resonance

Metal–dielectric films consist of a distribution of nanometer-sized metal grains randomly distributed on the surface of an insulating substrate.

The displacement of free electrons from their equilibrium position in each small particle, under an applied electromagnetic field, results in a noncompensated charge on the surface of the particle, leading to its polarization; this polarization, in turn, results in a restoring force that causes electron oscillations. Metallic particles are characterized by a dielectric constant $\epsilon$ with a negative real $\epsilon'$ part and a small imaginary $\epsilon''$ part, which depend on the wavelength $\lambda$. The polarizabilty $\alpha$ induced in each particle can be written as

$$\alpha = R^3 \frac{\epsilon(\lambda) - 1}{\epsilon(\lambda) + 2}, \tag{3}$$

where $R$ is the radius of the particle supposed spherical. For a given wavelength $\lambda_0$, the dielectric constant $\epsilon$ real can take the value $-2$, thus leading to a divergence of $\alpha$. Then, the electron oscillations experience a resonance around $\lambda_0$, which is known as "plasmon resonance" [6]. The associated frequency occurs at $\omega = \omega_p/\sqrt{3}$, where $\omega_p$ is the classical plasma frequency [39].

The light-induced oscillating dipoles of different particles interact with each other, forming collective optical excitations of the whole system. Numerical techniques described in detail in [38,40], allow us to calculate the spatial distribution of the local fields.

## 3.2   Shalaev–Sarychev Approach

Random media have no translational invariance symmetry and consist of very small structural particles which are much smaller than the wavelength $\lambda$ of the applied field, so that a quasi-static approximation can be used to describe the response of an individual particle. Typically, the particle size ranges from 10 to 30 nm in the semicontinuous gold-on-glass films considered here. If the skin effect in metal grains is small, a semicontinuous metal film can be considered a two-dimensional object placed in the $(x, y)$ plane. It is often scale-invariant within a certain interval of sizes, i.e., it looks self-similar on different scales. The key approach to describing such symmetry is the concept of fractals, as mentioned in the Sect. 2. The size of a whole system, in general, can be arbitrary with respect to the wavelength $\lambda$ of the light wave which propagates in the $z$ direction.

The original approach used by Shalaev and Sarychev is to represent the semicontinuous metal film in the optical spectral range, as a large random network made of capacitors $C$ and inductances $L$ in series with a weak resistance $R$ following [41,42]. The capacitance $C$ models the dielectric bridges between metal grains with the dielectric constant $\epsilon_d$, which is assumed equal to the dielectric constant of the substrate. The metallic grains are almost purely inductive elements $L$–$R$ in the high-frequency range considered here because losses in metal grains are small. The frequency $\omega$ of the incident wave is much larger than the relaxation rate $\omega_\tau = \tau^{-1}$ in this spectral domain. The metal is characterized by the Drude model following the dielectric function

$$\epsilon_m(\omega) = \epsilon_0 - (\omega_p/\omega)^2 \Big/ [1 + i\omega_\tau/\omega], \tag{4}$$

where $\epsilon_0$ is a contribution to $\epsilon_m$ due to interband transitions, $\omega_p$ is the plasma frequency, and $\omega_\tau = 1/\tau \ll \omega_p$ is the relaxation rate. The real part of the metal dielectric function is much larger than the imaginary part, and it is negative for frequencies $\omega$ less than the "renormalized" plasma frequency defined as

$$\omega_p^* = \omega_p/\sqrt{\epsilon_0}. \tag{5}$$

Thus, the metal conductivity is almost purely imaginary, and, as mentioned, metal grains can be thought of as $L$–$R$ elements, with the active component much smaller than the reactive one. It is worth pointing out that the authors use this $L$–$R$–$C$ lattice representation only for illustrative purposes; their calculations are general and do not actually rely on this representation, as shown by $Fyodorov$ in a recent paper [43].

The local conductivity $\sigma(r)$ of the film takes either the "metallic" values $\sigma(r) = \sigma_m$ in metallic grains or the "dielectric" values $\sigma(r) = -i\omega\epsilon_d/4\pi$ outside the metallic grains. The vector $r = (x, y)$ has two components in the plane of the film. In this case, the local field $E(r)$ can be represented as

$$E(r) = -\nabla\phi(r) + E_e(r), \tag{6}$$

where $E_e(r)$ is the applied (macroscopic) field and $\phi(r)$ is the potential of the fluctuating field inside the film. Ohm's law allows us to express the current density $j(r)$ at point $r$,

$$j(r) = \sigma(r)\left[-\nabla\phi(r) + E_e(r)\right]. \tag{7}$$

The current conservation law, $\nabla \cdot j(r) = 0$, takes the following form:

$$\nabla \cdot \{\sigma(r)\left[-\nabla\phi(r) + E_e(r)\right]\} = 0. \tag{8}$$

The authors solve the Laplace equation to find the fluctuating potential $\phi(r)$ and the local field $E(r)$ induced in the film by the applied field $E_e(r)$. When the wavelength of the incident electromagnetic wave is much larger than all spatial scales of a semicontinuous metal film, the applied field $E_e$, i. e., the field of the incident wave, is constant in the film plane: $E_e(r) = E^{(0)}$. Using the nonlocal conductivity introduced in [36], the local field $E(r)$ induced by the applied field $E_e(r)$ can be obtained. Provided that the local field $E(r)$ is known, the effective conductivity $\sigma_e$ can be obtained from the definition

$$\langle j(r) \rangle = \sigma_e E^{(0)}, \tag{9}$$

where the symbol $\langle \cdots \rangle$ denotes the average over the entire film. For further consideration of nonlinear effects, the authors suppose that the applied field has the following form:

$$E_e(r) = E^{(0)} + E_f(r), \tag{10}$$

where $E^{(0)}$ is the constant linearly polarized field and fluctuating field $E_f(r)$ may arbitrarily change over the film surface but its averaged value $\langle E_f(r) \rangle$

is collinear with $\boldsymbol{E}^{(0)}$. Then the average current density $\langle\boldsymbol{j}\rangle$ is also collinear with $\boldsymbol{E}^{(0)}$ in the macroscopically isotropic films considered here, and this expression is given by

$$\langle\boldsymbol{j}(\boldsymbol{r})\rangle = \boldsymbol{E}^{(0)}\left[\sigma_{\mathrm{e}} + \frac{\langle\sigma(\boldsymbol{r})\left[\boldsymbol{E}(\boldsymbol{r})\cdot\boldsymbol{E}_{\mathrm{f}}(\boldsymbol{r})\right]\rangle}{E^{(0)2}}\right] . \tag{11}$$

Thus, the average current induced in a macroscopically isotropic film by a nonuniform external field $\boldsymbol{E}_{\mathrm{e}}(\boldsymbol{r})$ can be expressed in terms of the fluctuating part $\boldsymbol{E}_{\mathrm{f}}(\boldsymbol{r})$ of the external field and the local field induced in the film by the constant part $\boldsymbol{E}^{(0)}$ of the external field.

The authors use a numerical model for calculating the effective conductivity of semicontinuous metal films and finding the field distributions. Among the various efficient methods [32,33], the authors chose the real space renormalization group method (RSRG), which allowed them to compute the full distribution of the local fields. Though not exact, this method leads to numerical results very close to the exact ones on a two-component composite. Typical distribution of the local-field intensities on a semicontinuous film at the percolation threshold is shown in Fig. 7.

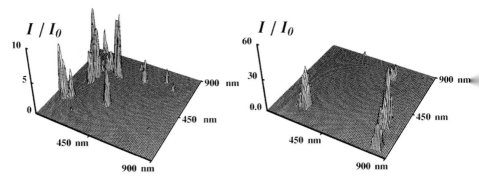

**Fig. 7.** Calculated distribution of local-field intensities for two wavelengths, $\lambda =$ 710 nm (*left*) and 780 nm (*right*)

## 3.3   Anderson Localization

When discretized on a square lattice, (8) acquires the form of Kirchhoff's equation characterized by the Hamiltonian $H$ with off-diagonal elements $H_{ij} = -\epsilon_{ij}$ and diagonal elements $H_{ii} = \sum_j \epsilon_{ij}$, where $j$ refers to the nearest neighbors of $i$ and $\epsilon_{ij}$ take values $\epsilon_{\mathrm{m}}$ and $\epsilon_{\mathrm{d}}$ for metal and dielectric bonds, respectively. It was suggested in [65] that the operators of such a type acting on a lattice be called Kirchhoff Hamiltonians (KH). Off-diagonal entries of such a Hamiltonian assume random values $\pm 1$ for directly connected nodes. This property makes KH similar to a tight-binding Hamiltonian describing the motion of a quantum particle in a disordered lattice with an off-diagonal disorder. Such a model was used by *Anderson* [44] to show "that at low enough densities no diffusion at all can take place", which is now called the theory of Anderson localization.

A scaling theory of local-field fluctuations and optical nonlinearities, which maps the Anderson localization problem, was simultaneously developed for random metal–dielectric composites near a percolation threshold [34]. The theory predicts that in the optical and infrared spectral ranges, the local fields are very inhomogeneous and consist of sharp peaks representing localized surface plasmons. The local fields exceed the applied field by several orders of magnitude, resulting in giant enhancements of various optical phenomena. The developed theory quantitatively describes enhancement in percolation composites for an arbitrary nonlinear optical process. Details of this theory are presented in the preceding contribution.

### 3.4   The Percolation Regime

Now, we can summarize the important points as follows. Because of the self-similarity of percolating clusters, every size of resonating cluster is present in a sample, and then every interaction length is active, giving rise to every plasmon mode frequency. Because of strong disorder in a thin granular metal film, these plasmon–polariton modes are strongly localized within a certain arrangement of metallic clusters, which resonates at a certain frequency. This strongly localized mode takes energy from the electromagnetic field, and this increases its amplitude dramatically, as long as the balance of loss and profit of energy is not reached. When this happens in a wavelength range of low intrinsic absorption of energy for both components (metal and vacuum), the process produces giant local enhancement of the electromagnetic field in very small areas called "hot spots". Because gold and silver poorly absorb in the infrared range, the absorption, as has been pointed out (see Sect. 2.3) in this range in the percolation regime, is due to the excitation of the hot spots. This is the same energy accumulation process which produces giant fields within a Fabry–Perot cavity and then allows about 100 % of light transmission at the resonating frequency, although the reflectance of both mirrors is close to 1. With the granular metal system, the main difference is that it appears whatever the frequency of the incident light because every frequency can find some special sites for its resonance modes.

The hot spots are then responsible for the existence of surface-enhanced Raman scattering (SERS), which would not be visible without locally enhanced electromagnetic fields. In fact, because the scattered light and the incident light are at frequencies very close to each other, the same local resonance mode acts for both of the wavelengths together in the same area. Then, if the field amplitude is locally enhanced by a factor of $10^2$, the scattered intensity can be enhanced by a factor of $10^8$, which makes the Raman scattering visible. Figure 8 shows the granular metal efficiency for ZnTPP

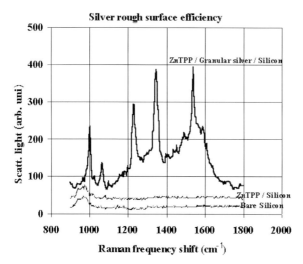

**Fig. 8.** Granular silver efficiency for SERS on ZnTPP adsorbed molecules. The *x* axis gives the Raman frequency shift when the sample is lighted by the 576-nm ray of an argon laser. The lowest curve gives the intensity of the light scattered by the bare silicon substrate. The intermediate curve (slightly shifted for visibility) gives the intensity of the light scattered by a single layer of ZnTPP adsorbed on the silicon substrate. The highest curve gives the scattered intensity for ZnTPP molecules adsorbed on the top of the silver percolating granular film deposited on the silicon substrate

(zinc tetraphenyl porphyne) molecules adsorbed by the Langmuir–Blodgett technique at the surface of a percolating silver film. This efficiency has been studied all over the percolation regime [35,36], for different metal filling factors $p$.

In accordance with the theoretical model presented in the preceding paper by Sarychev and Shalaev, it is found first that granular metallic films are efficient all along the percolation regime, and second that there exists a slight minimum of efficiency at the percolation threshold. This last point is shown in Fig. 9, which represents a histogram of the integrated scattered light for different values of $p$ on both sides of $p_c$. There is a small drop in the efficiency close to the percolation threshold $p_c \approx 0.6$. One should expect a second maximum of efficiency when increasing the filling factor above $p_c$. However, this is not shown here, because the growing process induces rearrangement of the metal grains for high values of $p$. This makes the high $p$ samples not easily comparable to the percolating samples.

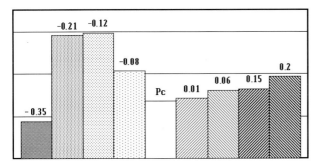

**Fig. 9.** Histogram of the SERS efficiency for the peaks at $1350\,\mathrm{cm}^{-1}$. There is a small drop in efficiency close to the percolation threshold $p_c \approx 0.6$. One should expect a second maximum of efficiency when increasing the filling factor above $p_c$. However, this is not shown here because the growing process induces rearrangement of the metal grains for high values of $p$. This makes the high $p$ samples not easily comparable to the percolating samples

## 4    Experimental Observation of "Hot Spots" Using a Scanning Near-Field Optical Microscope

Near-field optics has demonstrated its ability to break the diffraction barrier in terms of resolution and has recently become a powerful tool for investigating the physical properties of submicroscopic objects [45,46]. Usually, subwavelength sources or detectors (often a metal coated optical fiber that has a small aperture) are used close to the sample surface to get such resolution [47]. Another near-field optical method is the apertureless technique, which uses a sharp dielectric [48] or metallic [49] tip instead of a fiber. We have developed a new scanning near-field optical microscopy (SNOM) technique, which has the unigueness of using a metallic tip without an aperture as a probe [50].

### 4.1    Near-Field Versus Far-Field Imaging

Let us take a familiar example dealing with basic optics to recall the physical limitations of the classical microscope. A slit (or a hole) is illuminated by a monochromatic plane wave (wavelength $\lambda$) parallel to the slit plane; see Fig. 10.

A suitable optical system makes an image of the slit on a screen or a detector. When the slit width $a$ is much larger than the diffraction angle $\alpha$, the objective of the microscope collects the light and allows imaging of the slit. If $a$ decreases, $\alpha$ increases and fills up the half space when $a$ becomes smaller than $\lambda$. So, if we still reduce $a$, the diffraction angle can no longer increase, and any perfect optical system (diffraction limited system) used to image the slit will not be able to differentiate the slit's sizes when $a = \lambda/2$ or $\lambda/10$ or $\lambda/100$. The information dealing with the object structure does not propagate

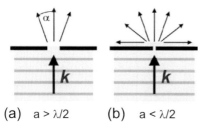

(a)  a > λ/2        (b)  a < λ/2

**Fig. 10.** Diffraction by a slit: **a** is the slit width, λ is the wavelength

but "stays" at the sample level in the near field (a distance of about $\lambda/2\pi$). The wave-vector module of these evanescent waves can increase along the slit surface to values larger than in air.

The aim of scanning optical near-field optical microscopy is to reveal optical information associated with subwavelength structures, which are mainly or totally contained in the evanescent field at distances from a sample surface much smaller than the wavelength. This subject has attracted considerable attention in the past fifteen years. Various experimental and theoretical approaches can be found in [45,51,52].

## 4.2   Fibers or Tips?

How can we reach the near field of the sample under investigation? After the very realistic approach suggested by *Synge* [53] in the 1930s and the pioneering work of *Lewis* [54] and *Pohl* [55], one can say that most of the instruments built for scanning optical near-field microscopy (SNOM) use a metal coated tapered optical fiber with a nanometric hole at the tip apex. In this widely adopted approach, only a tiny fraction of the light coupled in the fiber is emitted by the aperture because such systems squeeze the electromagnetic field in the tapered region and act like metallic waveguides above their cutoff frequency. So there is a trade-off between resolution, which is higher for small holes and small angles of the tapered part, and efficiency, which is larger for large holes and large angles. Typical resolution is around 50 nm (commercial instruments claim better than 100 nm), and efficiency is in the range $10^{-3}$ to $10^{-6}$ of the power injected in the fiber when the tip acts as a nanosource. Despite their low light throughput, tapered fibers are unique when one tries to make highly localized sources and/or detectors. Among the other drawbacks of the fiber approach, which are of importance for infrared spectroscopists, is the limited transmission range of the widely used silica fibers. Other fibers can be used in the infrared [56], but the tapering process is far from being optimized, and only micron range resolution has been reported.

In the application reported below, as in a recent application in biology [58], the spatial resolution required was not obtainable with optical fibers. The use of a tip such as the apex of a near-field probe (atomic force or tunneling) which will scatter the local field (propagation and evanescent waves) has proven to be very powerful in both resolution and efficiency [48,49,50]. To understand the physical origin of these properties, let us recall that in the very near field of the tip, electromagnetic fields behave like electrostatic ones and that a metallic tip exhibits a larger field and larger enhancement when the radius of curvature is smaller [57].

In our case, the tip is made of tungsten metal with a complex dielectric constant relatively flat in the spectral range investigated in our experiments (UV to near-infrared). It is conical, as can be seen in Fig. 11. The field at short distance $r$ from the tip apex ($kr \ll 1$) can be written as [57,59]

$$\boldsymbol{E} = k(kr)^{\nu-1} \sin\beta \left( \boldsymbol{u}_r + \frac{\boldsymbol{u}_\theta}{\nu} \frac{\partial}{\partial\theta} \right) a(\theta_0, \theta, \alpha) , \tag{12}$$

where $a$ is a function of the angle of incidence $\theta_0$, the angle of observation $\theta$, and the semiangle of aperture $\alpha$ of the tip cone. $\boldsymbol{u}_r$ and $\boldsymbol{u}_\theta$ are the unit vectors in spherical coordinates; $\nu$ is positive and smaller than 1, and it depends on $\alpha$. The other parameters are the wave vector $k = \omega/c$ and the angle of polarization $\beta$ of the incident wave.

In an elegant theoretical approach presented recently, a general expression for the signal detected in SNOM is derived based on the reciprocity theorem of electromagnetism [60]. The component of the electrical field in the direction of an analyzer at the detector position is given by an exact expression in the experimental situation. The detected field is an overlapping integral between the experimental field and a term proportional to the response function of the instrument [52]. The response function is proportional to the derivative of the reciprocal field. To determine an approximate and practical description of this reciprocal field, the authors use the expression for $\boldsymbol{E}$ given in (12). With this model, their approach provides a versatile and useful tool to analyze experimental results, such as the polarization effect and spectral response, and to identify the key parameters.

Using a metallic tip does not require any kind of propagation through dielectric materials, and the setup can be used over a very wide range of wavelengths. Although developing the relation between the properties of the

**Fig. 11.** SEM image of a tungsten tip

sample in the near field of the optical probe and the detected far field remains a difficult task [62], we will show below that we can gain useful information, which cannot be obtained by other techniques. Very recently, a new procedure to detect the near-field response even in the presence of strong background scattering has shown that simultaneous amplitude and phase nanoscale imaging are possible [61].

## 4.3   The Setup

Our scanning near-field optical microscope (SNOM) works simultaneously as an atomic force microscope (AFM) [50]. We describe below the transmission-mode scheme, shown in Fig. 12.

The SNOM and AFM probe is an apertureless metallic tip made from a tungsten wire etched by electrochemical erosion. The typical radius of curvature of the conical tip end is around 20 nm, sometimes less than 10 nm (measured by scanning electron microscopy SEM) (Fig. 11). The extremity of the tip is bent by $90°$ ($\approx 1$ to $2\,\mathrm{mm}$) before sealing the arm to a piezoelectric transducer that can excite the cantilever perpendicularly to the sample surface. The frequency of vibration (3 to $12\,\mathrm{kHz}$ depending on the length of the arm) is close to the resonant frequency of the cantilever, and the amplitude of vibration is chosen to be close to $50\,\mathrm{nm}$. The metallic tip vibrates above the sample as in the atomic force microscopy (AFM) tapping-mode technique. The vibrational amplitude is detected and monitored optically. A feedback system keeps the amplitude of vibration constant during scanning of the sample surface via a piezoelectric translator attached to the piezoelectric transducer. The error signal gives a topographical signal of the surface of the sample, the AFM signal. The tip and the sample are set below a commercial microscope (Olympus BH). The Ti/sapphire source is sent to a first micro-

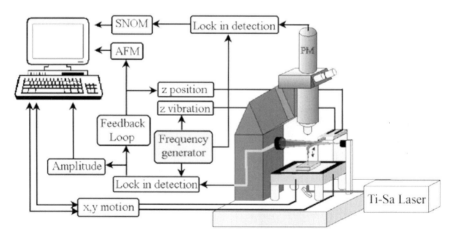

**Fig. 12.** Experimental SNOM setup

scope objective ($20\times$, $f = 10$, NA $= 0.4$) which focuses the light on the surface of the sample. The state of polarization of the incident light can be controlled externally. The collection is axially symmetrical above the sample and is done by a second microscope objective (for example $50\times$, $f = 6.9$, NA $= 0.6$). The detector is a photomultiplier.

The vibration of the tip modulates the near-zone field on the sample surface. The tip periodically scatters the electromagnetic field near the sample surface and radiates it in the far field. The locally scattered electromagnetic field of the near-field region is detected by using lock-in detection at the output of the photomultiplier. This near-field optical signal, the SNOM signal, has at least the optical resolution of the tip's end [64]. The dc signal coming from transmitted light and its residual modulation due to the modulation of the cantilever is rejected by a black screen inserted in the collected flux before the observation. It allows working almost in a dark-field configuration and also improves the dynamic of the detector. As the tip scatters the near-field optical signal of the sample surface in any direction, the signal collected in this dark-field configuration is mainly this scattered optical signal. The experiments presented below were made in the dark-field configuration.

## 4.4   Experimental Observation of "Hot Spots"

Direct experimental observation of localized surface plasmon modes in semicontinuous metal films was reported in [65]. The observed near-field images of a percolation film were in qualitative agreement with theoretical predictions and numerical simulations. However the optical excitations were localized in 100-nm areas, which is significantly smaller than $\lambda$ but larger than the lateral size of the grains. Typical images for two different wavelengths are displayed in Fig. 13.

To observe the localized fields in random semimetallic films at the scale of a grain's size, we improved the setup used in previous experiments.

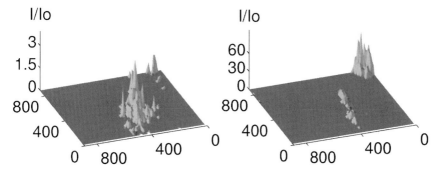

**Fig. 13.** First experimental observation of the "hot spots": SNOM images in a percolation gold-on-glass film for two wavelengths, $\lambda = 714\,\mathrm{nm}$ (*left*) and $770\,\mathrm{nm}$ (*right*)

### 4.4.1  AFM Measurements

Semicontinuous metal films are formed of metallic grains, whose typical sizes are of the order of 10–20 nm, deposited on a glass substrate. The end of the tip is of the same order of magnitude as the grain size, as can be seen in Fig. 11. Because the adhesive coefficient of gold grains on glass is poor, the tapping mode can induce the grains to slip. The first step is then to reduce the interaction between the tip and the sample to a minimum. This can be achieved by using low-frequency vibration of the cantilever and careful adjustment of the contact interaction. A typical AFM image of a semicontinuous film close to the percolation threshold is shown in Fig. 14. It can be seen that the roughness of the surface is of the order of 10 nm and the shape resembles that observed in the TEM image in Fig. 3. Due to convolution with the tip size, the mean height of the metallic grains is not directly related to the roughness. A better determination of the height of the grains is achieved by scanning the surface of a more dilute sample. This is shown in Fig. 15 where the scan of the surface of the sample (20-Å in mass thickness) is displayed. It can be seen that the typical height of metallic granules is of the order of 20 nm. TEM measurements indicate a lateral size of about 10 to 30 nm, in good agreement with these AFM measurements.

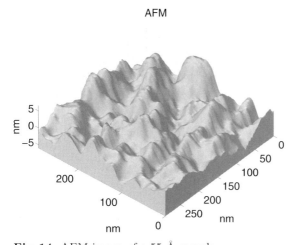

**Fig. 14.** AFM image of a 55-Å sample

### 4.4.2  SNOM Observations

SNOM and AFM data were recorded simultaneously on all samples investigated. We already mentioned that granular metallic films are efficient all along the percolation regime in enhancing the Raman scattered signal (see

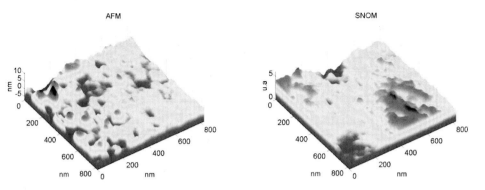

**Fig. 15.** AFM and SNOM images of 20-Å mass thickness sample

Fig. 9). This is clearly put in evidence by recording the SNOM signal on the 20-Å sample using a slightly different device (illumination by total internal reflection). The experimental results shown in Fig. 15 depict local enhancement of the fields not correlated with the topography of the sample.

New software was used to acquire and average the data, which resulted in a better signal-to-noise ratio using the improved setup described in the previous paragraph.

The SNOM experimental data were processed slightly. The "pseudoslope" resulting from imperfections in the feedback loop and, most importantly, from a drift of the whole setup was removed. The average slopes in the $x$ and $y$ directions of the image were determined using a least-squares fit. The coordinates of the plane of inclination were defined by removing these average slopes, and the removing plane was brought back to the $z = 0$ plane. We recall that the SNOM detected signal in our SNOM technique is proportional to the amplitude of the modulated field, and consequently we squared the detected signal to find the local-field intensity.

We also performed the near-field spectroscopy by parking the tip at different points of the sample surface and varying the wavelength. The nanostructures at different points resonate at different $\lambda$, as can be seen in Fig. 16. Again, there is qualitative agreement between theory and experiment. The spectra consist of several peaks $\approx 10$ nm in width, and they depend markedly on the spatial location of the point where the near-field tip is parked. Strong evidence of plasmon-mode localization comes from the fact that different spectra are observed for two points spaced apart by only 100 nm.

In Fig. 17, we show the near-field image of the local-field distribution on the surface of a percolating gold-on-glass film at the wavelength $\lambda = 800$ nm. Clearly our apparatus allows one to resolve even the most localized modes. The resolution of one pixel is 16 nm, close to the mean size of gold grains and about two orders of magnitude smaller than the optical wavelength. The intensity of the "hot spots" can exceed the applied field by at least two orders of magnitude even on one single localized mode.

**spectroscopy on two points 100 nm apart**

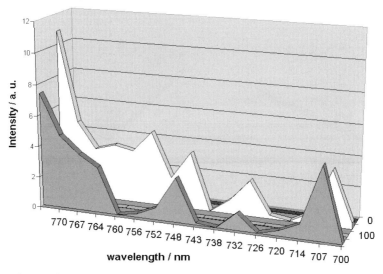

**Fig. 16.** Spectroscopy on two points

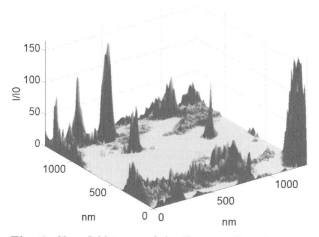

**Fig. 17.** Near-field image of the "hot-spots" at the wavelength $\lambda = 800\,\text{nm}$

# 5    Conclusion

The characterization of semicontinuous 2-D metal–dielectric films, using either nonoptical or linear optical methods, has been conducted for different stages of the filling factor, starting with low values, crossing the percolation transition, and ending with a homogeneous thin film. The near-field imaging and spectroscopy of random metal–dielectric films near percolation suggest localization of optical excitations in small nanometer-scale "hot spots," which represent very large local fields. The spatial positions of the spots strongly depend on the light frequency. The observed patterns of the localized modes and their spectral dependences agree with theoretical predictions and numerical simulations. Their existence is responsible for the observation of surface enhanced Raman scattering, which would not be visible without them. From the fact that the local-field enhancement is large, it can be anticipated that nonlinear processes of the $n$th order could exist because they are proportional to the local fields at the $n$th power. This opens a fascinating possibility for nonlinear near-field spectroscopy of single nanoparticles and molecules.

**Acknowledgment**

We gratefully acknowledge the efficient work of many colleagues and students who were at the origin of the results presented in this review. We would like especially to acknowledge Bruno Bérini who made the samples; Xavier Quélin, Steve Liberman, and Cyril Desmarest at the Versailles Saint-Quentin University; Samuel Grésillon who played the key role in observing the "hot-spots" for the first time using the krypton laser in the Optics laboratory; Lionel Aigouy who is at the origin of the new developments in SNOM; Yannick de Wilde who joined the SNOM team recently; Sébastien Ducourtieux who increased the signal-to-noise ratio dramatically in the AFM-SNOM equipment. We have benefited from stimulating discussions with Remi Carminati, Juan Porto, and Jean Jacques Greffet. The stimulating environment of the GDR "Optique du Champ Proche" of the CNRS is highly appreciated. J.C.R. is indebted to Claude Boccara who suggested using a metallic tip to detect optical near-field signals about fifteen years ago and to Philippe Gleyses who built the first setup to test the idea.

# References

1. S. Ducourtieux, S. Gresillon, A. C. Boccara, J. C. Rivoal, X. Quelin, P. Gadenne, V. P. Dratchev, W. D. Bragg, V. P. Safonov, V. A. Podolskiy, Z. C. Ying, R. L. Armstrong and V. M. Shalaev: Percolation and fractal composites: Optical studies, J. Nonlin. Opt. Phys. Mater. **9**, 105–116 (2000)
2. A. K. Sarychev, V. A. Shubin, V. M. Shalaev, Anderson localization of surface plasmons and nonlinear optics of metal-dielectric composites, Phys. Rev. B **60**, 16389–16408 (1999)

3. S. Gresillon, H. Cory, J. C. Rivoal, A. C. Boccara:Transmission-mode apertureless near-field microscope: Optical and magneto-optical studies, J. Opt. Soc. A **1**, 178–184 (1999)

4. P. Gadenne, Modifications of the optical and electrical properties of thin Au films as a function of structure during deposition, Thin Solid Films **57**, 77 (1979)

5. Y. Yagil, M. Yosephin, D. J. Bergman, G. Deutscher, P. Gadenne, Scaling theory for the optical properties of semi-continuous metal films; Phys. Rev. B **43**, 11342 (1991).

6. J. C. Maxwell-Garnett, Philos. Trans. R. Soc. **203**, 395 (1904); **205**, 237 (1905)

7. D. Bruggeman, Ann. Phys. **24**, 636 (1935); R. W. Cohen, G. D. Cody, M. D. Coutts, B. Abeles, Optical properties of granular silver and gold films, Phys. Rev. B **8**, 3689 (1973); C. G. Granqvist, O. Hunderi, Optical properties of ultrafine gold particles, Phys. Rev. B **16**, 3513 (1977); and Optical properties of $Ag - SiO_2$ cermet films: A comparison of effective-medium theories, Phys. Rev. B **18**, 2897 (1978; F. Brouers, J. P. Clerc, G. Giraud, Z. A. Randriamanantany, Dielectric and optical properties close to percolation threshold II, Phys. Rev. B **47**, 666 (1993); O. Hunderi, Effective Medium Theory and Nonlocal effects for Superlattices, J. Wave-Mater. Interactions **2**, 29–39 (1987); P. Sheng, Phys. Rev. Lett. **45**, 60 (1979)

8. J. C. Garland, D. B. Tanner (Eds.), *Electrical Transport and Optical Properties of Inhomogeneous Media*, Conf. Proc. ETOPIM1 (American Institute of Physics, New York 1978); J. Lafait, D. B. Tanner (Eds.), ETOPIM2 Physica A (North Holland Amsterdam 1987); L. Mochan, R. Barreira (Eds.), ETOPIM3 Physica A (North Holland Amsterdam 1994); A. N. Lagarkov and A. K. Sarychev (Eds.), ETOPIM4 Physica A 241, (North Holland Amsterdam 1997); ETOPIM5 Ping Sheng and L. H. Tan (Eds.) Physica B **279** (North Holland Amsterdam 2000)

9. B. B. Mandelbrot, *The Fractal Geometry of Nature* (Freeman, New York 1982)

10. R. F. Voss, R. B. Laibovitz, E. L. Alessandrini, Fractal (Scaling) clusters in thin gold films near the percolation threshold, Phys. Rev. Lett. **49**, 1441 (1982)

11. A. Kapitulnik, G. Deutscher, Percolation characteristics in discontinuous thin films of Pb, Phys. Rev. Lett. **49**, 1444 (1982)

12. S. Blacher, F. Brouers, A. K. Sarychev, A. Ramsamugh, P. Gadenne, Relation between morphology and alternating current electrical properties of granular metallic films close to percolation threshold, Langmuir **12**, 183 (1996); A. K. Sarychev, V. A. Shubin, V. M. Shalaev, Percolation-enhanced nonlinear scattering from metal-dielectric composites, Phys. Rev. E **59**, 7239 (1999); A. K. Sarychev, V. M. Shalaev, Electromagnetic field fluctuations and optical non-linearities in metal-dielectric composites, Phys. Rep. **335** (2000)

13. M. I. Stockman, V. M. Shalaev, M. Moskvits, R. Botet, T. F. George, Enhanced Raman Scattering by fractal clusters: scale-invariant theory, Phys. Rev. B **46**, 2821–2840 (1992); V. A. Markel, L. S. Muratov, M. I. Stockman, T. F. George, Theory and numerical simulations of optical properties of fractals, Phys. Rev. B **43**, 8133–8198 (1991)

14. F. Didier, J. Jupille, Surf. Sci. **314**, 378 (1994) and **307–309**, 587 (1994); D. Martin-Gagnot, F. Creuset, J. Jupille, Y. Borensztein, P. Gadenne, 2D and 3D silver adlayers on $TiO_2$ (110) surfaces, Surf. Sci. **377**, 958 (1997)

15. T. Lopez-Rios, Y. Borensztein, G. Vuye, Roughening of Al surfaces by Ag deposits studied by differential reflectivity, Phys. Rev. B **30**, 659 (1984)

16. Y. Borensztein, M. Roy, R. Alameh, Threshold and linear dispersion of the plasma resonance in thin films, Europhys. Lett. **31**, 311 (1995); C. Beita, Y. Borensztein, R. Lazzari, J. Nieto, R. G. Barrera, Substrate-induced multipolar resonances in supported free electron metal spheres, Phys. Rev. B **60**, 6018 (1999)

17. J. P. Clerc, G. Giraud, S. Alexander, E. Guyon, Conductivity of a mixture of conducting and insulating grains: Dimensionality effects, Phys. Rev. B **22**, 2489 (1980); M. Gadenne, P. Gadenne, Electrical and optical properties of $AuAl_2O_3$ films: dimensionality effects, Physica A **157**, 344 (1989)

18. H. Kiessig, Ann. Phys. **10**, 769 (1931); W. Umrath; Z Angew. Phys. **22**, 406 (1967)

19. J. P. Straley, Critical phenomena in resistor networks, J. Phys. C **9**, 783 (1976)

20. S. Kirkpatrick, Models of disordered materials, in *Condensed Matter* Les Houches Summer School, R. Balian, R. Maynard, G. Toulouse (Eds.), (North Holland Amsterdam 1979); A. L. Efros, B. I. Schlovskii, Critical behaviour of conductivity and dielectric constant near the metal-non metal transition threshold, Phys. Status Solidi B **76**, 475 (1976)

21. D. Stauffer, A. Aharony, *Introduction to Percolation Theory*, 2nd ed. (Taylor, London 1991)

22. A. Beghdadi, A. Constans, P. Gadenne, J. Lafait, Optimum image processing for morphological study of granular films; Rev. Phys. Appl. **21**, 73 (1986); A. Beghdadi, J. Lafait, P. Gadenne, A. Constans, O. Bouet, Thin film morphology; Acta Stereologica **6**, 809 (1987)

23. D. J. Bergman, D. Stroud, Physical properties of macroscopically inhomogeneous media, Solid State Phys. **46**, 147 (1992)

24. S. Blacher, F. Brouers, P. Gadenne, J. Lafait, Morphological analysis of discontinuous thin films on various substrates, J. App. Phys. **74**, 207 (1993)

25. P. Gadenne, Croissance de couches minces d'or: Proprietes optiques, conductivite electrique et etude morphologique, Dissertation, Chap. 5, Univ. Paris 6 Univ. (1986); P. Gadenne, A. Beghdadi, J. Lafait, Optical cross-over analysis of granular gold films at percolation, Opt. Commun. **65**, 17 (1988)

26. F. Abeles, J. Opt. Soc. Am. **47**, 473–482 (1957)

27. Y. Yagil, P. Gadenne, C. Julien, G. Deutscher, Scaling theory for the optical properties of semi-continuous metal films, Phys. Rev. B **43**, 11342, (1991)

28. P. Gadenne, Y. Yagil, G. Deutscher, In situ measurements of the optical properties of gold films near the percolation threshold, Physica A **157**, 279 (1989); P. Gadenne, Y. Yagil, G. Deutscher, Transmittance and Reflectance in situ measurements of semicontinuous gold films, J. Appl. Phys. **66**, 3019 (1989)

29. R. Lazzari, J. Jupille, Y. Borensztein; Appl. Surf. Sci. **142**, 451 (1999)

30. M. L. Theye, Investigation of optical properties of Au by means of thin semi-transparent films, Phys. Rev. B **2**, 3060 (1970)

31. P. B. Johnson, R. W. Christy, Optical properties of the noble metals, Phys. Rev. B **6**, 4370 (1972)

32. P. J. Reynolds, W. Klein, H. E. Stanley, A real space renormalization group for site and bond percolation, J. Phys. C **10**, L167 (1977)

33. A. K. Sarychev, Zh. Eksp. Teor. Fiz. **72**, 1001 (1977) [Sov. Phys. JETP **45**, 524 (1977)]

34. A. K. Sarychev, V. A. Shubin, V. M. Shalaev, Anderson localization of surface plasmons and nonlinear optics of metal-dielectric composites, Phys. Rev. B **60**, 16389–16408 (1999)

35. P. Gadenne, D. Gagnot, M. Masson, Surface enhanced resonant Raman scattering induced by silver thin films close to the percolation threshold, Physica A **241**, 161 (1997); P. Gadenne, F. Brouers, V. M. Shalaev, A. K. Sarychev, Giant Stokes fields on semicontinuous metal films, J. Opt. Soc. Am. B **15**, 68 (1998)

36. F. Brouers, S. Blacher, A. N. Lagarkov, A. K. Sarychev, P. Gadenne, V. M. Shalaev, Theory of giant Raman scattering from semicontinuous metal films, Phys. Rev. B **55**, 13234 (1997)

37. V. M. Shalaev, Electromagnetic properties of small-particles composites, Phys. Rep. **272**, 61–137 (1996)

38. V. M. Shalaev, A. K. Sarychev, Nonlinear optics of random metal-dielectric films, Phys. Rev. B **57**, 13265–13288 (1998)

39. R. P. Feynman, R. B. Leighton, M. Sands, in *The Feynman Lectures on Physics,* Vol.2 (Addison-Wesley, 1965)

40. V. M. Shalaev, *Nonlinear Optics of Random Media*, Springer Tracts Mod. Phys. **158** (Springer, Berlin, Heidelberg 2000)

41. D. J. Bergman, D. Stroud, Physical properties of macroscopically inhomogeneous media, Solid State Phys. **46**, 147 (1992)

42. J. P. Clerc, G. Girard, J. M. Laugier, J. M. Luck, Adv. Phys. **39**, 191 (1990)

43. Y. V. Fyodorov, Fluctuations in random $RL$-$C$ network: Non-linear $\sigma$-model description; JETP Lett. **70**, 743 (1999)

44. P. W. Anderson, Absence of diffusion in certain random lattices, Phys. Rev. **109**, (1958)

45. D. W. Pohl, D. Courjon (Eds.), Near-Field Optics (Kluwer Academic, Dordrecht 1993)

46. M. A. Pasler, J. P. Moyer, Near-Field Optics (Wiley-Interscience, New York 1996)

47. E. Betzig, J. K. Trautman, Near-Field Optics: Microscopy, spectroscopy and surface modification beyond the diffraction limit, Science **257**, 189–195 (1992)

48. F. Zenhausern, Y. Martin, H. K. Wickramasinghe, Scanning interferometric aperturless microscopy: Optical imaging at 10 Å resolution, Science **269**, 1083–1085 (1995)

49. Y. Inouye, S. Kawata, Near-Field scanning optical microscope with a metallic tip, Opt. Lett. **19**, 159–161 (1994)

50. R. Bachelot, P. Gleyses, A. C. Boccara, Near field optical microscope using local perturbation of a diffraction spot, Opt. Lett. **20**, 1924–1926 (1995)

51. C. Girard, A. Dereux, Near-field optics theories, Rep. Prog. Phys. **59**, 657–699 (1996)

52. J. J. Greffet, R. Carminati, Image formation in near-field optics, Prog. Surf. Sci. **56**, 133–237 (1997)

53. E. H. Synge, A suggested method for extending microscopic resolution into the ultramicroscopic region, Philos. Mag. **6**, 356–362 (1928)

54. A. Lewis, M. Isaacson, A. Harootunian, A. Muray, Development of a 500 Å spatial resolution light microscope, Ultramicroscopy **13**, 227–232 (1984)

55. D. W. Pohl, W. Denk, M. Lanz, Image recording with resolution $\lambda/20$, Appl. Phys. Lett. **44**, 651–653 (1984)

56. A. Piednoir, C. Licoppe, F. Creuset, Imaging and local spectroscopy with a near field optical microscope, Opt. Commun. **129**, 414–422 (1996)

57. H. Cory, A. C. Boccara, J. C. Rivoal, A. Lahrech, Electric field intensity variation in the vicinity of a perfectly conducting conical probe: Application to near-field microscopy, Microwave Opt. Technol. Lett. **18**, 120 (1998)

58. E. J. Sanchez, L. Novotny, X. S. Xie, Near-Field Fluorescence Microscopy based on two-photon excitation with metal tip, Phys. Rev. Lett. **82**, 4014–4017 (1999)

59. L. Aigouy, F. X. Andreani, C. Boccara, J. C. Rivoal, J. A. Porto, R. Carminati, J. J. Greffet, R. J. C. Megy, Near-field spectroscopy using an incoherent ligth source, Appl. Phys. Lett. **76**, 397–399 (2000)

60. J. A. Porto, R. Carminati, J. J. Greffet, Theory of electromagnetic fiels imaging and spectroscopy in scanning near-field optical microscopy, J. Appl. Phys. **88**, 4845–4850 (2000)

61. R. Hillenbrand, F. Keilmann, Complex optical constants on a subwavelength scale, Phys. Rev. Lett. **85**, 3029 (2000)

62. L. Novotny, Allowed and forbidden light in near-field optics. II- Interacting dipolar particles; J. Opt. Soc. Am. A **14**, 105–113 (1997)

63. L. Aigouy, A. Lahrech, S. Gresillon, H. Cory, A. C. Boccara, J. C. Rivoal, Polarization effects in aperturless scanning near-field optical microscopy: An experimental study, Opt. Lett. **24**, 187–189 (1999)

64. S. Gresillon, S. Ducourtieux, A. Lahrech, L. Aigouy, J. C. Rivoal, A. C. Boccara, Nanometer scale aperturless near field microscopy, Appl. Surf. Sci. **164**, 118–123 (2000)

65. S. Gresillon, L. Aigouy, A. C. Boccara, J. C. Rivoal, X. Quelin, C. Desmaret, P. Gadenne, V. A. Shubin, A. K. Sarychev, V. M. Shalaev, Experimental observation of localized excitations in random metal-dielectric films, Phys. Rev. Lett. **82**, 4520–4523 (1999)

# SERS and the Single Molecule

Martin Moskovits[1], Li-Lin Tay[2], Jody Yang[2], and Thomas Haslett[2]

[1] Department of Chemistry and Biochemistry, University of California,
Santa Barbara, CA 93106, USA
mmoskovits@ltsc.ucsb.edu

[2] Deparment of Chemistry, University of Toronto
Toronto M5S 3H6, Canada

**Abstract.** Surface Enhanced Raman spectroscopy (SERS) was discovered in 1978 and has grown to become a significant surface diagnostic and analytical technique. It has also launched a wide variety of investigations into the electromagnetic, and especially the optical, properties of nanostructured disordered materials. A number of phenomena contribute to SERS including adsorbate resonances as well as new resonances (such as metal to molecule charge transfer transitions) that result form the formation of the adsorbate-to-surface bonds, or other adsorbate-metal interactions. Chief among the contributions to SERS, however, is the enhancement of the optical fields in the vicinity of the nanoparticles constituting the SERS-active system. The field enhancement is especially high when highly localizable resonances such as surface plasmons are excited. Aggregates and assemblies of nanoparticles (of appropriate materials) can, in turn, manifest unusually enhanced SERS by virtue of particle-particle interactions. For example, while the SERS enhancement in the vicinity of single silver nanoparticles rarely exceed $10^4$, the Raman spectrum of molecules located in the interstitial volume between two closely-spaced nanoparticles can be enhanced some 10 orders of magnitude when the two particles approach each another to within molecular dimensions and the system is excited at an appropriate wavelength. Other aggregates can Show similar levels of enhancement at special locations within the aggregate. Large fractal aggregates form a special class of enhancing aggregates. Illuminating such aggregates, in general, results in a highly inhomogeneous distribution of enhancement over the body of the aggregate with electromagnetic hot Spots where the Raman enhancement can reach or slightly exceed 10 orders of magnitude. Moreover, such hot Spots can be excited with a broad range of wavelengths (although the pattern of hot spots is critically wavelength dependent). Recently, reports have been published suggesting SERS enhancements upwards of $10^{14}$, sufficient for *single molecule* SERS detection. Although the cause of such huge enhancements was at first mysterious, we suggest that these observations result from the aforementioned electromagnetic effects in aggregates combined with either intramolecular or metal-to-molecule (or molecule-tometal) resonances. We also Show that the purported optical pumping of vibrationally excited states by such intense SERS transitions is spurious.

## 1 Introduction

Surface-enhanced Raman Scattering spectroscopy (SERS) is now a well-established phenomenon that has been extensively studied for over two dec-

V. M. Shalaev (Ed.): Optical Properties of Nanostructured Random Media,
Topics Appl. Phys. **82**, 215–226 (2002)
© Springer-Verlag Berlin Heidelberg 2002

ades by several dozen groups [1]. Its attributes have attracted the attention of physicists, chemists, and engineers, and its potential as an analytical tool for detecting molecules at the subpicomolar level has kept SERS a vibrant field of inquiry. More recently, the prospect of detecting SERS from single adsorbed molecules [2,3,4] has invigorated the field enormously. Some aspects of the interpretation of those observations have created some controversy. For example, Nie reported the existence of *individual* "hot" silver and gold particles capable of SERS enhancements in excess of $10^{14}$ [2]. Although originally the authors of [2] concluded that the hot particles were individual particles of low aspect ratio, they have softened that conclusion recently in the light of more recent observations by *Käll* [2] and *Brus* [4] and now suggest that the hot "particles" might, in fact, be small compact clusters. The intriguing properties of the hot particles include very rapid saturation as a function of surface coverage by adsorbate, implying that only a small fraction of the surface of a hot particle manifests such large enhancements. Only a small fraction of a given ensemble of particles is "hot." The excitation spectra associated with the large enhancements narrowly peak at particle-size-dependent wavelengths. The aspect ratio of the particles, moreover, does not seem to correlate well with their giant enhancement abilities.

Kneipp and co-workers report an elegant series of experiments from which she argues two novel effects, both implying very large enhancements, again exceeding $10^{14}$. These large enhancements are reckoned in two ways. First, Kneipp observes unusually intense anti-Stokes Raman emission [3], implying a strongly non-Boltzmann population in vibrationally excited states of the adsorbed molecules studied. She claims that this is indicative of optical pumping by the Raman process itself. Normally Raman is orders of magnitude too weak to produce such pumping. A simple analysis of these observations suggests that for such an effect to be observed, the overall emission cross section must be enhanced by some 15 orders of magnitude, or alternatively, the local radiative intensity must be increased by an equivalent extent, or some of both mechanisms must occur whose total contribution produces a 15 order of magnitude increase over ordinary Raman. There are no known phenomena that would boost the Raman cross section so extraordinarily. If, on the other hand, one ascribes the effect to field enhancement, the resultant local field would not be tolerated by ordinary samples.

In another series of elegant experiments, Kneipp measures SERS from colloidal silver aggregates dosed with so small a quantity of adsorbate that, on average, only a single adsorbate molecule resides on any given silver aggregate [3]. A statistical analysis of the time series of the observed Raman intensities suggests that essentially every adsorbate molecule introduced into the system can be accounted for in the resulting Poisson statistics of the Raman signal suggesting that, in every case, the single adsorbate molecule riding on a silver aggregate occupies a high-enhancement location. This contrasts with Nie's observation that very few particles are capable of ultrahigh enhancement. Recently, *Käll* and co-workers [2] reported single molecule SERS from hemoglobin adsorbed on a silver colloid. In most other respects their observations paralleled those of Nie and members of his group. However, Käll's group concluded that the hot particles were dimers or small clusters of colloidal silver particles and that most of the enhancement could be accounted for by the very strong fields that, it is thought, exist in the interstices or sharp clefts in the dimers. *Michaels* et al. [4] repeated some of the experiments of Nie on silver and obtained phenomenologically the same results. Additionally, Michaels et al. measured the resonant Rayleigh spectra of the particles and concluded that intense Rayleigh and SERS scattering did not correlate.

Although some of the above observations appear to challenge the conventional understanding of SERS, in fact, they accord with our present understanding of SERS.

## 2    Results and Discussion

As it is currently understood, SERS is primarily a phenomenon associated with the enhancement of the electromagnetic field surrounding small metal (or other) objects optically excited near an intense and sharp (high $Q$), dipolar resonance such as a surface plasmon polariton. The enhanced reradiated dipolar fields excite the adsorbate, and, if the resulting molecular radiation remains at or near resonance with the enhancing object, the scattered radiation will again be enhanced (hence, the most intense SERS is really frequency-shifted elastic scattering by the metal). Under appropriate circumstances, the field enhancement will scale as $E_L^4$, where $E_L$ is the local optical field. A great deal of early SERS literature dealt with this phenomenon [5]. For particles with regions of very high curvature (ellipsoidal or rod-shaped particles), the enhancement near those sharp regions can be very much greater than for spherical or near-spherical particles.

Another contribution to the observed SERS intensity is generally referred to as "chemical enhancement." This arises from the fact that many adsorbates bind sufficiently strongly to the SERS-active surface that the Raman scatterer cannot properly be construed to be the adsorbate alone, but rather an adsorbate–surface complex, more or less analogous to a metal–ligand or perhaps cluster–ligand coordination complex. As a result, the Raman cross section of the scatterer might be increased in much the same way as the Raman cross section of ligand vibrational modes is often increased in coordination complexes over that of the free ligand. The creation of a surface complex might, moreover, lead to resonances in the visible region of the spectrum, even for colorless adsorbates, due to metal-to-molecule or molecule-to-metal transitions owing to the convenient location of the Fermi energy of most metals in an energy region intermediate between the energy of the HOMO and the LUMO of many molecules. These new states can contribute resonantly to the Raman cross section of the surface complex, increasing its magnitude further.

Although most workers in the field recognize these two contributions to SERS of which the electromagnetic contribution is the dominant (indeed, the contribution that defines the salient features of SERS, after all, all metal/ligand systems can, in principle, engage in "chemical enhancement," yet SERS is a phenomenon that is robust with only a few metals and with systems comprised predominantly of nanoparticles), the antipathy of some groups to the electromagnetic (EM) contribution [6] has obscured the origin of SERS over the past two decades. However, attempts to account for the major aspects of SERS primarily in terms of chemical contributions have not been successful.

A more subtle SERS mechanism (also often referred to as chemical enhancement), proposed by *Otto* [6] and *Persson* [7], suggests that the SERS enhancement results from the interaction of chemisorbed molecules with ballistic electrons that arise through plasmon excitations.

An EM model in terms of single particles is, in most cases, a poor model for real SERS-active systems. Most SERS-active systems are actually assemblies, sometimes very large assemblies, of coupled nanoparticles. The EM fields associated with such assemblies have been approximated in a variety of ways: gratings, fractal aggregates, small compact clusters, and periodic superlattices. For compact or periodic systems of particles, one has shown that the EM field strength at interstitial locations in the aggregate can be greatly increased over the field surrounding a single constituent particle in the aggregate [8]. Interstitial sites might also correspond to the chemically most active surface sites. An adsorbate molecule landing on a randomly rough surface might remain strongly bound to such chemically active sites after diffusing on the surface.

Many SERS-active systems, such as colloid clusters that have aggregated via so-called cluster–cluster aggregation or randomly rough surfaces produced by restricted diffusion on cold surfaces possess scaling or fractal symmetry [9]. *Stockman* [10] and *Shalaev* [11] and co-workers predicted a number of novel optical properties for such aggregates resulting from their fractality. The approach taken by those groups is to consider that the cluster or the surface is composed of a large number of individual particles each interacting through dipole–dipole coupling. The dipolar EM fields in the vicinity of such surfaces will be those corresponding to the normal modes of the interacting dipoles. (This problem is isomorphic with vibrational excitation, except that here we are dealing with plasmon "oscillations.")

Among the robust optical features of these fractal systems that distinguish them from the optical behavior of compact clusters are (1) the optically allowed normal modes span a broad range of wavelengths, whereas in compact or periodic clusters, most modes would not be dipole active. Hence, the absorption spectrum of a fractal aggregate (or surface) is, in general, broad, whereas that of a compact aggregate is much narrower. (2) Many of the normal modes will be highly localized in regions of the aggregate that are small with respect to the overall size of the aggregate and the wavelength of the exciting light. Although some normal modes of some compact clusters will be localized, this would be an accidental feature of those compact clusters. In general, for compact or periodic aggregates, the normal modes are delocalized over all or most of the volume of the cluster. (3) Most optical properties of fractal clusters, such as their absorption and their SERS excitation spectra, rapidly approach a size-independent form which will also be independent of the cluster's specific shape, so long as it belongs to the same class of fractal.

For compact clusters, size-independence is approached slowly, and the absorption spectrum will depend critically on cluster geometry. Of these three properties, the second — excitation localization — is the most intriguing. What it implies is that for essentially all fractal clusters, many of the normal modes will be so localized that the enhancement resulting from the interaction of all of the particles comprising the cluster will be concentrated in very small regions of the cluster, in so-called hot spots, where the enhancement can be many orders of magnitude higher than the average enhancement. Local enhancements $\sim 10^{11}$ have been predicted in these hot spots [11]. Because the normal mode patterns corresponding to even closely located excitation wavelengths can be quite different, the hot spots will not, in general, reflect local geometric features of the cluster. Hence, for some normal modes, the hot spots will correspond to interstitial surface sites, for others to hilltops. When the aggregate is dosed with very few molecules, as one might do in attempting to measure SERS from single molecules, an aggregate might seem "cold" or "hot" depending on whether the normal mode has a hot spot where the molecule happens to be for the excitation wavelengths used. If the molecule binds strongly to the surface, then, in general, few aggregates will seem hot. Contrariwise, a molecule that is mobile can diffuse in and out of the hot spot. It may even be trapped in the hot spot by the strong dipolar field gradients at the periphery of the hot spots.

The situation is more unpredictable for compact clusters. Because compact clusters have few normal modes for which the fields are highly concentrated, only few in an ensemble of compact clusters will appear hot and only for a narrow range of wavelengths. However, one can, in principle, find some sites, normally interstitial sites, where the electromagnetic enhancement could be such that, locally, SERS intensities $\sim 10^{11}$ are encountered. We believe that a great deal of the giant enhancements reported by Nie and Kneipp and others are due to the field-concentrating properties of aggregates. It is also clear, however, that another effect contributes as well to raise the observed enhancement to the reported $\sim 10^{14}$. We note that almost all of the results reported so far have been with molecules with a rich spectrum in the visible region of the spectrum. The added enhancement may be due to some form of resonance.

Many of the salient features of the EM theory of fractals have been corroborated using near-field microscopy and spectroscopy [12]. For example, it has been shown that the EM fields near the surfaces of self-affine, randomly rough surfaces produced by collapsing colloidal metal aggregates under the influence of gravity, are highly localized. More recently, we were able to measure good quality near-field SERS spectra from which one could construct near-field Raman maps of SERS transitions in which the hot spots are also visible. For example, Fig. 1 shows a scanning electron micrograph of a sample produced by allowing colloidal silver particles to aggregate then settle gravitationally on a glass plate. The silver aggregate was exposed to phthalazine. The near-field SERS spectrum measured at the surface of that sample with a sharpened optical fiber tip with a resolution < 100 nm is shown in Fig. 1. Maps of the SERS intensities of a number of Raman bands were obtained by manually moving the tip to lattice points over a $3 \times 3$ μm grid, measuring the near-field SERS spectrum at each point, fitting a Lorentzian under the band in each spectrum, and then plotting the result. A shear-force topographic image was taken before and after each set of measurements. The resulting maps are shown in Fig. 1. Hot spots with dimensions < 1 μm are clearly visible. We have also shown that the excitation spectra measured by probing the near field in very small regions of such samples display a number of narrow lines corresponding to the normal modes of surface plasmon excitation in which the region of the aggregate being probed participated.

Careful experiments in which the Stokes and anti-Stokes intensities in the SERS spectra of several molecules were measured over five orders of magnitude of laser excitation and showed ($i$) that although large anti-Stokes intensities were observed (for dye molecules), as previously reported, the anti-Stokes intensity depended linearly on laser power over many orders of magnitude; ($ii$) adsorbed small, colorless molecules, though showing intense SERS spectra, showed no unusual anti-Stokes intensities; and ($iii$) a careful analysis of the physics of the proposed Raman pumping would imply local radiative intensity (i. e., $\langle |E_{\mathrm{L}}|^2 \rangle$) enhancements $\sim 10^{14}$ which would result in a staggering expected SERS enhancement $\sim 10^{28}$.

Let us look at this problem in a little more detail. Following the method of Kneipp and co-workers [3], the anti-Stokes/Stokes ratios of various bands of toluene were measured using the same 830-nm laser excitation and instrument settings to take into account both the thermal population ratio and the wavelength dependence of the instrument response. Dividing the measured crystal violet anti-Stokes/Stokes ratio at wave number $\nu_{\mathrm{m}}$ by the ratio measured for a band of toluene with similar wave number produces the function $K(\nu_{\mathrm{m}})$ which should then be a wavelength and instrument-independent measure of the factor by which the anti-Stokes/Stokes ratio exceeds the expected thermal value. For Rhodamine 6G and crystal violet adsorbed on aggregated silver, values of $K$ were found (as with the authors of [3]) that greatly exceeded unity, ranging from 3 to 50. Unlike in [3], we were never able to discover a laser intensity range where the anti-Stokes intensity depended quadratically on the incident laser intensity, even at the highest laser fluxes below which the sample remained undamaged. Moreover, we found $K$ values

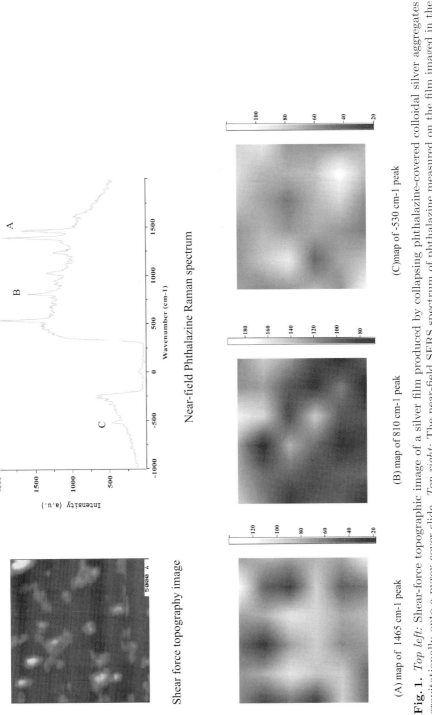

Shear force topography image

Near-field Phthalazine Raman spectrum

(A) map of 1465 cm-1 peak

(B) map of 810 cm-1 peak

(C)map of -530 cm-1 peak

**Fig. 1.** *Top left:* Shear-force topographic image of a silver film produced by collapsing phthalazine-covered colloidal silver aggregates gravitationally onto a pyrex cover slide. *Top right:* The near-field SERS spectrum of phthalazine measured on the film imaged in the previous panel. A, B, C: near-field Raman intensity maps of the SERS bands shown in the previous spectrum measured over a $3 \times 3$ μm portion of the colloidal silver film

greatly exceeding unity even at the very lowest laser powers used. Both of these observations seriously challenge the notion that the inordinately large anti-Stokes to Stokes ratios derive from optical pumping of vibrationally excited states by Raman transitions, as proposed in [3].

# 3  Interpretation

To propose a plausible alternative explanation, we begin by deriving an expression for $K$ which explicitly allows for the presence of a resonant contribution to the Raman process and hence for differing Stokes and anti-Stokes cross sections, $\sigma_S$ and $\sigma_A (cm^{-2})$, of the surface-adsorbed molecule. In addition, we consider the simplest case of Stokes and anti-Stokes SERS field enhancement, given by $G(\nu_i - \nu_m) = \beta_i^2 \beta_S^2$ and $G(\nu_i + \nu_m) = \beta_i^2 \beta_A^2$, where $\nu_i$ and $\nu_m$ are, respectively, the frequencies of the incident radiation and vibration $m$ and the parameters $\beta_i$, $\beta_S$, and $\beta_A$ denote the field enhancement at the incident, Stokes, and anti-Stokes wavelengths. This approach is more detailed than that of [3], where all of the enhancement, it was assumed, belongs to the effective cross section and is not associated with the field. This assumes implicitly that the Stokes and anti-Stokes cross sections are equivalent, which, in general, is not so, and that the entire magnitude of the SERS enhancement is effective in pumping population from the ground to the vibrationally excited state. In fact this is not the case. The electromagnetic contribution to the SERS enhancement, it is understood, is due to two resonances: the first is the resonance of the incident laser field with a surface plasmon polariton mode of the SERS-active system; the second is the resonance of the Raman-scattered field with the surface plasmon. The second process is clearly an electromagnetic effect that, in principle, does not involve the adsorbed molecule and, hence, would not contribute to the pumping of vibrationally excited molecular states. (This is not strictly true. One could postulate that the Raman-shifted field is involved in a variety of stimulated processes. However, we have considered this possibility and found that the inclusion of stimulated processes leads to strongly nonlinear dependences of the Raman signals on laser intensity. Accordingly, we will not discuss such processes further.) Since only the $\beta_i^2$ part of the field enhancement participates in any Raman pumping process and the $\beta_S^2$ and $\beta_A^2$ terms contribute only to the observed Stokes and anti-Stokes intensities without contributing to the pumping, the field enhancements associated with SERS, even taking into account the real possibility of the existence of hot spots in the aggregated samples where the field-intensity exceeds the average by four or five orders of magnitude, cannot account for the levels of field enhancement required to produce the anti-Stokes/Stokes ratios reported. Moreover, ascribing the effect to an inordinate increase in cross section introduces a new complexity in the mechanism for so unusual a cross-sectional increase that far exceeds what is encountered in even the most efficient resonance Raman processes and certainly exceeds those occurring near metal surfaces where radiative rates are normally increased. Even should a more detailed and complex model for the field enhancement be considered such as those proposed by *Stockman* and *Shalaev* [10,11], the point remains that only a small fraction of the SERS enhancement can be associated with the enhancment of the field that can play a role in populating the excited vibrational state.

An expression for $K(\nu_{\mathrm{m}})$ derived for a simple twostate model in terms of the lifetime of the excited vibrational state $\tau(\mathrm{s})$ and the incident light intensity $n_{\mathrm{L}}$ (photons cm$^{-2}$s$^{-1}$) is easily derivable. The expression, which assumes that a steady state is established in the vibrational state populations by the exciting field, the various optical transitions, and nonradiative depopulation processes, is

$$K(\nu_{\mathrm{m}}) = \left(\frac{\sigma_{\mathrm{A}}\beta_{\mathrm{A}}^2}{\sigma_{\mathrm{S}}\beta_{\mathrm{S}}^2}\right)\left[\frac{\sigma_{\mathrm{S}}(\nu_{\mathrm{m}})\tau(\nu_{\mathrm{m}})\beta_{\mathrm{i}}^2 n_{\mathrm{L}}e^{h\nu_{\mathrm{m}}/k_{\mathrm{B}}T}+1}{\sigma_{\mathrm{A}}(-\nu_{\mathrm{m}})\tau(\nu_{\mathrm{m}})\beta_{\mathrm{i}}^2 n_{\mathrm{L}}+1}\right]. \tag{1}$$

This expression reduces to that used in [3] in the limit of $\sigma_{\mathrm{A}}\tau n_{\mathrm{L}} \ll 1$ by setting $\sigma_{\mathrm{S}} = \sigma_{\mathrm{A}}$ and setting all values of $\beta$ to unity since all of the enhancement is combined in the cross section. This expression suggests a number of points: (1) $K$ approaches unity for small values of $n_{\mathrm{L}}$ and nonresonant excitation processes (i. e., $\sigma_{\mathrm{S}} = \sigma_{\mathrm{A}}$), provided that $\beta_{\mathrm{A}} \approx \beta_{\mathrm{S}}$; (2) the Stokes and anti-Stokes cross sections need not be the same under resonance conditions; therefore $K$ need not be unity even if the field enhancement and/or $n_{\mathrm{L}}$ is small; and (3) if we consider a value of $\nu_{\mathrm{m}} = 1200\,\mathrm{cm}^{-1}$, for example, then $e^{h\nu_{\mathrm{m}}/k_{\mathrm{B}}T} \approx 400$, indicating that if the Raman pumping mechanism is operating, there exists only a small range of incident radiation intensities before saturation is achieved. Therefore, it is unreasonable to truncate the denominator to unity as in [3] since by doing so, the correct saturation behavior of $K$ will not be realized. Exploring the expression further and taking typical parameter values from spectra of crystal violet: $n_{\mathrm{L}} = 10^{-24}\,\mathrm{cm}^{-2}\mathrm{s}^{-1}$, $\tau = 10^{-11}\mathrm{s}$, and $K(1174\,\mathrm{cm}^{-1}) = 8$; setting $\sigma_{\mathrm{S}} = \sigma_{\mathrm{A}}$ and $\beta_{\mathrm{A}} \approx \beta_{\mathrm{S}}$ and assuming that Raman pumping is the dominant mechanism for the nonunity value of $K$ leads to the condition $\sigma_{\mathrm{S}}\beta_{\mathrm{i}}^2 \approx 10^{-15}\,\mathrm{cm}^2$. If we further assume a large resonant Raman cross section of $10^{-25}\,\mathrm{cm}^2$, a local-field intensity at the molecule, this results in a prediction of incident field intensities of the order $\beta_{\mathrm{i}}^2 n_{\mathrm{L}} \approx 10^{34}\,\mathrm{cm}^{-2}\mathrm{s}^{-1}$, or over $10^{15}\,\mathrm{W/cm}^2$. Clearly, the adsorbate would not survive such fields intact, not to mention the fact that we assumed a value of $\beta_{\mathrm{i}}^2 \approx 10^{10}$ which is several orders of magnitude larger than most existing estimates of the average field-intensity enhancement. Any attempt to reduce the value of the local-field intensity would imply a proportional increase in the Raman cross section, resulting in resonant enhancements greater than $10^{10}$ which is already uncommonly high. It is worth noting that the power taken above is at the higher end of the powers used in our study. If we go to the lower powers, where nonunity values of $K$ are still observed, a further four to five orders of magnitude must be assumed for the value of $\sigma_{\mathrm{S}}\beta_{\mathrm{i}}^2$. None of these scenarios is as plausible as the possibility that the Stokes and anti-Stokes cross sections or field enhancements or both differ by relatively small factors and that no pumping is in fact taking place.

To probe further the origin of the nonunity values of $K$, SERS spectra of nondye adsorbates, including nitropyridine, phthalazine, benzoic acid, and pyrazine were investigated. In all of those cases, $K$ was $1 \pm 0.5$, implying that the anti-Stokes signal arises (within experimental error) from purely thermal effects. Within the context of our explanation, this observation also implies that for this last set of adsorbate, there is no resonance effect and $\sigma_S = \sigma_A$. This is further evidence that the origin of the unusual anti-Stokes intensities observed with the dyes is somehow related to unequal Raman cross sections for the Stokes and anti-Stokes processes, likely arising from the effects of resonance. One should note that neither of the dye molecules used has a significant absorption cross section at the excitation wavelength used (830 nm). Hence, the resonance postulated must involve a new resonance involving the adsorbate–surface complex. Such a resonance does not require the adsorbate to be a dye, i.e., to possess absorptions in the visible in its unadsorbed state. Nevertheless, dye molecules do possess states conveniently located in energy from which resonant transitions for the surface complex might be more readily manufactured. This is the likely reason why resonance is observed for adsorbed dyes, even when excited with a wavelength that is not resonant for the dye in solution and not observed for the series of colorless molecules referred to above when they are adsorbed on silver.

# 4   Conclusion

From this, we conclude that a better explanation for the observed intense anti-Stokes emission can be framed in terms of EM SERS enhancements coexisting with resonance Raman enhancement. The latter can result in unequal Raman cross sections for Stokes and anti-Stokes transitions [13]. This does not require one to propose the occurrence of optical pumping of excited vibrational states by the surface-enhanced Raman transitions.

Therefore, we believe that the results reported by Nie and Kneipp so far can be accounted for by a combination of strong local EM excitation in colloidal metal aggregates (generally small compact aggregates) with added contributions from resonance enhancements. The fact that, so far, the reports have focused largely on adsorbed dye molecules is, in our view, not a coincidence. One needed the opportunity for additional resonance enhancement that those systems provide, on top of electromagnetic enhancement. Because AFM images tend to show the particles as "compact", we propose that the particles involved are either compact colloidal aggregates or "failed fractals" [14]. Such aggregates would possess few normal plasmon modes that would have highly localized and concentrated fields (hot spots). Because of the high localization of the fields, only a few aggregates would manifest a SERS spectrum under circumstances where only one or very few adsorbate molecules decorate a given cluster. Only those clusters in which the adsorbed molecule is coincidentally bound at a location on the cluster corresponding to a hot spot would show inordinate SERS intensities. This explains the rarity of hot particles and the sharpness of the excitation spectrum for a given group of hot particles. This also accounts for the fact that as one continues to add adorbate to a sample of particles, progressively more particles become "hot." Adding adsorbate increases the probability that an ad-molecule eventually binds to a location on the aggregate where the field is concentrated. This also explains the fact that for a given particle, the SERS intensity saturates rapidly and then becomes roughly independent of further coverage. Once the surface sites within the hot spots on an aggregate are populated, further surface coverage would not add significantly to the SERS intensity.

According to this explanation, localization of the EM field due to particle–particle interactions in an aggregate results in a SERS enhancement of the order of $10^{11}$ [11]; the remaining enhancement arises from resonance Raman and other so-called chemical enhancement effects. The fact that resonant Rayleigh and SERS scattering do not correlate also accords well with this view. Rayleigh scattering is a coherent process that depends on averaging a two-point field correlation over the entire volume of the aggregate. Rayleigh scattering is phase-dependent, and the variation in the field's phase and amplitude from point to point on the aggregate is important. Raman is an incoherent, phase-independent process in which the enhancement is defined by the one-point correlation: $< |E(r)|^4 >$. The very intense fields that exist at hot spots would dominate this average [15]. Hence the aggregates that show intense Rayleigh scattering need not be those manifesting the most intense SERS signals or vice versa. Our explanation also does away with the implied requirement that there are very special and rare adsorption sites on a particle where the adsorbate molecule binds chemically. Although we recognize the known fact that some surface sites can be more reactive than others, the very early saturation of the SERS signal with added adsorbate [4] suggests that surface sites suitable for chemisorption are very rare indeed. There is no reason to suppose that silver aggregates possess such rare, highly reactive adsorption sites. The model we propose does not require such highly site-selective surface chemistry.

# References

1. M. Moskovits: Rev. Mod. Phys. **57**, 783 (1985); R. K. Chang: Ber. Bunsen Ges. Phys. Chem. **91**, 296 (1987)
2. J. T. Krug, G. D. Wang, S. R. Emory, S. M. Nie: J. Am. Chem. Soc. **121**, 9208 (1999); S. R. Emory, W. E. Haskins, S. M. Nie: J. Am. Chem. Soc. **120**, 8009 (1998); W. A. Lyon, S. M. Nie: Anal. Chem. **69**, 3400 (1997); S. M. Nie:

Emery SR, Science **275**, 1102 (1997); H. Xu, E. J. Bjerneld, M. Käll, L. Börjesson: Phys. Rev. Lett. **83**, 4357 (1999)

3. K. Kneipp, Y. Wang, H. Kneipp, I. Itzkan, R. R. Dasari, M. S. Feld: Phys. Rev. Lett. **76**, 2444 (1996); K. Kneipp, H. Kneipp, I. Itzkan, R. R. Dasari, M. S. Feld: Chem. Phys. **247**, 155 (1999); K. Kneipp, H. Kneipp, R. Manoharan, E. B. Hanlon, I. Itzkan, R. R. Dasari, M. S. Feld: Appl. Spectrosc. **52**, 1493 (1998); K. Kneipp, H. Kneipp, R. Manoharan, I. Itzkan, R. R. Dasari, M. S. Feld: J. Raman Spectrosc. **29**, 743 (1998); K. Kneipp, H. Kneipp, V. B. Kartha, R. Manoharan, G. Deinum, I. Itzkan, R. R. Dasari, M. S. Feld: Phys. Rev. E **57**, R6281 (1998); K. Kneipp, Y. Wang, H. Kneipp, L. T. Perelman, I. Itzkan, R. Dasari, M. S. Feld: Phys. Rev. Lett. **78**, 1667 (1997)

4. A. M. Michaels, M. Nirmal, L. E. Brus: J. Am. Chem. Soc. **121**, 9932 (1999)

5. M. Kerker, D. Wang, H. Chew: Appl. Opt. **19**, 4159 (1980)

6. A. Otto: in *Light Scattering in Solids IV*, M. Cardona, G. Gundtherodt (Eds.) (Springer Berlin, Heidelberg 1984) pp. 289–418

7. B. N. Persson: Chem. Phys. Lett. **82**, 561 (1981)

8. P. K. Aravind, A. Nitzan, H. Metiu: Surf. Sci. **110**, 189 (1981); N. Liver, A. Nitzan, J. I. Gersten: Chem. Phys. Lett. **111**, 449 (1984); A. Wirgin, T. López-Ríos: Opt. Commun. **48**, 416 (1984); N. Garcia, G. Diaz, J. J. Saenz, C. Ocal: Surf. Sci. **143**, 342 (1984); P. A. Kneipp, T. L. Reinecke: Phys. Rev. B **45**, 9091 (1992); F. J. García-Vidal, J. B. Pendry: Phys. Rev. L **77**, 1163 (1996)

9. D. A. Weitz, M. Oliveria: Phys. Rev. Lett. **52**, 1433 (1984)

10. M. I. Stockman: Phys. Rev. E **56**, 6494 (1997)

11. V. M. Shalaev, R. Botet, J. Mercer, E. B. Stechel: Phys. Rev. B **54**, 8235 (1996); E. Y. Poliakov, V. M. Shalaev, V. A. Markel, R. Botet: Opt. Lett. **21**, 1628 (1996); V. M. Shalaev, A. K. Sarychev: Phys. Rev. B **57**, 13265 (1998); S. Grésillon, L. Aigouy, A. C. Boccara, J. C. Rivoal, X. Quelin, C. Desmarest, P. Gadenne, V. A. Shubin, A. K. Sarychev, V. M. Shalaev: Phys. Rev. Lett. **82**, 4520 (1999)

12. V. A. Markel, V. M. Shalaev, P. Zhang, W. Huynh, L. Tay, T. L. Haslett, M. Moskovits: Phys. Rev. B **16**, 8080 (1999); P. Zhang, T. Haslett, C. Douketis, M. Moskovits: Phys. Rev. B **57**, 15513 (1998); D. P. Tsai, J. Kovacs, Z. Wang, M. Moskovits, V. Shalaev, J. S. Suh, R. Botet: Phys. Rev. Lett. **72**, 4149 (1994)

13. T. L. Haslett, L. Tay, M. Moskovits: J. Chem. Phys. **113**, 1641 (2000)

14. A. M. Michaels, J. Yang, L. Brus, unpublished

15. V. M. Shalaev, personal communication

# Nonlinear Raman Probe of Single Molecules Attached to Colloidal Silver and Gold Clusters

Katrin Kneipp[1,2], Harald Kneipp[2], Irving Itzkan[1],
Ramachandra R. Dasari[1], Michael S. Feld[1], and Mildred S. Dresselhaus[1,3]

[1] Massachusetts Institute of Technology
Cambridge, MA 02139, USA
kneipp@usa.net
[2] Technical University Berlin
10623 Berlin, Germany
[3] Office of Science, Department of Energy
Washington DC 2058, USA

**Abstract.** We review surface-enhanced linear and nonlinear Raman scattering experiments on molecules and single wall carbon nanotubes attached to colloidal silver and gold clusters. Surface-enhanced hyper-Raman scattering and surface-enhanced anti-Stokes Raman scattering from pumped vibrational levels are studied as two-photon excited Raman processes where the scattering signal depends quadratically on the excitation laser intensity. The experimental results are discussed in the framework of strongly enhanced electromagnetic fields predicted for such cluster structures in so-called "hot spots." The electromagnetic enhancement factors for Stokes, pumped anti-Stokes, and hyper-Raman scattering scale as theoretically predicted, and the field strengths in the hot spots, it is inferred, are enhanced of the order of $10^3$. From our experiments we claim a very small density of hot spots (0.01 % of the cluster surface) and lateral confinement of the strong field enhancement within domains that can be as small as 10 nm.

Effective cross sections of the order of $10^{-16}\,\mathrm{cm}^2$ and $10^{-42}\,\mathrm{cm}^4\,\mathrm{s}$ for Stokes and pumped anti-Stokes scattering, respectively, are adequate for one- and two-photon Raman spectroscopy of single molecules.

## 1 Introduction

During the last decade, detecting and characterizing single molecules, including artificial large molecules, such as nanotubes or quantum dots, using laser spectroscopy became a matter of growing interest [1,2,3]. In general, since the spectroscopic signal is proportional to the number of molecules that contribute to the signal, single molecule spectroscopy means dealing with very low signals. Therefore, methods for enhancing optical signals are essential for developing single molecule spectroscopy. Exciting opportunities for enhancing spectroscopic signals exist, when the target species is attached to metallic nanostructures. In the very close vicinity of such structures, local optical fields can be strongly enhanced when they are in resonance with the collective excitation of conduction electrons, also called surface plasmon resonances. Particularly high field enhancement seems to exist for ensembles of

V. M. Shalaev (Ed.): Optical Properties of Nanostructured Random Media,
Topics Appl. Phys. **82**, 227–247 (2002)
© Springer-Verlag Berlin Heidelberg 2002

metallic nanoparticles exhibiting fractal properties, such as colloidal silver or gold clusters formed by aggregation of colloidal particles or island films of those metals [4,5,6]. Plasmon resonances in such structures occur over a relatively wide distribution of frequencies and simultaneously, they tend to be spatially localized and enhanced in small areas, so-called "hot spots" whose dimensions can be smaller than tenths of a wavelength [7]. Very favorable conditions for single molecule spectroscopy exist when the target molecule is attached to a hot spot and can be probed by the high local optical fields. Moreover, the strong lateral confinement of the field enhancement provides an additional opportunity for selecting a single species. Hot areas on cluster structures provide particularly attractive opportunities for nonlinear probes of single molecules, where the signal depends on the enhanced field intensities raised to a power of two or greater.

Because of its high information content on chemical structure, Raman scattering is a very promising technique for single molecule spectroscopy. The information about a molecule is considerably increased when nonlinear Raman techniques are applied in addition to "normal" linear Raman scattering. For example, hyper-Raman scattering follows different selection rules than normal Raman scattering, and therefore it can probe so-called "silent modes" that are forbidden in normal Raman scattering and in infrared absorption [8,9]. The general disadvantage of Raman spectroscopy and also of nonlinear Raman spectroscopic techniques is the extremely small cross section that makes, for example, the use of hyper-Raman scattering nearly impossible as a practical "spectroscopic tool." Linear Raman scattering cross sections fall between $10^{-30}$ cm$^2$ and $10^{-25}$ cm$^2$ per molecule; the larger values occur under favorable resonance Raman conditions; hyper-Raman scattering has cross sections of approximately $10^{-65}$ cm$^4$ s/ photon.

In spite of these small cross sections, recently, single molecule Raman spectra have been measured by several groups (for an overview see [10]). The experiments are based on the phenomenon of a strongly increased Raman signal from molecules attached to metallic nanostructures, the effect of so-called "surface-enhanced Raman scattering" (SERS) [11,12,13]. The unexpectedly high Raman scattering signal from molecules attached to a metal substrate with nanometer-scaled structure or "roughness" [14,15,16] might be one of the most impressive effects for demonstrating the interesting optical properties of metallic nanostructures, which occur due to resonances with the plasmon excitations in the metal. In particular, strong enhancement has been observed when molecules are attached to colloidal silver or gold clusters or to island structures of those metals. The high local optical fields in the hot spots of such cluster structures provide a rationale for the high enhancement level, which is necessary for nonresonant single molecule Raman spectroscopy [17].

According to the electromagnetic field enhancement model, nonlinear Raman effects should be surface enhanced to a greater extent than "normal" Raman scattering. In agreement with this hypothesis, for molecules attached to colloidal silver clusters, we obtained Raman scattering and hyper-Raman scattering at nearly the same signal level since the much stronger field enhancement for the nonlinear effect compensated for its smaller cross section [18,19].

In this article, we review surface-enhanced linear and nonlinear Raman scattering experiments. In particular, we want to discuss the feasibility of applying nonlinear Raman scattering to single molecules located in the hot areas of a metal cluster structure and the potential of SERS experiments performed on large artificial molecules such as single wall carbon nanotubes. Surface-enhanced hyper-Raman scattering and "pumped" anti-Stokes Raman scattering are selected as nonlinear or two-photon excited Raman probes where the Raman scattering signal depends quadratically on the excitation laser intensity. Our experimental findings in surface-enhanced Stokes, anti-Stokes, and hyper-Raman scattering as well as in surface-enhanced Raman scattering from single wall carbon nanotubes are discussed in the framework of a strong local electromagnetic field enhancement in the hot spots of colloidal metal clusters.

# 2    Surface-Enhanced Linear and Nonlinear Raman Scattering

In this section, we briefly introduce surface-enhanced linear and nonlinear Raman scattering and discuss experiments when the target species is attached to colloidal silver or gold clusters.

## 2.1    Experimental

Figure 1 shows the schematic of an experimental setup for surface-enhanced linear and nonlinear Raman spectroscopy on molecules attached to colloidal silver and gold clusters [20]. The colloidal cluster–target molecule complex is provided in aqueous solution or "dry" on a glass slide. Raman scattering is excited using an argon-ion laser pumped Ti:sapphire laser operating in the near infrared (NIR). The Raman light can be observed at the Stokes and anti-Stokes side of the NIR excitation laser. Hyper-Raman scattering can be measured at the Stokes side of the second harmonic of the NIR excitation laser in the near-ultraviolet region.

Surface-enhanced Stokes and anti-Stokes Raman spectra are excited in the CW mode of the Ti:sapphire laser using $10^4$–$10^6$ W/ cm$^2$ intensities. For surface-enhanced hyper-Raman studies, the laser is used in the mode-locked picosecond regime to achieve excitation intensities of about $10^7$ W/ cm$^2$. Grating spectrographs were used to disperse the scattered light. Surface-enhanced Stokes and anti-Stokes Raman light (shifted relative to the NIR excitation) and surface-enhanced hyper-Raman light (shifted relative to the second harmonic of the NIR light) can be measured simultaneously in the

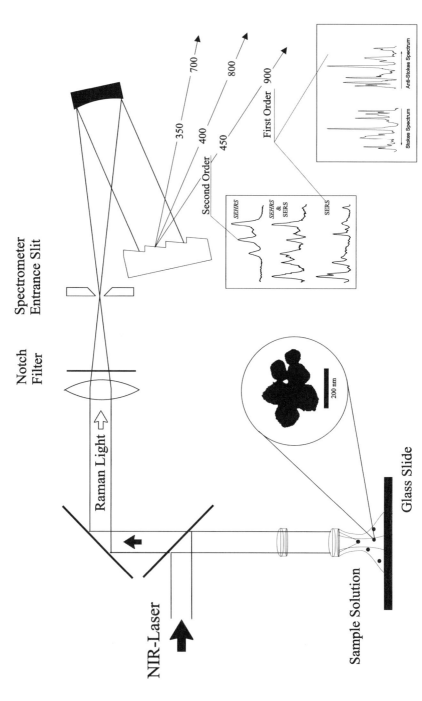

**Fig. 1.** Schematic experimental setup for surface-enhanced linear and nonlinear Raman spectroscopy

same spectrum using the first and second diffraction order of the spectrograph. This allows direct measurement of the ratio between SEHRS and SERS signal power [18,19]. The relative wavelength sensitivity of the system for Stokes Raman, compared to anti-Stokes Raman and hyper-Raman scattering, was determined by using Stokes and anti-Stokes Raman spectra of benzene measured at NIR excitation and its second harmonic.

## 2.2   Surface-Enhanced Linear Raman Scattering

Raman signals can be enhanced by more than 10 orders of magnitude, when the molecule is attached or in the very close vicinity of silver or gold structures in dimensions of tens of nanometers. In analogy to "normal" Raman scattering, the number of Stokes photons per second $P_S^{\mathrm{SERS}}$ in surface-enhanced Raman scattering can be written as

$$P_S^{\mathrm{SERS}} = N_0 \sigma_s^{\mathrm{SERS}} n_L , \tag{1}$$

where $\sigma_s^{\mathrm{SERS}}$ describes an effective cross section of the surface-enhanced Raman process and $S$ denotes the Stokes scattering. $n_L$ is the photon flux density of the excitation laser which is equal to the incoming laser field $|E^{(0)}(\nu_L)|^2$ divided by $h\nu_L$ and by the focus area. $E(\nu)$ is the field strength, and $\nu_L$ and $\nu_S$ are the laser and the Stokes frequencies with

$$\nu_S = \nu_L - \nu_M$$

or

$$\nu_{aS} = \nu_L + \nu_M . \tag{2}$$

$\nu_M$ is the molecular vibrational frequency and minus and plus stand for Stokes and anti-Stokes scattering, respectively. $N_0$ is the number of molecules in the vibrational ground state, that are involved in the SERS process.

It is generally agreed that more than one effect contributes to the enhancement of the Raman signal [11,12,13]. The enhancement mechanisms are roughly divided into so-called "electromagnetic" field enhancement effects and "chemical first layer" effects. The latter effects include enhancement mechanism(s) of the Raman signal, which can be explained in terms of specific interactions, i. e., electronic coupling between molecule and metal [21], resulting in a larger Raman cross section $\sigma_{\mathrm{ads}}^{\mathrm{RS}}$ compared to that of the molecule without coupling to the metal $\sigma_{\mathrm{free}}^{\mathrm{RS}}$. The electromagnetic field enhancement arises from an enhanced local optical field at the place of the molecule, described by field enhancement factors $A(\nu)$. Then, the SERS cross section can be written as

$$\sigma_s^{\mathrm{SERS}} = \sigma_{\mathrm{ads}}^{\mathrm{RS}} |A(\nu_L)|^2 |A(\nu_S)|^2 , \tag{3}$$

where

$$|A(\nu)|^2 = \frac{|E(\nu)|^2}{|E^{(0)}(\nu)|^2} . \tag{4}$$

$E(\nu)$ are the local optical fields (laser and the scattered field) and $E^{(0)}(\nu)$ are the same fields in the absence of the metal nanostructures.

The SERS enhancement factor $G_{\mathrm{SERS}}$ for Stokes scattering is determined by the ratio of the effective SERS cross section $\sigma_S^{\mathrm{SERS}}$ to the "normal" Raman cross section $\sigma_{S,\mathrm{free}}^{\mathrm{RS}}$,

$$G_{\mathrm{SERS}} = \frac{\sigma_{\mathrm{ads}}^{\mathrm{RS}}}{\sigma_{\mathrm{free}}^{\mathrm{RS}}} |A(\nu_{\mathrm{L}})|^2 |A(\nu_{\mathrm{S}})|^2 . \tag{5}$$

The first term in formula (5), $\sigma_{\mathrm{ads}}^{\mathrm{RS}}/\sigma_{\mathrm{free}}^{\mathrm{RS}}$, describes the "chemical" enhancement effect. Chemical SERS enhancement factors may contribute factors of 10 to 1000 to the total SERS enhancement [21,13]. The second term describes local-field enhancement effects.

The electromagnetic contribution to SERS enhancement strongly depends on the morphology of the metal nanostructures and on the dielectric constants of the metal (for an overview, see [11,12,13]). The field enhancement exhibits particularly exciting properties for fractal metallic nanostructures [22,5,6,4,23] and can reach 12 orders of magnitude for colloidal silver and gold cluster structures [24,25,26].

In the following, we discuss some experimental observations on SERS performed on colloidal silver and gold clusters. Figures 2a and 2b display Stokes and anti-Stokes SERS spectra of crystal violet attached to isolated colloidal gold spheres in aqueous solution. Figure 3a shows the electron micrographic view of one gold sphere and the extinction spectrum of the aqueous solution of many isolated spheres. NaCl was added to the aqueous solution to induce aggregation of the spheres, but spectra in Figs. 2a and 2b were measured in the first minutes after addition of the salt. No changes in the extinction spectrum of the colloidal solution were observed, indicating that no aggregation occurred during this time. Therefore, the electromagnetic SERS enhancement should be mainly related to isolated gold spheres of about 60 nm in diameter. On the other hand, if there is any additional "chemical" enhancement related to NaCl induced "active sites" [27], that effect should already exist. After several minutes, when changes in the extinction spectrum from curve $a$ to curve $b$ in Fig. 3 indicated the formation of colloidal gold clusters (see also electron micrographs in Fig. 3), the SERS Stokes signal strongly increased (Fig. 2c). Now a strong anti-Stokes spectrum also occurs, as shown in Fig. 2d. Particularly, higher frequency modes appear at unexpectedly high signal levels in the anti-Stokes spectrum. This behavior indicates a very high SERS enhancement since molecules that are "pumped" to the first excited vibrational levels due to the strong Raman process now contribute to the anti-Stokes signal in addition to the thermally excited molecules [24]. We discuss this "pumped" anti-Stokes Raman scattering as a nonlinear Raman process in more detail later. Here, we use the anti-Stokes scattering only for a rough estimate of the SERS enhancement factor. In the stationary case as in our CW experiments, SERS cross sections are inferred from anti-Stokes to

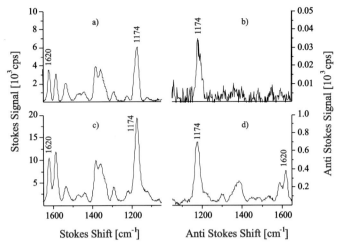

**Fig. 2.** Stokes (**a, c**) and anti-Stokes (**b, d**) SERS spectra of crystal violet on isolated colloidal gold spheres (**a, b**) and on colloidal gold clusters (**c, d**)

Stokes SERS signal ratios $P_{aS}^{SERS}/P_S^{SERS}$ normalized to the ratio in a normal Raman experiment $P_{aS}^{RS}/P_S^{RS}$ (Boltzmann population) according to [24]:

$$\frac{P_{aS}^{SERS}/P_S^{SERS}}{P_{aS}^{RS}/P_{aS}^{RS}} = \sigma^{SERS}\tau_1 e^{\frac{h\nu_M}{kT}} n_L + 1 \,, \tag{6}$$

where $\tau_1$ is the lifetime of the first excited vibrational state assumed to be of the order of 10 ps [28]. T is the sample temperature (300 K), and $h$ and $k$ are the Planck and Boltzmann constants, respectively [30].

In the anti-Stokes SERS spectrum measured from crystal violet on isolated gold spheres, the higher frequency anti-Stokes bands do not appear due to their weak thermal population. The anti-Stokes to Stokes signal ratio of the $1174\,cm^{-1}$ crystal violet SERS line measured in Fig. 2a,b is in agreement with the ratio measured for the $1211\,cm^{-1}$ line of toluene, which represents the Boltzmann population of the vibrational levels. That means that no Raman pumping can be observed for the smaller enhancement factors of isolated colloidal gold spheres since the effective Raman cross section is not large enough for measurably populating the first vibrational levels. According to formula (6), in SERS experiments on colloidal gold clusters, effective SERS cross sections on the order of $10^{-16}\,cm^2$ per molecule must be operative, corresponding to total enhancement factors of about $10^{14}$ for a nonresonant Raman process in order to explain the observed anti-Stokes spectra.

Similar total enhancement factors of about $10^{14}$ were found for colloidal silver clusters. Figure 4 shows surface-enhanced Stokes and anti-Stokes Raman spectra of the DNA base adenine on colloidal silver clusters displaying the strong Raman line of the adenine ring breathing mode at $735\,cm^{-1}$ and lines in the $1330\,cm^{-1}$ region. As an indication of a very strong SERS effect,

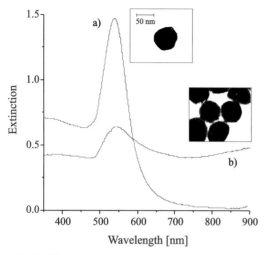

**Fig. 3.** Electron micrographs and extinction curves of isolated colloidal gold spheres and of colloidal clusters

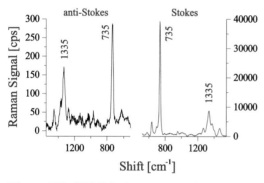

**Fig. 4.** NIR-SERS Stokes and anti-Stokes spectra of adenine measured on a silver cluster of about 8 μm. Parameters: 100 mW, 830 nm excitation with 1 μm spot size and 1 s collection time

higher frequency lines at the anti-Stokes side appear at relatively high signal levels, and effective Raman cross sections of the order of $10^{-16}$ cm$^2$/molecule can also be inferred for adenine [31]. Adenine has absorption bands in the ultraviolet. Therefore, at 830 nm excitation, no molecular resonance Raman effect will contribute to the large Raman cross section observed.

SERS enhancement factors of $10^{14}$, as they are inferred from vibrational pumping, are also confirmed in single molecule Raman experiments. Figure 5a shows 100 Raman spectra measured in time sequence from a sample, which contains an average of 0.6 crystal violet molecules in the probed volume. Single target molecules are attached to colloidal silver clusters (see magnification glass in Fig. 1) and move into and out of the probed volume due to

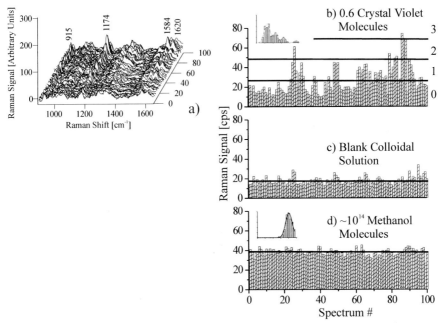

**Fig. 5.** Single molecule SERS spectra for confirming SERS enhancement factors of the order of $10^{14}$ (see text for explanation). (**a**) 100 SERS spectra collected from an average of 0.6 crystal violet molecules; (**b**) Peak heights of the 1174 cm$^{-1}$ line; (**c**) background signal; (**d**) peak heights of a Raman line measured from $\sim 10^{14}$ methanol molecules. The *horizontal* lines display the thresholds for one-, two- and three-molecule signals (**b**), the average background (**c**), and the average $10^{14}$-molecule signal (**d**). The *insets* in (**b**) and (**d**) show the Poisson and Gaussian statistics of single molecule and many molecule Raman signals, respectively. (Reprinted with permission from [25]. Copyright 1997 American Institute of Physics)

Brownian motion. Figure 5b displays the signal strengths of the 1174 cm$^{-1}$ crystal violet peak in the 100 measurements showing a Poisson distribution for seeing just one, two, three, or, very likely zero molecules in the probed volume. Figure 5c shows the background signal at 1174 cm$^{-1}$ when no crystal violet is in the sample. For comparison, Fig. 5d shows the signal of the 1030 cm$^{-1}$ methanol Raman line, which comes from about $10^{14}$ methanol molecules in the probed volume. Methanol does not show SERS enhancement. A comparison between Figs. 5b and 5d shows that SERS signals of a single crystal violet molecule appear at the same level as the normal Raman signal of $10^{14}$ methanol molecules, confirming a SERS enhancement factor of the order of $10^{14}$ compared to a nonresonant Raman process [25].

## 2.3    Surface-Enhanced Raman Scattering from Single Wall Carbon Nanotubes

The theory predicts that extremely strong field enhancement is confined within very small areas, smaller than the wavelength of the light [7], but still large compared to the size of a molecule. Therefore, a single molecule SERS experiment performed on a "normal" molecule provides no information about the dimension of a hot spot. This can be different for large artificial molecules, such as single wall carbon nanotubes. A single wall carbon nanotube (SWNT) is a graphene sheet rolled up into a seamless cylinder to form a high aspect ratio (length/diameter) one-dimensional (1-D) macro-molecule, with a cylinder typically from 1 to 3 nm in diameter and a few microns long. SWNTs can be semiconducting or metallic, depending on the nanotube geometry. Raman spectroscopy provides a sensitive probe to distinguish semiconducting from metallic nanotubes through measurements on the tangential $G$-band feature near $1580 \, \text{cm}^{-1}$. The variation in tube diameters results in small changes in the Raman frequencies for the tangential $G$ band from tubes of different diameters [32]. Therefore, Raman spectra measured from a bundle of nanotubes, normally consisting of tubes with a diameter distribution, show inhomogeneously broadened Raman lines.

In our experiment, small bundles of single wall carbon nanotubes are attached to a fractal colloidal silver cluster [26]. When nanotubes are in contact with the cluster, the Raman signal can be enhanced by more than 10 orders of magnitude. Simultaneously, the inhomogeneous line width of the Raman line can strongly decrease. Figure 6 shows SERS spectra in the region of the tangential $G$ band of semiconducting tubes measured from a bundle of tubes on a fractal silver cluster. The line width of the Raman band changes from place to place on the cluster, and in very rare cases, it becomes about $10 \, \text{cm}^{-1}$. This value is very close to the theoretical homogeneous line width, suggesting that only a very small number of tubes, maybe even a single tube, just in contact with the cluster at a "hot spot," contributes to the SERS spectrum. Single wall carbon nanotubes have diameters between 1 and 3 nm; selecting a "few" tubes, or even a single tube, requires field confinement within less than 10 nm.

## 2.4    Pumped Anti-Stokes Raman Scattering

Anti-Stokes Raman scattering starts from the first excited vibrational levels and is proportional to the number of molecules in the first excited vibrational state $N_1$. This number $N_1$ relative to the number of molecules in the vibrational ground state $N_0$ is determined by the Boltzmann factor. As briefly discussed above, a strong surface-enhanced Raman Stokes process with an effective cross section $\sigma_S^{\text{SERS}}$ populates the first excited vibrational levels. Depopulation of these levels is determined by anti-Stokes scattering and by the vibrational lifetime $\tau_1$. Figure 7 shows a schematic of

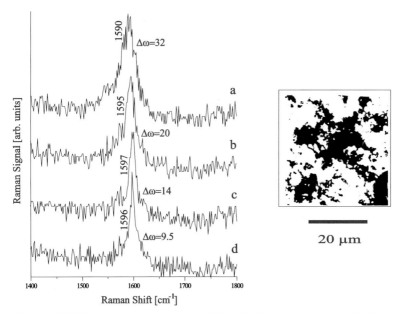

**Fig. 6.** SERS spectra of the tangential band of semiconducting single wall carbon nanotubes in contact with fractal colloidal silver clusters. Spectra measured with 1 μm spot size from different places on the cluster show different line widths of the Raman band (see text)

these processes and the appropriate rate equation for the number of molecules in the first excited vibrational state. In a weakly saturating intensity regime, $\exp(-h\nu_M/kT) \leq \sigma_S^{SERS}\tau_1 n_L \ll 1$ which in our experiments is between about $5 \times 10^5\,\mathrm{W/cm^2}$ and $1 \times 10^7\,\mathrm{W/cm^2}$, the anti-Stokes signal $P_{aS}$ and the Stokes signal $P_S$ can be estimated according to [24]

$$P_{aS}^{SERS} = (N_0 e^{-\frac{h\nu_M}{kT}} + N_0 \sigma_S^{SERS}\tau_1 n_L)\sigma_{aS}^{SERS} n_L \,, \tag{7}$$

$$P_S^{SERS} = N_0 \sigma_S^{SERS} n_L \,. \tag{8}$$

The second term in the equation for the anti-Stokes power $P_{aS}^{SERS}$ describes a quadratic dependence on the excitation intensity. This nonlinear anti-Stokes scattering

$$P_{aS,nl}^{SERS} = N_0 \sigma_{aS,nl}^{SERS} n_L^2 \tag{9}$$

can be described by an effective two-photon cross section

$$\sigma_{aS,nl}^{SERS} = \sigma_S^{SERS} \sigma_{aS}^{SERS} \tau_1 \,. \tag{10}$$

Figure 8 shows plots of anti-Stokes and Stokes signal powers of crystal violet on colloidal gold clusters versus excitation laser intensities. The lines indicate

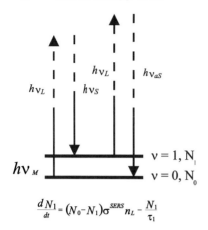

$$\frac{dN_1}{dt} = (N_0 - N_1)\sigma^{SERS} n_L - \frac{N_1}{\tau_1}$$

**Fig. 7.** Schematic Stokes and anti-Stokes Raman scattering

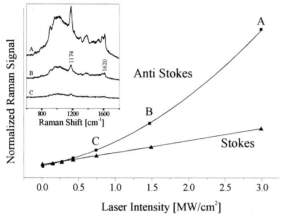

**Fig. 8.** Surface-enhanced Stokes (▲) and anti-Stokes (■) Raman scattering signals of the $1174\,\mathrm{cm}^{-1}$ line of crystal violet on colloidal gold clusters plotted vs 830 nm CW excitation intensity. The inset shows selected anti-Stokes spectra (A: $3\,\mathrm{MW/cm^2}$, B: $1.4\,\mathrm{MW/cm^2}$, and C: $0.7\,\mathrm{MW/cm^2}$). The Rayleigh background is suppressed by a notch filter up to about $900\,\mathrm{cm}^{-1}$. (Reprinted with permission from [20]. Copyright 2000 SPIE)

quadratic and linear fits to the experimental data, displaying the predicted quadratic and linear dependence. The Stokes signal $P_\mathrm{s}^{SERS}$ always remains linearly dependent on the laser intensity. This behavior is different from nonlinear coherent anti-Stokes Raman scattering, which is generated by nonlinear coupling of the anti-Stokes and Stokes fields. In that case, the Stokes power also becomes nonlinearly dependent on the excitation laser intensity.

Assuming a SERS cross section of approximately $10^{-16}\,\text{cm}^2$ and a vibrational lifetime of the order of 10 picoseconds, effective two-photon cross sections can be inferred at about $10^{-43}\,\text{cm}^4\,\text{s}$.

This provides a two-photon excited Raman probe at a cross section more than seven orders of magnitude larger than typical cross sections for two-photon excited fluorescence [33]. The large effective cross section can be explained by the nature of the process, which is a two-photon process exploiting the vibrational level as a real intermediate state.

Analogously to (3), we can split chemical and electromagnetic enhancement and write an expression for the effective surface-enhanced cross section for pumped anti-Stokes scattering:

$$\sigma_{\text{aS,nl}}^{\text{SERS}} = (\sigma_{\text{ads}}^{\text{RS}})^2 \tau_1 |A(\nu_{\text{L}})|^4 |A(\nu_{\text{S}})|^2 |A(\nu_{\text{aS}})|^2 \,. \tag{11}$$

The enhancement factor of this nonlinear effect is [34]

$$G_{\text{aS,nl}}^{\text{SERS}} = \left( \frac{\sigma_{\text{ads}}^{\text{RS}}}{\sigma_{\text{free}}^{\text{RS}}} \right)^2 |A(\nu_{\text{L}})|^4 |A(\nu_{\text{S}})|^2 |A(\nu_{\text{aS}})|^2 \,. \tag{12}$$

Because of its very high cross section, this two-photon effect can be observed at relatively low excitation intensities using CW lasers. For instance, at excitation photon flux densities of $10^{24}$ photons/ $\text{cm}^2\,\text{s}$, the anti-Stokes signal appears at a signal level 20 times lower than the Stokes signal [35]. This confirms cross sections of the order of $10^{-42}\,\text{cm}^4\,\text{s}$ for the two-photon process.

Surface-enhanced pumped anti-Stokes Raman spectroscopy provides all the advantages of two-photon spectroscopy, such as a linear increase of the spectroscopic signal relative to the background for increasing excitation intensities. For illustration, the inset in Fig. 7 shows anti-Stokes spectra together with the Rayleigh background. Moreover, anti-Stokes spectra are measured at the high energy side of the excitation laser, which is free from fluorescence, since two-photon excited fluorescence appears at much higher frequencies. The two-photon process inherently confines the volume probed by surface-enhanced anti-Stokes Raman scattering compared to that probed by one-photon surface-enhanced Stokes scattering [20]. Similar effects of confinement of the probed volume are known from two-photon excited fluorescent detection of single molecules [36].

## 2.5   Surface-Enhanced Hyper-Raman Scattering (SEHRS)

Hyper-Raman scattering (HRS) is a spontaneous nonlinear process, i.e., different molecules independently scatter light due to their hyperpolarizability and generate an incoherent Raman signal shifted relative to the second harmonic of the excitation laser. The number of surface-enhanced hyper-Raman Stokes photons $P^{\text{SEHRS}}$ can be written as

$$P^{\text{SEHRS}} = N_0 \sigma^{\text{SEHRS}} n_{\text{L}}^2 \,, \tag{13}$$

where $\sigma^{\mathrm{SEHRS}}$ is the effective cross section for the surface-enhanced hyper-Raman process.

Hyper-Raman scattering can follow symmetry selection rules different from one-photon Raman scattering, and therefore HRS can probe vibrations that are forbidden in Raman scattering and also in infrared absorption [8,9]. The "normal" hyper-Raman cross section $\sigma^{\mathrm{HRS}}$ is extremely small, of the order of $10^{-65}\,\mathrm{cm^4\,s}$. HRS can be enhanced analogously to normal Raman scattering by a "chemical" effect and by enhancement of the optical fields when the molecule is attached to metallic nanostructures. Then, the surface-enhanced hyper-Raman cross section is

$$\sigma^{\mathrm{SEHRS}} = \sigma_{\mathrm{ads}}^{\mathrm{HRS}} |A(\nu_{\mathrm{L}})|^4 |A(\nu_{\mathrm{HS}})|^2 , \tag{14}$$

where $\sigma_{\mathrm{ads}}^{\mathrm{HRS}}$ describes an enhanced hyper-Raman cross section compared to that of a "free" molecule [37]; the $A(\nu)$ describe the enhancement of the optical fields. We can write an enhancement factor for SEHRS as

$$G_{\mathrm{SEHRS}} = \frac{\sigma_{\mathrm{ads}}^{\mathrm{HRS}}}{\sigma_{\mathrm{free}}^{\mathrm{HRS}}} |A(\nu_{\mathrm{L}})|^4 |A(\nu_{\mathrm{HS}})|^2 . \tag{15}$$

Strong surface-enhancement factors can overcome the inherently weak nature of hyper-Raman scattering and surface-enhanced hyper-Raman spectra, and surface-enhanced Raman spectra can appear at comparable signal levels [18,19]. This is demonstrated in the middle spectrum in Fig. 9, which displays surface-enhanced hyper-Raman and Raman signals of crystal violet on colloidal silver clusters measured in the same spectrum (see also Fig. 1).

Taking into account the different sensitivity of the Raman system in the near-infrared and blue regions, the SEHRS signal is a factor of 100 weaker than the SERS signal. The experimental ratios between SERS and SEHRS intensities can be combined with the corresponding estimated "bulk" intensity ratio between RS and HRS scattering for the applied $10^7\,\mathrm{W/cm^2}$ excitation intensity (about $10^8$ [9]) to infer a ratio of about $10^6$ between surface-enhancement factors of hyper-Raman scattering and Raman scattering. Combining this ratio with NIR-SERS enhancement factors of crystal violet on colloidal silver clusters of the order of $10^{14}$, total surface-enhancement factors of hyper-Raman scattering on crystal violet adsorbed on colloidal silver clusters of the order of $10^{20}$ can be inferred [38].

The enormous total enhancement factor for HRS, six orders of magnitude more than for "normal" RS, in principle, can be discussed in terms of a strong increase of the hyperpolarizibility due to interaction between the molecule and the metal electrons or by a strong field enhancement.

Figure 10 shows hyper-Raman spectra of crystal violet on colloidal silver clusters measured at different excitation wavelengths. The hyper-Raman signals decrease strongly with decreasing excitation wavelength. No hyper-Raman spectrum was measured at 760 nm excitation and below. This can be explained in the framework of a "field-enhancement model" where the

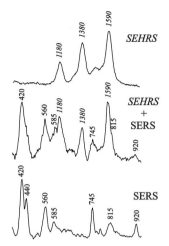

**Fig. 9.** SEHRS and SERS signals of crystal violet on colloidal silver clusters measured in the same spectrum using $10^7$ W/cm$^2$ NIR excitation (middle trace, see also Fig. 1). In the upper and the lower traces, SEHRS and SERS spectra are differentiated by placing a NIR absorbing filter in front of the spectrograph or by switching off the mode-locked regime of the Ti:sapphire laser, respectively

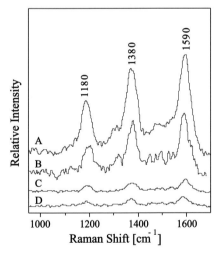

**Fig. 10.** Surface-enhanced hyper-Raman spectra of crystal violet on fractal colloidal silver clusters measured at 833 nm (A), 815 nm (B), 798 nm (C), and 785 nm (D). (Reprinted with permission from [38], Copyright 1999 Elsevier)

total enhancement benefits from enhancement of the laser and the scattering field (15). In general, field enhancement decreases for decreasing wavelengths. When the excitation wavelengths change from 830 to 750 nm, the wavelengths of the hyper-Raman fields change between 450 and 390 nm where field enhancement decreases rapidly [6]. At 750 nm excitation, the hyper-Raman Stokes field does not excite any eigenmodes on the colloidal silver cluster. This missing enhancement for the scattering field reduces the total enhancement to a level which is not sufficient to compensate for the extreme weakness of the hyper-Raman effect, and therefore we do not detect a hyper-Raman signal. This experimental finding supports the important role of electromagnetic enhancement in surface-enhanced hyper-Raman scattering.

The appearance of surface-enhanced hyper-Raman and Raman scattering at comparable scattering powers suggests that surface-enhanced hyper-Raman scattering is a spectroscopic technique that can be applied to single molecules.

## 3    Discussion

Extremely strong enhancement factors are observed for surface-enhanced linear and nonlinear Raman effects on colloidal silver and gold clusters at near-infrared excitation. Table 1 shows the effective cross sections observed in our experiments. For comparison, the table also shows typical cross sections for non-surface-enhanced optical processes.

The order of magnitude for the cross sections for "normal" linear SERS is confirmed in several experiments performed by other groups [39,40,41,42] (see Sect. 2.2), but most of these experiments, except [42], benefit from additional molecular resonance Raman enhancement.

In general, the strong enhancement of the Raman signal includes *electromagnetic* and *chemical* contributions. However, the experimental finding, that extremely large SERS enhancement is always related to colloidal silver or gold *clusters* [24,25,26,31,35,38,40,41,42,43,44] is an important indication that electromagnetic field enhancement plays a dominant role.

Figure 11 shows a typical fractal colloidal silver cluster structure used in our surface-enhanced linear and nonlinear NIR Raman experiments and the extinction spectrum of the aqueous solution of such clusters. The colloidal silver clusters have a fractal dimension of $1.63 \pm 0.05$ [45], in good agreement with values found for colloidal gold clusters [46]. Strongly enhanced

**Table 1.** Representative "normal" and surface-enhanced linear and nonlinear effective cross sections per molecule

| | |
|---|---|
| Resonance Raman | $10^{-16}\,\mathrm{cm}^2$ |
| Fluorescence | $10^{-16}\,\mathrm{cm}^2$ |
| Two-photon fluorescence | $10^{-50}\,\mathrm{cm}^4\,\mathrm{s/photon}$ |
| Hyper-Raman | $10^{-65}\,\mathrm{cm}^4\,\mathrm{s/photon}$ |
| Pumped anti-Stokes Raman | $10^{-71}\,\mathrm{cm}^4\,\mathrm{s/photon}$ |
| Surface-enhanced NIR-Raman | $10^{-16}\,\mathrm{cm}^2$ |
| Surface-enhanced pumped anti-Stokes Raman | $10^{-43}\,\mathrm{cm}^4\,\mathrm{s/photon}$ |
| Surface-enhanced hyper-Raman | $10^{-45}\,\mathrm{cm}^4\,\mathrm{s/photon}$ |

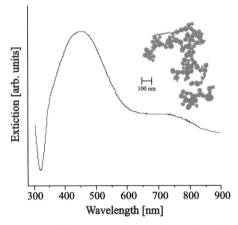

**Fig. 11.** Electron micrographs and extinction spectra of colloidal silver cluster structures used in the experiments

and highly localized fields, which are predicted for fractal colloidal cluster structures in so-called "hot spots" [5,6], can provide a rationale for the observed enhancement level. Moreover, plasmon resonances in a fractal cluster cover a broad frequency range from visible to the near infrared. This makes fractal clusters attractive for enhancing nonlinear effects, which very often involve optical fields at widely separated frequencies.

Note that for "normal" linear SERS, smaller clusters can show enhancement factors comparable to those obtained for fractal colloidal clusters. We also performed single molecule experiments on relatively small clusters 150–300 nm in size, formed by only 10–30 individual colloids (see for example [35]). Other authors have achieved single molecule sensitivity in SERS even for smaller clusters containing three to five colloidal particles [40,41,42]. Theoretical estimates for two particles in close contact show particularly strong electromagnetic enhancement at interparticle sites which result in electromagnetic SERS enhancement factors up to $10^{11}$ to $10^{12}$ [47], i. e., the same order of magnitude as SERS enhancement predicted for the hot spots on a fractal cluster [5,6]. But observation of surface-enhanced hyper-Raman scattering and strong vibrational pumping effects were possible only for cluster sizes of approximately 1 μ or larger. In particular, using small compact clusters, we did not observe surface-enhanced hyper-Raman scattering. This may be because the small compact clusters do not have such a broad plasmon resonance making it impossible that the laser and the hyper-Raman field can benefit simultaneously from plasmon resonance enhancement.

The surface of a fractal colloidal cluster structure shows a very inhomogeneous field distribution. This theoretical result was confirmed by near-field measurements [17]. SERS experiments performed on fractal cluster structures using relative high concentrations of the target molecule (about one monolayer on the fractal surface) also show that only a very small fraction (0.01 %) of the molecules available on the surface of the cluster can be involved in the Raman process at an extremely high enhancement level [24]. That means that total SERS enhancement factors of 14 orders of magnitude are available only at a few places on the cluster (about 0.01 % of the cluster surface).

SERS spectra performed on single wall carbon nanotubes show that the field enhancement on a fractal cluster is strongly confined within domains that can be as small as 10 nm.

Table 2 summarizes our experimental findings on the enhancement factors for nonresonant linear and nonlinear Raman scattering performed on fractal colloidal silver structures at near-infrared excitation. The first column shows the experimentally observed total enhancement factors [48].

We assume a "chemical" enhancement factor of about 100. This is the approximate order of magnitude we need to fill the gap between experimentally observed total SERS enhancement factors for small "NaCl activated" spheres [31] and electromagnetic estimates for such spheres [49]. The "chemical contribution" to the total enhancement should be preserved when colloidal particles form clusters.

Based on (5), (12), and (15), we can separate the "chemical" contribution to SERS enhancement. The third column in Table 2 then shows the value ascribed to electromagnetic enhancement. The table also shows the contributions to the electromagnetic enhancement factor from enhancing all optical fields involved in the process [50].

The estimate of the field-enhancement factors $A(\nu)$ for a fractal cluster is described in [6]. In general, these field-enhancement factors must be averaged over the cluster [6], but the numbers in the first column are experimental data taken from (single) molecules on "hot" spots, or the data are at least dominated by the Raman signals coming from these molecules [51]. Therefore, we believe that our experiments benefit from field-enhancement factors available in the hot spots of the cluster and do not represent average values.

In general, hot spots for different plasmon frequencies might be located at different places on the cluster, and the hot spot for a special Raman process is always a compromise. However, for small Raman shifts, both laser and scattered fields are very close, resulting in relative optimum field-enhancement conditions. Field enhancement for Stokes SERS and pumped anti-Stokes SERS should behave as a constant field-enhancement factor $A(\nu)$ raised to

**Table 2.** Enhancement factors of nonresonant linear and nonlinear Raman effects on fractal colloidal silver structures inferred from experiments using NIR excitation[*]

|  | Total | Field effect | |
| --- | --- | --- | --- |
| Raman scattering | $10^{14}$ | $10^{12}$ | $|A(\nu_{\mathrm{L}})|^2 * |A(\nu_{\mathrm{S}})|^2$ |
| Hyper-Raman scattering | $10^{20}$ | $10^{18}$ | $|A(\nu_{\mathrm{L}})|^4 * |A(\nu_{\mathrm{HRS}})|^2$ |
| Pumped anti-Stokes | $10^{28}$ | $10^{24}$ | $|A(\nu_{\mathrm{L}})|^4 * |A(\nu_{\mathrm{aS}})|^2 * |A(\nu_{\mathrm{S}})|^2$ |

[*] Assuming a factor of 10–100 "chemical" enhancement

the power of four and eight, respectively (see last column in Table 2). This agrees with our experimental results for a field-enhancement factor $A(\nu)$ of the order of $10^3$. The experimentally observed enhancement factor for hyper-Raman scattering also scales quite well in this schema, at least for a long excitation wavelength (830 nm), where the hyper-Raman field is still in resonance with plasmon excitations of the silver cluster. This is surprising, since it is unlikely to find such an optimum spot for the large shift between the frequencies of laser field and hyper-Raman Stokes field, where all fields are enhanced by a factor $10^3$ [52].

The importance of an enhancement of all fields is demonstrated by turning off the SEHRS effect when the frequency of the scattered field becomes too high to meet any plasmon resonance in the fractal (Fig. 10).

Effective cross sections of the order $10^{-16}\,cm^2$ and $10^{-42}\,cm^4\,s$ for Stokes and pumped anti-Stokes scattering , respectively, allow one- and two-photon Raman spectroscopy of single molecules. Moreover, the different surface-enhanced linear and nonlinear Raman experiments on molecules and single wall carbon nanotubes on colloidal silver and gold clusters provide insight into the dimensions and the nature of the field enhancement of colloidal metal clusters.

# References

1. M. J. Wirth, Chem. Rev. **99**, 2843 (1999)
2. X. S. Xie, J. K. Trautman, Annu. Rev. Phys. Chem. **49**, 441 (1998)
3. S. Nie, R. N. Zare, Annu. Rev. Biophys. Biomol. Struct. **26**, 567 (1997)
4. V. M. Shalaev, Phys. Rep. **272**, 61 (1996)
5. V. M. Shalaev, Fractal Nano-Composites: Giant Local-Field Enhancement of Optical Responses, in: M. Bertolotti, C. M. Bowden, C. Sibilia (Ed.): *Nanoscale Linear and Nonlinear Optics* Proc. V. 560, (American Institute Physics, Melville, New York 2001)
6. V. M. Shalaev, *Nonlinear Optics in Random Media* (Springer, Berlin, Heidelberg 2000)
7. V. P. Safanov, V. M. Shalaev, V. A. Markel, Yu. E. Danilova, Ni. N. Lepeshkin, W. Kim, S. G. Raution, R. L. Armstrong, Phys. Rev. Lett. **80**, 1102 (1998)
8. V. N. Denisov, B. N. Mavrin, V. B. Podobedov, Phys. Rep. **151**, 1 (1987)
9. L. D. Ziegler, Raman Spectrosc. **21**, 769 (1990)
10. K. Kneipp, H. Kneipp, I. Itzkan, R. R. Dasari, M. S. Feld, Chem. Rev. **99**, 2957 (1999)
11. A. Otto, Surface-enhanced Raman scattering: 'classical' and 'chemical' origins, in *Light Scattering in Solids IV. Electronic Scattering, Spin Effects, SERS and Morphic Effects*, M. Cardona, G. Güntherodt (Eds.) (Springer, Berlin, Heidelberg 1984) p. 289
12. M. Moskovits, Rev. Mod. Phys. **57**, 783 (1985)
13. A. Campion, P. Kambhampati, Chem. Soc. Rev. **27**, 241 (1998)
14. M. Fleischman, P. J. Hendra,, A. J. McQuillan, Chem. Phys. Lett. **26**, 123 (1974)

15. D. L. Jeanmaire, R. P. V. Duyne, J. Electroanal. Chem. **84**, 1 (1977)
16. M. G. Albrecht, J. A. Creighton, J. Am. Chem. Soc. **99**, 5215 (1977)
17. V. A. Markel, V. M. Shalaev, P. Zhang, W. Huynch, L. Tay, T. L. Haslett, M. Moskovits, Phys. Rev. B **59**, 10903 (1999)
18. H. Kneipp, K. Kneipp, F. Seifert, Chem. Phys. Lett. **212**, 374 (1993)
19. K. Kneipp, H. Kneipp, F. Seifert, Chem. Phys. Lett. **233**, 519 (1995)
20. K. Kneipp, H. Kneipp, I. Itzkan, R. R. Dasari, M. S. Feld, Near-infrared surface-enhanced Raman spectroscopy of biomedically relevant single molecules on colloidal silver and gold clusters, SPIE Proc. **3922**, 49 (2000)
21. A. Otto, I. Mrozek, H. Grabhorn, W. Akeman, J. Phys. Cond. Matter **4**, 1143 (1992)
22. M. I. Stockman, V. M. Shalaev, M. Moskovits, R. Botet, T. F. George, Phys. Rev. B **46**, 2821 (1992)
23. E. Y. Poliakov, V. M. Shalaev, V. A. Markel, R. Botet, Opt. Lett. **21**, 1628 (1996)
24. K. Kneipp, Yang Wang, H. Kneipp, I. Itzkan, R. R. Dasari, M. S. Feld, Phys. Rev. Lett. **76**, 2444 (1996)
25. K. Kneipp, Yang Wang, H. Kneipp, L. T. Perelman, I, Itzkan, R. R. Dasari, M. S. Feld, Phys. Rev. Lett. **78**, 1667 (1997)
26. K. Kneipp, H. Kneipp, P. Corio, S. D. M. Brown, K. Shafer, J. Motz, L. T. Perelman, E. B. Hanlon, A. Marulci, G. Dresselhaus, M. S. Dresselhaus, Phys. Rev. Lett. **84**, 3470 (2000)
27. P. Hildebrandt, M. Stockburger, J. Phys. Chem. **88**, 5935 (1984)
28. W. Kaiser, J. P. Maier, A. Selmeier, in *Laser Spectroscopy IV*, H. Walther, E. Rothe (Eds.) (Springer, Berlin, Heidelberg 1979)
29. T. L. Haslett, L. Tay, M. Moskovits, Chem. Phys. **113**, 1641 (2000)
30. Deviations of the anti-Stokes to Stokes signal ratios from those expected from a Boltzmann population can also be observed when the resonance Raman conditions for anti-Stokes and Stokes scattering are different. Since the resonance Raman effect depends on resonance conditions of both excitation laser and the scattered light, the anti-Stokes scattering may benefit in some special circumstances from some preresonance Raman effect. This can occur, in particular, at excitation wavelengths in the NIR for molecules with absorption bands in the visible. In that case, the anti-Stokes to Stokes signal ratio also shows deviations from the Boltzmann ratio [29]. But this effect is independent of the excitation laser intensity. An indication for vibrational pumping is also a linear increase of the anti-Stokes to Stokes signal ratio or a quadratic dependence of the anti-Stokes signal power vs. the excitation laser intensity (see Sect. 2.4).
31. K. Kneipp, H. Kneipp, V. B. Kartha, R. Manoharan, G. Deinum, I. Itzkan, R. R. Dasari, M. S. Feld, Phys. Rev. E **57**, R6281 (1998)
32. M. S. Dresselhaus, P. Eklund, Adv. Phys. **49**, 705 (2000)
33. J. Mertz, C. Xu, W. W. Web, J. Opt. Soc. Am. B **13**, 481 (1996)
34. The enhancement factor is determined relative to a fictive process, the population of excited vibrational states by spontaneous Raman scattering. For a typical nonresonant Raman cross section of $10^{-30}$ cm$^2$ and vibrational lifetimes of about 10 ps, laser intensities on the order of $10^{20}$ W/cm$^2$ would be required to generate a measurable vibrational population
35. K. Kneipp, H. Kneipp, G. Deinum, I. Itzkan, R. R. Dasari, M. S. Feld, Appl. Spectrosc. **52**, 175 (1998)

36. J. Mertz, C. Xu, W. W. Webb, Opt. Lett. **20**, 2532 (1995)
37. J. T. Golab, J. R. Sprague, K. T. Carron, G. C. Schatz, R. P. van Duyne, J. Chem. Phys. **88**, 7942 (1988)
38. K. Kneipp, H. Kneipp, I. Itzkan, R. R. Dasari, M. S. Feld, Chem. Phys. **247**, 155 (1999)
39. S. Nie, S. R. Emory, Science **275**, 1102 (1997)
40. H. Xu, E. J. Bjerneld, M. Kåll, L. Borjesson, Phys. Rev. Lett. **83**, 4357 (1999)
41. M. Michaels, M. Nirmal,, L. E. Brus, J. Am. Chem. Soc. **121**, 9932 (1999)
42. E. Bjernheld, P. Johansson,, M. Kaell, private communication (2000)
43. K. Kneipp, Exp. Technik Phys. **36**, 161 (1988)
44. K. Kneipp, Y. Wang, R. R. Dasari, M. S. Feld, Appl. Spectrosc. **49**, 780 (1995)
45. K. Güldner, R. Liedtke, K. Kneipp, H. J. Eichler, Morphological Study of Colloidal Silver Clusters Employed in Surface-Enhanced Raman Spectroscopy (SERS), Europhys. Conf. Abstr., H. D. Kronfeldt (Ed.) (European Physical Society, Berlin 1997) Vol. 21 C, p. 128
46. D. A. Weitz, M. Oliveria, Phys. Rev. Lett. **52**, 1433 (1984)
47. H. Xu, J. Aizpurua, M. Kaell, P. Apell, Phys. Rev. E **62**, 4318 (2000)
48. It is worth noting that the total enhancement factor for normal Raman scattering was derived from vibrational pumping and independently, from a very straightforward experiment by comparing the nonenhanced Raman signal from $10^{14}$ methanol molecules and the surface-enhanced Raman signal from one molecule attached to a colloidal silver cluster. Also the ratio between SEHRS and SERS signals is measured in a manner in which experimental errors can be excluded (see Sect. 2.5)
49. D. S. Wang, M. Kerker, Phys. Rev. B **24**, 1777 (1981)
50. K. Kneippy, Exp. Technik Phys. **38**, 3 (1990)
51. In single molecule experiments, we detected between 70 and 80 % of the molecules [25,31]. This implies that it is very likely that molecules attach to the cluster just at the "hot" spots
52. There is also a possibility that hyper-Raman scattering benefits from a larger "chemical" enhancement effect than Raman scattering, which may compensate for the more unfavorable conditions for field enhancement [37]

# Electromagnetic Response of Ferromagnetic Cermet: Superparamagnetic Transition

Mireille Gadenne

Laboratoire d'Optique des Solides, Université Pierre et Marie Curie
Paris VI, 4 Place Jussieu, 75252 Paris Cedex 05, France
paga@ccr.jussieu.fr

**Abstract.** Magnetic interactions between light and matter are known to be very weak and are nearly always neglected. However, in well-separated monodomain magnetic particles, the local interaction between the magnetic field of light and the magnetic moments does not vanish and has to be taken into account. After recalling the necessary concepts about the magnetism of nanosized superparamagnetic particles, we present the theory (an extension of Onsager's theory) modeling the local interaction between the magnetic field and the magnetic medium, leading to an averaged description in terms of effective dielectric and magnetic permeability. Then, we apply the model in the infrared range and show the expected main effects; we use physical realistic parameters describing nanosized ferromagnetic particles (nickel) embedded in an insulating matrix (aluminum nitride) and look at the influence of particle size and temperature on the variations of permeability versus frequency. These results are compared with the first experiments.

## 1 Introduction

Interaction between the magnetic field of an electromagnetic wave and a magnetic medium is scarcely taken into account and generally the permeability is supposed to be equal to the permeability of vacuum $\mu_0$. In this review, we are interested in a medium whose permeability may be different from $\mu_0$, and we look at its influence on the optical response in the infrared range.

We shall consider light as a probe for the study of the relaxation of superparamagnetic particles, and the period of the light will then be taken as the unit of measurement time.

First, we recall the definitions relative to the magnetic properties of the elementary particle and present superparamagnetism and the fundamental relaxation time involved in switching the magnetic moments in a two-level system. Then, we build a model of an inhomogeneous medium including monodomain grains, and determine an effective magnetic permeability for such a mixture which takes into account both processes: induction and superparamagnetic relaxation, when the medium is submitted to the field of electromagnetic radiation.

Afterward, in the framework of our model, we study the influence of magnetic permeability on the optical properties by calculating the transmission,

V. M. Shalaev (Ed.): Optical Properties of Nanostructured Random Media,
Topics Appl. Phys. **82**, 249–273 (2002)
© Springer-Verlag Berlin Heidelberg 2002

reflection, and absorption of a thin homogeneous film equivalent to a hetero-geneous cermet film. Then, we look at the influence of the main parameters on the superparamagnetism–ferromagnetism transition and predict the conse-quences for optical absorption. Finally, we compare the theoretical calculation to the first experimental results.

## 2    Description of the Materials Studied

### 2.1    Cermet

The "cer-(amic) met-(al)" is an inhomogeneous medium constituted of an insulating matrix with nanoscale inclusions of metal; the matrix is quasi-amorphous, actually composed of microcrystals. The volumetric concentra-tion in metal (or filling factor) $f$ is much smaller than the critical value corre-sponding to the percolation threshold. The material behaves macroscopically as an insulator and for simplicity we shall suppose that the metallic inclusions are regularly shaped (almost spheroidal) and dilute in the matrix, with weak electrical interactions. These inclusions are made of classical "ferromagnetic metals" which have a magnetic moment even in a zero applied magnetic field. According to Weiss, an internal interaction due to an exchange field lines up electrons spins and magnetic moments in a volume called a magnetic domain. Macroscopically, a ferromagnetic crystal appears as a collection of regularly oriented domains called Weiss domains, delimited by a thick border, called a Bloch wall, where the magnetization direction progressively changes from one domain to its neighbor.

Transition metals are optically "bad" metals, in the sense they have an intricate band structure and an interband absorption edge in infrared range, situated at very low energy. In the near-infrared range, the free-electron model (or Drude model) cannot be used [1]. And when it becomes useful for energy lower than the interband absorption edge, the relaxation time is very short ($h/\tau_e \approx 0.5\,\text{eV}$ instead of 0.05 for a noble metal) [1].

### 2.2    Definition of a Ferromagnetic Mono-Domain Particle

When the size of the metallic grains is very small, the dipolar interaction be-tween crystallites inside a grain decreases, becomes negligible in comparison with the exchange energy, and there is no thermodynamic need to separate the grain into several domains. Anyway, the thickness of the Bloch wall should become of the same order of magnitude as the size of the grain itself. So, in agreement with experimental results [2,3], we are allowed to consider that each nanosized particle has uniform magnetization whether or not the ap-plied field exists. The associated permanent magnetic moment in each grain cannot be cancelled; when no field is applied, its orientation could be ran-dom but because of anisotropic energy, there are privileged directions called easy magnetization directions. When a static magnetic field is applied, the direction of the permanent moment changes, following the field.

The critical size, under which the particle becomes monodomain, is obtained by comparing the free energy of a uniform magnetization distribution to the energy of a repartition in domains [4]. On one hand, if the anisotropy is very small, the energy of a spherical monodomain particle, with magnetization $M$ and radius $a$, is $F_m = \frac{1}{2}\left(\frac{4\pi}{3}\right)^2 a^3 M^2$. On the other hand, the necessary energy to build a Bloch wall between two domains is $F = F_p \pi a^2$ where $F_p$ is the wall energy per unit area. For a grain to be small enough to be a monodomain, its radius has to be smaller than a critical value $a_{mono}$. We obtain this value by writing that the wall energy has to be larger than the energy $F_m$, which gives

$$a_{mono} = \frac{9F_p}{8\pi M^2}.$$

Considering the numerical values for nickel given by [4] ($M = 480\,\text{G}$ and $F_p \approx 4\,\text{erg}\,\text{cm}^{-3}$), the critical radius is $a_{mono} \approx 30\,\text{nm}$. The reader may look at [5] for a more rigorous calculation.

## 3    Magnetic Anisotropy

When subjected to an external magnetic field, the magnetization of a macroscopic ferromagnetic crystal depends on the direction of the applied field. It is known that there exist preferential directions for magnetization, called easy axes. The corresponding anisotropic energy can be expressed by using the angles between magnetization and lattice vectors of the crystal unit cell. A rigorous theory was developed by *Van Vleck* [6].

In the particular case of fcc nickel crystals, for instance, the easy axis for magnetization is [111], and the expression of anisotropic energy becomes

$$E_K = k(\sin^2 2\theta + \sin^4 \theta \sin^2 2\Phi), \tag{1}$$

where $\theta$ is the polar angle and $\Phi$ the azimuthal angle between magnetization and easy axis [111] and the thermal variations of $K$ follow an exponential law [7]

$$K = K_0 \exp(-\alpha T^2). \tag{2}$$

The numerical values which we shall consider in this chapter are $K_1 = 5 \times 10^4\,\text{erg}\,\text{cm}^{-3}$ for $T = 300\,\text{K}$ and $K_1 = 3 \times 10^5\,\text{erg}\,\text{cm}^{-3}$ for $T = 200\,\text{K}$. These values provide $\alpha = 3.5 \times 10^{-5}\,\text{K}^{-2}$ and $K_0 = 1.258 \times 10^6\,\text{erg}\,\text{cm}^{-3}$.

When ferromagnetic crystals are of micro- or nanoscopic size, crystalline anisotropy is not the only factor to be taken into account. For instance, when inclusions are ellipsoidal, the direction of magnetization is determined mainly by the shape and is oriented along the longer dimension of the grain; this is called "shape anisotropy." Moreover, in nanosized inclusions, the organization of atoms on the surface plays a very important role and becomes predominant. The surface anisotropic energy can be written as

$$E_\sigma = K_\sigma \cos^2 \theta_s \, , \tag{3}$$

where $\theta_s$ is the angle between the magnetic moment and the normal to the surface. The variations of these two last anisotropic energies versus temperature are generally considered negligible but still have to be investigated.

## 3.1  Relaxation Time

A magnetic grain in a monodomain particle is considered to have a macrospin with a single magnetic moment per grain in one direction related to the easy magnetization axis. We can suppose that the assembly of such particles is governed by the two-level model due to *Néel* [5].

   According to this model (Fig. 1), a monodomain particle has two stable levels and may switch between two opposite orientations, if the thermal activation allows the transition through the potential barrier. A relaxation time $\tau$ is phenomenologically introduced to describe the competition between anisotropy that aligns the moments and thermal agitation that gives a random orientation, and the way thermodynamical equilibrium is reached. Making the approximation that anisotropy is a small perturbation, the expression of energy is simplified and can be written as $KV \sin^2 \theta$ [8] where $\theta$ is the angle between the easy magnetization axis and the applied magnetic field; this energy has two minima ($\theta = 0$ and $\theta = \pi$). There are two orientations for the moment and two corresponding values of magnetization. Under thermal agitation, some moments may change their orientation with a probability given by Néel's law $\nu = \nu_0 \exp(-\frac{KV}{k_b T})$ which is a consequence of the more general Arrhénius law for relaxation time $\tau = \tau_0 \exp(\frac{V}{k_b T})$.

   Here the prefactor $\tau_0$ is taken as a constant but depends on the material and is still studied, leading to controversial numerical results, according to authors [8,9,10].

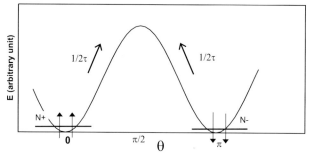

**Fig. 1.** Superparamagnetic two-level model. $\theta$ is the angle between the magnetic moment and the easy magnetization axis

We emphasize here because it will be studied at the end of this review, that relaxation time depends strongly on the anisotropic constant $K$, the volume of the particles $V$, and also the temperature $T$.

## 3.2  Magnetization and Coercive Field

The way the magnetic moment of the particle follows the applied field depends on the kind of dominating anisotropy, the value, and the direction of the applied field [11].

It is possible to determine the coercive field $H_c$ which is the minimum field able to switch the magnetization of one grain. Studying the variations of $H_c$ of an assembly of particles for various concentrations does not give trivial results [12]. In our case of very low concentration, $H_c$ is almost zero when the temperature is in the range $T = 200$ to $300\,\mathrm{K}$ as shown in Fig. 2 corresponding to experimental SQUID measurements on a thin cermet Ni–AlN film.

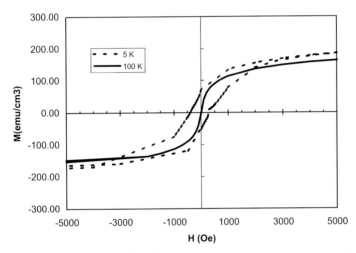

**Fig. 2.** Experimental SQUID magnetization measurement of a Ni–AlN film (thickness $d = 1$ μm, filling factor $f = 0.2$); $t_m$ can be regarded as infinite. The hysteresis cycle area decreases when temperature increases, and all the monodomain particles have superparamagnetic behavior when temperature $T$ is above $100\,\mathrm{K}$

## 3.3  Relation between Measuring Time, Critical Volume, and Blocking Temperature $T_b$

When studying the magnetization switch in a metallic grain, we have to compare the time $t_m$ necessary for observation and the relaxation time $\tau$. Two extreme situations may happen:

- either $t_m \ll \tau$, then the particle cannot reach equilibrium during measurements, the moments cannot switch, and the medium seems to be blocked in the initial state.
- or $t_m \gg \tau$, then the medium reaches equilibrium, and every monodomain particle behaves as a paramagnetic medium.

So, for a given value of $t_m$, two approaches may be followed:

When temperature $T$ is defined, it is possible to determine a critical volume $V_{cr}$ below which grains can switch and then behave as superparamagnetic.

$$V_{cr} = \frac{k_b T}{K} \ln\left(\frac{t_m}{\tau_0}\right) = \frac{k_b T}{K} \ln\left(\frac{1}{\omega\tau_0}\right).$$ (4)

When the volume of particles is determined, there exists a blocking temperature above which the medium is superparamagnetic, given by the relation,

$$T_B = \frac{KV}{k_b \ln\left(\frac{t}{\tau_0}\right)} = \frac{KV}{k_b \ln\left(\frac{1}{\omega\tau_0}\right)}.$$ (5)

As an example, we show evidence for superparamagnetic behavior in Fig. 2. In a SQUID, the measuring time is much larger than the relaxation time, and the magnetic moments may follow the applied magnetic field if thermal agitation is large enough to prevent blocked states. There is competition between anisotropic energy and thermal energy. When $T = 5\,\mathrm{K}$, the moments are blocked, and there is a classical hysteresis cycle for magnetization, but when $T$ is larger (here $T = 100\,\mathrm{K}$ and above), thermal energy liberates the moments that can follow the applied magnetic field as a paramagnetic medium.

When a weak, high-frequency magnetic field is applied, it can be considered a perturbation, and a linear response is obtained. (If this assumption cannot be made, see [13,14,15,16])

The alternative applied field is considered a probe, and its frequency corresponds to the time of measurement $t_m = 1/\omega$.

Using a Debye function, the linear response for susceptibility of the medium is given by

$$< \chi(\omega) >= \frac{Nm^2}{3k_b T} \frac{1}{1 - i\omega\tau}.$$ (6)

For a mixture of grains in a nonmagnetic matrix of metal concentration $f$ and magnetization per volume unit $m_0$,

$$\chi(\omega) = \frac{\bar{m}_0^2 fV}{3k_b T} \frac{1}{1 - i\omega\tau}.$$ (7)

When the real and imaginary parts can be separated, the usual variation versus $\omega$ is observed:

$$\chi'(\omega) = \frac{\bar{m}_0^2 f V}{3 k_{\mathrm{b}} T} \frac{1}{1 + \omega^2 \tau^2},$$

$$\chi''(\omega) = \frac{\bar{m}_0^2 f V}{3 k_{\mathrm{b}} T} \frac{\omega \tau}{1 + \omega^2 \tau^2}. \tag{8}$$

If the time $t_{\mathrm{m}}$ is larger than the relaxation time $\tau$, the response does not depend on the frequency and is almost the static response of a superparamagnetic system,

$$\chi'_{(\omega=0)} = \frac{m_0^2 f V}{3 k_{\mathrm{b}} T}$$

and

$$\chi'' = 0. \tag{9}$$

If $t_{\mathrm{m}}$ is smaller than $\tau$, the moments cannot follow the variation of the field, and the system is blocked.

The transition between these two behaviors is the range of frequency (when $t_{\mathrm{m}} \approx \tau$) we are interested in, since the imaginary part is maximum and the real part shows up as an inflexion point.

Using the linear response theory, we define magnetic permeability as

$$\mu = 1 + 4\pi\chi \tag{10}$$

that will be needed in optics.

## 4    Interaction with an Electromagnetic Wave: Theoretical Approach

The behavior of inhomogeneous media, when subjected to an external static or variable magnetic field is studied in some of other reviews in this book. In this section, we limit ourselves to the study of the response of material to an electromagnetic wave. We suppose that the wavelength is much larger than the grain size and much larger than the average intergrain separation. Then, the quasi-static approximation for the electromagnetic field surrounding metallic grain is valid.

Starting from the Maxwell equations (c.g.s.-u.e.m. system)

$$\nabla \cdot \boldsymbol{E} = 0,$$

$$\nabla \times \boldsymbol{E} = -\frac{\mu}{c} \frac{\partial \boldsymbol{H}}{\partial t},$$

$$\nabla \cdot \boldsymbol{H} = 0,$$

$$\nabla \times \boldsymbol{H} = \frac{\varepsilon}{c} \frac{\partial \boldsymbol{E}}{\partial t} + \frac{4\pi\sigma}{c} \boldsymbol{E}, \tag{11}$$

the propagation equation in a conducting medium is obtained:

$$\nabla^2 \boldsymbol{E} = \frac{\varepsilon\mu}{c^2}\frac{\partial^2 \boldsymbol{E}}{\partial t^2} + \frac{4\pi\sigma\mu}{c^2}\frac{\partial \boldsymbol{E}}{\partial t}, \tag{12}$$

where $E, H, \sigma, \varepsilon, \mu$ are, respectively, the electric field, the magnetic field, the conductivity, the dielectric function and the magnetic permeability.

In the case studied, when scattering of the radiation can be neglected, inhomogeneous nanocermets behave as homogeneous isotropic media and can then be characterized by effective dielectric function $\varepsilon$ and magnetic permeability $\mu$. Doing so, we replace the local-field distribution by an effective equivalent mean field. In fact, we shall see that the nonlocal electrical field is valid in the range of filling factor and wavelength studied because interactions are not the dominant problem. However, we shall consider the local magnetic field distribution, as needed for establishing the mean field equivalence when it is necessary.

## 4.1  Study of the Effective Dielectric Function

As the metallic inclusions are in a very low concentration range ($f < 0.2$), we consider that the particles are far enough from each other to prevent any dipolar or multipolar interaction. It is possible to calculate the effective dielectric function $\varepsilon$ using an effective medium theory, and it is well known that the most simple Maxwell Garnett theory is efficient enough with this assumption. The grains are not connected to each other and are totally embedded in an insulating matrix. The sample is globally insulating. In this approach, inclusions are subjected to a mean external applied field, and the behavior of the whole medium is that of the matrix. Far below the percolation threshold, the effective dielectric function is given by the relation,

$$f\frac{\varepsilon_i - \varepsilon_m}{L\varepsilon_i + (1-L)\varepsilon_m} = \frac{\varepsilon - \varepsilon_m}{L\varepsilon + (1-L)\varepsilon_m}, \tag{13}$$

where $\varepsilon_i, \varepsilon_m, \varepsilon$ are, respectively, the dielectric function of the insulator, the metal, and the effective medium. We use the expression found in reference [17] where depolarizing factor $L$ is included to take into account the morphology of inclusions. If we assume that these metallic inclusions are spherical, the depolarizing factor $L$ is equal to 1/3.

## 4.2  Study of Effective Magnetic Permeability of a Cermet

Usually, in optical studies concerning isotropic nonmagnetic media, because the components of the mixture do not have any permanent magnetic moments, the relative permeability remains real and constant equal to 1 as it is in vacuum. But for inhomogeneous media with ferromagnetic particles, namely, cermets with ferromagnetic metal inclusions, the permeability is no longer equal to 1; it becomes a complex quantity in an adequate spectral range of electromagnetic radiation.

Moreover, to predict the magnetic permeability, it is not possible to use the Maxwell Garnett approximation because of the presence of permanent magnetic moments: using the Clausius–Mossotti relation leads to a polarization divergence when the density of permanent dipoles increases, which is not realistic. To prevent this "catastrophe," it is better to follow the Onsager approach [18] established for permanent electric dipoles. The expression of a local magnetic field can be established and extended [19] to an inhomogeneous medium, including permanent magnetic dipoles.

### 4.2.1    Onsager's Theory

Let us assume that the metallic grains are very small and the dipolar interaction negligible in regard to exchange interactions. In a quasi-static magnetic field, the magnetization of the composite, in which the inclusions have permanent magnetic moments, is due to the alignment of the permanent ferromagnetic moments and also to the induced magnetic moment resulting from eddy currents.

If we consider a single magnetic monodomain particle in a quasi-static magnetic field $\boldsymbol{H}$, its magnetization has two terms

$$\boldsymbol{m} = m_0\boldsymbol{u} + \beta\boldsymbol{F}, \tag{14}$$

where $m_0$ is the ferromagnetic moment of the particle, $\boldsymbol{u}$ is the unit vector in the direction of the permanent moment, $\beta$ is the complex magnetic polarizability of the particle, and $\boldsymbol{F}$ is the local magnetic field, which is different from the applied field $\boldsymbol{H}$.

Onsager analyzes the local field $\boldsymbol{F}$ as composed of two terms: the cavity field component $\boldsymbol{G}$ and the reaction field component $\boldsymbol{R}$. Both components are obtained by solving boundary-value problems. Because of the permanent moment of a single particle, an inhomogeneous field is created, which induces magnetic moments in every particle around the first one: the orientation of the permanent moment of the neighboring particles is modified, creating the reaction field $\boldsymbol{R}$. If we consider that the anisotropy is small, the reaction field can be considered proportional to the permanent moment. To find its expression, it is possible to solve Laplace equation $\Delta\Psi = 0$ for a magnetic moment in the center of a sphere of radius $a$ surrounded by a magnetic medium with permeability $\mu$. Using boundary conditions for $\Psi$ and its derivative, the solutions

$$\Psi_1 = \sum_{n=0}^{\infty}\left(A_n r^n + \frac{B_n}{r^{n+1}}\right)P_n(\cos\theta) \qquad \text{when} \quad r > a$$

$$\Psi_2 = \sum_{n=0}^{\infty}\left(C_n r^n + \frac{D_n}{r^{n+1}}\right)P_n(\cos\theta) \qquad \text{when} \quad r < a \tag{15}$$

become

$$\Psi_1 = \frac{3}{2\mu + 1} \frac{m}{r^2} \cos\theta,$$

$$\Psi_2 = \frac{m}{r^2} \cos\theta - \frac{2(\mu - 1)}{2\mu + 1} \frac{m}{a^3} r \cos\theta. \tag{16}$$

The local Onsager fields are

$$F_1 = \frac{3\mu}{2\mu + 1} H_m \qquad \text{when} \quad r > a,$$

$$F_2 = H_m + \frac{2(\mu - 1)}{(2\mu + 1)a^3} m \qquad \text{when} \quad r < a, \tag{17}$$

where

$$\|H_m\| = \frac{m\sqrt{1 + 3\cos^2\theta}}{r^3}. \tag{18}$$

The reaction field is

$$R = \frac{2(\mu - 1)}{(2\mu + 1)a^3} m. \tag{19}$$

Moreover, we have to determine the perturbation of the applied field due to a cavity in a medium of permeability $\mu$ around the particle. We have, then, to solve the same Laplace equation with other boundary values.

Instead of $H$, the field becomes

$$G = \frac{3\mu}{2\mu + 1} H, \tag{20}$$

called the cavity field.

The local field which interacts with the particle is the Onsager field

$$F = G + R. \tag{21}$$

This expression given here for a spherical particle can be generalized [20].

### 4.2.2   Calculation of Effective Permeability

Now, we consider an assembly of spherical ferromagnetic monodomain particles ($N$ is the number of particles per unit volume). Each particle has a magnetic moment randomly oriented when there is no applied field $H$ (we suppose that anisotropy is negligible).

The polarizability of one particle is given by

$$\beta = \frac{4\pi a^3}{3} \bar{\beta} \tag{22}$$

if $\bar{\beta}$ is the polarizability per unit volume. The total magnetization (per unit volume) is

$$\boldsymbol{M} = N \langle \boldsymbol{m} \rangle \tag{23}$$

and is related to the applied field by

$$\boldsymbol{M} = \frac{\mu - 1}{4\pi} \boldsymbol{H}, \tag{24}$$

which provides the following equation for a static applied field:

$$\frac{(\mu - 1)}{4\pi} \boldsymbol{H} = \frac{\bar{m}_0 \langle \boldsymbol{u} \rangle 3(2\mu + 1) + 9\mu\bar{\beta}\boldsymbol{H}}{3(2\mu + 1) - 8\pi(\mu - 1)\bar{\beta}}. \tag{25}$$

The magnetic permanent moment per unit volume is $\bar{m}_0$; it corresponds to the saturation magnetization. According to Onsager's model, the thermodynamic averaged $\boldsymbol{u}$ (in every direction) is

$$\langle \boldsymbol{u} \rangle = \frac{\mu \bar{m}_0 4\pi a^3 \boldsymbol{H}}{k_{\mathrm{b}} T [3(2\mu + 1) - 8\pi(\mu - 1)\bar{\beta}]}. \tag{26}$$

For an oscillating field $\boldsymbol{H}$, the averaged $\boldsymbol{u}$ has to be multiplied by a Debye factor

$$\Gamma(\omega) = \frac{1}{1 - i\omega\tau}, \tag{27}$$

where $\tau$, the orientation relaxation time previously defined (Sect. 3.1), follows the Arrhénius law [21]

$$\tau = \tau_0 \exp\left(\frac{KV}{k_{\mathrm{b}}T}\right). \tag{28}$$

Finally, the self-consistent equation which gives the permeability is

$$\frac{(\mu - 1)}{4\pi} = \frac{9\mu(2\mu + 1)\Gamma(\omega)\left(\frac{4\pi a^3 \bar{m}_0^2}{3k_{\mathrm{b}}T}\right)}{\left[3(2\mu + 1) - 8\pi(\mu - 1)\bar{\beta}\right]^2} + \frac{9\mu\bar{\beta}}{3(2\mu + 1) - 8\pi(\mu - 1)\bar{\beta}}. \tag{29}$$

In our case of random distribution of very small metallic particles, bearing a magnetic dipolar moment, the expression has to be modified to take into account the volume fraction $f$ occupied by the metallic particles in the medium. The polarizability per unit volume $\bar{\beta}$ becomes $f\bar{\beta}$, and the relation giving the effective magnetic relative permeability becomes

$$\frac{(\mu - 1)}{4\pi} = \frac{9f\mu(2\mu + 1)\Gamma(\omega)\left(\frac{4\pi a^3 \bar{m}_0^2}{3k_{\mathrm{b}}T}\right)}{\left[3(2\mu + 1) - 8\pi(\mu - 1)f\bar{\beta}\right]^2} + \frac{9f\mu\bar{\beta}}{3(2\mu + 1) - 8\pi(\mu - 1)\bar{\beta}f}. \tag{30}$$

This equation is a third-order polynomial where only one solution is physically correct:

$$A\mu^3 + B\mu^2 + C\mu + D = 0, \tag{31}$$

where

$$A = 4(9 + 16\pi^2 f^2 \bar{\beta}^2 - 24\pi f \bar{\beta}),$$
$$B = -24\pi f [\bar{\beta}(3 - 4\pi f \bar{\beta}) + 3X\Gamma(\omega)],$$
$$C = -3\{32\pi^2 \beta^2 f^2 + 12\pi f (X\Gamma(\omega) + 3\beta) + 9\},$$
$$D = -[9 + 16\pi\bar{\beta}f(3 + 4\pi\bar{\beta})f],$$

and                                                                    (32)

$$X = \frac{4\pi a^3 \bar{m}_0^2}{3k_{\mathrm{b}}T}.$$                  (33)

If the permanent magnetic moment $m_0 = 0$, the quantity $X$ cancels, and this expression becomes the Clausius–Mossotti relation.

# 5  Numerical Results and Influence of Various Parameters

## 5.1  Superparamagnetism

To calculate the magnetic permeability, we first assumed that the temperature is responsible for equilibrium (Boltzmann statistical study) in the presence of a static magnetic field. Then this assumption was supposed to be valid also when the applied field alternates with a period larger than the relaxation time of the magnetic moments, so that the system can equilibrate and follow the field. That means that the medium behaves as a paramagnetic medium but with a moment larger than paramagnetic atoms.

This is a manifestation of superparamagnetic behavior.

The variations are governed both by the magnetic relaxation time $\tau$, which is included in the Debye function, and the magnetic polarizability (eddy currents). The terms $X$ and $\tau$ depend on the filling factor, the radius of the particles, and the temperature.

With this model, it is possible to describe the whole response of the inhomogeneous medium to an electromagnetic wave in a very large spectrum. As both the polarizability and the Debye function depend on frequency, the response of the system is governed by induced magnetization and by the maximum of the Debye function, whatever the intensity of the magnetic field.

Let us look at some results of calculation and predict the influence of various parameters. Equation (30) is solved numerically. The right solution is found by keeping only the positive determination of the real part $\mu'$ and the imaginary part $\mu''$ of $\mu$.

## 5.2  Influence of the Filling Factor

Because we use Onsager's model, there is no longer a catastrophe in Fig. 3, even in the range of large filling factors. Anyway, the large values of filling

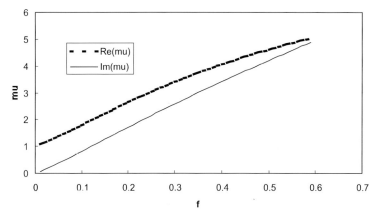

**Fig. 3.** Variations in real and imaginary parts of $\mu$ versus ferromagnetic metal filling factor $f$ for $\lambda = 500\,\mu\mathrm{m}$ which corresponds to wave number $\sigma = 20\,\mathrm{cm}^{-1}$ according to Onsager's model

factor will not be studied here because interactions are definitely supposed to be weak. The variations of permeability for a wavelength belonging to the range of interest ($\lambda = 500\,\mu\mathrm{m}$) versus filling factor $f$ are regularly monotonic and increasing. For $f = 0$, the values are $\mu' = 1$ and $\mu'' = 0$, as expected.

### 5.3 Variations of Magnetic Permeability versus Frequency

The results of calculation using formula (30) are presented in Fig. 4. In the high optical frequency range ($> 10^{15}\,\mathrm{Hz}$), because of eddy currents and Lenz's law, the real part of the magnetic permeability becomes smaller than one (the susceptibility is negative, and the behavior is diamagnetic), and the imaginary part shows a maximum corresponding to the maximum of the magnetic polarizability of an elementary spherical particle. In the low-frequency range ($\approx 10^{12}\,\mathrm{Hz}$), the dominant feature is the superparamagnetic relaxation, responsible for a maximum in the imaginary part $\mu''$, when the optical frequency corresponds to the relaxation time for superparamagnetic behavior and for a large variation in the value of the real part $\mu'$ (from 1 to 4 or 5).

### 5.4 Influence of the Particle Size

As can be presumed from the expression for the relaxation time, the exponent is proportional to the volume of the particle and consequently, the maximum of the imaginary part $\mu''$ of $\mu$ shifts towards smaller frequencies when the radius of the spherical particles increases. In the meantime, the maximum becomes higher (Fig. 5).

In actual material, the assembly of particles is not monodisperse, and generally it is necessary to consider a distribution $D(r)$ of particle size. The

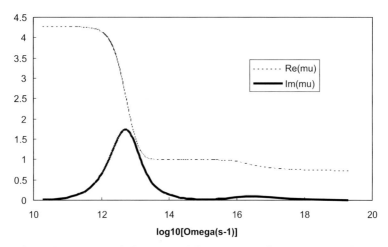

**Fig. 4.** Variations of the permeability $\mu$ versus frequency according to (30). The radius is 5 nm, temperature is $T = 300\,\mathrm{K}$, relaxation time $\tau_0 = 10^{-13}\,\mathrm{s}$, and the constant of anisotropy is $K = 5 \times 10^4\,\mathrm{erg\,cm}^{-3}$

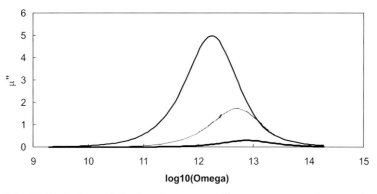

**Fig. 5.** Variations of the imaginary part $\mu''$ versus energy when varying the radius of the metallic inclusions according to the effective medium theory ($a = 3$, 5 and 7 nm). The maximum value increases with the radius of inclusions

equation that gives $\mu$ becomes

$$\frac{\mu - 1}{4\pi} = \frac{9\mu(2\mu + 1)}{k_{\mathrm{b}}T} \int_0^\infty \frac{D(r)\Gamma(\omega, r)f\bar{m}_0^2\left(\frac{4\pi r^3}{3}\right)\mathrm{d}r}{\left[3(2\mu + 1) - 8\pi f\bar{\beta}(r)(\mu - 1)\right]^2}$$
$$+ 9\mu \int_0^\infty \frac{D(r)f\bar{\beta}(r)\mathrm{d}r}{\left[3(2\mu + 1) - 8\pi f\bar{\beta}(r)(\mu - 1)\right]} \tag{34}$$

The curves are modified, and the variations are wider and smoother. We can deduce from Fig. 6 that the optical effect we are looking for will be attenuated because of the non-mono-dispersion of the size of the metallic grains.

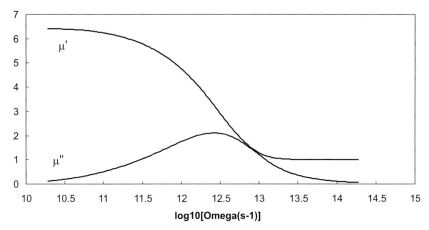

**Fig. 6.** Influence of the dispersion of the size of particles: the log-normal size distribution $D(r)$ is centered on $< r >= 5\,\mathrm{nm}$ with $\sigma = 1.3$

## 5.5 Influence of Temperature on the Variation in $\mu$ versus Frequency

When the temperature decreases, the relaxation time increases, the curve describing the variations of $\mu$ shifts toward the low frequencies (higher time period of the electromagnetic field), and the amplitude of the maximum increases.

Moreover the magneto-crystalline anisotropy depends on temperature according to an exponential law

$$K = K_0 e^{-cT^2} . \tag{35}$$

Consequently, when the temperature decreases, the exponent in the expression of the relaxation time increases, and the maximum of the $\mu$ curve happens at lower frequency (100 times less frequency between $300\,\mathrm{K}$ and $200\,\mathrm{K}$) (Fig. 7).

When anisotropy is supposed to be mainly due to shape and surface for very small and irregular shaped particles we have seen that the energy barrier is independent of temperature $T$. In this case, we observe (Fig. 8) that the relaxation time does not vary so much as for magneto-crystalline anisotropy and that there is only a small shift of the permeability value when temperature varies.

In Fig. 9, we show the variations of magnetic permeability versus temperature for a given frequency (the frequency in the figure is $\omega = 3.7 \times 10^{12}\,\mathrm{s}^{-1}$).

We observe a maximum in temperature which can be related to the blocking temperature, the border between the superparamagnetic relaxation, and the blocked state.

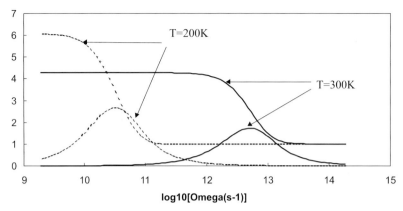

**Fig. 7.** Magnetocrystalline anisotropy: the anisotropic constant $K_1$ depends on temperature. Variations of permeability $\mu$ versus frequency for two different temperatures: $T = 300\,\text{K}$, $K_1 = 5 \times 10^4\,\text{erg}\,\text{cm}^{-3}$ and $T = 200\,\text{K}$, $K_1 = 3 \times 10^5\,\text{erg}\,\text{cm}^{-3}$, according to [7]

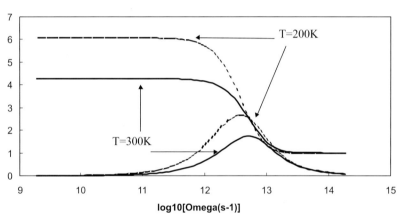

**Fig. 8.** Shape and surface anisotropy: the anisotropic constant does not depend on temperature; variations in $\mu$ versus frequency for two temperatures

## 6　Consequences on Optical Absorption

It is necessary to test the validity of our assumptions in regard to actual materials which could be ferrofluids, powders of metallic grains and insulating oxide mixed together, or metallic grains in a polymer matrix. Our own approach consists of studying the optical properties of thin cermet films deposited onto transparent substrates. For instance, these samples can be classically prepared by radio-frequency cosputtering on intrinsic silicon substrates, and the measurements are performed with a fast fourier transform spectrometer (FFTS).

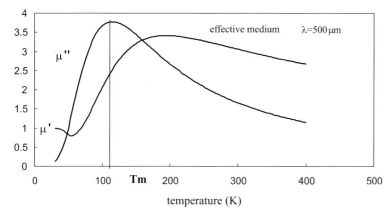

**Fig. 9.** Variations of real and imaginary parts of magnetic permeability calculated with effective medium, versus temperature for $\lambda = 500\,\mu\mathrm{m}$ (the period of light is $T_{\mathrm{w}} = 1.67 \times 10^{-12}\,\mathrm{s}, \omega = 3.7 \times 10^{12}\,\mathrm{s}$)

## 6.1    Expression of Reflectance and Transmittance of a Thin Magnetic Film

For that purpose, we have to rewrite the expressions of the reflectance and transmittance of a thin film when the effective magnetic permeability is no longer equal to one, but has a complex value. These optical parameters (reflectance and transmittance) for an effective medium containing ferromagnetic particles are explicit functions of both the optical thickness

$$\varphi = \frac{2\pi d}{\lambda}\sqrt{(\varepsilon\mu)_{\mathrm{eff}}} \tag{36}$$

and the admittance

$$g = \sqrt{\left(\frac{\varepsilon}{\mu}\right)_{\mathrm{eff}}}. \tag{37}$$

Following *Lamb* et al. [22], we assume that there is no coupling between the effective values of $\varepsilon$ and $\mu$. We consider that the product $(\varepsilon\mu)_{\mathrm{eff}} = \varepsilon_{\mathrm{eff}}\mu_{\mathrm{eff}}$, and we define

$$\beta = kd = \frac{2\pi d}{\lambda}\sqrt{\varepsilon_{\mathrm{eff}}\mu_{\mathrm{eff}}} = \varphi' - i\varphi'' \tag{38}$$

as the optical thickness of the film and

$$g = \sqrt{\frac{\varepsilon_{\mathrm{eff}}}{\mu_{\mathrm{eff}}}} = g' - ig'' \tag{39}$$

as the admittance.

If these quantities are known, it becomes possible to calculate the absorptance $A = 1 - R - T$ for a thin film by replacing complex $\varepsilon$ and $\mu$ in the complete expressions of $R$ and $T$ (including $\varphi$ and $g$) [23]. As it is unusual in optics calculations, we shall present the whole development.

Let us consider [23] an isotropic thin film characterized by the complex effective functions and its thickness $d$.

Let $A(z)$ be the field matrix vector in a plane $z$. We define

$$A(z) = \begin{vmatrix} E(z) \\ H(z) \end{vmatrix} = M(d) A(z+d) \tag{40}$$

with

$$M(d) = \begin{vmatrix} m_{11} & m_{12} \\ m_{21} & m_{22} \end{vmatrix} = \begin{vmatrix} \cos\varphi & (i/g)\sin\varphi \\ ig\sin\varphi & \cos\varphi \end{vmatrix}. \tag{41}$$

We can notice that the physical solutions for the optical thickness and the admittance should have a positive imaginary part because of the absorbing nature of the medium.

So we can express

$$g' = \frac{1}{\sqrt{2}} \sqrt{ \frac{\varepsilon'\mu' + \varepsilon''\mu''}{\mu'^2 + \mu''^2} + \sqrt{\frac{\varepsilon'^2 + \varepsilon''^2}{\mu'^2 + \mu''^2}} },$$

$$g'' = \frac{1}{\sqrt{2}} \sqrt{ -\frac{\varepsilon'\mu' + \varepsilon''\mu''}{\mu'^2 + \mu''^2} + \sqrt{\frac{\varepsilon'^2 + \varepsilon''^2}{\mu'^2 + \mu''^2}} }, \tag{42}$$

and

$$\varphi' = \frac{2\pi d}{\lambda} \frac{1}{\sqrt{2}} \sqrt{ (\varepsilon'\mu' - \varepsilon''\mu'') + \sqrt{(\varepsilon'^2 + \varepsilon''^2)(\mu'^2 + \mu''^2)} },$$

$$\varphi'' = \frac{2\pi d}{\lambda} \frac{1}{\sqrt{2}} \sqrt{ -(\varepsilon'\mu' - \varepsilon''\mu'') + \sqrt{(\varepsilon'^2 + \varepsilon''^2)(\mu'^2 + \mu''^2)} }. \tag{43}$$

Given the amplitude of the incident wave (taken as a unit), we have to find the complex amplitudes of the reflected and transmitted waves. They can be defined as

$$r = \rho \exp(i\delta_r) \tag{44}$$

and

$$t = \tau \exp(i\delta_t - 2i\pi n_0 d/\lambda), \tag{45}$$

respectively.

Solving the matrix (41) and using the continuity of the tangential components of the field across the first and the last surfaces of the film,

$$r = \frac{g_0(m_{11} + g_s m_{12}) - (m_{21} + g_s m_{22})}{g_0(m_{11} + g_s m_{12}) + (m_{21} + g_s m_{22})},$$

$$t = \frac{2g_0 \exp(2i\pi n_0 d/\lambda)}{g_0(m_{11} + g_s m_{12}) + (m_{21} + g_s m_{22})}, \tag{46}$$

where $g_0, g_s$ are, respectively, the values of $g$ for the ambient and the substrate. As we suppose that they are nonabsorbing media and without permanent magnetic moment, these quantities are real: $g_0 = \sqrt{\varepsilon_0} = n_0$, and $g_s = \sqrt{\varepsilon_s} = n_s$. Separating real and imaginary parts, we obtain the module of the amplitude of the reflection and transmission coefficients, $|\rho|$, and $|\tau|$, and we can express the energy factors reflectance $R = \rho\rho^* = |r|^2$ and transmittance $T = \frac{n_s}{n_0}\tau\tau^* = \frac{n_s}{n_0}|t|^2$:

$$R = \frac{abe^{2\varphi''} + cde^{-2\varphi''} + 2r\cos 2\varphi' + 2s\sin 2\varphi'}{bde^{2\varphi''} + ace^{-2\varphi''} + 2t\cos 2\varphi' + 2u\sin 2\varphi'},$$

$$T = \frac{16n_0 n_s \left(g'^2 + g''^2\right)}{bde^{2\varphi''} + ace^{-2\varphi''} + 2t\cos 2\varphi' + 2u\sin 2\varphi'}, \qquad (47)$$

with

$$\begin{aligned}
\frac{a}{d} &= (g' \mp n_0)^2 + g''^2 \\
\frac{b}{c} &= (g' \pm n_s)^2 + g''^2 \\
\frac{r}{t} &= \left(n_0^2 + n_s^2\right)\left(g'^2 + g''^2\right) - \left(g'^2 + g''^2\right)^2 - n_0^2 n_s^2 \mp 4n_0 n_s g''^2 \\
\frac{s}{u} &= 2g''(n_s \mp n_0)\left(g'^2 + g''^2 \pm n_0 n_s\right).
\end{aligned}$$

Taking the limit where the material is no longer magnetic ($\mu' = 1$, and $\mu'' = 0$), we obtain the same expressions for the admittance and the optical thickness as the classical Abelès formulas [23].

Now, it is possible to predict the variations of thin cermet films in the whole optical range from the near-UV region (0.4 μm) up to the microwave region (500 μm). In next section, we propose, as an example, the calculation of the optical reflectance $R$ and transmittance $T$ under normal incidence in the optical range up to $\lambda = 1000$ μm of cermet thin films, including very small ferromagnetic monodomain particles of nickel (grain size $r < 10$ nm) embedded in an amorphous matrix of aluminum nitride or aluminum oxide deposited on a double-sided optically polished silicon substrate.

## 6.2    Application to Cermet: Curves in the Spectral Range Corresponding to the Relaxation Time $\tau_0$

### 6.2.1    Comments on Calculated Curves

Let us sum up the main points of our theoretical study. We consider in Fig. 10 the curves of real and imaginary parts of permeability versus frequency on a logarithmic scale using the effective medium model discussed previously. The initial relaxation time $\tau_0 = 10^{-13}$ s. Considering anisotropy as mainly magnetocrystalline and according to [7], we take $K_1 = 5 \times 10^4$ erg cm$^{-3}$ when $T = 300$ K and $K_1 = 3 \times 10^5$ erg cm$^{-3}$ when $T = 200$ K. This means a large variation of barrier energy with temperature. When the temperature $T = 300$ K, we can notice that in the high frequency range, larger than

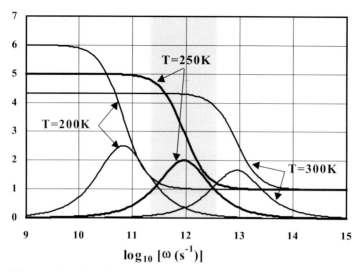

**Fig. 10.** Real and imaginary parts of permeability calculated with the effective Onsager model for three temperatures. The *hatched window* represents the optical measurement range performed with a FFTS

$5 \times 10^{13}\,\mathrm{s}^{-1}$, the real part $\mu'$ is equal to one and the imaginary part $\mu''$ is close to zero: the material is transparent. The frequency is too high to be followed by the magnetic moment. This behavior looks like the behavior of a magnetically ordered crystal, although there is no organization in domains here. On the contrary, when frequency decreases in the very far infrared, below $10^{12}\,\mathrm{s}^{-1}$, the permeability increases ($\mu' \approx 4.3$ and $\mu''$ can reach 2.5), and the material becomes absorbing: as the period of light is larger than the relaxation time, the moments follow the electromagnetic field, consuming some energy to change the direction of the moments. The maximum absorbed energy in the incident electromagnetic wave corresponds to the maximum of the imaginary part of $\mu$, which is a kind of resonance.

When the temperature decreases, we have seen that the curves shift toward a lower frequency. For $T = 200\,\mathrm{K}$, the maximum of $\mu''$ happens when the frequency is lower than $10^{11}\,\mathrm{s}^{-1}$. The sample is always transparent in the optical range ($10^{12}$, $10^{15}\,\mathrm{s}^{-1}$) and becomes absorbing only in the microwave range. The transition region between $T = 200\,\mathrm{K}$ and $T = 300\,\mathrm{K}$ is particularly interesting as it corresponds to the change in behavior between superparamagnetic absorbing material and blocked-state transparent material. In Fig. 10 we have pointed out the range corresponding to the optical measurements we can reach with a fast fourier transform spectrometer ($10\,\mu\mathrm{m} < \lambda < 1000\,\mu\mathrm{m}$). We observe that in this hatched window a sample can change its behavior just by heating from 200 to 300 K.

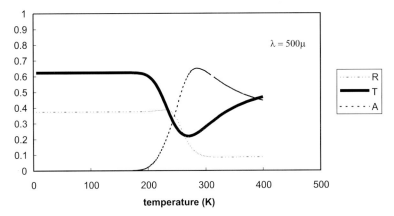

**Fig. 11.** Calculated variations of reflectance $R$, transmittance $T$, and absorption $A$ versus temperature. The wavelength is 500 µm, and the film thickness is 25 µm. Anisotropy is supposed to be purely magnetocrystalline

The optical consequence is obvious in Fig. 11 where the calculated variations of reflectance, transmittance, and absorptance are presented versus temperature for a cermet film made of nickel aluminum nitride (thickness $d = 25$ µm, $\lambda = 500$ µm). Between 5 and 200 K, the film is almost transparent; above 200 K, the behavior changes and the transmission (which is the right parameter for studying this phenomenon) decreases strongly; the transmittance minimum happens at $T \approx 250$ K. Finally, when the temperature continues to increase, the transmission slowly increases, which confirms what was predicted by the variations of $\mu'$ and $\mu''$ shown in Fig. 9.

In crystalline anisotropy, the variation of the barrier energy versus temperature $K(T)V$ is mainly responsible for these results. But in "shape" or "surface" anisotropy, the barrier energy is supposed to be independent of temperature. For the calculations presented in Fig. 12, we take $K_1$ constant and equal to $5 \times 10^4 \, \text{erg cm}^{-3}$. The shape of the curves of optical properties becomes completely different. In particular, the minimum of transmission happens at a much lower temperature than previously. The temperature corresponding to the transmission minimum seems to be very sensitive to the kind of anisotropy for a given film thickness, filling factor and wavelength. The transmission measurement in a well-defined size grain sample may give an idea of the kind of anisotropy (which is always a hard problem to solve).

### 6.2.2    Two Examples of Experimental Results

In Figs. 13, 14, and 15 are shown the measurements performed with a FFT spectrometer on two different cermet films: the matrix is either alumina (Fig. 13) or aluminum nitride (Figs. 14 and 15). Heating was very slow between 5 K and 300 K. All of the influence of temperature on substrate and matrix has been taken off by differential measurements.

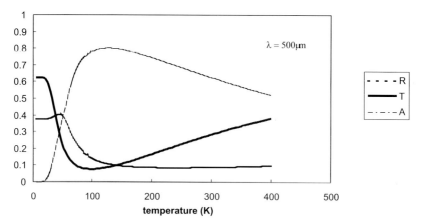

**Fig. 12.** Calculated variations of reflectance $R$, transmittance $T$, and absorption $A$ versus temperature. The wavelength is 500 μm, and the film thickness is 25 μm. Anisotropy is considered as purely shape and surface anisotropy without any strong dependence on temperature

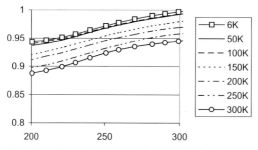

**Fig. 13.** Transmittance of a Ni–Al$_2$O$_3$ film versus wavelength. The film thickness is $d = 1$ μm and the filling factor $f = 0.2$

In Fig. 13, we observe that transmission increases with wavelength and that there is a shift of transmission beginning between 100 and 150 K. The shift goes on regularly when the temperature increases, corresponding to increasing absorbed energy when more and more particles become superparamagnetic.

In Fig. 14, we show the transmittance of a Ni–AlN film with $d = 1$ μm and $f = 0.2$. We can notice that the curves superimposed in the small wavelength region ($\lambda = 50$ μm), where permeability $\mu' = 1$ and $\mu'' = 0$, are shifted with a change in the slope of transmission between 5 and 100 K from $\lambda \geq 80$ μm, up to 350 μm. The curves corresponding to 100, 200, and 300 K are very close to each other, which is significant, as accurate measurements are difficult in the far-infrared range. The blocking temperature corresponding to the measuring time $10^{-12}$ s could be between 5 and 100 K in Fig. 14, whereas it seems to

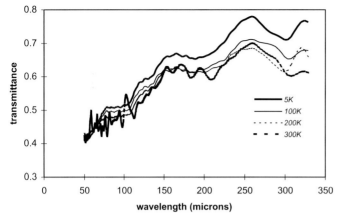

**Fig. 14.** Experimental results. Transmittance of a Ni–AlN film versus wavelength. The thickness of the film is $d = 1$ μm and the filling factor is $f = 0.2$

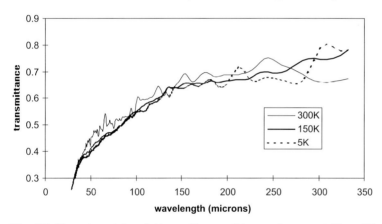

**Fig. 15.** Experimental variations of transmittance of a cermet film with same filling factor and thickness but without granular structure

be a little higher for the alumina matrix in Fig. 13, probably because of the mean size of particles.

In an actual assembly of particles, there is a distribution of size and shape and consequently, a distribution of barrier energy. For this reason the change between blocked states and the superparamagnetic state cannot be so sudden, experimentally. For wavelengths between 200 μm and 250 μm, other particles becomes superparamagnetic, and there is a new shift between transmission at $T = 100$ K and transmission at $T = 200$ K.

As a proof of the superparamagnetic behavior in the previous figure, we show in Fig. 15, the same measurement on a film with the same thickness and composition but where the nickel atoms are mostly dispersed in the matrix of aluminum nitride without forming nanosized ferromagnetic metallic clusters, as confirmed by TEM micrographs. For this "amorphous" film, the measurement in the SQUID does not present any evidence of superparamagnetic behavior. In this case, we notice that the transmission is not sensitive to temperature and that the only variations are due to the usual noise in far-infrared measurements.

# 7   Conclusion

In this contribution, we have pointed out that it is not always justifiable to neglect interaction between the magnetic field of an electromagnetic wave and a magnetic medium. This interaction induces local effects, very often averaged and weak, but sometimes measurable when they are cooperative. It is the case for ferromagnetic nanosized inclusions where exchange energy is negligible.

The experimental results shown here are only the first stammering ones. The author hopes that this work will raise other experimental studies, where the various parameters will be isolated, controlled, and optimized.

Practically, the numerical calculation shows that granular ferromagnetic metals could be temperature-tunable absorbers for far-infrared and microwave radiation.

### Acknowledgment

The author wants to thank Pr. V. Shalaev for giving her the opportunity to present this atypical work in this book. She wants to say how much she feels in debt to Pr. P. Sheng and Pr. P. Gadenne for fruitful discussions, to the students Dr. J. Plon and Dr. C. Desmarest, who worked on the subject for several years, and to C. Naud who performed the optical measurements.

# References

1. M. Gadenne, J. Lafait, J. Phys. (Paris) **47**, 1405–1410 (1986)
2. W. C. Elmore, Phys. Rev. **54**, 1092 (1938); **60**, 593 (1941)
3. A. H. Morrish, S. P. Yu, J. Appl. Phys. **26**, 1049 (1955)
4. C. Kittel, Phys. Rev. **70**, 965 (1946)
5. L. Néel, Comp. rend. Acad. Sci. (Paris) **224**, 1448 (1947); Comp. Rend. Acad. Sci. (Paris) **228**, 664 (1949); J. Phys. Soc. Jpn. **17**, B-1, 676 (1962)
6. J. H. Van Vleck, Phys. Rev. **52**, 1178 (1937)
7. R. M. Bozorth, *Ferromagnetism* (Van Nostrand, New York 1951)
8. L. Bessais, L. Ben Jaffel, J. L. Dormann, *Studies of Magnetic Properties of Fine Particles*, J. L. Dormann, D.Fiorani (Eds.) (Elsevier, Amsterdam 1992)
9. A. Aharoni, Phys. Rev. B **7**, 1103 (1971)
10. D. H. Jones, K. K. P. Strivastava, Phys. Rev. B **34**, 7542 (1986)
11. S. Shrikman, D. Treves, *Magnetism,* Vol. 3, Micromagnets (Academic, New York 1963)

12. S. H. Liou, L. C. Chien, App. Phys. Lett. **52**, 512 (1988)
13. J. L. Dormann, D. Fiorani, E. Tronc, Adv. Chem. Phys. **98**, (1996) Chap. 4
14. W. F. J. Brown, J. Appl. Phys. **34**, 1319 (1963)
15. H. Pfeiffer, Phys. Stat. Solidi A **118**, 295 (1990); Phys. Stat. Solidi A **122**, 377 (1990)
16. S. Dattagupta, *Relaxation Phenomena in Condensed Matter Physics* (Academic, New York 1987)
17. R. W. Cohen, G. D.Cody, M. D. Coutts, B. Abeles, Phys. Rev. B **8**, 3689 (1973)
18. L. Onsager, J. Am. Chem. Soc. **58** 1486 (1936)
19. P. Sheng, M. Gadenne, J. Phys. Condens. Matter **4**, 9735–9740 (1992)
20. C. J. F. Böttcher, *Theory of Electric Polarization* (Elsevier, Amsterdam 1973)
21. G. Xiao, S. H. Liou, A. Levy, J. N. Taylor, C. L. Chien, Phys. Rev. B **34**, 7573 (1986)
22. W. Lamb, D. M. Wood, N. W. Ashcroft, Phys. Rev. B **21**, 2248–2266 (1978)
23. F. Abelès, *Optics of Thin Films* (Van Heel, Amsterdam 1967)

# Manipulating Light with a Magnetic Field

Bart A. van Tiggelen[1] and Geert L. J. A. Rikken[2]

[1] CNRS/Laboratoire de Physique et Modélisation des Milieux Condensés
Université Joseph Fourier, Maison des Magistères,
B.P. 166, 38042 Grenoble, France
tiggelen@belledonne.polycnrs-gre.fr

[2] Grenoble High Magnetic Field Laboratory, Max Planck Institut für
Festkörperforschung/CNRS, B.P. 166, 38042 Grenoble, France

**Abstract.** We review our theoretical and experimental work done on light propagation and scattering in magnetic fields.

## 1 Introduction

For more than one century, we have known that Maxwell's equations provide a complete description of the propagation of classical electromagnetic waves. For applications in daily life, it has become customary to describe the interaction of matter on a macroscopic level, i. e., without worrying about individual atoms, but looking only at charge distributions on scales large compared to the atomic scale. Microscopic charges and currents are described by the polarization density vector $\boldsymbol{P}$ and the magnetization $\boldsymbol{M}$. It is important to realize that this description is only approximate. Cases are known for which the macroscopic Maxwell equations seem to break down since they do not predict the observed behavior [1,2,3]. Macroscopically, it is still possible to consider a charge density $\rho$ and a current density $\boldsymbol{J}$, but we will focus on dielectric materials for which both of them vanish.

A solution of Maxwell's equations becomes feasible when so-called *constitutive* relations are put forward that relate the microscopic parameters $\boldsymbol{P}$ and $\boldsymbol{M}$ to the macroscopic electromagnetic fields $\boldsymbol{E}$ and $\boldsymbol{B}$. Constitutive relations are subject to symmetry relations [4]. For instance $\boldsymbol{P}$ is, like the electrical field $\boldsymbol{E}$, a polar (parity-odd) vector that changes sign upon space inversion. On the other hand, the magnetic field $\boldsymbol{B}$ is a pseudovector, invariant under a space inversion, but variant upon time-reversal. One symmetry allowed, a constitutive relation for the polarization density $\boldsymbol{P}$ could be [4]

$$\boldsymbol{P} = \chi_0 \boldsymbol{E} + \chi_1 \partial_t \boldsymbol{E} \times \boldsymbol{B} + \chi_2 (\boldsymbol{B} \cdot \boldsymbol{B})\boldsymbol{E} + \chi_3 (\boldsymbol{E} \cdot \boldsymbol{B})\boldsymbol{B} + \cdots . \tag{1}$$

For simplicity, we have adopted an isotropic medium so that all constitutive parameters $\chi_n$ are scalars and not second-rank tensors. In the above equation, many other terms are possible, and we have — for future use — just collected the terms linear in the electrical field and without time derivatives of the magnetic field. They are still nonlinear in the magnetic field, which complicates

V. M. Shalaev (Ed.): Optical Properties of Nanostructured Random Media,
Topics Appl. Phys. **82**, 275–301 (2002)
© Springer-Verlag Berlin Heidelberg 2002

the solution of the macroscopic Maxwell equations. A dramatic simplification occurs if the magnetic field $\boldsymbol{B}$ is in fact a low-frequency, external field $\boldsymbol{B_0}$ that is orders of magnitude larger than that of the electromagnetic field itself. In that case, the constitutive relation linearizes to

$$\boldsymbol{P}(\omega) = \boldsymbol{\chi}(\omega, \boldsymbol{B_0}) \cdot \boldsymbol{E}(\omega) \,, \tag{2}$$

with the electrical susceptibility tensor,

$$\chi_{ij}(\omega, \boldsymbol{B}) = \chi_0 \delta_{ij} + \chi_1 \omega \, i\epsilon_{ijk} B_k + \chi_2 B^2 \delta_{ij} + \chi_3 B_i B_j \,. \tag{3}$$

We have inserted harmonic waves with frequency $\omega$. The dielectric tensor is defined as $\boldsymbol{\varepsilon} \equiv 1 + \boldsymbol{\chi}$. The constitutive relation above is purely linear in the electromagnetic field $\boldsymbol{E}$ and demonstrates that magneto-optics can be considered a particular case of nonlinear optics. If we disregard the existence of the microscopic magnetization $\boldsymbol{M}$, an "unwarrantable refinement at optical frequencies" according to *Landau, Lifshitz* and *Pitaevskii* [5], Maxwell's equations and the constitutive equation (2) for $\boldsymbol{P}$ can be combined to give one linear "Helmholtz equation",

$$\boldsymbol{\nabla} \times \boldsymbol{\nabla} \times \boldsymbol{E}(\omega, \boldsymbol{r}) + \frac{\omega^2}{c_0^2} \boldsymbol{\chi}(\boldsymbol{r}, \boldsymbol{B_0}) \cdot \boldsymbol{E}(\omega, \boldsymbol{r}) = \frac{\omega^2}{c_0^2} \boldsymbol{E}(\omega, \boldsymbol{r}) \,. \tag{4}$$

The fascinating analogy of this equation to the Schrödinger eigenvalue equation has frequently been emphasized [6,7,8] and is often of great use in finding its solutions, using results from quantum mechanics. As in Schrödinger's theory, conservation of electromagnetic energy is guaranteed when the electromagnetic "potential" $\boldsymbol{\chi}$ is a hermitan operator, i.e., $\chi_{ij} = \bar{\chi}_{ji}$. This happens when all coefficients $\chi_n$ in (3) are real-valued.

## 2 Magneto-Optics of Homogeneous Media

Local, homogeneous media are characterized by a susceptibility that is independent of $\boldsymbol{r}$. Equation (4) can now easily be solved upon inserting plane waves,

$$\boldsymbol{E}(\omega, \boldsymbol{r}) = \boldsymbol{e}(\omega, \boldsymbol{k}) \exp(\mathrm{i}\boldsymbol{k} \cdot \boldsymbol{r}) \,, \tag{5}$$

with a yet unknown polarization vector $\boldsymbol{e}$. The wave equation (4) reduces to the so-called "Fresnel equation",

$$\det\left[ \boldsymbol{k}^2 - \mathbf{kk} - \frac{\omega^2}{c_0^2} - \boldsymbol{\chi}(\boldsymbol{B_0})\omega^2 \right] = 0 \,, \tag{6}$$

which provides the complex dispersion law $\omega(\boldsymbol{k})$.

The different terms in (3) correspond to different well-known magneto-optical effects. Let us first concentrate on the term $\chi_2$ in (3). We easily obtain the dispersion law given by

$$n^2 \frac{\omega^2}{c_0^2} \approx \left( \boldsymbol{k} \pm \frac{V c_0}{n} \boldsymbol{B}_0 \right)^2, \tag{7}$$

where we dropped small terms quadratic in the magnetic field and introduced the complex index of refraction $n \equiv \sqrt{1 + \chi_0}$, as well as the Verdet constant $V \equiv \frac{1}{2} \chi_1 \omega^2$. If $\chi_0$ and $\chi_1$ are both real-valued, (7) locates the "constant-frequency" surfaces in $\boldsymbol{k}$ space around two spheres, translated over a distance $VB/n$ from the origin along and opposite to the magnetic field (Fig. 1). Their separation is largest when the $\boldsymbol{k}$ vector is parallel to the magnetic field (the so-called "Faraday" geometry) and vanishes when they are mutually orthogonal (the "Voigt" geometry). The two spheres lift the degeneracy of the two states of circular polarization $\pm$, resulting in two different group velocities for different circular polarizations. For linearly polarized light in the Faraday geometry, this leads to a rotation of the polarization vector along the magnetic field over an angle $VBr$, with $r$ the distance of propagation. This is called the Faraday effect. If $\chi_0$ has an imaginary part, the light will be absorbed. A nonzero value of $\mathrm{Im}\,\chi_1$ implies different absorption for different states of circular polarization. This is called magnetic circular dichroism (MCD).

The terms involving $\chi_2$ and $\chi_3$ in (3) are quadratic in the magnetic field and generate a uniaxial symmetry in the dielectric tensor. The resulting linear birefringence is called the Cotton–Mouton effect. This effect is often much smaller than the Faraday effect and has a crucial difference from the Faraday effect. Only the Faraday effect satisfies the relations,

$$\varepsilon_{ij}(\boldsymbol{B}_0) = \varepsilon_{ji}(-\boldsymbol{B}_0) \neq \varepsilon_{ji}(\boldsymbol{B}_0). \tag{8}$$

The equality is a general consequence of the time-reversal symmetry of matter + magnet [5]; the minus-sign is due to the fact that the magnetic field is

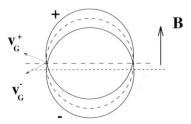

**Fig. 1.** Dispersion law for the Faraday effect. The degeneration of the two circular states $\pm$ of polarization is lifted. Their constant frequency surface is shifted along or opposite to the magnetic field. They remain degenerate only for wave vectors perpendicular to the magnetic field, but the group velocity for both modes is deflected along the magnetic field

a pseudovector. The inequality implies that the Faraday effect, contrary to the Cotton–Mouton effect, breaks time-reversal symmetry only in the subsystem of matter. This aspect gives the Faraday effect an important role in the more general context of light propagation in the presence of broken symmetries.

## 2.1    Magnetodeflection of Light

The bending of electrons in a magnetic field caused by the Lorentz force lies at the very base of many electronic phenomena in disordered metals and semiconductors, such as the Hall effect and the magnetic suppression of weak localization. The question naturally arises whether light propagating in non-scattering, homogeneous media is also deflected by a static transverse magnetic field. Some aspects of the deflection of light by a magnetic field have already been discussed in [5]. Those effects, as has been estimated, are very small [9], which may explain why, to our knowledge and surprise, they have never been observed until recently. A different kind of magnetodeflection has been reported in absorbing homogeneous media [10]. The question whether light is bent by magnets in homogeneous media has recently been raised again by 't Hooft and Van der Mark [11]. This topic is part of a much broader discussion on the properties of macroscopic electromagnetic fields inside dielectrics, that still yields new results, despite its long history [12,13].

In a nonabsorbing medium, the direction of wave propagation is unambiguously given by the group velocity $v_G = d\omega/dk$. The absence of absorption also guarantees that $v_G$ is parallel to the Poynting vector $S = (c_0/4\pi) E \times B$, a theorem that is left as an exercise in [5]. From (7), for the group velocity in the presence of the Faraday effect,

$$v_G^{\pm} = \frac{c_0}{n}\,\widehat{k} \pm \frac{V c_0^2}{\omega n^2}\,B_0\,.\tag{9}$$

On the basis of this equation, the optical energy flow can be deflected by a magnetic field. We note that the deflection is only in the direction of the magnetic field and no "magnetotransverse" term, perpendicular to both magnetic field and wave vector is present.

From Maxwell's equations, we can calculate the Poynting vector $(c_0/4\pi)$ $E \times B$,

$$S_{\pm}(B_0) \propto \mathrm{Re}\,n\widehat{k} \pm \mathrm{Re}\left(\frac{V c_0 \overline{n}}{\omega n}\right) B_0 + \mathrm{Im}\left(\frac{V c_0 \overline{n}}{\omega n}\right)\widehat{k}\times B_0\,.\tag{10}$$

In this expression, we have allowed for absorption and magnetic circular dichroism. They cause differences between the directions of the Poynting vector and the group velocity. In particular, the Poynting vector contains a magneto-transverse component $\widehat{k}\times B_0$.

In nonlocal media, the electromagnetic current density is known to be different from the Poynting vector [5,14,15]. In media with local response, the Poynting vector is widely accepted as the current density. The conflict above with the group velocity puts the validity of the Poynting vector as the energy flow at stake in absorbing media. Arguments based on energy conservation and macroscopic Maxwell equations show that the energy flow of an electromagnetic wave is given by the more general expression

$$\widetilde{S} \equiv \frac{c_0}{4\pi} E \times B + \nabla \times T, \tag{11}$$

where $T$ is some vector field to be determined [16]. The ambiguity in the energy flow, reflected by the existence of the second term in (11), is well appreciated in standard textbooks on electrodynamics [4,16,17,18] but is always discarded.

## 2.2   Bending of Light by Magnetic Fields

To experimentally test the various predictions for the direction of energy flow, we determined the deflection of light upon propagation in several homogeneous dielectrics in a transverse magnetic field [19]. The setup is shown schematically in Fig. 2. A light beam of a given polarization state is normally incident on the sample, placed in a transverse magnetic field, alternating at 8 Hz. The transmitted light is detected by two-quadrant split photodiodes, whose interconnecting axis can be directed along the $B_0$ axis or the $k \times B_0$ axis. The difference between the photodiode signals represents a magnetic-field-induced lateral displacement of the beam after passage through the sample. From (10), it is clear that this displacement can be determined by subtracting the $B_0$ axis displacement signals for left- and right-circularly polarized light, according to

$$\frac{S_+(B_0) - S_-(B_0)}{S} = \frac{2V c_0}{n\omega} B_0. \tag{12}$$

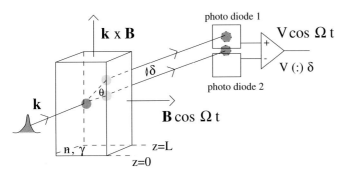

**Fig. 2.** Schematic setup of the deflection experiment. As shown, the displacement of the beam in the $k \times B$ direction is detected. By rotating the photodiode assembly over 90° along $k$, the $B$ axis displacement is detected. Taken from *Rikken* and *Van Tiggelen* [19]

Figure 3 confirms the displacement *along* the magnetic field, predicted by (9). Note that the observed deflection is only of the order of $10^{-7}$ radians. Our estimated inaccuracy is well below this value. This is the *first* experimental observation of deflection of light in a magnetic field. The good agreement with theory also gives us confidence in our experimental setup.

A comment on these conclusions was published by *'t Hooft, Nienhuis,* and *Paaschens* [20]. They argued that *exactly* in Voigt geometry, the eigenmodes are both linearly polarized, none of which suffer from magnetodeflection. They explained our experiments by a "misalignment" of only $10^{-7}$ mrad of the magnetic field. Their considerations are correct but do not affect the magnetodeflection predicted by (12) [21]. Our He Ne laser has a nearly diffraction-limited angular divergence of $10^{-3}$ mrad. The photodiodes measure the deflection of the *intensity weighted* wave vector average. Therefore, only a fraction of $10^{-4}$ of the light flux propagates within the critical range. This fraction will not be deflected. The remainder behaves conform to our description using circularly polarized eigenmodes. For our experiment, the improved deflection theory of *'t Hooft* et al., therefore, would introduce a relative correction in the deduced deflection angle of the beam of $10^{-4}$, which is far below the experimental relative uncertainty of this angle ($\pm 5 \times 10^{-2}$).

We have also carefully looked for the magnetotransverse bending present in the Poynting vector (10) and proportional to the magnetic circular dichroism Im $V$. Experimentally, we found no significant deflection at the level of two orders of magnitude below the theoretical prediction of (10) [19]. To our

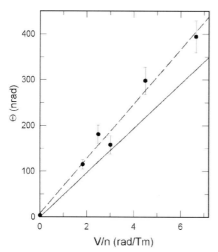

**Fig. 3.** $\widehat{\boldsymbol{B}}$ axis displacement $\delta$ versus sample length $L$ [laser wavelength 632.8 nm, sample material is Plexiglas ($n = 1.49$, $V = 4.5$ mrad/Tm) and magnetic field strength $B = 0.48$T.] *Dashed line* is a linear fit to the data points. *Inset* shows the dependence of the deflection angle $\theta = \delta / L$ on magnetic field strength, also for Plexiglas. *Dashed line* is a linear fit to the data points

knowledge, this is the first time that the widely accepted Poynting vector definition, its direction in particular, is experimentally proven to be incorrect and that the more general relation (11) must be invoked. The absence of a magnetotransverse deflection in homogeneous absorbing media emphasizes the different impact of a magnetic field on absorbing and scattering media; as will be shown below, the latter exhibits such a reflection.

It seems indeed possible to find a choice for $T$ that reconciles the energy flow as expressed by (11) with the group velocity, which in turn is consistent with our experimental observation. We emphasize that this choice is not necessarily unique. Although we have not clarified the physical significance of such remarkable, albeit unavoidable, choices for $T$, these findings contribute to the ongoing discussion on the interpretation of the Poynting vector as the flow of electromagnetic energy [23].

# 3   Magneto-Optics of Heterogeneous Media

## 3.1   Single Magneto-Mie Scattering

To undertake a study of magneto-optics of heterogeneous media, one is obliged first to understand the magneto-optics of one small spherical particle. Already in the 1970s, *Ford* and *Werner* [24] made an extensive study of magneto-Mie scattering. The small perturbation of the Faraday effect, of the order of $10^{-4}$, to the standard Mie problem [25], justifies a perturbation theory linear in the magnetic field, quite similar to the standard treatment of the Zeeman effect in atomic orbitals.

The first-order magneto-optical change in the differential cross section $d\sigma/d\Omega(\boldsymbol{k} \to \boldsymbol{k}')$ can be guessed from symmetry arguments. Let us consider a Mie sphere made of a dielectric constant given by (3). Being a scalar, the magneto cross section linear in $\boldsymbol{B}_0$ must be proportional to either $\boldsymbol{k} \cdot \boldsymbol{B}_0$, $\boldsymbol{k}' \cdot \boldsymbol{B}_0$ or $\det(\boldsymbol{k}, \boldsymbol{k}', \boldsymbol{B}_0)$. Being pseudoscalars, the first two options are parity-forbidden. The *only* expression allowed by symmetry and linear in the magnetic field is

$$\frac{1}{\sigma_{\text{tot}}}\frac{d\sigma}{d\Omega}(\boldsymbol{k} \to \boldsymbol{k}'; \boldsymbol{B}_0) = F_0(\theta) + F_1(\theta)\det(\boldsymbol{k}, \boldsymbol{k}', \boldsymbol{B}_0)\,. \tag{13}$$

This cross section also obeys the reciprocity principle

$$\frac{d\sigma}{d\Omega}(\boldsymbol{k} \to \boldsymbol{k}'; \boldsymbol{B}_0) = \frac{d\sigma}{d\Omega}(-\boldsymbol{k}' \to -\boldsymbol{k}; -\boldsymbol{B}_0)\,. \tag{14}$$

$F_0(\theta)$ is the phase function of the conventional Mie problem [25] and — by rotational symmetry — depends only on the angle $\theta$ between $\boldsymbol{k}$ and $\boldsymbol{k}'$. For the same reason, $F_1$ can depend only on the angle $\theta$. In [26], we developed a method for calculating $F_1(\theta)$. For a Rayleigh scatterer, the Born approximation can be used which leads to $F_1(\theta) \sim (V/k)\cos\theta$.

It is well known that for applications in multiple scattering, the anisotropy of the cross section is important [6]. For Mie spheres, this anisotropy is quantified by the "anisotropic factor" $\langle \cos\theta \rangle$, which is $\cos\theta$ averaged over $F_0(\theta)$. This factor discriminates forward $(\cos\theta > 0)$ from backward $(\cos\theta < 0)$ scattering. In magneto-Mie scattering, a second anisotropy shows up that discriminates "upward" from "downward" scattering (Fig. 4). If the magnetic field is perpendicular to both the incident and outgoing wave vectors, (13) predicts a difference between upward and downward flux, both defined with respect to the magnetotransverse direction $\boldsymbol{k} \times \boldsymbol{B}_0$. The magnetoanisotropy $\eta_1$ of one scatterer can be quantified exactly as the normalized difference between total flux upward and total flux downward. An easy calculation yields [26]

$$\eta_1 \equiv 2\pi \int_0^\pi \mathrm{d}\theta \sin^3\theta F_1(\theta).\tag{15}$$

If $\eta_1 \neq 0$, we shall speak of "magnetotransverse light scattering." No magnetotransverse anisotropy survives for one Rayleigh scatterer, as can easily be checked by filling in $F_1 \sim \cos\theta$. At least two Rayleigh scatterers are required to generate a net effect [26], or, alternatively, a Mie sphere with finite size (Fig. 5).

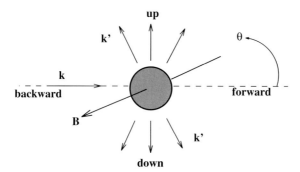

**Fig. 4.** Scattering geometry for magnetotransverse scattering from a Mie sphere, given an incident plane wave from the left

### 3.2 Multiple Magnetoscattering of Light

In principle, when the medium is heterogeneous, light is scattered and concepts like "Faraday rotation" become ill-defined. Even in a random medium, it seems possible to define a dielectric constant associated with the "effective medium". When small particles of the heterogeneous medium are magnetoactive, the effective medium is undoubtedly magnetoactive as well. The wave will undergo a Faraday effect on its way from one scatterer to the other. Since

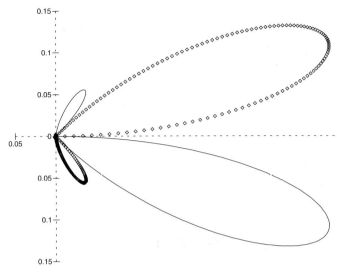

**Fig. 5.** Polar plot of the phase function associated with the magneto cross section of Mie scatterers with size parameter $x = 2\pi a/\lambda = 5$. *Solid line* denotes positive magneto cross section, *symbols* denote negative magneto cross section. Magnetic field is directed perpendicular to the plot. Taken from *Lacoste* et al. [26]

the typical distance between two scattering events is typically the mean free path $\ell$, we expect a typical Faraday rotational angle of

$$\alpha \approx V_{\text{eff}} B\ell \cos\theta , \tag{16}$$

with $V_{\text{eff}}$ some effective medium Verdet constant and $\theta$ the angle between the wave vector and the magnetic field. If the matrix is inert and the scatterers have a Verdet constant $V$ and volume fraction $f$, the choice $V_{\text{eff}} \approx fV$ does not *a priori* seem unreasonable.

Alternatively, the Faraday effect can be understood as a magnetically induced phase shift $\phi = \sigma V_{\text{eff}} B\ell \cos\theta$ of a wave with circular polarization $\sigma$. Since multiple scattering tends to randomize the state of circular polarization, the magnetic field induces a net zero phase shift of a wave. This excludes the Faraday effect as a direct phenomenon in multiple scattering. Nevertheless, novel effects in multiple scattering can exist that have the Faraday effect at their origin. The typical fluctuation of the magnetically induced phase shift in multiple scattering, $\sqrt{\langle\phi^2\rangle} \approx |V_{\text{eff}} B\ell|$, is likely to be a parameter quantifying the impact of Faraday effect in multiple scattering, just like the parameter $\omega_c \tau$ (the cyclotron motion executed by an electron during one mean free time) is known to quantify magnetic effects in electronic transport. Three magneto-optical scattering phenomena are now known to exist: the photonic Hall effect, photonic magnetoresistance, and the magnetic supression of coherent backscattering.

The influence of a magnetic field on coherent backscattering will not be discussed here. Experimental work by *Maret* et al. demonstrated that interference phenomena in multiple scattering are suppressed by an external magnetic field due to the broken time-reversal symmetry [27]. The universal parameter governing the suppression is indeed found to be $V_{\text{eff}} B_0 \ell$ [28], as confirmed by numerical simulations [29] and calculations with point scatterers [30] as well as Mie spheres [31].

The physics of multiple scattering can best be understood by adopting a diffusion picture for the multiply scattered light. This familiar picture asserts that the average local current density $\boldsymbol{J}$ is due to a gradient in electromagnetic energy density $\boldsymbol{\nabla}\rho$. The so-called Fick's law reads [6],

$$\boldsymbol{J}(\boldsymbol{r}) = -\boldsymbol{D} \cdot \boldsymbol{\nabla}\rho(\boldsymbol{r}), \tag{17}$$

with $D_{ij}$ the diffusion tensor. Because the current and the gradient are both parity-odd vectors, $\boldsymbol{D}$ must be parity-even. Fick's law breaks time-reversal symmetry since $\boldsymbol{D}$ is believed to be a positive definite tensor, and only the current changes sign upon time-reversal. This symmetry breaking is due to the ensemble-average that has been assumed implicitly in (17) and makes macroscopic transport phenomena irreversible.

The impact of a magnetic field on light diffusion can be understood qualitatively by realizing that the magnetic field is a pseudovector. Onsager relations apply to transport coefficients and thus also to the diffusion tensor,

$$D_{ij}(\boldsymbol{B}_0) = D_{ji}(-\boldsymbol{B}_0), \tag{18}$$

so that,

$$\begin{aligned} D_{ij}(\boldsymbol{B}_0) = {} & D_0 \delta_{ij} + D_{\text{H}}(B_0)\epsilon_{ijk}(B_0)_k \\ & + D_\perp \left(\delta_{ij} B_0^2 - (B_0)_i (B_0)_j\right) + D_\parallel (B_0)_i (B_0)_j. \end{aligned} \tag{19}$$

The first term is just the ordinary, isotropic diffusion tensor. The second term induces an energy current perpendicular to the energy gradient and the magnetic field, i. e., is magnetotransverse. In analogy to a similar effect for electrons in disordered semiconductors, we will refer to this as the *photonic Hall Effect* (PHE). The last two terms are quadratic in the magnetic field and make the current along the magnetic field different from that perpendicular. We will call this *photonic magnetoresistance* (PMR). Similar effects occur in the so-called Beenakker–Senftleben effect, describing the influence of a magnetic field on the thermal conductivity of paramagnetic and diamagnetic gases [32]. In the following subsections, we discuss the relation of these macroscopic effects to their microscopic equivalents discussed earlier, and report on their experimental verification.

## 3.3    Theory of Magnetodiffusion

One of the aims of multiple scattering theory is to establish a link between the macroscopic and the microscopic world. More specifically, we want to understand, qualitatively and quantitatively, the diffusion tensor (19) from the magneto cross section of one particle.

A rigorous theory for the photonic Hall effect was recently developed by *Lacoste* and *Van Tiggelen* [33,34]. Using a microscopic transport theory, a relation was found between the photonic Hall effect of one Mie particle, quantified by the parameter $\eta_1$ in (15), and the magnetotransverse diffusion $D_H$ defined in (19),

$$\frac{D_H B_0}{D_0} = \frac{1}{2} \frac{\eta_1}{1 - \langle \cos\theta \rangle} = \frac{\pi}{1 - \langle \cos\theta \rangle} \int_0^\pi d\theta \sin^3\theta F_1(\theta). \tag{20}$$

Such a relation is intuitively reasonable, provided that single scattering is the basic building block of multiple scattering, which is true when the mean free path is much bigger than the wavelength. This was assumed in deriving (20). Perhaps less intuitive is the presence of the average cosine $\langle \cos\theta \rangle$ of the scattering cross section in (20). This parameter can be very close to one, as for large dielectric spheres [25], which greatly amplifies the relative importance of the photonic Hall effect. This factor was absent in an earlier calculation, which used Rayleigh point scatterers [35] but turned out to be important to come to a quantitative agreement between theory and experiments. Note that, by (20), $D_H/D_0$ should be independent of the scatterer concentration, provided it is small enough to ignore cluster effects.

In view of the familiar relation $D = \frac{1}{3} v_E \ell^*$ of the diffusion constant in terms of the speed of light and the mean free path $\ell^*$ [6], one can write the ratio $D_H B_0/D_0$ as a ratio of two transport mean free paths. This is more convenient for later purposes. This ratio can be used to define what we call the "photonic Hall angle". In electronic transport, this would be $\sigma_{xy}/\sigma_{xx}$ with $\sigma_{ij}$ the conductivity tensor, directly proportional to the diffusion tensor [36].

In Fig. 6, we show $\ell_\perp^*/\ell^*$ as a function of the size parameter, for a contrast in the index of refraction of $m \equiv n_S/n_m = 1.128$, corresponding to $CeF_3$ in glycerol. For particles sizes around 4 µm, we calculate $\ell_\perp^*/\ell^* = +0.06VB\lambda$ which, for $V = -1100\,\text{rad/Tm}$ (at temperature T = 77 K) and vacuum wavelength $\lambda_0 = 0.457$ µm yields $\ell_\perp^*/\ell^* = -2 \times 10^{-5}/\text{T}$. The experimental value is $\ell_\perp/\ell^* \approx -1.1 \pm 0.3 \times 10^{-5}/\text{T}$ for a 10 vol. % suspension [22].

A theory for the photonic magnetoresistance, i.e., the diffusion coefficients $D_\perp$ and $D_\parallel$ in (19), has so far been developed only for pointlike scatterers with Verdet constant $V$ and volume fraction $f$ [37], for which the prediction is

$$\frac{D_\perp}{D_0} = -\frac{12}{5}(fVB_0\ell)^2$$

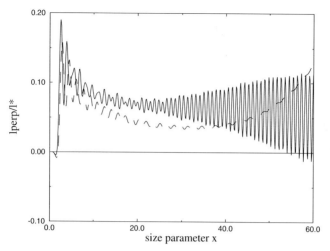

**Fig. 6.** Ratio $D_H/D_0$ (photonic Hall angle) of magnetotransverse diffusion to isotropic diffusion, in units of the dimensionless parameter $2\pi V B/k$, as a function of size parameter $x = ka$ ($a$ is the particle radius). *Solid line* corresponds to Mie spheres with index of refraction $m = 1.28$; *dashed line* is the Rayleigh–Gans approximation, valid as $m \to 1$. Taken from *Lacoste* and *Van Tiggelen* [33]

and

$$\frac{D_\parallel}{D_0} = -\frac{6}{5}\left(fVB_0\ell\right)^2 . \tag{21}$$

This calculation always implies a negative magnetoconductance (positive magnetoresistance) proportional to the typical fluctuations in the magnetically induced phase shift between two collisions. This property is expected to be valid in general. In the dilute regime, the ratios $D_{\perp,\parallel}/D_0$ are expected to be *independent* of the concentration of the scatterers.

## 3.4   Experiments on Magnetodiffusion

For experimental observation of the photonic Hall effect, scatterers with a large Verdet constant are required. This can be found in materials containing large concentrations of rare-earth ions like $Ce^{3+}$, $Ho^{3+}$, or $Dy^{3+}$. In these paramagnetic materials, the Verdet constant is inversely proportional to temperature and can thus be further enhanced by cooling.

The photonic magnetotransport phenomena were measured by phase-sensitive detection of the magnetically induced changes in the scattered and transmitted intensities. An alternating magnetic field $B_0(t) = B_0 \cos \Omega t$ with $B_0 \approx 1\,\text{T}$ and $\Omega \approx 40\,\text{Hz}$ was applied perpendicularly to the illuminating and collecting light guides. Monochromatic illumination was provided by an argon ion laser or with an incandescent lamp in combination with a narrow-band interference filter. The scattered light was measured with silicon photo-diodes or photomultipliers outside the magnetic field region (see Fig.7). Both linear and quadratic magnetic field responses can be measured this way at the fundamental and second-harmonic frequencies $\Omega$ and $2\Omega$, respectively. The mean free path $\ell^*$ of the light in these scattering media was determined by measuring their optical transmission $T$ by an integrating sphere and using $T = 1.6\, \ell^*/L$, where $L$ is the sample thickness [38].

To deduce the various diffusion coefficients from our measurements, we have to know their relation to the current emerging from our cylindrical sample. This engineering part of the experiment was examined by solving the diffusion equation, with diffusion constant $D_0$ for a cylinder geometry with radius $R$ and length $L$, with radiative boundary conditions at the surface and on the sides [38]. We adopted a source at a depth $x_0 \approx \ell$ at one side of the cylinder with a radial profile $J_{in}(R)$ across the output of the multimode fiber used in the experiment. The magnetotransverse current can be calculated from this solution using Fick's law (17). These calculations demonstrated the following relation for the normalized photonic Hall effect:

$$\eta \equiv \frac{I_{up} - I_{down}}{\frac{1}{2}(I_{up} + I_{down})} = F\left(\frac{L}{R}\right)\frac{\ell_H}{R}. \tag{22}$$

The function $F$ depends only on the ratio of the length and width of the cylinder, not on the mean free path. For the experimental value $L/R = 2.6$, we estimated $F \approx 5$. A measurement of $\eta$ thus gives direct access to the magneto-transverse transport mean free path $\ell_H$.

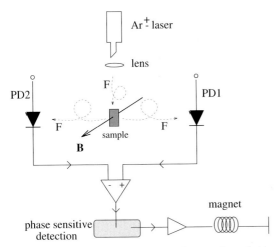

**Fig. 7.** Schematic setup for the observation of the photonic Hall effect. F = optical fiber, PD = photo-diode, $\boldsymbol{B}$ is the magnetic field

The role of sample geometry for the photonic magnetoresistance is easier to obtain since a slab geometry and an integrating sphere suffice. The total transmission coefficient, normalized to the one measured without magnetic field, is obtained from

$$\frac{\Delta T(\boldsymbol{B}_0 = \|, \perp)}{T} = \frac{D_{\|,\perp} B_0^2}{D_0} . \tag{23}$$

All geometry-dependent factors cancel since they are the same with and without the magnetic field.

### 3.4.1  Normal Photonic Hall Effect

The photonic Hall effect is qualified by the difference of transverse photon flux $\Delta I_\perp = I_L - I_R$, normalized by the transversely scattered intensity $I = \frac{1}{2}(I_L + I_R)$. This ratio should be proportional to the ratio of the magneto-transverse mean free path and the transverse dimension, as stated by (22). We estimate a systematic error of at least a factor of 2 in *attributing* values to $\ell_H$ and $\ell^*$ on the basis of Fick's law. Another uncertain factor is the broad size distribution of the scatterers.

The results shown in Fig. 8 confirm the predicted linear magnetic field dependences of this quantity. By normalizing it by the magnetic field, we obtain the photonic Hall effect per Tesla which is characteristic for a given scattering sample. Figure 9 shows the temperature dependence of the photonic Hall effect per tesla. As the only temperature-dependent parameter in the scattering process is the Verdet constant, the observed linear dependence on the

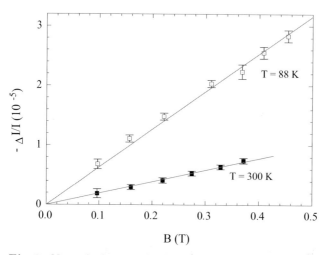

**Fig. 8.** Normal photonic Hall effect vs. magnetic field, showing linear behavior. Sample was EuF$_2$ in resin, observed with light of wavelength $\lambda = 457$ nm. Taken from *Rikken* et al. [42]

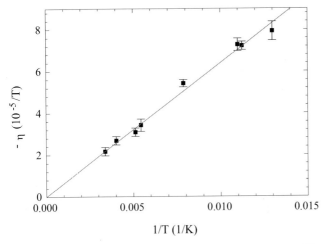

**Fig. 9.** Photonic Hall effect (per tesla) vs. temperature. The plot confirms the $1/T$ behavior predicted for paramagnetic scatterers. Sample was $EuF_2$ in resin, observed with light of wavelength $\lambda = 457\,nm$. Taken from *Rikken* et al. [42]

inverse temperature confirms the linear relation between $\eta$ and the Verdet constant. Figure 10 shows the mean free path $\ell^*$ and $\eta$ as a function of the volume fraction of scatterers. We see that for large $f$, where $\ell^*$ becomes much smaller than the sample thickness and we are entering the multiple scattering regime, the Hall angle seems to become independent of $f$. The value for the magnetotransverse scattering length that we deduce at that point, using (22), is $\ell_H/B = -1\,nm/T$, i.e., $\ell_H/\ell^* \approx -1.1 \times 10^{-5}/T$, which is in good agreement with the Mie theory [33] discussed earlier. The inset shows the normalized photonic Hall effect for several different scatterers, as a function of the Verdet constant of the scatterers. The sign of the magnetotransverse photon flux was deduced from the phase angle of the lock-in. The linear relation that is observed, including the sign, cannot simply be explained by the linearity with the Verdet constant because the index of refraction also varies considerably among the different samples. Nevertheless, our magneto-Mie theory for $CeF_3$ theory reproduces this linearity for the photonic Hall effect assuming polydisperse samples containing $ZnS$, $Al_2O_3$, $TiO_2$, $CeF_3$, and $EuF_2$ [34]. The calculated photonic Hall effect changes rapidly as a function of particle size and can even change sign, but on average, the sign of the photonic Hall effect reflects the sign of the Verdet constant. The estimated anisotropic factor for our sample equals $\langle\cos\theta\rangle \approx 0.9$. Therefore, the amplification factor $1/(1 - \langle\cos\theta\rangle)$ for the photonic Hall effect is significant and improves theoretical predictions considerably for Rayleigh scatterers [35]. The pertinent role of the sign of the photonic Hall effect makes the analogy with the behavior of electrons and holes in the electronic Hall manifest again.

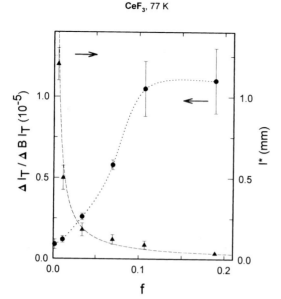

**Fig. 10.** Photonic Hall effect (per tesla) vs. volume fraction $f$ of CeF$_3$ particles at a temperature $T = 77$ K. Also plotted is the transport mean free path as a function of volume fraction (*left vertical axis*) which decays basically as $1/f$ (*dashed*). Taken from *Rikken* and *Van Tiggelen* [21]

In this analogy, the Verdet constant of the scatterer takes over the role of electronic charge.

### 3.4.2   Photonic Hall Effect in Inverted Media

So far, we considered media with magnetoactive scatterers in a passive matrix. One might argue that in electronic magnetotransport, the effect of the magnetic fields occurs mainly between the scattering events. This raises the question whether a photonic Hall effect exists in "inverted" media, consisting of passive scatterers in a magnetoactive matrix.

The inverted medium is mathematically much more difficult to handle, and therefore no theoretical description is available at present. *Martinez* and *Maynard* [29] studied the inverted medium using a Monte-Carlo simulation but did not treat magnetodiffusion. Intuitively, there is no reason to believe that the photonic Hall effect differs from that in normal media. For volume fractions around $f \approx 50\%$, it is in fact impossible to discriminate between scatterers and matrix.

In the experiments [39], we used two different matrices, a saturated aqueous dysprosium chloride solution and a dysprosium nitrate glass. These are transparent apart from a few narrow and weak $4f$–$4f$ transitions of the

$Dy^{3+}$ ion. The following values were found for the optical parameters at a wavelength of 457 nm: The refractive index of the solution $n_m = 1.44$, the Verdet constant $V = -19$ rad/T m at 300 K, the refractive index of the glass $n_m = 1.53$, and its Verdet constant $V = -59$ rad/Tm at 300 K. Scattering samples were prepared by adding $Al_2O_3$ particles to the matrix. The average size of these particles was 1 μm with a 50% size dispersion. The refractive index of these scatterers was $n_s = 1.72$. Their volume fraction $f$ ranged between 0.5 and 10%.

We measured $\ell^*$ as a function of the volume fraction $f$ of the solution samples and found that over a large range of volume fractions, $\ell^* \propto f^{-1}$. Figure 11 confirms the predicted linear magnetic field dependence of the photonic Hall angle. It is important to note that the sign of the photonic Hall effect in inverted media with paramagnetic scatterers is the same as that obtained in normal media, where the magneto-optical activity is concentrated in paramagnetic scatterers. The temperature dependence is implied in Fig. 11. As the only temperature-dependent parameter in the scattering process is the Verdet constant, the observed linear dependence on the inverse temperature confirms the linear relation between the normalized photonic Hall effect and the Verdet constant of the matrix, which identifies the Faraday effect between the scatterers as its physical origin.

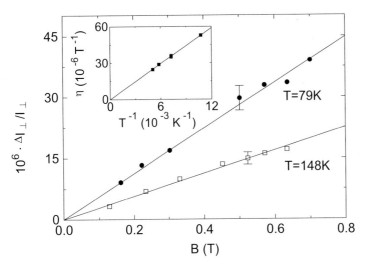

**Fig. 11.** Inverted photonic Hall effect vs. the magnetic field and the temperature (*inset*). As for the normal photonic Hall effect, the inverted photonic Hall effect is linear in the magnetic field and inversely proportional to temperature. Sample was $Al_2O_3$ particles with volume fraction $f = 4.2\%$ imbedded in a dysprosium nitrate glass. Wavelength of the light was $\lambda = 457$ nm. Taken from *Düchs* et al. [39]

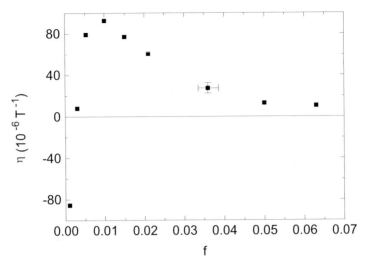

**Fig. 12.** Inverted photonic Hall effect vs. volume fraction of inert $Al_2O_3$ scatterers in a dysprosium nitrate glass, measured at a temperature $T = 77\,\mathrm{K}$. Taken from *Düchs* et al. [39]

The effect of scatterer concentration on the photonic Hall effect is shown in Fig. 12 for the glass samples. Upon decreasing the scatterer volume fraction, an increase in the normalized photonic Hall effect is obtained, until at a low volume fraction, the photonic Hall effect decreases and even changes sign. The increasing photonic Hall effect with decreasing volume fraction can be understood as a mean free path effect; Fig. 13 shows the photonic Hall effect as a function of the mean free path of the high volume fraction samples in the top panel of Fig. 12. It shows the linear dependence of the photonic Hall effect on the mean free path as long as this is much smaller than the sample dimensions, i. e., in the multiple scattering regime. For smaller volume fractions, i. e., longer mean free paths, this no longer holds and for the lowest volume fractions, one enters the single scattering regime which apparently gives an opposite sign for the magnetotransverse photon flux. The volume fraction at which this turnover occurs is higher for the glass samples because the refractive index difference between scatterer and matrix is smaller for these samples; so the mean free path is longer than for a solution sample at the same scatterer concentration.

For "normal" media, our experiments have shown empirically that the normalized photonic Hall effect in the multiple scattering regime is proportional to $V_{\mathrm{eff}}B\ell^*$ for the range of concentrations studied, where $V_{\mathrm{eff}} = fV_{\mathrm{s}}$ [22]. We now propose the empirical expression for arbitrary two-component media in the multiple scattering regime:

$$\eta = G[fV_{\mathrm{s}} + (1 - f)V_{\mathrm{m}}]\ell^* . \qquad (24)$$

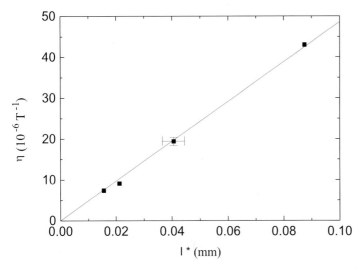

**Fig. 13.** Inverted photonic Hall effect vs. mean free path, measured in the multiple scattering regime (i.e., relatively large $f$ in Fig. 12). Sample was the same as in Fig. 12. Taken from *Düchs* et al. [39]

The factor $G$ can depend only on the shape of the sample, but this was not studied. For the experiments done so far, the value $G \approx 1 \times 10^{-2}$ covers both the normal and the inverted photonic Hall effect.

### 3.4.3  Photonic Hall Effect in Absorbing Media

Absorption is a specific property of classical waves and has no equivalent in charge transport. Therefore, we conducted experiments on the photonic Hall effect in the presence of absorption [40].

We first investigated the role of absorption in the scatterers. To that end, we used particles of $HoF_3$, obtained by chemical precipitation, and put them in a transparent, inert homogeneous matrix. The particles had an average radius $r \approx 0.5$ μm and a broad size distribution between $r = 0.2$ μm and $r = 5$ μm, that was determined using scanning electron microscopy. The $Ho^{3+}$ ions have a narrow $4f$–$4f$ transition $^5I_8 \rightarrow ^5 F_4$ around $\lambda \approx 534$ nm. By varying the wavelength of the diffusing light over a few tens of nanometers, we were able to scan across the absorption band, thereby strongly varying the imaginary part of the index of refraction $\kappa_s$, without significantly affecting the other optical parameters. The real part of the refractive index is $n_s = 1.6$ outside the absorption peak, and it varies only by 0.004 around the absorption peak.

The absorption spectrum in Fig. 14 was measured with a UV-vis spectrometer at room temperature from a thin slab of $HoF_3$ powder dispersed in an index-matched resin at 23 vol%. The absorption maximum blueshifts from $\lambda = 534\,\text{nm}$ at $T = 300\,\text{K}$ toward $\lambda = 529\,\text{nm}$ at $T = 85\,\text{K}$. From the measurements carried out at $T = 300\,\text{K}$ on samples of different thickness, we concluded that the imaginary part of the refractive index is $\kappa_\text{s} \approx 0.0012$ at $\lambda = 535\,\text{nm}$ and $T = 300\,\text{K}$ and the absorption length in the multiple scattering samples is $L_\text{a} \approx 31 \pm 6$ µm at $\lambda = 535\,\text{nm}$. We assumed that these values stay the same at $T = 85\,\text{K}$ when corrected for the blueshift. Other optical parameters undergo only minor changes: the Verdet constant of $HoF_3$ is $\text{Re}\,V_\text{s} \approx 400\,\text{rad/Tm}$ and was seen to vary by only 20 % across the absorption band (Fig. 14a). The magnetic circular dichroism $\text{Im}\,V_\text{s}$ is estimated to be smaller than $1\,\text{rad/Tm}$ [41] in this spectral range. We estimate a variation $\Delta\ell^*/\ell^* \leq 0.25$ in the mean free path across the absorption band due to the change of the complex index of refraction $n_\text{s} + i\kappa_\text{s}$ with $\lambda$. The resin matrix had a refractive index $n_\text{m} \approx 1.566$ at $589\,\text{nm}$. Samples were prepared by mixing $HoF_3$ powder at a volume fraction of $f = 23\,\%$ with the liquid resin, followed by curing. We measured a transport mean free path $\ell^* \approx 70 \pm 26$ µm in the transparent spectral region of $HoF_3$.

We observed again a linear magnetic field dependence of $\Delta I_\perp/I_\perp$. This linear behavior was independent of the amount of absorption. Figure 15 contains the main result: It shows the normalized photonic Hall effect $\eta$ as a function of wavelength around the absorption maximum. Outside the absorption band, we obtain negative values for $\eta$. These values agree in sign and magnitude with the results obtained in [22] and [42] for similar paramagnetic

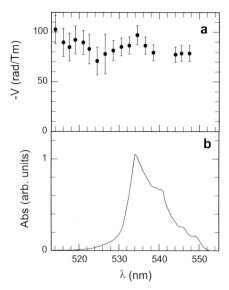

**Fig. 14.** Verdet constant (**a**) and absorption profile (**b**) for $HoF_3$ powder in resin, both at room temperature, around a wavelength of $535\,\text{nm}$. Taken from *Wiebel* et al. [40]

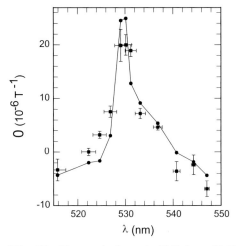

**Fig. 15.** Observed photonic Hall for a HoF$_3$ powder scanned along the absorption profile. The *solid line* is the prediction of Mie theory. Taken from *Wiebel* et al. [40]

scatterers. As the absorption increases, the photonic Hall effect $\eta$ decreases to zero and rapidly changes sign around the absorption band center. The sign change of $\eta$ seems to occur at the wavelength where $L_a$ roughly equals $\ell^*$. In the center of the absorption band, we obtain values for the photonic Hall effect that are roughly four times larger than those measured in the transparent region of HoF$_3$. At this wavelength, the absorption length $L_a \approx 31$ μm is smaller than $\ell^*$ and much smaller than the sample dimensions. As the maximum value of the photonic Hall effect coincides with the wavelength of maximum absorption, we conclude that the observed changes in the photonic Hall effect are dominated by absorption effects. Variations of other relevant optical parameters would have led to only minor changes in the photonic Hall effect.

To investigate the impact of absorption in the medium outside the scatterers, we prepared samples of similar, but transparent CeF$_3$ scatterers in the same resin matrix that was made absorbing by dissolving an organic dye into it. Values up to $\kappa_m \approx 0.001$ were obtained this way, i.e., comparable to the values of $\kappa_s$ for the absorbing scatterers in the transparent matrix. We found *no significant variation* of the normalized photonic Hall effect $\eta$ with increasing $\kappa_m$. Quite surprisingly, we see that the role of absorption in the matrix is very different from that of absorption in the magnetoactive scatterers.

The Mie theory developed by us to calculate the photonic Hall effect of spherical scatterers of arbitrary size [26,33] allows for the inclusion of absorption inside the scattering particles but does not allow for absorption in the matrix. In the calculation, we assume a complex refractive index $n_s + i\kappa_s(\lambda)$ and use $\kappa_s(\lambda)$ obtained from the absorption spectra in Fig. 14, corrected for the low temperature blueshift. We took Re $V_s = 400$ rad/Tm and $\ell^* = 70$ μm and completely disregarded $\mathrm{Im}\, V_s$. The geometry factor $F$ in (22) was kept constant, although we estimate theoretically that this factor may increase by perhaps a factor of 2 in the range of absorption covered in this experiment. The broad size distribution of the scatterers was taken into account by averaging our numerical results over particle sizes between 0.2 and 5 µm. The results were found quite independent of the exact choice of the particle size distribution. As shown in Fig. 15, they show good agreement with the experimental results. Note that no adjustable parameters are used, as all of them were determined experimentally. The good agreement suggests that the observed wavelength dependence of the photonic Hall effect $\eta$ is mainly due to the effect of absorption on the magneto cross section of one Mie particle. Our numerical study can be summarized by the simple relation

$$\eta = (\alpha + \beta\kappa_s)V_s B_0 \ell^* \,, \tag{25}$$

where $\alpha$ and $\beta$ are parameters that may depend on sample geometry, refractive index contrast, size- and shape-distribution of the scatterers, etc., but not on the Verdet constant, absorption of the matrix, or magnetic field and concentration.

### 3.4.4    Photonic Magnetoresistance

Now that the analogy between the electronic and the photonic Hall effect seems to be well established, one can wonder whether electronic magnetoresistance also has a photonic equivalent. The photonic magnetoresistance was measured by performing phase-sensitive detection at $2\Omega$ on the magnetically induced change of the transmitted intensity and normalizing by the total transmitted intensity. In this experiment, the magnetic field was aligned perpendicularly to the direction of transmission (see inset, Fig. 16). This ratio is then taken equal to the ratio of the magnetoresistive and normal diffusion coefficients, as expressed by (23). Based on the outcome (21) for Rayleigh scatterers, we used the relation,

$$\frac{\Delta T}{T} \sim -(fVB_0\ell^*)^2 \,, \tag{26}$$

as a starting point for our experiments.

Figure 16 shows our results for the transmission modulation at $2\Omega$ as a function of the square of the magnetic field amplitude. Both curves, measured for different particle sizes and for different volume fractions, reveal a quadratic field dependence. From the lock-in phase, we deduce that for these samples, the transmission decreases with increasing magnetic field, i. e., $\Delta I/I \propto -B^2$.

As in the photonic Hall effect, the explicit dependence of the transmission modulation on the Verdet constant can be conveniently studied by varying the temperature. Figure 17 shows the relative transmission modulation at $2\Omega$

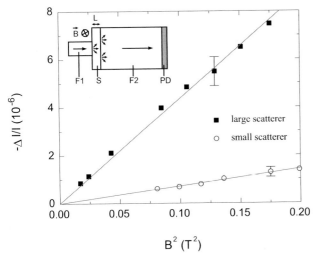

**Fig. 16.** Transmission modulation at $2\Omega$ versus the square of the magnetic field strength. Temperature is 105 K. Full symbols represent samples with an average particle diameter of 2 μm and a volume fraction $f = 17\%$. Open symbols are for an average particle diameter of 0.4 μm and a volume fraction $f = 10\%$. *Lines* are fit to the data points. The *inset* shows a schematic setup. Light is guided through an optical fiber F1 (diameter 1 mm) to the sample S, which consists of EuF$_2$ powder in a polymer disk. Forwardly scattered light is collected by a light guide F2 and detected by a silicon photodiode PD outside the magnetic field region. The magnetic field is directed perpendicularly to the plane of the drawing. Taken from *Sparenberg* et al. [43]

as a function of the square of the inverse temperature at constant magnetic field. The observed linear dependence confirms the quadratic dependence of the photonic magnetoresistance on $V$.

On the basis of (26), we expect that the magnetoresistance should become independent of $f$, once we are in the diffusive regime. This is demonstrated in Fig. 18: for high volume fractions, where $\ell^*$ is smaller than the sample thickness, the observed transmission modulation is independent of $f$. For low volume fractions, the mean free path is no longer small compared to the sample thickness, and diffusion theory should break down. Experimentally, we see that at low concentrations, the transmission modulation is proportional to the concentration, which is consistent with what may be anticipated from low order scattering events. In the diffusive regime, the data shown in Fig. 18 all fall on the same curve when plotted against the square of $\ell^*$. This indicates that that the proportionality factor in (26) is independent of scatterer size.

We conclude that (26) qualitatively covers our observations of photonic magnetoresistance. This identifies the Faraday effect as the underlying mechanism. Now, we come to a quantitative comparison. Using relation (23), we obtain the value $\Delta D_\perp / D_0 \approx -3 \times 10^{-5} \mathrm{T}^{-2}$ for the 0.4 μm particles and

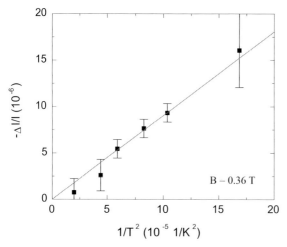

**Fig. 17.** Transmission modulation (magnetoconductance) at $2\Omega$ versus the inverse square of the temperature. Sample is $EuF_2$ powder in a polymer disk, subject to a magnetic field of $B = 0.36$ T. The $EuF_2$ particles have an average diameter of 0.4 μm and volume fraction $f = 17\%$. *Solid line* is a linear fit to the data points. Taken from *Sparenberg* et al. [43]

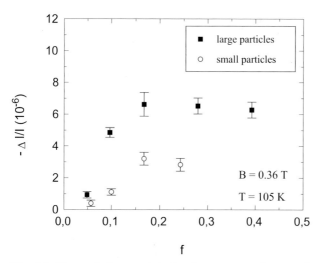

**Fig. 18.** Transmission modulation (magnetoconductance) at $2\Omega$ versus the volume fraction of the $EuF_2$ scatterers, observed at a wavelength $\lambda = 457$ nm. Full symbols represent samples with an average particle diameter of 2 μm, and open symbols denote an average particle diameter of 0.4 μm. Taken from *Sparenberg* et al. [43]

$\Delta D_\perp/D_0 \approx -7 \times 10^{-5} \mathrm{T}^{-2}$ for the 2- μm particles at 105 K. It is important to note that the theory of [37] is valid only for monodisperse Rayleigh scatterers. The small particles, $d \approx 0.4$ μm, are in the Rayleigh–Gans scattering regime [25], since $|n_\mathrm{s} - n_\mathrm{m}| \ll 1$ and $2(m-1)x \approx 0.4$, where $x \equiv n_\mathrm{m}k\,d/2$ is the so-called size parameter and $m \equiv n_\mathrm{s}/n_\mathrm{m}$. The Rayleigh–Gans theory for spheres [25] gives, for $m = 1.051$ and $d = 0.4$ μm, the value $f \cdot \ell^* = 25$ μm. The difference from our measurement $f \cdot \ell^* \approx 8.3$ μm is sufficiently small to be attributed to systematic errors and polydispersity. By the absence of a Rayleigh–Gans theory for magnetodiffusion, we shall compare our experimental results to the prediction for monodisperse Rayleigh scatterers, that follows directly from (21):

$$\frac{\Delta D_\perp}{D_0} = -\frac{243}{5}m^2\frac{(Vd)^2}{(m^2-1)^4x^8}. \tag{27}$$

For the observed range of diameters 0.3 μm $< d <$ 0.5 μm, (27) provides a theoretical range $-2\times10^{-6}\mathrm{T}^{-2} > \Delta D_\perp/D_0 > -4\times10^{-5}\mathrm{T}^{-2}$, which agrees in sign and order of magnitude with the experimental value $\Delta D_\perp/D_0 \approx -3\times 10^{-5}\mathrm{T}^{-2}$ at 105 K. The 2- μm scatterers are no longer in the Rayleigh–Gans regime, and a prediction on the basis of Rayleigh scattering theory is even less justifiable. Nevertheless, we emphasize that the scaling relation (26) is found to apply to all samples that we investigated, which strongly suggests that the Faraday effect is the universal mechanism of photonic magnetoresistance.

### 3.4.5   Conclusions and Acknowledgments

This work has attempted to review our theoretical and experimental studies of the magneto-optics of random media. A magnetic field breaks time-reversal symmetry in light propagation; the most well-known manifestation of this breaking is the Faraday effect. Our study therefore contributes to the more general study of waves in complex media [44] with broken spatial or temporal symmetries, like human tissue and seismic media, or highly disordered media. At the same time, since the magnetic field enables us to manipulate the phase shifts of the multiply scattered light waves externally, magneto-optical studies are in many ways the "photonic equivalents" of the successful mesoscopic studies of charge transport in a magnetic field.

Most of this work was carried out in close collaboration with our students Anja Sparenberg, David Lacoste, Georg Düchs, Sabine Wiebel, Anne Napierala, and Cornelius Strohm whom we would like to thank for their efforts and enthusiasm. It is a pleasure to thank Profs. Roger Maynard and Peter Wyder for their continuous interest and support.

### Note Added in Proof

Magneto-transverse scattering has recently also been seen for surface-plasmon polaritons (G. Düchs, A. Sparenberg, G. L. J. A. Rikken, P. Wyder, Phys. Rev. Lett. **87**, 127902-1 (2001)

# References

1. J. P. Gordon: Phys.Rev. A **8**, 14 (1973)
2. I. Brevik: Phys. Rep. **52**, 133 (1979)
3. R. Peierls: *More Surprises in Theoretical Physics*, (Princeton University Press, Princeton 1991)
4. J. D. Jackson: *Classical Electrodynamics* (Wiley, New York 1975)
5. L. D. Landau, E. M. Lifshitz, L. P. Pitaevskii: *Electrodynamics of Continuous Media* (Pergamon, Oxford 1984)
6. Ping Sheng: *Introduction in Wave Scattering, Localization and Mesoscopic Physics* (Academic, San Diego 1995)
7. Research Group POAN (Ed): *New Aspects of Electromagnetic and Acoustic Wave Diffusion* (Springer, Berlin, Heidelberg, 1998)
8. A. Lagendijk, B. A. van Tiggelen: Phys. Rep. **270**, 143 (1996)
9. N. B. Baranova, B. Ya. Zel'dovich: JETP Lett. **509**, 681 (1994)
10. R. Schlesser, A. Weis: Opt. Lett. **14**, 1015 (1992)
11. G. W. 't Hooft, M. B. van der Mark: Nature **381**, 27 (1996)
12. D. F. Nelson: Phys. Rev. Lett. **76**, 4713 (1996)
13. Y. Jiang, M. Liu: Phys. Rev. Lett. **77**, 1043 (1996)
14. M. F. Bishop, A. A. Maradudin: Phys. Rev. B **14**, 3384 (1976)
15. A. Puri, J. L. Birman: Phys. Rev. A **27**, 1044 (1983)
16. M. Born, E. Wolf: *Principles of Optics* (Pergamon, Oxford 1980)
17. R. P. Feynman, R. B. Leighton, M. Sands: *The Feynman Lectures on Physics*, Vol. II, Sect. 27–4 (Addison-Wesley, Reading 1979)
18. J. A. Stratton: *Electromagnetic Theory* (McGraw-Hill, New York 1941) Sect. 2–19
19. G. L. J. A. Rikken, B. A. van Tiggelen: Phys. Rev. Lett. **78**, 847 (1997)
20. G. W.'t Hooft, G. Nienhuis, J. C. J. Paaschens: Phys. Rev. Lett. **80**, 1114 (1998)
21. G. L. J. A. Rikken, B. A. van Tiggelen: Phys. Rev. Lett. **80**, 1115 (1998)
22. G. L. J. A. Rikken, B. A. van Tiggelen: Nature **381**, 54 (1996)
23. I. Campos, J. L. Jiménez: Eur. J. Phys. **13**, 117 (1992)
24. G. W. Ford, S. A. Werner: Phys. Rev. B **18**, 6752 (1978)
25. H. C. van de Hulst: *Light Scattering by Small Particles* (Dover, New York 1981)
26. D. Lacoste, B. A. van Tiggelen, G. L. J. A. Rikken, A. Sparenberg: J. Opt. Soc. Am. A **15**, 1636 (1998)
27. F. Erbacher, R. Lenke, G. Maret: Europhys. Lett. **21**, 551 (1993)
28. R. Lenke: Diffusion Multiple de la Lumière: Déstruction de la Rétrodiffusion Cohérente par la Rotation Faraday Magnéto-optique. Ph.D. Thesis, Université Joseph Fourier, Grenoble (1994)
29. A. S. Martinez, R. Maynard: Phys. Rev. B **50**, 3714 (1994)
30. F. C. MacKintosh, S. John: Phys. Rev B **37**, 1884 (1988)
31. D. Lacoste, B. A. van Tiggelen: Phys. Rev. E **61**, 4556 (2000)
32. J. J. M. Beenakker, G. Scoles, H. F. P. Knaap, R. M. Jonkman: Phys. Lett. **2**, 5 (1962)
33. D. Lacoste, B. A. van Tiggelen: Europhys. Lett. **45**, 721 (1999)
34. D. Lacoste: Diffusion de la Lumière dans les Milieux Magnéto-optiques ou Chiraux. Ph.D. Thesis, Université Joseph Fourier, Grenoble (1999)
35. B. A. van Tiggelen: Phys. Rev. Lett. **75**, 422 (1995)

36. N. W. Ashcroft, N. D. Mermin: *Solid State Physics* (Holt, New York 1976) App. E
37. B. A. van Tiggelen, R. Maynard and Th.M. Nieuwenhuizen: Phys. Rev. E **53**, 2881 (1996)
38. J. H. Li, A. A. Lisyansky, T. D. Cheung, D. Livdan, A. Z. Genack: Europhys. Lett. **22**, 675 (1993)
39. G. Düchs, A. Sparenberg, G. L. J. A. Rikken, B. A. van Tiggelen: Phys. Rev. E. **62**, 2840 (2000)
40. S. Wiebel, A. Sparenberg, G. L. J. A. Rikken, D. Lacoste, B. A. van Tiggelen: Phys. Rev. E. **62**, 8636 (2000)
41. C. Görller-Walrand, H. Peeters, Y. Beyens, N. de Moitié-Neyt, M. Behets: Nouv. J. Chim. **4**, 715 (1980)
42. G. L. J. A. Rikken, A. Sparenberg, B. A. van Tiggelen: Physica B **246, 247**, 188 (1998)
43. A. Sparenberg, G. L. J. A. Rikken, B. A. van Tiggelen: Phys. Rev. Lett. **79**, 757 (1997)
44. J. P. Fouque (Ed.): *Wave Diffusion in Complex Media* (Kluwer, Dordrecht 1999)

# Random Lasers with Coherent Feedback

Hui Cao

Department of Physics and Astronomy, Materials Research Center
Northwestern University, Evanston, IL 60208-3112, USA
h-cao@northwestern.edu

**Abstract.** We have demonstrated lasing with resonant feedback in active random media. Recurrent light scattering provides coherent feedback for lasing. A detailed experimental study of laser emission spectra, spatial distribution of laser intensity, dynamics, and photon statistics of random lasers with coherent feedback is presented. The fundamental difference and transition between a random laser with resonant feedback and a random laser with nonresonant feedback are illustrated. We have achieved spatial confinement of laser light in micrometer-sized random media. The optical confinement is attributed to disorder-induced scattering and interference. Using the finite-difference time-domain method, we simulate lasing with coherent feedback in active random media.

## 1 Introduction

Optical scattering in a random medium may induce a phase transition in the photon transport behavior [1]. When the scattering is weak, the propagation of light can be described by a normal diffusion process. With an increase in the amount of scattering, recurrent light scattering events arise. Interference between the counterpropagating waves in a disordered structure gives rise to the enhanced backscattering, also called weak localization [2,3]. When the amount of scattering is increased beyond a critical value, the system makes a transition into a localized state. Light propagation is inhibited due to interference in multiple scattering [4,5,6,7,8,9]. This phenomenon is called Anderson localization of light. It is an optical analog to Anderson localization of electrons in solids [10].

Apart from the remarkable similarities, there are striking differences between electron transport and photon transport in a disordered medium. For example, the number of electrons is always conserved, whereas the number of photons may not be. In an amplifying random medium, a photon may induce the stimulated emission of a second photon. A fascinating phenomenon, which would never occur in an electronic system, is the lasing action in a disordered gain medium. Such lasers are called random laser. There are two kinds of random lasers: one has nonresonant (incoherent) feedback, the other has resonant (coherent) feedback. I will first introduce the two kinds of random lasers and explain the difference between them. Then, I will focus on our study of random lasers with resonant feedback.

V. M. Shalaev (Ed.): Optical Properties of Nanostructured Random Media,
Topics Appl. Phys. **82**, 303–328 (2002)

## 2    Two Kinds of Random Lasers

Lasing with nonresonant feedback occurs in the diffusive regime [11,12,13,14]. In a disordered medium, light is scattered and undergoes a random walk before leaving the medium. In the presence of gain, a photon may induce the stimulated emission of a second photon. When the gain length is equal to the average length of light path in the medium, the probability that a photon generates second photon before leaving the gain medium approaches one. Thus the photon density increases. From the theoretical point of view, the solution to the diffusion equation, including optical gain, diverges [15]. This phenomenon is similar to neutron scattering in combinations of nuclear fission.

When optical scattering is strong, light may return to a scatterer from which it is scattered before, and thereby form a closed loop path. When the amplification along such a loop path exceeds the loss, laser oscillation could occur in the loop which serves as a laser resonator. The requirement that the phase shift along the loop is equal to a multiple of $2\pi$ determines the oscillation frequencies. This is a random laser with coherent feedback [16,17,18]. Of course, the picture of a closed loop is intuitive but naive. The light may come back to its original position through many different paths. All of the backscattered light waves interfere and determine the lasing frequencies. Thus, a random laser with coherent feedback is a randomly distributed feedback laser. The feedback is provided by disorder-induced scattering.

Experimentally, we have observed the difference in the two kinds of random lasers and the transition between them [19]. The random medium we used in our experiment is a laser dye solution containing nanoparticles. The advantage of the suspension is that the gain medium and the scattering elements are separated. Thus, we can independently vary the amount of scattering by particle density and the optical gain by dye concentration.

Experimentally, rhodamine 640 perchlorate dye and zinc oxide (ZnO) particles are mixed in methanol. The ZnO particles have a mean diameter of 100 nm. To keep the particles from clustering, the solution, contained in a flask, is shaken in an ultrasonic cleaner for 20 minutes right before the photoluminescence experiment. The frequency-doubled output ($\lambda = 532$ nm) of a mode-locked Nd:YAG laser (10 Hz repetition rate, 25-ps pulse width) is used as pump light. The pump beam is focused by a lens (10 cm focal length) onto the solution contained in a 1 cm × 1 cm × 3 cm cuvette at nearly normal incidence. Emission in the direction $\sim 45°$ from the normal of the cell front window is collected by a fiber bundle and directed to a 0.5-m spectrometer with a cooled CCD detector array.

By changing the ZnO particle density in the solution, we continuously vary the amount of scattering. Figure 1 shows the evolution of the emission spectra with the pump intensity when the ZnO particle density is $\sim 2.5 \times 10^{11}$ cm$^{-3}$. The dye concentration is fixed at $5 \times 10^{-3}$ M. When the pump intensity exceeds a threshold, drastic spectral narrowing occurs. As shown in the insets of Fig. 1, when the incident pump pulse energy exceeds $\sim 3$ µJ, the emission line width is quickly reduced to $\sim 5$ nm; meanwhile, the peak intensity increases much more rapidly with the pump power because optical scattering by the ZnO particles increases the path length of the emitted light inside the gain region. As the pump power increases, the gain length is reduced. Eventually the gain length at frequencies near the maximum of the gain spectrum approaches the average path length of photons in the gain regime. Then, a photon generates a second photon by stimulated emission before leaving the gain medium; thus the photon density increases dramatically. The sudden increase of emission intensity at frequencies near the maximum of the gain spectrum results in drastic narrowing of the emission spectrum. This process is lasing with nonresonant feedback.

Next, we keep the same dye concentration and increase the ZnO particle density to $1 \times 10^{12}$ cm$^{-3}$. Figure 2 plots the evolution of the emission spectra with pump intensity. We can see that the phenomenon becomes very different in strong scattering. When the incident pump pulse energy exceeds 1.0 µJ, discrete peaks emerge in the emission spectrum. The line width of these peaks is less than 0.2 nm, which is more than 50 times smaller than the line width of the amplified spontaneous emission (ASE) below the threshold. When the pump intensity increases further, more sharp peaks appear. As shown in the inset of Fig. 2, when the pump intensity exceeds the threshold where discrete peaks emerge in the emission spectrum, the emission intensity increases much more rapidly with the pump power. Hence, lasing occurs in the random cavities formed by recurrent light scattering. The phase relationship of the backscattered light determines the lasing frequencies. Laser emission from the random cavities results in discrete narrow peaks in the emission spectrum. Because the ZnO particles are mobile in the solution, the frequencies of the lasing modes change from pulse to pulse. The emission spectra in Figs. 1 and 2 are taken for a single pump pulse. When the pump power increases further, the gain exceeds the loss in more random cavities. Laser oscillation in these cavities gives additional peaks in the emission spectrum.

Next, we study the transition from lasing with nonresonant feedback to lasing with resonant feedback. Figure 3 shows the evolution of the emission spectra with the pump intensity when the ZnO particle density is $\sim 5 \times 10^{11}$ cm$^{-3}$. As the pump power increases, a drastic spectral narrowing occurs first. Then at higher pump intensity, discrete narrow peaks emerge in the emission spectrum. Because the amount of scattering in the solution is between the previous two cases, there is some but not a large probability that a photon is scattered back to the same scatterer from which it is scattered before. In other words, the random cavities formed by recurrent scattering are quite lossy. The pump intensity required to reach the lasing threshold in these cavities is high. Thus, the pump intensity first reaches the threshold where the gain length near the maximum of the gain spectrum becomes equal to the average path length of photons in the excitation volume. A significant

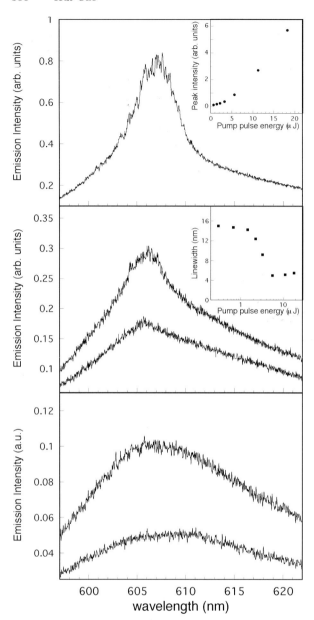

**Fig. 1.** Emission spectra when the incident pump pulse energy is (from bottom to top) 0.68, 1.5, 2.3, 3.3, 5.6 μJ. The ZnO particle density is $\sim 2.5 \times 10^{11}$ cm$^{-3}$. The *upper inset* is the emission intensity at the peak wavelength versus the pump pulse energy. The *lower inset* is the emission line width versus the pump pulse energy

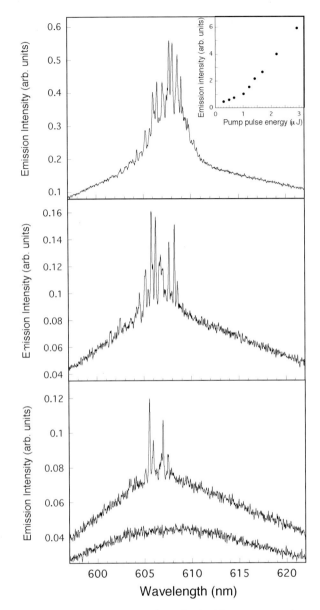

**Fig. 2.** Emission spectra when the incident pump pulse energy is (from bottom to top) 0.68, 1.1, 1.3, 2.9 μJ. The ZnO particle density is $\sim 1 \times 10^{12} \, \mathrm{cm}^{-3}$. The *inset* shows the emission intensity versus the pump pulse energy

spectral narrowing and a sudden increase of peak emission intensity occur, similar to what happens in Fig. 1. Then, the pump intensity reaches a second threshold where the gain exceeds the loss in some random cavities. Lasing oscillation occurs in these cavities, adding discrete peaks to the emission spectrum. However, the number of lasing modes in Fig. 3 is less than that in Fig. 2 under similar pump power. When the gain length and excitation volume are the same, a fewer number of scatterers leads to weaker optical scattering. Hence, the number of random cavities where the lasing threshold can be reached is smaller.

Therefore, there are two kinds of lasing processes in an active random medium, and they correspond to two lasing thresholds [19]. From the ray optics point of view, lasing with nonresonant feedback corresponds to the instability of light amplification along *open* trajectories in a random medium, and lasing with resonant feedback corresponds to the instability of light amplification along *closed* paths formed by recurrent scattering. An alternative and perhaps more accurate explanation for random lasers is based on quasi-states. Quasi-states are the eigenmodes of the Maxwell equations in a finite-sized random medium. When photons in a quasi-state reach the boundaries of the random medium, they are either reflected back to the medium or transmitted into the air. The transmitted photons are lost, and the reflected photons may enter other quasi-states. Hence, the decay of a quasi-state results from both light leakage through the boundaries and energy exchange with other quasi-states. When $kl > 1$ ($k$ is a wave vector, $l$ is the transport mean free path), the average decay rate of a quasi-state is larger than the average frequency spacing of adjacent quasi-states. Hence, the quasi-states are spectrally overlapped, giving a continuous emission spectrum.

In weak scattering, the quasi-states decay fast, and they are strongly coupled. Because of photon exchange among quasi-states, the loss of a set of interacting quasi-states is much lower than the loss of a single quasi-state. In an active random medium, when the optical gain for a set of interacting quasi-states at maximum gain reaches the loss of these coupled quasi-states, the total photon number in these coupled states builds up. The drastic increase of photon number at the frequency of maximum gain results in a significant spectral narrowing, as shown in Fig. 1.

With an increase in the amount of optical scattering, the dwell time of light in the random medium increases, and the mixing of the quasi-states is reduced. Hence, the decay rates of the quasi-states decrease. When the optical gain increases, it first reaches the threshold for lasing in a set of coupled quasi-states at maximum gain. As the optical gain increases further, it exceeds the loss of a quasi-state that has a long lifetime. Then, lasing occurs in a single quasi-state. The spectral line width of the quasi-state is reduced dramatically above the lasing threshold. A further increase of optical gain leads to lasing in more low-loss quasi-states. Laser emission from these quasi-states gives discrete peaks in the emission spectrum, as shown in Fig. 3.

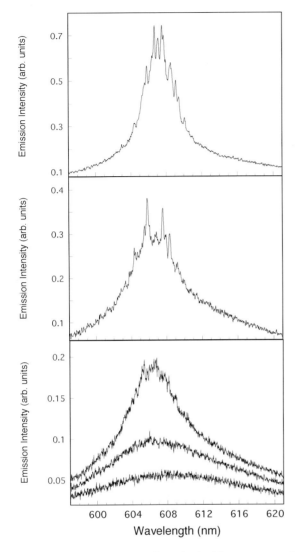

**Fig. 3.** Emission spectra when the incident pump pulse energy is (from bottom to top) 0.74, 1.35, 1.7, 2.25, and 3.4 μJ. The ZnO particle density is $\sim 6 \times 10^{11}\,\mathrm{cm}^{-3}$

When the scattering strength increases further, the decay rates of the quasi-states and the coupling among them continue decreasing. Because of the wide distribution of the decay rates of quasi-states, the threshold gain for lasing in individual low-loss quasi-states becomes lower than the threshold gain for lasing in the coupled quasi-states at maximum gain. Thus, lasing with resonant feedback occurs first, as shown in Fig. 2.

There has been much study of random lasers with nonresonant feedback [20,21]. Next, I will focus on our investigation of random lasers with resonant feedback.

# 3    Random Lasers with Resonant Feedback

In this section, I will present a quantitative study of random lasers with resonant feedback. To confirm that the coherent feedback is indeed provided by light scattering, we measured the dependence of the lasing threshold on the transport mean free path.

Because of the sedimentation of particles from the solution, the suspension was not suitable for quantitative measurement. Hence, we switched to a solid random medium and fabricated PMMA films containing rhodamine 640 perchlorate dye and $TiO_2$ particles. The average size of $TiO_2$ particles was 400 nm. By varying the $TiO_2$ particle density in the polymer film, we changed the scattering length. We fabricated a series of PMMA films with the same dye concentration but different particle density. The $TiO_2$ particle density in the polymer films varied from $8 \times 10^{10}$ to $6 \times 10^{12}$ cm$^{-3}$. The lasing threshold in these films was measured under identical conditions. To characterize the transport mean free path in these films, we conducted coherent backscattering experiments [2,3]. The output from a He:Ne laser was used as the probe light since its wavelength is very close to the emission wavelength of rhodamine 640 perchlorate dye. To avoid absorption of the probe light, we fabricated PMMA films which contained only $TiO_2$ particles but not the dye. From the angular width of the backscattering cone, we estimated the transport mean free path $l$, after taking into account the internal reflection [22].

Figure 4 plots the incident pump pulse energy at the lasing threshold versus the transport mean free path. The dye concentration in the polymer film is fixed at $5 \times 10^{-2}$ M. With an increase of the $TiO_2$ particle density in the polymer film, the transport mean free path decreases, and the lasing threshold also decreases. The strong dependence of the lasing threshold on the transport mean free path clearly illustrates the important contribution of scattering to lasing. With an increase in the amount of optical scattering, the feedback provided by scattering becomes stronger. The quasi-states of the random medium have lower loss. Thus, the lasing threshold is reduced. Through curve fitting, we found that the incident pump intensity at the lasing threshold is proportional to the square root of the transport mean free path. Figure 5 shows the number of lasing modes in the samples with different scattering lengths at the same pump intensity. The stronger the scattering, the more lasing modes emerge, because in a random medium with stronger scattering strength, there are more low-loss quasi-states. Hence, when the optical gain is fixed, lasing occurs in more quasi-states. An interesting feature in Figs. 4 and 5 is that when the transport mean free path approaches the optical wavelength, the lasing threshold pump intensity drops quickly, and

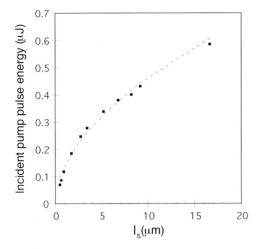

**Fig. 4.** The incident pump pulse energy at the lasing threshold versus the transport mean free path $l_s$ in PMMA films. The *dashed line* is the fitted curve represented by $P_{th} = 0.13/l_s^{0.53}$

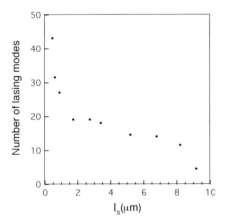

**Fig. 5.** The number of lasing modes as a function of the scattering length $l_s$ in PMMA films. The incident pump pulse energy is 1.0 μJ

the number of lasing modes increases dramatically. Therefore, the regime of $l \sim \lambda$ is important to both fundamental physics and practical application. Next, I will discuss random lasers in the regime $l \leq \lambda$. To achieve such a short scattering mean free path, we used ZnO powder and polycrystalline thin films to increase the contrast of refractive indexes in the binary random medium.

ZnO films are deposited on sapphire or amorphous fused silica substrates by laser ablation. A pulsed KrF excimer laser (248 nm) is used to ablate a hot pressed ZnO target in an ultrahigh vacuum chamber. A detailed description of the growth apparatus and growth procedure can be found in [23]. We took transmission electron microscopy (TEM) images of the ZnO films. From the plane view TEM image, the ZnO film consists of many irregularly-shaped grains whose sizes vary from 30 to 130 nm. The cross-sectional TEM image reveals that the grains have straight sidewalls that are perpendicular to the substrate. Hence, the polycrystalline film is a 2-D random medium. The in-plane randomly oriented polycrystalline grain structure results in strong optical scattering in the plane of the film [16,24]. The optical confinement in the direction perpendicular to the film is achieved through index guiding, similar to 2-D photonic crystals.

ZnO powder is a 3-D random medium [17,25]. The ZnO nanoparticles are synthesized by a precipitation reaction. The process involves hydrolysis of a zinc salt in a polyol medium. Through the process of electrophoresis, ZnO powder films are made on ITO-coated substrates. The film thickness varies from a few to 50 μm. Figure 6a is a scanning electron microscope (SEM) image of the ZnO particles. The average particle size is about 50 nm.

We characterized the transport mean free path $l$ in the coherent backscattering experiment [2,3]. ZnO has a direct band gap of 3.3 eV. To avoid absorption, the frequency-doubled output ($\lambda = 410$ nm) of a mode-locked Ti:Sapphire laser (76 MHz repetition rate, 200 fs pulse width) was used as

**(a)**

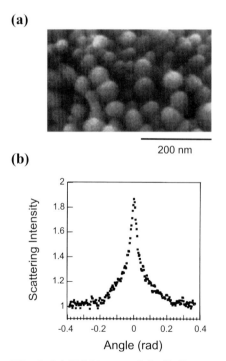

200 nm

**(b)**

**Fig. 6.** (a) SEM image of the ZnO nanoparticles. (b) Measured backscattering cone from the ZnO powder film

the probe light. Figure 6b shows the measured backscattering cone of the ZnO powder film. From the angle of cusp, we estimated that $l \approx 0.5\lambda$, after taking internal reflection into account [22].

In the photoluminescence experiment, the ZnO powder film is optically pumped by the fourth harmonic ($\lambda = 266\,\text{nm}$) of a mode-locked Nd:YAG laser. The pump beam is focused to a $\sim 20$-µm spot on the film surface with normal incidence. Electrons in the valence band absorb pump photons and jump to the conduction band. They subsequently relax to the bottom of the conduction band before radiative decay. The spectrum of emission from the powder film is measured by a spectrometer with 0.13-nm spectral resolution. At the same time, the spatial distribution of the emitted light intensity in the film is imaged by an ultraviolet (UV) microscope onto a UV-sensitive charge-coupled device (CCD) camera. The amplification of the microscope is about 100 times. The spatial resolution is around 0.24 µm. A band-pass filter is placed in front of the microscope objective to block the pump light.

Figure 7 shows the measured emission spectra and spatial distribution of emission intensity in a ZnO powder film at different pump intensities. At low pump intensity, the spectrum consists of a single broad spontaneous emission peak. Its full width at half maximum (FWHM) is about 12 nm (Fig. 7a). As shown in Fig. 7b, the spatial distribution of the spontaneous emission intensity is smooth across the excitation area. Due to the variation in pump intensity over the excitation spot, the spontaneous emission in the center of the excitation spot is stronger. When the pump intensity exceeds a threshold, very narrow discrete peaks emerge in the emission spectrum (Fig. 7c). The FWHM of these peaks is about 0.2 nm. Simultaneously, bright tiny spots appear in the image of the emitted light distribution in the film (Fig. 7d). The sizes of the bright spots were between 0.3 and 0.7 µm. When the pump intensity is increased further, additional sharp peaks emerge in the emission spectrum. Correspondingly, more bright spots appear in the image of the emitted light distribution. The frequencies of the sharp peaks depend on the sample position. When we move the pump beam spot across the sample, the frequencies of the sharp peaks change. Figure 8 plots the spectrally integrated emission intensity as a function of pump intensity. A threshold behavior is clearly seen: above the pump intensity at which multiple sharp peaks emerge in the emission spectrum, the integrated emission intensity increases much more rapidly with pump intensity.

We also measured the temporal profile of the emission from the ZnO powder film with a Hamamatsu streak camera. The temporal resolution of the streak camera is 2 ps. The scattered pump light is blocked by the input optics of the streak camera.

Figure 9 shows the temporal evolution of emission below the lasing threshold, just above the lasing threshold, and well above the lasing threshold. Below the lasing threshold, the spontaneous emission decay time is 167 ps. When the pump intensity exceeds the lasing threshold, the emission pulse is

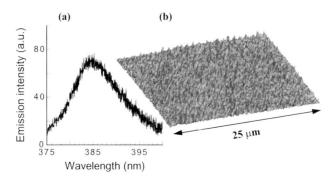

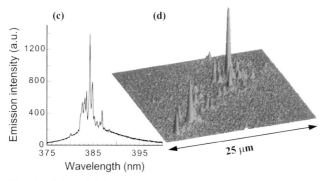

**Fig. 7.** (**a**) and (**c**) The measured spectra of emission from the ZnO powder film. (**b**) and (**d**) The measured spatial distribution of emission intensity in the film. The incident pump pulse energy is 5.2 nJ for (**a**) and (**c**) and 12.5 nJ for (**b**) and (**d**)

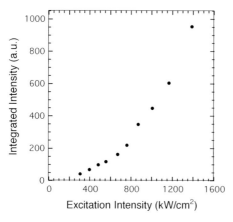

**Fig. 8.** Spectrally integrated intensity of emission from the ZnO powder film versus the excitation intensity

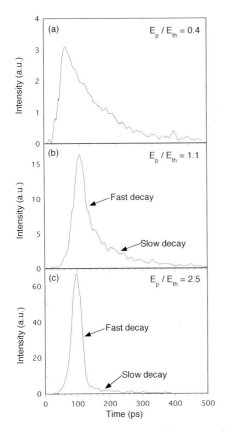

**Fig. 9.** Temporal evolution of emission from ZnO powder film. The incident pump pulse energy is (**a**) 2.8 nJ, (**b**) 5.5 nJ, and (**c**) 8.9 nJ

shortened significantly. By curve fitting, we find that the exponential decay of emission consists of a fast component and a slow component. As shown in Fig. 9b, the initial decay of emission is quite fast. The decay time is 27 ps. After $\sim 50$ ps, the fast decay is replaced by slow decay. The latter decay time is 167 ps, which is equal to the emission decay time below the lasing threshold. Since the sample is pumped by 20-ps pulses, the optical gain is transient. Laser oscillation occurs in a short time after each pump pulse. The strong laser emission depletes the population inversion quickly. When the optical gain is reduced below the loss, laser oscillation stops. Laser emission is replaced by spontaneous emission. Hence, the initial fast decay corresponds to laser emission, and the later slow decay is due to spontaneous emission. As the pump power is increased further, the initial laser emission becomes much stronger than the later spontaneous emission, as shown in Fig. 9c. When the pump power exceeds the threshold, the emission pulse is dramatically shortened from 200 to 30 ps. From the threshold behavior of the emission intensity, the emergence of very narrow spectral peaks, and the dramatic shortening of the emission pulses, we conclude that lasing has occurred in the ZnO powder film. Similar lasing phenomena have been observed in ZnO polycrystalline films and GaN powder.

The experimental fact that the bright spots in the emission pattern and the lasing modes in the emission spectrum always appear simultaneously suggests that the bright spots are related to the laser light. There are two possible explanations for the bright spots. One is that the laser light intensity at the locations of the bright spots is high. The other is that the laser light is not particularly strong at the locations of the bright spots. However, there are some efficient scattering centers at the locations of the bright spots, and thus, the laser light is strongly scattered. In the latter case, these scattering centers should also strongly scatter the spontaneously emitted light below the lasing threshold because scattering is a linear process. Hence, these bright spots should exist below the lasing threshold. However, there are no bright spots below the lasing threshold. Therefore, these bright spots are not caused by efficient scatterers, but by strong laser light in the medium.

Next we present an explanation for our experimental data. The short transport mean free path indicates very strong light scattering in the powder film. However, the transport mean free path obtained from the coherent backscattering measurement is an average over a large volume of the sample. Due to the local variation of particle density and spatial distribution, there exist small regions of higher disorder and stronger scattering. Light can be confined in these regions through multiple scattering and interference. For a particular configuration of scatterers, only light at certain wavelengths can be confined because the interference effect is wavelength sensitive. In a different region of the sample, the configuration of the scatterers is different, and thus, light at different wavelengths is confined. In other words, some quasi-states are spatially localized in small regions, and they have relatively long lifetimes. When the optical gain reaches the loss of such a quasi-state, lasing action occurs in the quasi-state. The lasing peaks in the emission spectrum illustrate the frequencies of the quasi-states, and the bright spots in the spatial light pattern exhibit the positions and shapes of the quasi-states.

Unlike conventional lasers with directional output, laser emission from random media can be observed in all directions. However, the laser emission spectra vary with the observation angle. Since different quasi-states have different output directions, lasing modes observed at different angles are different.

Although the data presented above indicate that lasing has occurred in the ZnO powder, the final proof of a laser comes from coherence and statistics measurement. Next, I will present our photon statistics measurement result which shows that indeed coherent light is generated from the disordered medium.

ZnO nanoparticles are cold pressed under a pressure of 200 MPa to form a pellet 2 mm thick. From the coherent backscattering experiment, we estimate that $l \sim 2.3\lambda$. Under the optical pumping of a mode-locked Nd:YAG laser, the emission from the ZnO pellet is collected by a lens and focused to the entrance slit of a 0.5-m Jarrell–Ash spectrometer. The output port of the spectrometer is connected to a Hamamatsu streak camera whose entrance slit is perpendicular to that of the spectrometer. The photocathode width of the streak camera gives an observable spectral window of 6.7 nm with a spectral resolution of 0.1 nm. Partial output of the pump laser goes directly to a fast photodiode whose output signal triggers the streak camera. A Peltier-cooled CCD camera, operating at $-50°$ C for reduced dark noise, is used to record the streak image. The streak camera operates in the photon counting mode. A threshold is set to eliminate the contribution of the dark-current noise. Thus, in the absence of an input signal, no photons are counted.

By combining the spectrometer with the streak camera, we are able to separate different lasing modes and measure the temporal evolution of each mode. Figure 10 is a 2-D image taken by the CCD camera. The horizontal axis is the time, and the vertical axis is the wavelength. When the pump power exceeds a threshold, discrete lasing modes appear in the spectrum. For different modes, lasing starts at different times and lasts for different periods of time. Such dynamic behavior is caused by different decay rates of the quasi-states that have lased.

Next, we measure the photon statistics of a single lasing mode. In the 2-D image shown in Fig. 10, we draw a rectangle, one of whose sides is wavelength interval $\delta\lambda$ and the other side is time interval $\delta t$. The number of photons inside this rectangle is counted for each pulse. After collecting photon count data for a large number of pulses, the probability $P(n)$ of $n$

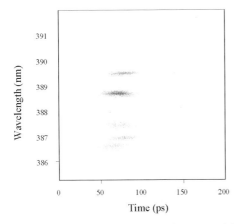

**Fig. 10.** The measured spectral-temporal image of the emission from a ZnO pellet. The incident pump pulse energy is 4.5 nJ

photons within $(\lambda, \lambda + \delta\lambda)$ and $(t, t + \delta t)$ is obtained. The frequency interval $\delta\nu$ of the counting area is calculated from $\lambda$ and $\delta\lambda$. When $\delta\nu \cdot \delta t \leq 1$, the counting area corresponds to a single electromagnetic (EM) mode. For coherent light, the photon number distribution $P(n)$ in a single mode satisfies Poisson distribution $P(n) = \langle n \rangle^n e^{-\langle n \rangle}/n!$, where $\langle n \rangle$ is the average photon number. For chaotic light, the photon number distribution $P(n)$ in a single mode satisfies the Bose–Einstein distribution $P(n) = \langle n \rangle^n/(1 + \langle n \rangle)^{n+1}$. Note that the above distribution holds only for a single mode. In multimode chaotic light, the photon number distribution approaches Poisson distribution. From $P(n)$, we obtain the normalized second-order correlation coefficient $G_2 = 1 + \left[ \langle (\Delta n)^2 \rangle - \langle n \rangle \right]/\langle n \rangle^2$. For the Poisson distribution, $G_2 = 1$. For the Bose–Einstein distribution, $G_2 = 2$.

We measured the photon statistics of coherent light and chaotic light to confirm the reliability of the spectrometer–streak camera setup for the photon statistics measurement. Then, we moved to the photon statistics measurement of random lasers with coherent feedback. The pump intensity is above the threshold where discrete spectral peaks appear, so that we can measure the photon statistics of a single peak. In the 2-D spectral-temporal image of ZnO emission, we pick up one of the brightest peaks and count the number of photons within the area of $(\lambda_0 - \Delta\lambda/2, \lambda_0 + \Delta\lambda/2)$ and $(t_0 - \Delta t/2, t_0 + \Delta t/2)$ for each pulse. $\lambda_0$ is the center wavelength of the spectral peak we choose, and $t_0$ is the time when the intensity of the emission pulse at $\lambda_0$ is maximum. $\Delta\lambda = 0.12\,\mathrm{nm}$, and $\Delta t = 3.9\,\mathrm{ps}$. The corresponding $\Delta\nu = 2.4 \times 10^{11}\,\mathrm{Hz}$, and hence, $\Delta\nu \cdot \Delta t = 0.95$. There are 100 CCD pixels within the area of $(\lambda_0 - \Delta\lambda/2, \lambda_0 + \Delta\lambda/2)$ and $(t_0 - \Delta t/2, t_0 + \Delta t/2)$. If two photons from the same pulse hit the same pixel, the count is one instead of two. Hence, such events cause error in the photon counting. To eliminate this kind of error, the photons that hit the pixels must be sparse enough that the probability of two photons hitting the same pixel is negligible. When the ZnO emission is strong, we use neutral density filters to attenuate the signal, so that the percentage of the pixels that are hit by the photons during each pulse is below 5 %.

Figure 11a shows the measured photon statistics at the threshold where discrete spectral peaks appear. From the data of $P(n)$, we calculate the count mean $\langle n \rangle$ and obtain the Bose–Einstein distribution. The measured photon number distribution is very close to the Bose–Einstein distribution for the same mean photon number. Using the data of $P(n)$, we also calculate $G_2 = 1.94$. As we increase the pump intensity, the photon statistics of ZnO emission starts deviating from the Bose–Einstein statistics. As shown in Fig. 11b, when the pump intensity is 1.5 times the threshold, the measured photon number distribution is between the Bose–Einstein distribution and the Poisson distribution for the same mean photon number. $G_2$ becomes 1.51. When the pump intensity is increased to three times the threshold, the photon number distribution of ZnO emission gets closer to the Poisson distri-

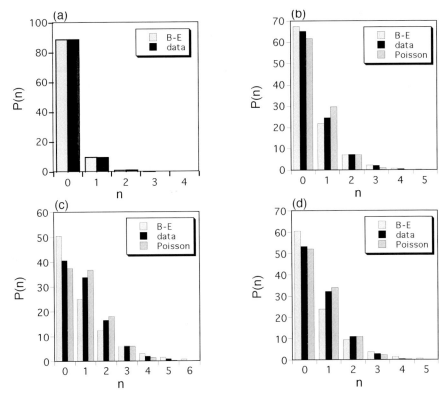

**Fig. 11.** The solid columns are the measured photon count distributions of the emission from the ZnO pellet. The *dotted* (*dashed*) columns are the Bose–Einstein (Poisson) distributions for the same count mean. The incident pump intensity is (**a**) 1.0, (**b**) 1.5, (**c**) 3.0, (**d**) 5.6 times the threshold intensity where discrete spectral peaks appear

bution (Fig. 11c). $G_2$ is reduced to 1.19. Eventually, when the pump intensity is 5.6 times the threshold, the photon number distribution is nearly identical to the Poisson distribution (Fig. 11d). The corresponding $G_2$ is 1.06.

Figure 12 shows the value of second-order correlation coefficient $G_2$ as a function of pump intensity. As the pump intensity increases, $G_2$ decreases gradually from 2 to 1. The error bar in Fig. 12 results from the finite number of samplings in the measurement. Figures 11 and 12 illustrate that the photon statistics of emitted light from a ZnO pellet changes continuously from Bose–Einstein statistics at the threshold to Poisson statistics well above the threshold. Therefore, the light field in the random medium has undergone a second-order phase transition.

The photon statistics of a random laser with resonant feedback is very different from that of a random laser with nonresonant feedback. The random laser with nonresonant feedback consists of many low-$Q$ modes that are

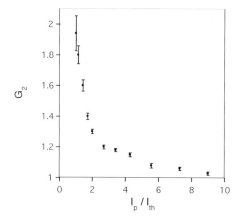

**Fig. 12.** The second-order correlation coefficient $G_2$ as a function of the ratio between the incident pump intensity $I_p$ and the threshold intensity $I_{th}$

strongly coupled. The fluctuation of the total number of photons in all modes of laser emission is smaller than that of blackbody radiation with the same number of modes [26]. The stabilization of the total photon number results from the gain saturation effect. However, the photon number distribution in each mode remains Bose–Einstein distribution even well above the threshold. Strong mode coupling prevents stabilization of the photon number in individual modes. In contrast, for a random laser with resonant feedback, the number of photons in each lasing mode is stabilized well above the threshold by the gain saturation effect. This phenomenon indicates very weak coupling among lasing modes. The difference in the photon statistics of the two kinds of random lasers originates from different scattering strength in the random medium. When optical scattering is weak, the modes are strongly coupled. As the amount of scattering increases, the interaction of the modes decreases. In fact, the decoupling of the modes is an indication of incipient photon localization. Although a random laser with resonant feedback is a multimode laser, it behaves like an ensemble of almost independent single-mode lasers.

## 4    Microlasers Made of Disordered Media

Disorder-induced optical scattering provides coherent feedback for lasing and also leads to spatial confinement of light in micrometer-sized volume. Utilizing this new physical mechanism of optical confinement, we fabricated microlasers with a disordered medium [27].

The micrometer-sized random material is made of ZnO nanocrystallites. Specifically, 0.05 mol of zinc acetate dihydrate is added to 300 ml diethylene glycol. The solution is heated to $160°$ C. As the solution is heated, more zinc acetate is dissociated. When the $Zn^{2+}$ concentration in the solution exceeds the nucleation threshold, ZnO nanocrystallites precipitate and agglomerate to form clusters. The size of the clusters can be controlled by varying the rate at which the solution is heated. The average size of ZnO nanocrystallites $\sim 50$ nm. The size of the clusters varies from submicron to a few microns [28].

The inset of Fig. 13 is the SEM image of a typical ZnO cluster. The size of the cluster is about 1.7 μm, and it contains roughly 20000 ZnO nanocrystallites. The ZnO cluster is optically pumped by the fourth harmonic of a mode-locked Nd:YAG laser. The pump light is focused by a microscope objective onto a single cluster. The spectrum of emission from the cluster is measured by a spectrometer with 0.13 nm spectral resolution. Simultaneously, the spatial distribution of the emitted light intensity in the cluster is imaged by a ultraviolet (UV) microscope onto a UV-sensitive CCD camera. A band-pass filter is placed in front of the microscope objective to block the pump light.

We performed optical measurement of the cluster shown in Fig. 13. At low pump power, the emission spectrum consists of a single broad spontaneous emission speak (Fig. 14a). Its FWHM is 12 nm. The spatial distribution of the spontaneous emission intensity is uniform across the cluster (Fig. 14b). When the pump power exceeds a threshold, a sharp peak emerges in the emission spectrum shown in Fig. 14c. Its FWHM is 0.22 nm. Simultaneously, a couple of bright spots appear in the image of the emitted light distribution in the cluster in Fig. 14d. The size of the bright spot is $\sim 0.3$ μm. When the pump power is increased further, a second sharp peak emerges in the emission spectrum (see Fig. 14e). Correspondingly, additional bright spots appear in the image of the emitted light distribution in Fig. 14f.

As shown in Fig. 13, above the pump intensity at which sharp spectral peaks and bright spots appear, the emission intensity increases much more

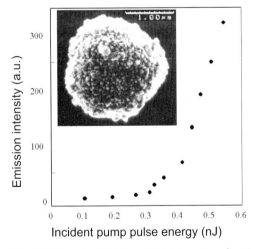

**Fig. 13.** Spectrally integrated intensity of emission from a ZnO cluster versus the incident pump pulse energy. The inset is the SEM image of a ZnO cluster

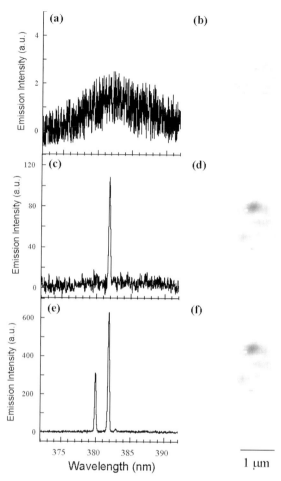

**Fig. 14.** (**a**), (**c**), and (**e**) The spectra of emission from the ZnO cluster shown in Fig. 1. (**b**), (**d**), and (**f**) The corresponding spatial distributions of emission intensity in the cluster. The incident pump pulse energy is 0.26 nJ for (**a**) and (**b**), 0.35 nJ for (**c**) and (**d**), and 0.50 nJ for (**e**) and (**f**)

rapidly with the pump intensity. These data suggest that lasing action has occurred in the micrometer-sized cluster. The incident pump pulse energy at the lasing threshold is $\sim 0.3$ nJ. Note that only $\sim 1\%$ of the incident pump light is absorbed. The rest is scattered.

Since the cluster is very small, optical reflection from the boundary of the cluster might have some contribution to light confinement in the cluster. However, the laser cavity is not formed by total internal reflection at the boundary. Otherwise, the spatial pattern of laser light would be a bright ring near the edge of the cluster [29]. We believe that 3-D confinement of laser light in the micrometer-sized ZnO cluster is achieved through disorder-induced scattering and interference. Since the interference effect is wavelength sensitive, only light at certain wavelengths can be confined in the cluster. To check that the main mechanism of optical confinement is not light reflection at the surface of the cluster, we chose some clusters with irregular shape and rough surface and repeated the above measurement.

Figure 15 presents the measurement result of a second cluster with irregular shape. It is slightly larger than 1 μm. Similar lasing phenomenon is observed in this cluster. The incident pump pulse energy at the lasing threshold is ~ 0.2 nJ. The FWHM of the emission spectrum narrows dramatically from 12 nm below the lasing threshold to 0.16 nm above the lasing threshold (Fig. 15b). After taking into account the instrumental broadening, the actual line width of the lasing mode is only 0.09 nm. Bright spots appear in the image of laser light distribution in the cluster. By adjusting the microscope objective, light distribution on a different plane inside the cluster is imaged onto the CCD camera. Figure 15c,d are the images of light distribution on two planes with different depths inside the cluster. Some bright spots appear in one image, but not in the other. This suggests that these bright spots are buried at different depths inside the cluster. Because lasing can occur in ZnO clusters with irregular shape and rough surfaces, we confirm that the optical

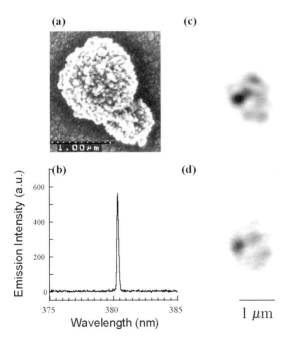

**Fig. 15.** (a) The SEM image of a second ZnO cluster. (b) The spectrum of emission from this cluster above the lasing threshold. The incident pump pulse energy is 0.27 nJ. (c) and (d) The spatial distribution of emission intensity in the cluster at the same pump power

confinement is not caused by reflection at the surface of the cluster, but by scattering inside the cluster.

Finally, I would like to compare the powder microlaser with other types of microlasers. Microlasers have important applications to integrated photonic circuits. Over the past decade, several types of microlasers have been developed. The key issue for a microlaser is to confine light in a small volume with dimensions on the order of an optical wavelength. In the vertical cavity surface emitting laser, light is confined by two distributed Bragg reflectors [30]. The microdisk laser utilizes total internal reflection at the edge of a high index disk to form whispering gallery modes [31]. In the recent demonstration of a two-dimensional photonic band-gap defect mode laser, lateral confinement of light is achieved by Bragg scattering in a two-dimensional periodic structure [32]. The fabrication of these microlasers requires expensive state-of-the-art semiconductor growth and microfabrication facilities. We demonstrate a new type of microlaser which is made of a disordered medium [27]. The optical confinement is achieved through disorder-induced scattering and interference. The fabrication of such a microlaser is much easier and cheaper than that of most microlasers.

## 5   Theoretical Modeling

In the last section, I will briefly describe our theoretical simulation of random lasers with coherent feedback. Several models have been set up in the theoretical study of the stimulated emission in an active random medium, e. g., the diffusion equation with gain [33,34], the Monte Carlo simulation [35], and the ring laser with nonresonant feedback [36]. However, these models cannot predict lasing with coherent feedback because the phase of the optical field is neglected. We take a different approach: we directly calculate the electromagnetic field distribution in a random medium by solving the Maxwell equations using the finite-difference time-domain (FDTD) method [37]. The advantage of this approach is that we can model the real structure of a disordered medium and calculate both the emission pattern and the emission spectrum [25,38].

In our model, ZnO particles are randomly positioned in space. The particle size is 50 nm. The random medium has a finite size, and it is surrounded by air. To model the random medium located in infinitely large space, we use the uniaxial perfect matched layer (UPML) absorbing boundary condition to absorb all of the outgoing light waves in the air [39]. Using the FDTD method, we solve the Maxwell curl equations

$$\frac{\partial \boldsymbol{H}}{\partial t} = -\frac{1}{\mu_0} \nabla \times \boldsymbol{E} \,,$$

$$\frac{\partial \boldsymbol{E}}{\partial t} = \frac{1}{\epsilon} \nabla \times \boldsymbol{H} - \frac{\sigma}{\epsilon} \boldsymbol{E} \,, \tag{1}$$

in the time domain. The randomness is introduced into the Maxwell equations through the dielectric constant $\varepsilon$, which varies spatially due to the random distribution of ZnO particles. We introduce optical gain by negative conductance [40]. The spectral gain profile of the dye solution is

$$\sigma(\omega) = -\frac{\sigma_0}{2}\left[\frac{1}{1+i(\omega-\omega_0)T_2} + \frac{1}{1+i(\omega+\omega_0)T_2}\right]. \tag{2}$$

$\sigma_0$ is related to the peak value of the gain set by the pumping level, and $T_2$ is the dipole relaxation time, which is inversely proportional to the spectral gain width.

In our simulation, a seed pulse, whose spectrum covers the ZnO emission spectrum, is launched in the center of the random medium at $t = 0$. When the optical gain is above the lasing threshold, the EM field oscillation builds up in the time domain. Using the discrete Fourier transform of the time domain data, we obtain the emission spectrum. Figure 16 shows the calculated emission spectrum and emission pattern for a specific configuration of scatterers. The size of the random medium is 3.2 µm. The filling factor of ZnO particles is 0.5. When the optical gain is just above the lasing threshold, the emission spectrum, shown in Fig. 16a, consists of a single peak. Figure 16b represents the light intensity distribution in the random medium. There are a few bright spots near the center. At the edge of the random medium, the light intensity is almost zero. To check the effect of the boundary, we change the spatial distribution of the scatterers near the edges of the random medium. We find that both the emission spectrum and the emission pattern remain the same. Their independence of the boundary condition indicates that the lasing mode is formed by multiple scattering and interference inside the disordered medium. When the optical gain is increased further, an additional lasing mode appears.

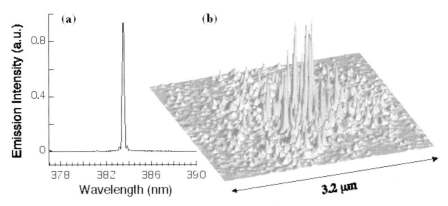

**Fig. 16.** (a) The calculated emission spectrum. (b) The calculated spatial distribution of emission intensity in the random medium

Our numerical simulation of random lasers with coherent feedback also confirms the second-order phase transition of the light field at the threshold. Below the lasing threshold, the seed pulse dies away from the random medium into the absorbing boundary layers. Only when the optical gain exceeds the threshold, the EM field builds up inside the random medium. Since the classical EM field represents the coherent part of a quantum field, our simulation result indicates that the quantum field in a random medium has no coherent part below the threshold; its coherent component appears only above the threshold.

# 6    Conclusion

In summary, we have observed lasing with resonant feedback in active random media. Recurrent light scattering provides coherent feedback for lasing. When the pump power exceeds the threshold, discrete lasing modes appear in the spectrum, the emission intensity increases suddenly, and the emission pulses are shortened dramatically. The photon statistics changes gradually from Bose–Einstein statistics at the threshold to Poisson statistics well above the threshold. Laser emission from random media can be observed in all directions.

In addition, we achieved spatial confinement of laser light in micrometer-sized random media. Since the transport mean free path is less than the optical wavelength, the optical confinement is attributed to the disorder-induced scattering and interference. Using the finite-difference time-domain method, we simulate lasing with coherent feedback in active random media. We find that the lasing modes are insensitive to the boundary conditions.

Finally, we illustrated very different lasing mechanisms for random lasers with resonant feedback and random lasers with nonresonant feedback. By varying the amount of scattering, we demonstrate the transition between the two kinds of random lasers.

**Acknowledgments**

I acknowledge the collaboration with R. P. H. Chang, S. T. Ho, P. Kumar, J. Y. Xu, Y. Ling, S.-H. Chang, A. Burin, H. C. Ong, E. W. Seelig, X. Liu, Y. G. Zhao, D. Z. Zhang, and J. Y. Dai. This work is supported partially by the United States National Science Foundation under Grant No. ECS-9877113, the David and Lucile Packard Foundation, and the Alfred P. Sloan Foundation.

# References

1. S. John, Localization of light (in dielectric microstructures), Phys. Today **44**, 32 (1991)
2. M. P. van Albada, A. Lagendijk, Observation of weak localization of light in a random medium, Phys. Rev. Lett. **55**, 2692 (1985)
3. P. E. Wolf, G. Maret, Weak localization and coherent backscattering of photons in disordered media, Phys. Rev. Lett. **55**, 2696 (1985)
4. A. Z. Genack, N. Garcia, Observation of photon localization in a three-dimensional disordered system, Phys. Rev. Lett. **66**, 2064 (1991)
5. R. Dalichaouch, J. P. Armstrong, S. Schultz, P. M. Platzman, S. L. McCall, Microwave localization by two-dimensional random scattering, Nature **354**, 53 (1991)
6. D. S. Wiersma, P. Bartolini, A. Lagendijk, R. Righini, Localization of light in a disordered medium, Nature **390**, 671 (1997)
7. F. J. P. Schuurmans, D. Vanmaekelbergh, J. van de Lagemaat, A. Lagendijk, Strongly photonic macroporous gallium phosphide networks, Science **284**, 141 (1999)
8. F. J. P. Schuurmans, M. Megens, D. Vanmaekelbergh, A. Lagendijk, Light scattering near the localization transition in macroporous GaP networks, Phys. Rev. Lett. **83**, 2183 (1999)
9. A. A. Chabanov, M. Stoytchev, A. Z. Genack, Statistical signatures of photon localization, Nature **404**, 850 (2000)
10. P. W Anderson, Absence of diffusion in certain random lattices, Phys. Rev. **109**, 1492 (1958)
11. N. M. Lawandy, R. M. Balachandran, A. S. L. Gomes, E. Sauvain, Laser action in strongly scattering media, Nature **368**, 436 (1994)
12. W. Sha, C.-H. Liu, R. Alfano, Spectral and temporal measurements of laser action of Rhodamine 640 dye in strongly scattering media, Opt. Lett. **19**, 1922 (1994)
13. C. Gouedard, D. Husson, C. Sauteret, F. Auzel, A. Migus, Generation of spatially incoherent short pulses in laser-pumped neodymium stoichiometric crystals and powders, J. Opt. Soc. Am. B **10**, 2358 (1993)
14. G. van Soest, M. Tomita, A. Lagendijk, Amplifying volume in scattering media, Opt. Lett. **24**, 306 (1999)
15. V. S. Letokhov, Generation of light by a scattering medium with negative resonance absorption, Sov. Phys. JETP **26**, 835 (1968)
16. H. Cao, Y. G. Zhao, H. C. Ong, S. T. Ho, J. Y. Dai, J. Y. Wu, R. P. H. Chang, Ultraviolet lasing in resonators formed by scattering in semiconductor polycrystalline films, Appl. Phys. Lett. **73**, 3656 (1998)
17. H. Cao, Y. G. Zhao, S. T. Ho, E. W. Seelig, Q. H. Wang, R. P. H. Chang, Random laser action in semiconductor powder, Phys. Rev. Lett. **82**, 2278 (1999)
18. S. V. Frolov, Z. V. Vardeny, K. Yoshino, A. Zakhidov, R. H. Baughman, Stimulated emission in high-gain organic media, Phys. Rev. B **59**, 5284 (1999)
19. H. Cao, J. Y. Xu, S.-H. Chang, S. T. Ho, Transition from amplified spontaneous emission to laser action in strongly scattering media, Phys. Rev. B **61**, 1985 (2000)
20. R. V. Ambartsumyan, N. G. Basov, P. G. Kryukov, V. S. Letokhov, Non-resonant feedback in lasers, in *Progress in Quantum Electronics*, J. H. Sanders, K. W. H. Stevens (Eds.) (Pergamon Press, London 1970) Vol. 1, Pt. 3, pp. 105–193

21. D. S. Wiersma, A. Lagendijk, Laser action in very white paint, Phys. World January, 33 (1997)
22. J. X. Zhu, D. J. Pine, D. A. Weitz, Internal reflection of diffusive light in random media, Phys. Rev. A **44**, 3948 (1991)
23. H. C. Ong, R. P. H. Chang, Effect of laser intensity on the properties of carbon plasmas and deposited films, Phys. Rev. B **55**, 13213 (1997)
24. H. Cao, Y. G. Zhao, H. C. Ong, R. P. H. Chang, Far-field characteristics of random lasers, Phys. Rev. B **59**, 15107 (1999)
25. H. Cao, J. Y. Xu, D. Z. Zhang, S.-H. Chang, S. T. Ho, E. W. Seelig, X. Liu, R. P. H. Chang, Spatial confinement of laser light in active random media, Phys. Rev. Lett. **84**, 5584 (2000)
26. R. V. Ambartsumyan, P. G. Kryukov, V. S. Letokhov, and Yu. A. Matveets, Statistical emission properties of a nonresonant feedback laser, Sov. Phys. JETP **26**, 1109 (1968)
27. H. Cao, J. Y. Xu, E. W. Seelig, R. P. H. Chang, Microlasers made of disordered media, Appl. Phys. Lett. **76**, 2997 (2000)
28. D. Jezequel, J. Guenot, N. Jouini, F. Fievet, Submicrometer zinc oxide particles: Elaboration in polyol medium and morphological characteristics, J. Mater. Res. **10**, 77 (1995)
29. H. Taniguchi, S. Tanosaki, K. Tsujita, H. Inaba, Experimental Studies on Output, Spatial, and Spectral Characteristics of a Microdroplet Dye Laser Containing Intralipid as a Highly Scattering Medium, IEEE J. Quant. Electron. **32**, 1864 (1996)
30. J. L. Jewell, J. P. Harbison, A. Scherer, Y. H. Lee, L. T. Florez, Vertical-cavity surface-emitting lasers: Design, growth, fabrication, characterization, IEEE J. Quant. Electron. **27**, 1332 (1991)
31. S. L. McCall, A. F. J. Levi, R. E. Slusher, S. J. Pearton, R. A. Logan, Whispering-gallery mode microdisk lasers, Appl. Phys. Lett. **60**, 289 (1992)
32. O. Painter, R. K. Lee, A. Scherer, A. Yariv, J. D. O'Brien, P. D. Dapkus, I. Kim, Two-dimensional photonic band-gap defect mode laser, Science **284**, 1819 (1999)
33. D. S. Wiersma, A. Lagendijk, Light diffusion with gain and random lasers, Phys. Rev. E **54**, 4256 (1996)
34. S. John, G. Pang, Theory of lasing in a multiple-scattering medium, Phys. Rev. A **54**, 3642 (1996)
35. G. A. Berger, M. Kempe, A. Z. Genack, Dynamics of stimulated emission from random media, Phys. Rev. E **56**, 6118 (1997)
36. R. M. Balachandran, N. M. Lawandy, Theory of laser action in scattering gain media, Opt. Lett. **22**, 319 (1997)
37. A. Taflove, *Computational Electrodynamics: The Finite-Difference Time Domain Method* (Artech House, Boston 1995)
38. X. Y. Jiang, C. M. Soukoulis, Time dependent theory for random lasers, Phys. Rev. Lett. **85**, 70 (2000)
39. Z. S. Sacks, D. M. Kingsland, R. Lee, J. F. Lee, A perfectly matched anisotropic absorber for use as an absorbing boundary condition, IEEE Trans. Antenna Propagation **43**, 1460 (1995)
40. S. C. Hagness, R. M. Joseph, A. Taflove, Subpicosecond electrodynamics of distributed Bragg reflector microlasers: Results from finite difference time domain simulations, Radio Sci. **31**, 931 (1996)

# Localization Phenomena in Elastic Surface Plasmon Polariton Scattering

Sergey I. Bozhevolnyi

Institute of Physics, Aalborg University
Pontoppidanstræde 103, 9220 Aalborg Øst, Denmark
sergey@physics.auc.dk

**Abstract.** Scattering of surface plasmon polaritons (SPPs) propagating along a metal film surface with random roughness is considered with the emphasis on the elastic (in-plane) SPP scattering driven by the resonantly excited SPP. The usage of scanning near-field optical microscopy (SNOM) for near-field imaging of SPP fields is discussed in detail, establishing a firm basis for the interpretation of experimentally obtained SNOM images. The near-field images, which exhibit spatially localized SPP intensity enhancement (bright spots) at rough metal surfaces, are presented and attributed to the phenomenon of strong (Anderson) localization of SPPs that occurs due to interference effects in multiple scattering caused by surface roughness. Several specific features of this phenomenon, viz., wavelength and angular dependence of the spatial location of bright spots, self-similarity of the surface topography, and statistical properties of SPP intensity distributions, are illustrated by the near-field optical images obtained experimentally.

## 1 Introduction

Light localization is one of the most fascinating wave phenomena in contemporary physics that challenges our understanding of basic concepts of wave scattering and, at the same time, opens up exciting avenues for the exploitation of localization effects in both basic and applied research. Essentially, localization of light as well as electron localization is an interference phenomenon related to multiple elastic scattering in random media [1]. When a wave propagates through a strongly scattering and nonabsorbing random medium, the mean free path is reduced due to constructive interference of waves scattered along the same path in opposite directions. In lower orders of multiple scattering, this effect results in a phenomenon of enhanced backscattering (also referred to as weak localization) [2]. With the increase of scattering, the constructive interference of backscattered waves eventually brings transport to a complete halt, i.e., the mean free path vanishes, and propagation no longer exists — the wave is captured in a "random" cavity (strong or Anderson localization). Strong localization is expected once the (elastic) scattering mean free path decreases below the light's wavelength. However, contrary to electron localization, this criterion is extremely difficult to satisfy for electromagnetic waves. In the latter case, the scattering potential is frequency-dependent, resulting in divergence of the mean free path in the

V. M. Shalaev (Ed.): Optical Properties of Nanostructured Random Media,
Topics Appl. Phys. **82**, 329–358 (2002)

limit of both low and high frequencies [1]. Direct experimental evidence of light localization in three dimensions has only recently been reported in strong scattering media of semiconductor powders whose transmission as a function of the sample thickness exhibited transition from linear to exponential behavior when the particle size became progressively smaller [3]. Indirectly, Anderson localization of light has been conjectured from the observation of spatial confinement of laser light in an active random medium consisting of ZnO nanoparticles [4].

The situation with localization changes dramatically in two-dimensional scattering geometry: light, as well as electrons, is localized with any degree of disorder, at least in the absence of absorption [1]. Qualitatively, it can be explained by the fact that a random walk is recurrent in two dimensions, and the total sojourn time becomes infinite. Consequently, the effective cross section of a single scatterer tends to infinity, drastically reducing the effective mean free path. Surface plasmon polaritons (SPP) are (quasi-) two-dimensional waves that can be excited at an interface between two media with opposite signs of the real parts of dielectric constants, typically, between a dielectric and a metal [5]. The associated electromagnetic fields are coupled to oscillating surface charge density distributions provided by the metal and decay exponentially with the distance from the interface that supports them. When the SPP interacts with surface features, one can distinguish between two scattering processes: scattering into free electromagnetic waves propagating away from the interface (inelastic SPP scattering) and scattering into other SPPs propagating along the interface (elastic SPP scattering). If the latter process that represents scattering in (quasi-) two-dimensional geometry, is dominating, SPP scattering by surface roughness should exhibit localization effects [6,7].

A conventional approach in investigations of the phenomenon of strong localization both in solid state physics and in optics is to study the transmission properties of a sample as a function of its thickness. However, this approach implies usage of average characteristics and does not explicitly reveal the fundamental role being played by the interference effects. Alternatively, instead of studying the transport properties of monochromatic electromagnetic waves, one can investigate the time-averaged energy density distribution of the field in the scattering random medium. It has been shown that localized waves correspond to eigenmodes of the system of equations that describes the self-consistent field acting on each scatterer [8]. The energy density distribution in the strongly scattering random medium should thereby represent regions with large (local) field enhancement. Consequently, in elastic SPP scattering at rough metal surfaces, strong localization should lead to spatially localized enhancement of the total (incident and scattered) SPP field. Indeed, for resonant SPP excitation at a relatively rough gold surface, subwavelength-sized and very bright spots have been observed in the corresponding optical images that were obtained by using scanning near-field optical microscopy (SNOM) [9,10]. Actually, it has been suggested [6] that strongly enhanced and localized SPP fields play the major role in the surface-enhanced Raman scattering (SERS) observed for rough metal surfaces [11]. Investigations of SPP localization are therefore fueled not only due to the interest in the basic physics of localization but also because of its contribution to SERS and other surface-enhanced optical phenomena.

Below, I review the main experimental results concerning localization phenomena in elastic SPP scattering caused by random roughness. Since direct observation of these phenomena is possible only with the help of SNOM, the first part is devoted to the technique of near-field mapping of SPP fields, discussing a number of important issues related to the interpretation of SNOM images. In the second part, the near-field optical images exhibiting spatially localized SPP intensity enhancement are presented, and various characteristic features of this phenomenon are discussed.

## 2    Near-Field Mapping of Surface Plasmon Polaritons

Different near-field techniques have been employed to investigate various SPP properties (an overview can be found in [9]). A photon tunneling SNOM (PT-SNOM) configuration, in which an uncoated fiber tip is used to probe an evanescent field of the light being totally internally reflected at the sample surface, seems to be the most suitable technique for local and unobtrusive probing of the SPP field [12]. It is generally accepted that, due to the relatively low refractive index of optical fiber, such a tip can be considered a passive probe of the electrical field intensity or, in our particular case, the SPP field intensity. However, even in this case, there are several very important features that should be borne in mind when interpreting the near-field optical images.

### 2.1    Elastic and Inelastic Scattering

The PT-SNOM configuration used to obtain experimental results presented here consists of a stand-alone near-field microscope (with an uncoated fiber tip) combined with a shear-force-based feedback system and an arrangement for SPP excitation in the usual Kretschmann configuration; it is described in detail elsewhere [9]. Schematic representation of SPP excitation and scattering and near-field imaging of the resulting SPP intensity distribution are shown in Fig. 1a. The SPP is excited at a surface of a thin metal film by p-polarized (electrical field is parallel to the plane of incidence) laser radiation directed on the surface at the resonant angle $\vartheta$, which provides the minimum in the angular dependence of the reflected light power (Fig. 1b). The resonant angle is determined by the phase matching condition: $\mathrm{Re}\,\beta = (2\pi/\lambda)n\sin\vartheta$, where $\beta$ is the SPP propagation constant, $\lambda$ is the light wavelength in air, and $n$ is the prism refractive index. The SPP propagation constant depends

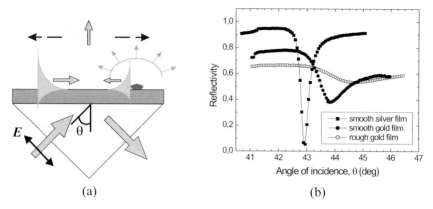

(a)                        (b)

**Fig. 1.** Schematic representation (**a**) of excitation (in the Kretschmann configuration), scattering, and near-field imaging of SPP fields. Polarization of the incident light exciting the SPP is also indicated. Angular dependencies (**b**) of the reflected light power measured for 45-nm-thick smooth silver film and 53-nm-thick smooth and rough gold films at the light wavelength of 633 nm

on the film characteristics and has a simple expression only for semi-infinite media on both sides of the planar air–metal interface:

$$\beta = \frac{2\pi}{\lambda} \sqrt{\frac{\varepsilon}{\varepsilon + 1}}, \tag{1}$$

where $\varepsilon$ is the film dielectric constant [5]. A film surface can be rough due to the fabrication process [9,10], or it can be made rough, e.g., if the film is evaporated on a sublayer of colloidal gold particles [12,13]. Any surface feature would scatter the excited SPP both elastically, i.e., in the surface plane, giving rise to a diverging cylindrical SPP, and inelastically, i.e., out of the surface plane, producing field components propagating away from the surface (Fig. 1a). In general, both elastic and inelastic scattering processes result in decreasing the efficiency of SPP excitation (Fig. 1b), which also depends on the film thickness and dielectric constant [5].

    It is very important to control the amount of inelastic SPP scattering when investigating SPP localization phenomena. First of all, even for a perfectly flat metal surface, the SPP propagation length $L$ is finite due to the internal damping: $L = [2\mathrm{Im}\,(\beta)]^{-1}$. For a thin metal film placed on the prism surface (Fig. 1a), the SPP propagation length is further reduced due to the coupling between the SPP and the field components propagating in the prism (the radiation damping) [5]. Inelastic SPP scattering by surface roughness contributes to radiation damping as well and thereby decreases the SPP propagation length even further [5,14]. This scattering may eventually bring the propagation length below the elastic scattering mean free path $l$, which depends on the average separation between and the strength of scatterers. In such a case, elastic SPP scattering is dominated by single scattering events and, consequently, the localization effects become insignificant [13,15]. It should be emphasized that the conditions for strong SPP localization include the requirement of multiple SPP scattering, i.e., $L \gg l$, and also that the surface roughness (disorder) is sufficiently strong, i.e., $l \sim \lambda$ [15].

Another important circumstance to be borne in mind is related to the influence of propagating (away from the surface) field components on the near-field mapping of SPP fields. If the contributions of propagating waves and SPPs in the optical signal detected are comparable, the resultant near-field optical image can be completely obscured by their interference. Actually, the contribution of propagating waves can even be enhanced during the detection process because they have lower spatial frequencies than SPPs. Therefore, when the SPP is resonantly excited, checking that the average optical signal decreases exponentially (and with the proper exponential factor) as the tip–surface distance increases is recommended [12]. Usually, if the angular dependence of the sample reflectivity exhibits a pronounced minimum corresponding to the resonant SPP excitation (Fig. 1b), the near-field optical signal contains mainly the SPP-related signal [12,13].

## 2.2   Topographical Artifacts

Let us consider the situation when the average optical signal does decrease exponentially with the increase of the tip–surface distance in accord with the evanescent decay of the SPP field intensity. Such a distance dependence of the optical signal indicates the possibility of inducing topographical artifacts in the near-field optical images obtained in the constant tip–surface distance mode [16,17]. When a fiber tip scans along the sample surface following its profile (e.g., by using shear-force feedback) and probing the optical field, the tip moves up and down with respect to the mean surface plane. Therefore, the optical signal detected necessarily contains a contribution related to the exponential decrease of the incident field SPP intensity with the distance from the mean plane. This contribution (when sufficiently strong) manifests itself in the near-field optical images as the features that are strongly correlated with the surface topography and thus are called topographical artifacts. In general, the influence of topographical artifacts can be discarded only when the surface features inspected are known to extend over the sensitivity window of the SNOM used [17].

One can evaluate the artifact contribution, at least of the kind described above, and take it into account when interpreting SNOM images by comparing the images recorded in constant (tip–surface) distance and constant height (when the distance to the mean surface plane is constant) modes [16,17]. Fortunately, in our experience, the near-field optical images that are obtained, when the resonantly excited SPP propagates along a rough surface, do not exhibit topographical artifacts. In other words, the only difference that is usually found between the images in constant distance and constant height modes is related to the displacement of the fiber tip when moving from contact with the sample surface [12,13]. This surprising fact has a simple yet rather profound explanation. When the surface is rough but supports the propagating SPP, the inevitable localization effects lead to strong and rapid variations of the total SPP intensity in the surface plane, variations that dominate over the topographical artifacts. Typically, localization effects result in near-field images that are not correlated with the topographical images taken simultaneously (Fig. 2), a feature which implicitly indicates that topographical artifacts are insignificant [13].

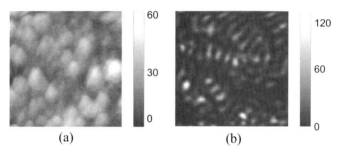

**Fig. 2.** (**a**) Typical gray-scale topographical and (**b**) near-field optical images $(4.4 \times 4.4 \, \mu m^2)$ obtained with the 80-nm-thick rough gold film and with the SPP resonantly excited at the light wavelength of 633 nm. Corresponding bar scales reflect (**a**) the depth (in nm) and (**b**) the detected light power (in pW)

## 2.3    Influence of a Probe

The aforementioned contribution to the topographical artifacts is of primary importance only if probe–object coupling is negligibly small. In general, the optical signal detected is determined by the self-consistent field at the site of the probe tip scanned along the sample surface. If the sample surface is strongly structured and the probe polarizability is large, the total self-consistent field can be significantly different from the field existing in the absence of the probe, i.e., from the measured field. The resultant optical image would then represent a map of the strength of probe–sample interactions rather than the intensity distribution of the measured field. One can regard the probe–sample interaction as yet another source of topographical artifacts. It is expected that this interaction can become significant when SNOM is used to image strongly enhanced and localized optical fields established in the process of multiple scattering [18].

One can argue that the larger the field enhancement (in the absence of a probe), the stronger the influence of the probe on the self-consistent field that is actually detected. It is possible to estimate the field enhancement in a system of dipolar scatterers (representing the sample) that would be notably influenced by a small probe sphere, whose response is represented by the dipolar (electric) polarizability $\alpha_p$. The main idea is that, if there is a resonant enhancement of the self-consistent field in the absence of the probe, then the field scattered by the probe (toward the sample) would be resonantly enhanced as well. Thereby, the probe contribution can eventually become comparable with the measured field. Introducing the field enhancement (at the site of the probe) factor $\Gamma$ and using the electrostatic (Rayleigh) expression for probe polarizability, the following estimate of the enhancement resulting in a strong perturbation can be obtained [18]:

$$\Gamma \frac{\mu_0 \alpha_p c^2}{4 \pi R^3} \propto 1 \,, \Rightarrow \Gamma \propto \left( \frac{R}{a_p} \right)^3 \frac{\varepsilon_p + 2}{\varepsilon_p - 1} \,, \tag{2}$$

where $a_p$ and $\varepsilon_p$ are the radius of the probe sphere and its dielectric constant, and $R$ is the probe–sample distance, or, more precisely, the smallest distance between the probe and one of the scatterers representing the sample. Considering, for example, a metal probe ($|\varepsilon_p| \gg 1$) in contact with a spherical scatterer of the same size, i.e., $R \sim 2a_p$, one obtains from (2) the field enhancement value $\Gamma \sim 10$. Such a field enhancement is similar to that observed experimentally in SPP scattering by surface roughness [10,13], as well as in light scattering by surface clusters of colloidal silver particles [19] and semicontinuous gold films [20].

The above estimate demonstrates that the problem of field perturbation by a probe becomes rather important in configurations with well-developed multiple scattering by surface features, especially for metal probes [20]. It is also clear that the probe influence rapidly decreases when the probe–sample distance increases (2), although probably at the expense of the resolution obtained. Therefore, by comparing the near-field optical images taken at different probe–surface distances, it is possible to deduce the extent of the field perturbation introduced by the probe. In fact, these images should become similar to each other at the distances for which the probe–sample coupling is negligibly small [18]. The reason is that dipole–dipole coupling scales as $R^{-6}$ (here $R$ is the dipole–dipole separation), whereas the field distributions at different distances from the sample differ only because of relatively soft filtering out of evanescent field components. Since the near-field optical images are usually obtained simultaneously with the topographical ones, the recommended procedure of imaging at different probe–sample distances would also help to distinguish the topographical artifacts discussed in the previous section.

## 2.4   Image Formation

Let us now assume that all sources of topographical artifacts (including perturbation by a probe) can be disregarded. There is still the issue of the relation between the near-field optical image and the measured field. It is intuitively appealing to think of an uncoated fiber tip as a detector of the electric near-field intensity [21] and, if needed, to account for its finite size by introducing the intensity transfer function [22]. Such an understanding has been widely (explicitly or implicitly) used in near-field optics. On the other hand, it is clear that, when dealing with coherent optical fields, one has to employ the transfer function that involves (complex) *amplitudes* of optical fields (but not their intensities). The appropriate treatment of image formation has only recently been developed [23]. Below, I present a simplified description of image formation [24], which is adapted to the special case of near-field mapping of SPP fields.

### 2.4.1 Basic Equations

Image formation in the PT-SNOM can be described as follows. The light collected by an uncoated fiber tip is propagating toward the output end of the fiber, where it is detected by a photodetector, as a combination of guided fiber modes [23]. For simplicity, I shall consider here only the constant height mode of operation, when a fiber tip is scanning at a constant height from the mean surface plane (Fig. 3a). Furthermore, let us take into account that most fibers are weakly guiding, i.e., the index difference between the fiber core and its cladding is very small. In such a case, the guided waves are approximately transverse electromagnetic (TEM), and the linearly polarized modes can be introduced [25]. I shall also assume that the fiber used is single-mode and the total field is composed only of (excited and scattered) SPPs that propagate over sufficiently long distances, i.e., $\mathrm{Re}\,\beta \gg \mathrm{Im}\,\beta$ (Sect. 2.1). The latter assumption means that all plane waves in the plane-wave decomposition of the total field [24] have the same magnitude of the wave-vector projection on the surface plane $\beta_1 = \mathrm{Re}\,\beta$. One can then reduce the plane-wave decomposition of the incident field $\boldsymbol{E}$ at the scanning plane $z = z_\mathrm{t}$ ($z_\mathrm{t}$ is the $z$ coordinate of

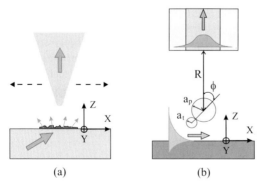

(a)                    (b)

**Fig. 3.** Schematic representation (**a**) of the PT-SNOM configuration and (**b**) model geometry used to calculate the detected signal related to SPP fields

the tip end) to the following expression [24,26]:

$$E(x, y, z_\mathrm{t}) = \frac{\beta_1 \exp(-\alpha z_\mathrm{t})}{2\pi} \int_0^{2\pi} F(\varphi) \exp\left[\mathrm{i}(x\beta_1 \cos\varphi + y\beta_1 \sin\varphi)\right] \mathrm{d}\varphi \,, \quad (3)$$

$$F_x(\varphi) = F_z(\varphi) r \cos\varphi \,,$$

$$F_y(\varphi) = F_z(\varphi) r \sin\varphi \,,$$

$$\alpha = \sqrt{\beta_1^2 - k^2} \,, r = \mathrm{i}\left(\sqrt{|\mathrm{Re}\,\varepsilon|}\right)^{-1} \,, \quad (4)$$

where $F(\varphi)$ represents the field components (at the metal surface) of the (plane) SPP propagating in the direction inclined at angle $\varphi$ with respect to the $x$ axis, $r$ is the ratio between the in-plane and perpendicular field components [5], $k = 2\pi/\lambda$, and $\alpha$ is the SPP penetration depth in air. A single-mode weakly guiding fiber actually supports two (degenerated) modes with orthogonal polarizations [25]. It is convenient to choose as a basis the modes that are linearly polarized along the $x$ and $y$ axes. Their amplitudes can then be related *linearly* to the plane SPP components [24]:

$$A(x_\mathrm{t}, y_\mathrm{t}, z_\mathrm{t}) = \frac{\beta_1 \exp(-\alpha z_\mathrm{t})}{2\pi}$$

$$\int_0^{2\pi} H(\varphi) \cdot F(\varphi) \exp\left[\mathrm{i}(x_\mathrm{t}\beta_1 \cos\varphi + y_\mathrm{t}\beta_1 \sin\varphi)\right] \mathrm{d}\varphi \,, \quad (5)$$

where $x_\mathrm{t}$ and $y_\mathrm{t}$ are the scanning coordinates of the probe tip, $A = (A_x, A_y)$ is composed of the fiber mode amplitudes, and $H(\varphi)$ is the vector-coupling coefficient that accounts for the contribution of each component of the plane SPP (propagating at angle $\varphi$) to the respective mode amplitude. The detected (with conventional detection schemes) signal is proportional to the total power carried by the fiber modes: $S \sim |A_x|^2 + |A_y|^2$. Note that by virtue of dealing with only one spatial frequency, $\beta_1$, the tip–surface distance dependence of the detected signal can be explicitly introduced.

It can be shown that if an uncoated fiber tip possesses axial symmetry with respect to the fiber axis and is oriented perpendicularly to the average surface plane, the cross-coupling coefficients $H_{xy}$ and $H_{yx}$ vanish and $H_{xx} = H_{yy} = H$ due to symmetry. In addition, $H_{xz} = h \cos\varphi$, and $H_{yz} = h \sin\varphi$ because it is only due to the SPP *propagation*, that the axial symmetry is broken, and thereby it is only the polarization in this direction that can be excited. Inserting these relations in (5) and taking into account (3), (4), one finds that the PT-SNOM image reproduces *only* the intensity distribution of the *in-plane* field components: $S \sim \left(|E_x|^2 + |E_y|^2\right)|H + hr^{-1}|^2$. Therefore, even though the perpendicular SPP component is being detected, the near-field image for the symmetrical detection configuration does not reflect its distribution. Note that this component can be considerably stronger than the in-plane one, since $r \to = 0$ when $|\varepsilon| \to = \infty$ (4). Here, it should be understood that these components are proportional to each other only for the plane SPP and there is no simple relation between them in arbitrary field distribution. Finally, if the fiber tip is asymmetrical, the perpendicular SPP component would also contribute to the near-field image to the extent determined by tip asymmetry.

### 2.4.2   Numerical Results

In this section, the above conclusions regarding near-field imaging of SPPs are illustrated by numerical results. Similarly to the previously reported numerical simulations [24], a combination of two small spherical scatterers of different size, which are in contact with each other and placed in front of a flat-faced single-mode fiber, were used to model an uncoated fiber tip (Fig. 3b). By changing the angle $\theta$ between the line connecting the spheres and the fiber axis, SPP detection with both symmetrical ($\theta = 0°$) and asymmetrical (e.g., $\theta = 45°$) configurations was simulated. The incoming SPP field was scattered by the probe spheres and projected on the fiber-mode fields (with a Gaussian radial distribution) polarized along the $x$ and $y$ axes. The detected signal was evaluated as a sum of the square magnitudes of the mode amplitudes. The dipole polarizabilities of the spherical scatterers were approximated by the electrostatic expression. The following parameters were used in the calculations: $\lambda = 633$ nm, the metal dielectric constant $\varepsilon = -16 + 0.6i$ (which is a typical constant for silver films at this wavelength [5,12]), the radius of the lower (upper) sphere $a_{t(p)} = 30(40)$ nm (Fig. 3b), their dielectric constant $\varepsilon_p = 2.25$, the distance between the fiber end and the upper sphere $R = 5000$ nm, the fiber-mode width $w = 1000$ nm, and the separation of the bottom of the lower sphere from the metal surface $d = 1$ nm.

Let us first consider the near-field mapping of the interference pattern for three plane SPPs propagating in directions with $\varphi = 0°, 90°$, and $45°$ and having different amplitudes, viz., $F_z(0°) = F_z(90°) = 1$ and $F_z(45°) = 0.1$. The total SPP intensity distribution reflects mainly the interference pattern corresponding to the perpendicular components of strong (orthogonal) SPPs (Fig. 4a). However, the symmetrical ($\theta = 0°$) detection configuration does not image this pattern at all (Fig. 4b) because it reproduces the intensity distribution of the in-plane SPP components. At the same time, the asymmetrical ($\theta = 45°$) configuration results in an image that is quite similar to the intensity distribution (see Fig. 4a,c). The reason is that the two tilted probe spheres act as a single anisotropic dipole scattering the perpendicular SPP components into fiber modes. It seems that using an asymmetrical tip might be even advantageous because it facilitates detecting the perpendicular SPP field components. On the other hand, such a tip also introduces asymmetry in detecting the in-plane components and can be considered a better choice only if these components are negligibly small in comparison with the perpendicular one.

Imaging of localized SPP fields that originate in the process of strong multiple scattering is much more complicated. As an example, let us consider the

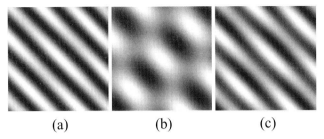

**Fig. 4.** (a) Linear gray-scale representations $(2 \times 2\,\mu m^2)$ of SPP intensity distribution for three plane SPPs, and the corresponding PT-SNOM images simulated for (**b**) symmetrical and (**c**) asymmetrical detection configurations. The contrast of images, i.e., the relative difference between maximum and minimum intensities, is (**a**) 86%, (**b**) 23%, and (**c**) 45%

simulation results for the near-field imaging of the interference pattern corresponding to a plane SPP and seven cylindrical SPPs with equal amplitude and randomly located origins. The phases of cylindrical waves are set by the phase of plane SPPs at their origins. The total intensity distribution exhibits a typical interference pattern with high contrast and bright spots (Fig. 5a). However, both near-field images simulated for the symmetrical (Fig. 5b) and asymmetrical (Fig. 5c) configurations look more similar to each other than to the SPP intensity distribution. This circumstance can be explained by the fact that, even with the asymmetrical configuration, only a small part of the perpendicular SPP field component is detected. However, one can still see that the image simulated with the asymmetrical configuration is somewhat closer to the total intensity distribution than the image for the symmetrical configuration. It should be also mentioned that, in many our experiments, the tip used was found to be asymmetrical as appeared from the topograph-

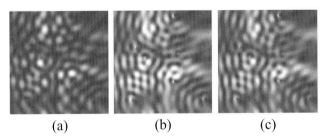

**Fig. 5.** Linear gray-scale representations $(4 \times 4\,\mu m^2)$ of SPP intensity distribution (a) for seven randomly situated point sources of cylindrical SPPs (of equal amplitude) interfering with a plane SPP and the corresponding PT-SNOM images simulated for (**b**) symmetrical and (**c**) asymmetrical detection configurations. The contrast of images is 100%

ical images [13,26] or from the measurements of the contrast of evanescent wave interference patterns [24]. This observation accounts for the fact that the experimental near-field images resemble rather well the SPP intensity distributions simulated numerically [13].

# 3    Localization of Surface Plasmon Polaritons

Localization of electromagnetic waves and, in particular, SPP localization have already been introduced as well as the PT-SNOM arrangement used for the near-field imaging of SPP fields (Fig. 1a). However, before presenting and discussing the main experimental results, I would like to briefly consider the underlying physics of the phenomenon of (coherent) multiple scattering, a phenomenon that is intimately related to the localization phenomena. Wave *localization* is the result of interference in *multiple* scattering [1] and, therefore, the characteristic features of localization originate from those of multiple scattering.

## 3.1    Single and Multiple Scattering

SPP scattering by surface inhomogeneities has been treated in many theoretical papers by using different approximations and limitations (a good overview can be found in [27]). In general, this problem is very complicated: even the modeling of SPP scattering by a circularly symmetrical defect requires elaborate numerical simulations [28]. The approximation of pointlike dipolar scatterers has been used recently to describe local excitation and scattering of SPPs [29] and, in a simplified form, to model elastic SPP scattering [13]. I shall adopt this approximation because it is rather suitable for the discussion of multiple scattering phenomena and their characteristic features. In the following, the main emphasis is put on the differences in the total intensity distribution of fields established in the processes of single and multiple scattering (Fig. 6).

Let us represent a scattering system by small spherical scatterers placed near the surface (e.g., of a multilayer system that includes a metal film). The particle–field interaction can then be treated in the electric dipole approximation with the optical response (to an incident electromagnetic field) of an individual sphere expressed in terms of the isotropic dipole polarizability. By assuming that the incident light is monochromatic with an angular frequency $\omega$, the *total* electrical field $\boldsymbol{E}(\boldsymbol{r}, \omega)$ at an arbitrary observation point can be written as follows [30]:

$$\boldsymbol{E}(\boldsymbol{r}, \omega) = \boldsymbol{E}_{\mathrm{in}}(\boldsymbol{r}, \omega) - \mu_0 \omega^2 \sum_{j=1}^{N} \alpha_j(\omega) \boldsymbol{G}(\boldsymbol{r}, \boldsymbol{r}_j, \omega) \cdot \boldsymbol{E}(\boldsymbol{r}_j, \omega) , \qquad (6)$$

where $\boldsymbol{E}_{\mathrm{in}}(\boldsymbol{r}, \omega)$ is the incident field, i.e, the field that would prevail in space if the scatterers were absent; $\boldsymbol{G}(\boldsymbol{r}, \boldsymbol{r}_j, \omega)$ is the appropriate field propagator

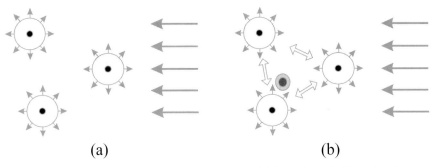

**Fig. 6.** Schematic representation of the regimes of (**a**) single and (**b**) multiple scattering

describing the field propagation from a source point $r_j$ to an observation point $r$; and $E(r_j, \omega)$ is the *self-consistent* field at the site of $j$th dipole scatterer with polarizability $\alpha_j(\omega)$. Note that several propagation channels can exist, e.g., reflection from the surface and propagation in the form of SPPs [29,30], and that, in general, the self-consistent field at the site of the scatterers is *different* from the incident field because it includes the fields scattered from *all* other scatterers.

The regime of *single* scattering means that the field at the site of the scatterers can be approximated by the incident field $E(r_j) = E_{\text{in}}(r_j)$, and thereby the total field can be adequately represented by the incident field and the primary scattered waves (Fig. 6a). This approximation can be used when the amplitudes of primary waves are too weak to produce appreciable secondary scattered waves. The amplitudes of scattered waves decrease rapidly with the distance from the scatterers and, therefore, the total field at a given point is completely determined by the relative positions of the nearest scatterers, i.e., by the *local environment* and the incident field in this environment. Since the strongest field decrease occurs in the near-field region (at subwavelength distances), the wave characteristics of the incident field (wavelength and propagation direction) are not important. Therefore, the total field is strongly influenced only by the polarization because the scattered field is anisotropic (even for isotropic scatterers) due to the anisotropy of the field propagator (6).

When the scattered waves are sufficiently strong, i.e., their amplitudes are comparable with the incident field, all *orders* of scattered waves become important and have to be taken into account (Fig. 6b). This corresponds to the regime of *multiple* scattering. The self-consistent field $E(r_j, \omega)$ at the site of the $j$th dipole scatterer has to be determined by directly resolving a set of self-consistent equations, which can be obtained from (6) by letting $r$ coincide in turn with each of the dipole positions:

$$\boldsymbol{E}(r_j, \omega) = \boldsymbol{E}_{\mathrm{in}}(r_j, \omega) - \mu_0 \omega^2 \sum_{i=1}^{N} \alpha_{\mathrm{i}}(\omega) \boldsymbol{G}(r_j, \boldsymbol{r}_{\mathrm{i}}, \omega) \cdot \boldsymbol{E}(\boldsymbol{r}_{\mathrm{i}}, \omega) \,. \tag{7}$$

The self-consistent field at the site of a scatterer is, thereby, determined by the relative positions of all scatterers and the distribution of the incident field over the whole system, i.e., by the *global environment*. In the regime of multiple scattering, the self-consistent field at the observation point is the result of the interference of all scattered waves and the incident field (6). This field depends, therefore, on the geometry of the whole scattering system and on the phase, amplitude, and polarization distribution of the incident field in the system.

The difference between single and multiple scattering results in the different (local and global) sensitivities of the scattered field distributions as discussed above. Another important consequence is the difference in the possible magnitude of the total field. In the regime of *linear* scattering considered here, the total field is proportional to the incident field in both cases. However, for single scattering, the total field can be significantly stronger than the incident field only when the observation point is very close to a scatterer, especially if the latter is resonantly excited. For multiple scattering, if all scattered waves interfere constructively (at some point), the total field can be enormously enhanced even without resonant excitation of individual scatterers [30]. Such a resonant enhancement is extremely sensitive to any changes in the scattering configuration and to the position of the observation point. This accounts for the fact that the field enhancement due to multiple scattering is often strongly localized in space, a feature that creates additional interest in multiple scattering phenomena.

Strong multiple scattering resulting in large local fields can be realized by using *strong* scatterers that interact *efficiently* with each other (7). One possibility (of enhancing the interaction) is to employ two-dimensional geometry since scattered waves decay relatively slowly in two dimensions. Elastic SPP scattering discussed here is an example of the realization of two-dimensional scattering. One can also realize two-dimensional dipole–dipole interaction via excitation of waveguide modes, i.e., via modes propagating in a planar waveguide with scatterers placed on its surface [31]. Another way is to compose nanometer-sized metallic particles very close to each other [32]. Strongly enhanced and localized optical fields have been observed in experiments with surface clusters of metallic particles [19] and discontinuous metal films [20]. However, one should not confuse this kind of field localization with Anderson localization of electromagnetic waves. In the latter case [19,20], the scattered waves are three-dimensional, whereas the structures are essentially two-dimensional. Therefore, this phenomenon, contrary to the phenomenon of Anderson localization (such as strong SPP localization), cannot be considered a transport phenomenon. Actually, in the limit of strongly rough metal surfaces, inelastic SPP scattering becomes dominant, and strong SPP localization transforms into the multiple scattering phenomenon that results in near-field images rather similar to those obtained with fractal surface clusters [26]. Finally, it should be mentioned that multiple scattering of light in a periodic array of scatterers renders the formation of *photonic* band gaps in such a system, opening, conceptually, new possibilities for manipulating electromagnetic waves [33].

## 3.2    Observation of Localized Surface Polaritons

One of the main differences between electron localization and localization of light is that the electromagnetic waves can be easily absorbed and, thereby, taken irrevocably out of the process of multiple scattering. This feature, together with the fact that realistic scattering systems are of limited sizes, transforms a seemingly sure case of strong (two-dimensional) SPP localization into a rather problematic one. It is clear that to realize the regime of multiple scattering, the SPP propagation length (inelastic mean free path) $L$ has to be larger than the elastic scattering mean free path $l$: $L > l$. The SPP propagation length at $\lambda = 633\,\mu m$ can be estimated (Sect. 2.1) by using the available optical constants [34] $\sim 25\,\mu m$ and $\sim 8\,\mu m$ for air–silver and air–gold interfaces, respectively. The elastic scattering mean free path depends on the spatial distribution of surface scatterers: $l \sim R^2/\sigma$, where $R$ is the average separation between scatterers and $\sigma$ is the elastic scattering cross section. For an individual scatterer, the elastic scattering cross section can be estimated from the contrast of the interference pattern between the excited (plane) SPP and the scattered (cylindrical) SPP [13]. The cross section $\sigma \sim 0.2\,\mu m$ has been obtained in this way for a wavelength-sized scatterer, a value which also agrees with theoretical considerations [28]. The overall conclusion, therefore, is that, when the separation $R$ is not larger than a few micrometers, the regime of multiple SPP scattering can be realized.

Multiple SPP scattering does not, however, warrant automatically strong localization. If the scatterers are not immediately adjacent to each other, the elastic scattering mean free path can extend over several micrometers, i.e., $l \gg \lambda$, which is the condition of weak disorder. It has been shown that, in weak disorder, the following condition should be satisfied for strong localization to occur [15]:

$$L \gg l \exp\left(\frac{4\pi l^2}{3\lambda^2}\right). \tag{8}$$

Exponential divergence of this condition with respect to the ratio $l/\lambda$ means that strong SPP localization can hardly be achieved in weak disorder, even when the condition of multiple scattering ($L > l$) is fulfilled. Contrary to that, enhanced backscattering of SPPs (weak SPP localization) can be observed in lower orders of multiple scattering, actually already in the regime of double scattering [13,15,35]. This effect is a precursor of strong localization and, for example, has also been observed and extensively studied in three dimensions [2].

In the following, I shall concentrate on the experimental results obtained for SPP scattering by sufficiently rough films [9,10,12,13,15]. The most convincing evidences of strong SPP localization have been obtained with gold films at the light wavelength of 633 nm. The SPP excitation exhibited usually well-pronounced resonance behavior with an average optical signal of $\sim 0.1$ nW, whose exponential decrease with the increase of the probe–surface distance indicated that the detected signal was primarily due to SPP intensity (Sect. 2.1). By comparing optical images obtained with shear-force feedback and in the constant height mode, it was established that the topographical artifacts were insignificant (Sect. 2.2) and that the field perturbation by an uncoated fiber probe was negligibly small (Sect. 2.3). Hereafter, the optical images obtained with shear-force feedback (in the constant distance mode) are shown along with the simultaneously recorded topographical images. All images are oriented so that the excited SPP propagates from the right side toward the left in the horizontal direction.

### 3.2.1   Near-Field Images

The regime of multiple SPP scattering leading to strong localization can be realized only if the scatterers are located very close to each other. The first (80-nm-thick gold) film that appeared suitable for observations of multiple SPP scattering [9,10] was evaporated in a relatively poor vacuum ($\sim 10^{-5}$ torr) on a cold substrate during a very short time interval ($\sim 0.1$ s). Topographical images of the film surface showed the typical island structure consisting of bumps with various heights (5–100 nm) and sizes (50–1000 nm) in the surface plane (Fig. 2a). Since the surface scatterers are practically adjacent to each other, it is plausible to assume that $R \sim \sigma \sim \lambda$ and, therefore, $L \gg l \sim \lambda$. In such a case, strong SPP localization leading to spatially localized SPP field enhancement should be expected. Indeed, near-field optical images exhibited well-pronounced bright spots related to spatially localized (within 150–250 nm) signal enhancement by up to 10 times (Fig. 2b). Note that the contrast of the typical near-field optical image is very high, indicating that the signal from bright spots (localized SPPs) is much stronger than the signal related to the resonantly excited SPP. The corresponding topographical image clearly shows the effect of convolution between the surface topography and the (asymmetrical) shape of a fiber tip. Actually, such an asymmetry may even facilitate faithful near-field imaging of SPP intensity distributions (Sect. 2.4.2).

Strong localization of SPPs has been also observed with artificially roughened gold films [12,13]. These films were fabricated according to the following recipe. A drop of diluted suspension containing colloidal gold particles (diameter $\approx 40\,\text{nm}$) was first deposited on the base of a glass prism and slowly dried in the air. A thin gold film was then thermally evaporated on this base. The corresponding topographical images showed a rough surface revealing a sublayer of randomly located gold particles and particle clusters (Fig. 7a). Again, the convolution effect could be seen in the elongated appearance of individual particles that revealed the asymmetrical shape of the fiber tip. The corresponding near-field optical images exhibited strongly localized and pronounced bright spots that were rather similar in appearance to those observed with the first gold film (cf. Figs. 7b and 2b). It is interesting to note that, even though the topographical images of the above films are distinctly different (cf. Figs. 2a and 7a), one can hardly distinguish these films by comparing only the optical images.

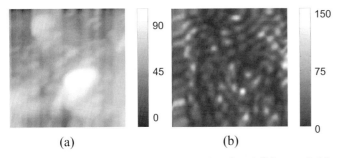

(a)                              (b)

**Fig. 7.** (a) Typical gray-scale topographical and (b) near-field optical images ($4.4 \times 4.4\,\mu\text{m}^2$) obtained with a 53-nm-thick gold film, which was evaporated on a sublayer of 40-nm-diameter gold spheres, and with the SPP resonantly excited at the light wavelength of 633 nm. Corresponding bar scales reflect (a) the depth (in nm) and (b) the detected light power (in pW)

### 3.2.2 Fourier Analysis

The SPP intensity distribution within an image area is determined by the excitation and scattering of SPPs in the *surrounding* area within the SPP propagation length. This circumstance may result in unexpected images, e.g., topographical images showing a featureless flat surface and optical images exhibiting a complicated interference pattern. It turned out that the analysis of spatial Fourier spectra of the optical images facilitates their interpretation, especially on account of the presence of localization effects [12,13,26]. Let us assume that the main contribution to the total field comes from *propagating* SPP field components, bearing in mind that the total field very close to a scatterer can be noticeably influenced by near-field components. Separating explicitly the incident and (total) scattered SPP fields and considering only the component perpendicular to the surface field (Sect. 2.4.1), one can write the following relation for the total field intensity distribution along the surface [26]:

$$I(x,y) \propto \left| E_0 \exp(i\beta_1 x) + \int_0^{2\pi} f(\varphi) \exp\left[i(x\beta_1 \cos\varphi + y\beta_1 \sin\varphi)\right] d\varphi \right|^2, \quad (9)$$

where $E_0$ is the amplitude of the excited SPP propagating along the $x$ axis and $f(\varphi)$ is determined by the amplitude $F_z(\varphi)$ of the appropriate plane wave component [cf. (3) and (9)]. The near-field signal detected is related to the field intensity in a nontrivial manner (Sect. 2.4), but the content of the spatial spectrum of the image and that of the intensity distribution should be similar.

In the regime of well-developed multiple SPP scattering leading to (strong) localization, the total scattered SPP field becomes dominant, and the intensity distribution can be approximated by the following expression [26]:

$$\begin{aligned}
I(x,y) &\propto \left| \int_0^{2\pi} f(\varphi) \exp\left[i(x\beta_1 \cos\varphi + y\beta_1 \sin\varphi)\right] d\varphi \right|^2 \\
&= \int_0^{2\pi} \int_0^{2\pi} f(\varphi) f^*(\phi) \exp\left[ix\beta_1(\cos\varphi - \cos\phi) \right. \\
&\quad \left. + iy\beta_1(\sin\varphi - \sin\phi)\right] d\varphi d\phi.
\end{aligned} \quad (10)$$

The corresponding spatial Fourier spectrum covers the whole area on the $(u,v)$ plane: $u(\varphi,\phi) = \beta_1(\cos\varphi - \cos\phi), v(\varphi,\phi) = \beta_1(\sin\varphi - \sin\phi)$. This area represents a filled circle with radius $2\beta_1$. It should be borne in mind that, for sufficiently rough surfaces (needed to ensure SPP localization), the observation point would always be close to one or several scatterers. In such a case, the spectral representation would contain the near-field components of scattered SPPs with high spatial frequencies, and the spectrum area would spread beyond the aforementioned circle.

The spatial spectra of near-field optical images that were obtained with rough gold films and exhibit subwavelength-sized bright spots represented indeed filled circles with the radius twice the SPP propagation constant [12,13,26]. Similar spectra were also found in the course of numerical simulations of the multiple SPP scattering regime [13]. The spatial spectra of the optical images presented here (Figs. 2b and 7b) represent filled circles with radius $\cong 2\beta_1$ as well (Fig. 8). Furthermore, in the logarithmic representation, these spectra extend beyond the circle due to the near-field components of scattered SPPs [12,13]. It should be noted that essentially the same result for the Fourier spectra can be obtained by relating the SPP localization to the electromagnetic energy density distribution in the form of bright spots. Indeed, a spatially localized (on the wavelength scale) intensity enhancement should result in a spatial Fourier spectrum concentrated within an area whose size is of the order of the wave number.

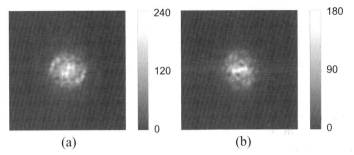

**Fig. 8.** Gray-scale representations of Fourier spectrum magnitude corresponding to the near-field optical images shown in (**a**) Fig. 2b and (**b**) Fig. 7b

### 3.2.3 Global Sensitivity

The global character of multiple scattering (Sect. 3.1) results in *global sensitivity* of the total (self-consistent) field intensity distribution, i.e., in sensitivity with respect to the global scattering geometry and incident field distribution. For strong SPP localization by surface roughness, the former means that the positions of bright spots in the optical images are not dictated by the local surface topography or, in other words, there is no correlation between the optical and topographical images [9,10]. This characteristic feature shows clearly in the images presented in Figs. 2 and 7. The correlation coefficients, which were calculated for the optical and topographical data corresponding to these images, are less than 0.01, confirming the conclusion based on visual perception. Another way of demonstrating the absence of correlation is to investigate the local topography at the positions of bright spots. It has been shown that the bright spots can be located at local maxima, minima, or slopes of the surface topography [9]. The global sensitivity with respect to the scatterers' locations was directly revealed in the experiments on local modification of surface topography. The topographic modification in a small region of the surface area resulted in drastic and global changes in the near-field optical image [13].

The sensitivity of localized SPPs with respect to the global distribution of the incident SPP field has been extensively studied in the course of experimental investigations [10,13] and also simulated numerically [13]. The phase distribution of the incident field along the surface is determined (for a given wavelength) by the angle of incidence $\theta$ (Fig. 1a). The angular sensitivity of the positions of bright spots in near-field images of SPP scattering is illustrated by the optical images recorded at the same places on the rough gold film for different angles of incidence of the excited SPP (Fig. 9). One can notice that the angular variation of $\delta\theta \sim 2°$ does not change the overall image appearance much, but it does strongly influence the brightness of spots, i.e., the degree of SPP field enhancement. Taking into account that the bright SPP spots are the result of interference of scattered SPPs, one can argue that,

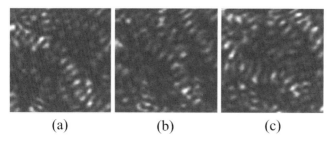

(a)                    (b)                    (c)

**Fig. 9.** Gray-scale near-field optical images ($4.4 \times 4.4 \, \mu m^2$) obtained in the surface region of the 80-nm-thick rough gold film shown in Fig. 2a and the SPP excited at the light wavelength of 633 nm at different angles of exciting beam incidence (Fig. 1a): (**a**) 51°, (**b**) 49°, and (**c**) 47°. The optical images are presented in the same scale corresponding to $\sim 5$–110 pW in the optical power detected

instead of the phase matching condition for the SPP excitation on a flat surface, a similar one exists for the excitation of a localized plasmon field on a rough surface. The observed angular width (within which a bright spot is pronounced) can be used to estimate the size $S$ of the surface area within which SPP scattering contributes to the formation of a particular bright spot. In the experiment considered [10], $n = 1.56$ and $S \sim \lambda/2\delta\theta n \cos\theta \sim 8 \, \mu m$, a value which actually corresponds to the SPP propagation length $L$ for smooth gold films.

The change in light wavelength influences both the phase distribution of the incident field along the surface and the field propagator, i.e., the electromagnetic interaction between scatterers (6). Considering only the phase distribution, one can estimate the wavelength variation $\delta\lambda$ needed to change the interference pattern as follows: $\delta\lambda \sim \lambda^2/L \sim 15 \, nm$ (for silver films and the light wavelength of 600 nm)[13]. It has been demonstrated for both silver and gold films that the near-field optical images corresponding to the resonant SPP excitation at two different wavelengths, viz., 594 and 633 nm, featured bright spots located at completely different positions (with respect to the same topography) [13]. Near-field imaging of SPP multiple scattering and localization at the surface of a gold film (evaporated on a sublayer of gold spheres) have been also carried out with a tunable Ti:Sapphire laser. The near-field optical images obtained for different light wavelengths in the same surface region show gradual redistribution of the total SPP field intensity (Fig. 10). It is seen that the wavelength variation of 10 nm is just at the limit of being sufficient to extinguish a bright spot, implying that the SPP propagation length can be evaluated as $L \sim 64 \, \mu m$. This value is a reasonable estimate for rough gold films and a light wavelength of 800 nm [5,36].

Overall, the enhancement ratio, the subwavelength sizes of the observed bright spots, the spectral content of the near-field optical images, and the global sensitivity of the positions of the bright spots (i.e., the fact that the positions of the bright spots do not correlate with the local surface topography

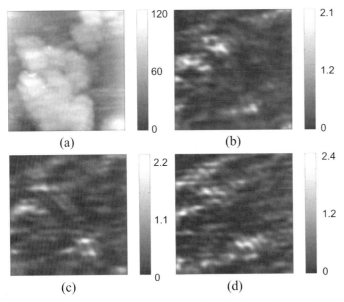

**Fig. 10.** (**a**) Gray-scale topographical and (**b, c, d**) near-field optical images (5 × 5 μm²) obtained with the 53-nm-thick gold film evaporated on a sublayer of gold particles (Fig. 7) and the SPP resonantly excited at different light wavelengths: (**b**) λ782, (**c**) 792, and (**d**) 805 nm. Corresponding bar scales reflect (**a**) the depth (in nm) and (**b, c, d**) the light power detected (in nW)

but depend on the incident angle and the wavelength of the exciting beam) provide altogether conclusive evidence of strong SPP localization caused by interference in multiple SPP scattering. Note that similar features of near-field optical images, e.g., similar modifications in the positions of bright spots with changes in the wavelength and angle of incidence, have been observed for localized dipolar excitations at fractal surface clusters of colloidal silver particles [19]. This similarity is related to the fact that both phenomena originate from interference in multiple light scattering and, moreover, become similar to each other in the limit of strongly corrugated metal surfaces (Sect. 3.1). Finally, it should be mentioned that coherent multiple SPP scattering by surface features of silver and gold films has been also observed with the help of pump–probe measurements in Kretschmann geometry [36]. These measurements revealed long exponential tails in the reflected light amplitude that were explained by using a two-dimensional diffusion model to describe the effect of multiple scattering.

## 3.3 Statistics of Surface Polariton Intensity Distributions

The topographical and near-field optical images demonstrating SPP scattering and localization by surface roughness are, in a sense, unique. Indeed, a small variation of the surface topography and/or parameters of the incident light drastically changes the self-consistent SPP field distribution and thereby the near-field optical image (Sect. 3.2.3). This circumstance makes the comparison of different samples and scattering configurations rather difficult. One can compare, of course, the sizes of bright spots and the corresponding values of signal enhancement, but it is clear that bright spots can be quite different for the same sample and scattering configuration. For example, resonant dipolar excitations in fractal surface clusters have localization lengths very different from the maximum size of a sample to the minimum roughness scale [37]. Below, the fractal surface characterization [38] is used to describe the topography of the rough gold film, and the near-field optical images corresponding to SPP localization are treated to determine the probability density function for intensity enhancement.

### 3.3.1  Fractal Surface Characterization

Characterization of randomly rough surfaces is a delicate issue. The conventional roughness characteristics, such as the root-mean-square roughness amplitude or the maximum peak-to-valley distance, though useful in some instances, fail to distinguish certain spatial frequency differences in surface profiles [39]. On the other hand, the regime of SPP scattering is very sensitive to the distances between and sizes of surface scatterers (Sect. 3.2), i.e., to the spatial frequency spectrum of the surface profile. Let us recall that surfaces with random roughness exhibit self-similarity, at least in an intermediate range of sizes, and that the light scattering by fractal structures results in localized resonant excitations [40]. Therefore, it is natural to suggest using fractal surface characterization for rough metal films that exhibit SPP localization. Note that the corresponding surface profiles are extended along the (average) surface plane but localized in the perpendicular direction. Moreover, the physical phenomenon of interest involves quasi-two-dimensional waves that propagate along the surface plane. These features indicate that the dimensions in the plane and normal to it should be scaled differently. Since the only relevant length dimensions are those of the SPP, one can use the SPP wavelength and penetration depth (in metal) to normalize the in-plane dimensions and the surface elevation, respectively [38]. Calculating the surface area $S$ as a function of the discretization step $d$ for experimentally measured surface profiles, one should obtain (for fractal structures) the dependence $S \sim d^{2-D}$, where $D$ is called the fractal (Hausdorff) dimension [41].

The corresponding dependencies calculated for four different surface regions of the rough gold film that was used to obtain the images presented in Figs. 2 and 9 are shown in Fig. 11 in double logarithmic scale [38]. The in-plane and normal dimensions were normalized with the values of 570 and 35 nm, respectively, representing the appropriate SPP characteristics at 633 nm for this film. Even though the individual dependencies $\ln[S(\ln d)]$ are

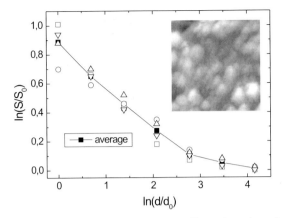

**Fig. 11.** Normalized surface area $S$ as a function of the discretization step $d$ for four different regions (*open symbols*) of the 80-nm-thick rough gold film and the average dependence (*filled squares connected with a solid line*). $S_0$ is the corresponding flat surface area and $d_0 = 40\,\text{nm}$ is the minimum discretization step. An inset shows a typical topographical image of $4 \times 4\,\mu\text{m}^2$

somewhat different (due to the limited number of points available), the average dependence clearly exhibits linear behavior in the range of sizes from 80–640 nm, and the slope corresponds to the (normalized) fractal dimension $D \approx 2.26$. The bend in the dependence for large values of $d$ is evidently related to the absence of micrometer-sized surface features on the sample surface (see inset in Fig. 11). It should be mentioned that similar logarithmic dependencies calculated for relatively smooth gold and silver films (that did not result in multiple SPP scattering) were essentially nonlinear in the whole range of sizes [38].

The circumstance that the film surface, at which SPPs were found localized, has a fractal structure allows one to consider the SPP localization from another point of view. In general, propagating waves appear in media that exhibit translational invariance, e.g., in homogeneous or periodic media. The SPP localization can be accounted for by the fact that the fractal surface structures do not possess lateral translational invariance and, therefore, propagating surface waves cannot be transmitted on it. Actually, the rough film considered here should be regarded as a self-affine film, i.e., an object whose scaling properties in one of the canonical directions is different from those in the other two. Self-affine films are often characterized by using the height–height correlation function [18,42] or the Hurst exponent [43]. However, the method described above seems especially suitable for characterizing metal film surfaces, whose roughness leads to multiple SPP scattering and eventually localization, because of the usage of SPP characteristics for normalization of coordinates in different directions.

### 3.3.2  Probability Distribution of Intensity Enhancement

The total SPP field intensity distribution established in the process of multiple scattering features regions with very different intensity levels, from bright spots representing large field intensities to rather weak background level. One can employ a statistical treatment of near-field optical images to characterize SPP intensity distribution and establish common features in the optical images obtained at different places and under different conditions. For example, the statistics of local field intensity distribution in fractal clusters has been theoretically analyzed, and the probability distribution function for intensity enhancement, it was found, exhibits a power-law dependence, which is a fingerprint of the scale invariance of the system [44]. In self-affine surfaces, the scale invariance is different for the in-plane and normal directions, a circumstance that can influence the statistics of the intensity distribution. Indeed, numerical simulations based on a rigorous integral formulation of wave scattering showed that rough metal surfaces possessing self-affine scaling behavior result in an exponential decrease of the probability density function (PDF) [43]. The PDF was also found very dependent on the fractality of the surface. However, the model geometry used in the calculations differs from the experimental configuration considered here in two significant instances. It is two-dimensional and contains only one (air–metal) interface implying that SPPs can be excited only via the incident field scattering by surface roughness [43]. These differences should be borne in mind when comparing theoretical and experimental results.

Let us assume that the detected signal is proportional to the intensity of the total SPP field, which is composed of resonantly excited and scattered SPP fields (Sect. 2.4). One of the main problems in treating experimental near-field optical images is to determine the reference signal level to be used for calculating the intensity enhancement. In the regime of strong multiple scattering leading eventually to strong SPP localization, scattered SPPs become dominant. Their constructive interference results in bright spots enhanced up to 10 times compared to the average level. In surface regions with destructive interference of scattered SPPs, one can expect to detect mainly an incident SPP field. Taking into account that the detected signal can be zero because of destructive interference of all SPP fields (and/or due to noise), one can argue that the enhancement should be related to the signal averaged within the surface regions showing the *weakest* signals. The size of such a region should be larger than the spatial resolution (to average noise) but smaller than the half of the SPP wavelength (to stay within the region of destructive interference). The procedure described is somewhat cumbersome but when used with care, allows one to obtain a more realistic estimate of the intensity enhancement than by using the signal averaged over the whole image area.

The corresponding PDFs, that were calculated by using the near-field optical images obtained with two different gold films are shown in Fig. 12. It

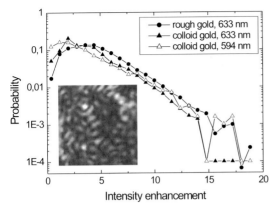

**Fig. 12.** Probability density function of the intensity enhancement calculated from the near-field optical images obtained for the 80-nm-thick rough gold film at the light wavelength of 633 nm (*filled circles*) and for the 53-nm-thick gold film, evaporated on a sublayer of colloidal gold particles, at the light wavelength of 633 nm (*filled triangles*) and 594 nm (*open triangles*). An inset shows a typical near-field optical $4 \times 4\,\mu m^2$ image

is seen that these PDFs exhibit exponential decay (over two decades) and resemble rather well the theoretical PDFs simulated for rough surfaces with large fractal dimensions [43]. Deviations from the exponential behavior appear only for large values of enhancement probably due to the limited amount of data available. It is interesting that the PDF slopes (in the logarithmic scale) for different samples and wavelengths are very similar to each other. This circumstance is yet to be elucidated because the PDF slope is influenced mainly by the value of intensity enhancement whose determination procedure used here is somewhat ambiguous. One should try to employ a direct method, e.g., by measuring the signal with the SPP excited along the flat surface of reference film. The most important result obtained is the demonstration of the *exponential* behavior of the PDF in strong SPP localization. At the same time, the PDFs, that were calculated for the images obtained from smooth silver and gold films [12,13] are completely different and distributed symmetrically around the average intensity level (Fig. 13). These near-field images are dominated by features corresponding to the regimes of single and double scattering [12,13]. Note that the PDFs shown in Fig. 13 are quite similar to the theoretical PDFs simulated for weakly rough surfaces with small fractal dimensions [43]. Overall, one can conclude that the exponential PDF is a fingerprint of the regime of well-developed multiple SPP scattering resulting eventually in strong SPP localization. Further investigations are needed to quantify the PDF slope and to link the slope value to the parameters of scattering configuration.

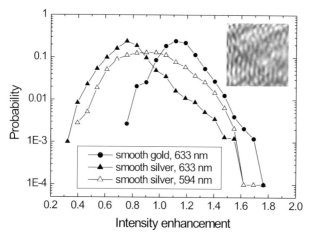

**Fig. 13.** Probability density function of the intensity enhancement calculated from the near-field optical images obtained with smooth gold (*filled circles*) and silver films (*open and filled triangles*). The *inset* shows a typical near-field optical image $4 \times 4\,\mu\mathrm{m}^2$

# 4  Conclusions

In this contribution, elastic (in-plane) SPP scattering by surface roughness was considered and the main experimental results concerning SPP localization were reviewed. Near-field imaging of SPP fields by using the collection SNOM configuration was discussed, and the general approach to image formation was developed. It was established that, in the regime of strong elastic SPP scattering and for an asymmetrical probe tip, near-field optical images approach the total SPP intensity distribution along the surface plane. The regimes of single and multiple scattering were qualitatively compared with an emphasis on the differences in the total field distributions and their sensitivity with respect to variations of the system parameters. The near-field images, which exhibit spatially localized SPP intensity enhancement (bright spots) on rough metal surfaces, were presented and used to illustrate the main features of the phenomenon of strong SPP localization: the appearance (size, enhancement ratio) of bright spots, the spectral content of the optical images, the global sensitivity of the positions of the bright spots, and the exponential behavior of the probability distribution of the intensity enhancement.

The phenomenon of strong SPP localization first attracted attention because of its implications for surface-enhanced optical phenomena, e.g., SERS [6]. Indeed, the intensity enhancement due to localized SPPs was directly demonstrated in various experiments reviewed here. One can also point out that SPP localization is readily accessible to SNOM techniques, i.e., to mapping of the SPP intensity distribution. The latter feature can be further exploited for detailed investigations of various fascinating phenomena related to multiple light scattering, e.g., laser action in disordered gain medium [45] and wavelength/angle-selective memory by interference of scattered light [46]. The first observations of these phenomena in three-dimensional media were reported recently, but one can foresee certain advantages in their realization by using surface electromagnetic waves, e.g., SPPs. It is worth mentioning that field enhancement at the metal surface due to the resonantly excited SPP reaches two orders of magnitude [5]. Finally, multiple SPP scattering in periodic structures can be used to realize an SPP band gap [47] and, consequently, to manipulate confined (in the surface plane) SPP beams in essentially the same manner as suggested for conventional photonic band gap structures [33]. Taking into account advances in SPP micro-optics by specially configured arrays of microscatterers [13,48], I believe that the successful implementation of channel waveguide structures based on the SPP band gap effect might be just around the corner.

## Acknowledgments

I am very thankful to my co-workers V. Coello, B. Vohnsen, A. V. Zayats, and I. I. Smolyaninov, who shared with me the excitement of watching the bright spots appearing on the monitor of our SNOM setup and finding out other salient features of strong SPP localization. I am also very grateful to the foundation "San Cataldos Venner" that granted my stay in its quiet residence located in the beautiful surroundings of southern Italy where this contribution was largely prepared.

# References

1. S. John, The localization of waves in disordered media, in *Scattering and Localization of Classical Waves in Random Media*, P. Sheng (Ed.) (World Scientific, Singapore 1990) pp. 1–96
2. M. P. van Albada, M. B. van der Mark, A. Lagendijk, Experiments on weak localization of light and their interpretation, in *Scattering and Localization of Classical Waves in Random Media*, P. Sheng (Ed.) (World Scientific, Singapore 1990) pp. 97–136
3. D. S. Wiersma, P. Bartolini, A. Lagendijk, R. Righini, Localization of light in a disordered medium, Nature **390**, 671 (1997)
4. H. Cao, J. Y. Xu, D. Z. Zhang, S.-H. Chang, S. T. Ho, E. W. Seelig, X. Liu, R. P. H. Chang, Spatial confinement of laser light in active random media, Phys. Rev. Lett. **84**, 5584 (2000)
5. H. Raether, *Surface Plasmons*, Springer Tracts Mod. Phys. **111** (Springer, Berlin, Heidelberg 1988)
6. K. Arya, Z. B. Su, J. L. Birman, Localization of the surface plasmon polariton caused by random roughness and its role in surface-enhanced optical phenomena, Phys. Rev. Lett. **54**, 1559 (1985)
7. J. A. Sánchez-Gil, A. A. Maradudin, Competition between Anderson localization and leakage of surface-plasmon polaritons on randomly rough periodic metal surfaces, Phys. Rev. B **56**, 1103 (1997)

356     Sergey I. Bozhevolnyi

8. M. Rusek, A. Orlowski, J. Mostowski, Localization of light in three-dimensional random dielectric media, Phys. Rev. E **53**, 4122 (1996)

9. S. I. Bozhevolnyi, I. I. Smolyaninov, A. V. Zayats, Near-field microscopy of surface-plasmon polaritons: Localization and internal interface imaging, Phys. Rev. B **51**, 17916 (1995)

10. S. I. Bozhevolnyi, B. Vohnsen, I. I. Smolyaninov, A. V. Zayats, Direct observation of surface polariton localization caused by surface roughness, Opt. Commun. **117**, 417 (1995)

11. R. K. Chang, T. P. Furtak (Eds.), *Surface-Enhanced Raman Scattering* (Plenum, New York 1982)

12. V. Coello, S. I. Bozhevolnyi, F. A. Pudonin, Imaging of surface plasmons with a near-field microscope, in *Optical Inspection and Micromeasurements II*, C. Gorecki (Ed.), SPIE Proc. **3098**, 536 (1997)

13. S. I. Bozhevolnyi, V. Coello, Elastic scattering of surface plasmon polaritons: modeling and experiment, Phys. Rev. B **58**, 10899 (1998)

14. M. Arnold, A. Otto, Notes on localization of surface-plasmon-polaritons, Opt. Commun. **125**, 122 (1996)

15. S. I. Bozhevolnyi, Localization phenomena in elastic surface plasmon polariton scattering caused by surface roughness, Phys. Rev. B **54**, 8177 (1996)

16. B. Hecht, H. Bielefeldt, Y. Inouye, D. W. Pohl, L. Novotny, Facts and artifacts in near-field optical microscopy, J. Appl. Phys. **81**, 2492 (1997); R. Carminati, A. Madrazo, M. Nieto-Vesperinas, J.-J. Greffet, Optical content and resolution of near-field optical images: influence of the operating mode, J. Appl. Phys. **82**, 501 (1997); C. Girard, D. Courjon, The role of scanning mode in near-field optical microscopy, Surf. Sci. **382**, 9 (1997)

17. S. I. Bozhevolnyi, Topographical artifacts and optical resolution in near-field optical microscopy, J. Opt. Soc. Am. B **14**, 2254 (1997)

18. S. I. Bozhevolnyi, Near-field optical microscopy of localized excitations on rough surfaces: influence of a probe, J. Microsc. **194**, 561 (1999)

19. P. Zhang, T. L. Haslett, C. Douketis, M. Moskovits, Mode localization in self-affine fractal interfaces observed by near-field microscopy, Phys. Rev. B **57**, 15513 (1998); S. I. Bozhevolnyi, V. A. Markel, V. Coello, W. Kim, V. M. Shalaev, Direct observation of localized dipolar excitations on rough nanostructured surfaces, Phys. Rev. B **58**, 11441 (1998); V. A. Markel, V. M. Shalaev, P. Zhang, W. Huynh, L. Tay, T. L. Haslett, M. Moskovits, Near-field optical spectroscopy of individual surface-plasmon modes in colloid clusters, Phys. Rev. B **59**, 10903 (1999)

20. S. Grésillon, L. Aigouy, A. C. Boccara, J. C. Rivoal, X. Quelin, C. Desmarest, P. Gadenne, V. A. Shubin, A. K. Sarychev, V. M. Shalaev, Experimental observation of localized optical excitations in random metal-dielectric films, Phys. Rev. Lett. **82**, 4520 (1999)

21. D. Van Labeke, D. Barchiesi, Probes for scanning tunneling optical microscopy: a theoretical comparison, J. Opt. Soc. Am. A **10**, 2193 (1993)

22. R. Carminati, J.-J. Greffet, Two-dimensional numerical simulation of the photon scanning tunneling microscope: concept of transfer function, Opt. Commun. **116**, 316 (1995)

23. J.-J. Greffet, R. Carminati, Image formation in near-field optics, Prog. Surf. Sci. **56**, 133 (1997)

24. S. I. Bozhevolnyi, B. Vohnsen, E. A. Bozhevolnaya, Transfer functions in collection scanning near-field optical microscopy, Opt. Commun. **172**, 171 (1999)

25. T. Okoshi, *Optical Fibers* (Academic, New York 1982)
26. S. I. Bozhevolnyi, Direct observation of strong localization of quasi-two-dimensional light waves, Ann. Phys. **8**, 717 (1999)
27. F. Pincemin, A. A. Maradudin, A. D. Boardman, J.-J. Greffet, Scattering of a surface plasmon polariton by a surface defect, Phys. Rev. B **50**, 15261 (1994)
28. A. V. Shchegrov, I. V. Novikov, A. A. Maradudin, Scattering of surface plasmon polaritons by a circularly symmetric surface defect, Phys. Rev. Lett. **78**, 4269 (1997)
29. L. Novotny, B. Hecht, D. W. Pohl, Interference of locally excited surface plasmons, J. Appl. Phys. **81**, 1798 (1997)
30. O. Keller, M. Xiao, S. Bozhevolnyi, Configurational resonances in optical near-field microscopy: a rigorous point-dipole approach, Surf. Sci. **280**, 217 (1993)
31. H. R. Stuart, D. G. Hall, Enhanced dipole–dipole interaction between elementary radiators near a surface, Phys. Rev. Lett. **80**, 5663 (1998)
32. V. M. Shalaev, Electromagnetic properties of small-particle composites, Phys. Rep. **272**, 61 (1996)
33. J. D. Joannopoulos, R. D. Meade, J. N. Winn, *Photonic Crystals* (Princeton University Press, Princeton, NJ 1995)
34. E. D. Palik, *Handbook of Optical Constants of Solids* (Academic, New York 1985)
35. S. I. Bozhevolnyi, A. V. Zayats, B. Vohnsen, Weak localization of surface plasmon polaritons: direct observation with photon scanning tunneling microscope, in *Optics at the Nanometer Scale*, M. Nieto-Vesperinas, N. Garcia (Eds.) (Kluwer, Dordrecht 1996) pp. 163–173
36. W. Wang, M. J. Feldstein, N. F. Scherer, Observation of coherent multiple scattering of surface plasmon polaritons on Ag and Au surfaces, Chem. Phys. Lett. **262**, 573 (1996)
37. M. I. Stockman, L. N. Pandey, T. F. George, Inhomogeneous localization of polar eigenmodes in fractals, Phys. Rev. B **53**, 2183 (1996)
38. S. I. Bozhevolnyi, B. Vohnsen, A. V. Zayats, I. I. Smolyaninov, Fractal surface characterization: implications for plasmon polariton scattering, Surf. Sci. **356**, 268 (1996)
39. L. Spanos, E. A. Irene, Investigation of roughened silicon surfaces using fractal analysis, J. Vac. Sci. Technol. A **12**, 2646 (1994)
40. V. A. Markel, L. S. Muratov, M. I. Stockman, T. F. George, Theory and numerical simulation of optical properties of fractal clusters, Phys. Rev. B **43**, 8183 (1991)
41. K. J. Falconer, *Fractal Geometry: Mathematical Foundations and Applications* (Wiley, Chichester 1990)
42. V. M. Shalaev, R. Botet, J. Mercer, E. B. Stechel, Optical properties of self-affine thin films, Phys. Rev. B **54**, 8235 (1996)
43. J. A. Sánchez-Gil, J. V. García-Ramos, Calculations of the direct electromagnetic enhancement in surface enhanced Raman scattering on random self-affine fractal metal surfaces, J. Chem. Phys. **108**, 317 (1998)
44. M. I. Stockman, L. N. Pandey, L. S. Muratov, T. F. George, Giant fluctuations of local optical fields in fractal clusters, Phys. Rev. Lett. **72**, 2486 (1994)
45. H. Cao, Y. G. Zhao, S. T. Ho, E. W. Seelig, Q. H. Wang, R. P. H. Chang, Random laser action in semiconductor powder, Phys. Rev. Lett. **82**, 2278 (1999)

46. A. Kurita, Y. Kanematsu, M. Watanabe, K. Hirata, T. Kushida, Wavelength- and angle-selective optical memory effect by interference of multiple-scattered light, Phys. Rev. Lett. **83**, 1582 (1999)
47. S. C. Kitson, W. L. Barnes, J. R. Sambles, Full photonic band gap for surface modes in the visible, Phys. Rev. Lett. **77**, 2670 (1996)
48. S. I. Bozhevolnyi, F. A. Pudonin, Two-dimensional micro-optics of surface plasmons, Phys. Rev. Lett. **78**, 2823 (1997)

# Multiple-Scattering Phenomena in the Second-Harmonic Generation of Light Reflected from and Transmitted Through Randomly Rough Metal Surfaces

Tamara A. Leskova[1], Alexei A. Maradudin[2], and Eugenio R. Méndez[2]

[1] Institute of Spectroscopy, Russian Academy of Sciences
Troitsk, 142092, Russia
[2] Department of Physics and Astronomy, and Institute for Surface and Interface Science, University of California
Irvine, CA 92697, USA
aamaradu@uci.edu
[3] División de Física Aplicada, Centro de Investigación Científica y de Educación Superior de Ensenada
Apartado Postal 2732, Ensenada, Baja California 22800, México

**Abstract.** Theories of multiple-scattering effects in the second-harmonic generation of light reflected from clean randomly rough metal surfaces and reflected from and transmitted through randomly rough metal surfaces in Kretschmann attenuated total reflection geometry are outlined. Both weakly rough and strongly rough surfaces are considered, the former by perturbative approaches, the latter by numerical simulations. Comparisons of theoretical results with experimental data for second-harmonic generation on clean random metal surfaces are presented.

## 1 Introduction

The interaction of randomness and nonlinearity is of interest in many branches of physics [1]. In this contribution, we describe one manifestation of this kind of interaction in optics, namely, multiple-scattering effects in the second-harmonic generation (SHG) of light reflected from and transmitted through randomly rough surfaces.

Experimental and theoretical studies of second-harmonic generation of light reflected from a planar metal surface go back at least three decades to the first experimental observation of the effect [2] and the first theoretical description of it [3]. At this early stage of theoretical studies [3], it was already established that this phenomenon is a surface effect. Consequently, it is very sensitive to modifications of the surface.

In the past several years, interest in second-harmonic generation on metal surfaces has been directed at generating it on random metal surfaces rather than on planar surfaces. This is due to the growing interest in the broader area of interference effects occurring in the multiple scattering of electromagnetic waves from, and their transmission through, randomly rough metal

V. M. Shalaev (Ed.): Optical Properties of Nanostructured Random Media,
Topics Appl. Phys. **82**, 359–442 (2002)
© Springer-Verlag Berlin Heidelberg 2002

surfaces and the related enhanced backscattering [5] and enhanced transmission [6] phenomena. It has been expected that nonlinear optical interactions on a randomly rough metal surface should also give rise to new features due to the interference of multiply-scattered electromagnetic waves.

Indeed, in the first theoretical study of the phenomenon, *McGurn* et al. [7] predicted on the basis of a perturbative calculation that enhanced second-harmonic generation of light on a weakly rough clean metal surface should occur not only in the retroreflection direction but also in the direction normal to the mean scattering surface. The peak in the direction normal to the mean surface was attributed to interference effects in the multiple scattering of surface plasmon polaritons of the fundamental frequency, excited by the incident light through the roughness of the surface, and the occurrence of the peak in the retroreflection direction was attributed to interference effects in the multiple scattering of surface plasmon polaritons of the harmonic frequency.

These predictions stimulated several experimental studies of second-harmonic generation in the scattering of light from random metal surfaces [8,9,10,11,12,13], and enhanced second-harmonic generation peaks in the direction normal to the mean surface and in the retroreflection direction were observed [8,9,10,11,12,13]. In these experiments, however, the scattering system was not a clean random interface between vacuum and a semi-infinite metal. To amplify the second-harmonic signal, the Kretschmann attenuated total reflection (ATR) geometry [14] was used. In this geometry, a thin metal film is deposited on the planar base of a prism through which the film is illuminated by p-polarized light. The back surface of the metal film is randomly rough. In the experiments of [8,9,11], the metal film was silver, and its randomly rough surface was in contact with a nonlinear quartz crystal, so that the nonlinear interaction occurred in the quartz crystal rather than on the significantly more weakly nonlinear silver surfaces. A well-defined peak of the second-harmonic generation in the direction normal to the interface was observed in transmission in [8]. When the experiment was carried out with long-range surface plasmon polaritons [15], peaks of the enhanced second-harmonic generation were detected in both the retroreflection direction [9] and in the direction normal to the mean surface in transmission [11]. In [10,12,13], attempts to detect the peaks of the enhanced second-harmonic generation at a silver film–vacuum interface were made. A well-defined peak in the direction normal to the mean surface was observed in transmission in [10] and [13], whereas only a broad depolarized background, but no peak in the direction normal to the mean surface, was observed in transmission in [12]. The peak of the intensity of the generated light in the direction normal to the mean surface observed in [8,10,11,13] was attributed to coherent nonlinear mixing of multiply-scattered contrapropagating surface plasmon polaritons. It is well known [16] that the symmetry of the surface nonlinear polarization of a clean planar metal surface forbids second-harmonic generation due to contrapropagating beams of surface plasmon polaritons. In [10] and [13], it was argued that the surface roughness breaks the symmetry and, as a result makes second-harmonic generation by contrapropagating beams of surface plasmon polaritons possible.

Recently, the first experimental studies of multiple scattering effects in the second-harmonic generation of light scattered from a clean one-dimensional random vacuum–metal interface were reported in a series of papers by O'Donnell and his colleagues [17,18,19]. It was found that for both weakly [17] and strongly [19] rough silver surfaces, a dip is present in the retroreflection direction in the angular dependence of the intensity of the second-harmonic light rather than the peak that occurs in scattering at the fundamental frequency. This result was in agreement with the results of rigorous numerical simulations of second-harmonic generation from such surfaces carried out by Leyva-Lucero et al. [20]. However, no peak or dip in the direction normal to the mean surface was observed in the experiments of [17,18,9].

In this contribution, we outline theories of multiple-scattering effects in the second-harmonic generation of light reflected from clean randomly rough metal surfaces and reflected from and transmitted through randomly rough metal surfaces in the Kretschmann ATR geometry. Both weakly rough and strongly rough surfaces are considered. However, only one-dimensional random surfaces will be treated because it is only for such surfaces that multiple-scattering effects can be calculated readily at the present time. The extension of the present work to two-dimensional random surfaces is a goal for the future. Comparisons of the theoretical results obtained with experimental data for second-harmonic generation on clean one-dimensional random metal surfaces will be made.

This review is organized as follows. The manner in which we characterize the random surfaces considered in this work is described in Sect. 2. In Sect. 3, theories of second-harmonic generation in scattering from clean, one-dimensional, randomly rough surfaces will be presented. Both weakly rough and strongly rough surfaces will be considered, and the mechanisms responsible for the features present in the retroreflection direction and in the direction normal to the mean surface, which are different for the two types of surfaces, will be discussed. Theories of second-harmonic generation in the reflection of light from, and in its transmission through, a thin metal film in the Kretschmann ATR geometry will be presented in Sect. 4, again for weakly rough and strongly rough one-dimensional random metal–vacuum interfaces. The conclusions drawn from these theoretical studies will be presented in Sect. 5.

## 2    Characterization of a Random Surface

All of the one-dimensional random surfaces with which we will be concerned here are defined by the equation $x_3 = \zeta(x_1)$. The surface profile function $\zeta(x_1)$ is assumed to be a single-valued function of $x_1$ that is at least twice differentiable and constitutes a stationary, zero-mean, Gaussian random process defined by

$$\langle \zeta(x_1)\zeta(x_1') \rangle = \delta^2 \, W(|x_1 - x_1'|). \tag{1}$$

In (1), the angle brackets denote an average over the ensemble of realizations of the surface profile function, and $\delta = \langle \zeta^2(x_1) \rangle^{\frac{1}{2}}$ is the rms height of the surface. In the Fourier integral representation of $\zeta(x_1)$,

$$\zeta(x_1) = \int_{-\infty}^{\infty} \frac{dk}{2\pi} \hat{\zeta}(k) \exp(ikx_1), \tag{2}$$

the Fourier coefficient $\hat{\zeta}(k)$ is also a zero-mean, Gaussian random process that is defined by

$$\langle \hat{\zeta}(k)\hat{\zeta}(k') \rangle = 2\pi\delta(k + k')\delta^2 g(|k|). \tag{3}$$

The function $g(|k|)$ in (3) is the power spectrum of the surface roughness and is the Fourier transform of the surface height autocorrelation function $W(|x_1|)$:

$$g(|k|) = \int_{-\infty}^{\infty} dx_1 W(|x_1|) \exp(-ikx_1). \tag{4}$$

In the numerical calculations whose results will be presented in this contribution, two forms of the power spectrum will be used. The first is the Gaussian form

$$g(|k|) = \sqrt{\pi}a \exp(-a^2 k^2/4). \tag{5}$$

The characteristic length $a$ in this expression is the transverse correlation length of the surface roughness. The second form of the power spectrum to be used is [17]

$$\begin{aligned}
g(|k|) = {} & \frac{\pi h_1}{k_{\max}^{(1)} - k_{\min}^{(1)}} \\
& \times \left[ \theta(k - k_{\min}^{(1)})\theta(k_{\max}^{(1)} - k) + \theta(-k - k_{\min}^{(1)})\theta(k_{\max}^{(1)} + k) \right] \\
& + \frac{\pi h_2}{k_{\max}^{(2)} - k_{\min}^{(2)}} \\
& \times \left[ \theta(k - k_{\min}^{(2)})\theta(k_{\max}^{(2)} - k) + \theta(-k - k_{\min}^{(2)})\theta(k_{\max}^{(2)} + k) \right],
\end{aligned} \tag{6}$$

where $\theta(z)$ is the Heaviside unit step function and $h_1 + h_2 = 1$. The manner in which $k_{\min}^{(1)}$, $k_{\max}^{(1)}$, $k_{\min}^{(2)}$, $k_{\max}^{(2)}$ are chosen will be described at the appropriate points in the text.

## 3    Clean Metal Surfaces

As was noted in the Introduction, the only experimental studies of second-harmonic generation in the reflection of light from one-dimensional random metal surfaces have been carried out on clean metal surfaces [17,18,9]. It is therefore appropriate to begin our discussion of multiple-scattering phenomena in the second-harmonic generation of light reflected from a random metal surface by considering just this case.

Thus, the scattering system we consider in this section is depicted in Fig. 1. It consists of vacuum in the region $x_3 > \zeta(x_1)$ (region I), and a metal characterized by an isotropic, frequency-dependent, complex dielectric function $\epsilon(\omega)$ in the region $x_3 < \zeta(x_1)$ (region II). It is illuminated from the vacuum by p-polarized light of frequency $\omega$, whose plane of incidence is the $x_1 x_3$ plane. In this case, it is convenient to work with the single nonzero component of the magnetic field in this system, $H_2(x_1, x_3|\omega)$.

In the absence of the surface roughness $[\zeta(x_1) \equiv 0]$, the system depicted in Fig. 1 supports a single surface electromagnetic wave — a surface plasmon polariton — whose dispersion relation is

$$\epsilon(\omega)\beta_0(q, \omega) + \beta(q, \omega) = 0. \tag{7}$$

In (31) $\beta_0(q, \omega) = [q^2 - (\omega/c)^2]^{\frac{1}{2}}$, with $\operatorname{Re}\beta_0(q, \omega) > 0$, $\operatorname{Im}\beta_0(q, \omega) < 0$, and $\beta(q, w) = [q^2 - \epsilon(\omega)(\omega/c)^2]^{\frac{1}{2}}$, with $\operatorname{Re}\beta(q, \omega) > 0$ and $\operatorname{Im}\beta(q, \omega) < 0$, are the inverse decay lengths of the electromagnetic field of the surface wave in the vacuum and the metal, respectively. The solutions of (7) for the wave numbers of the surface plasmon polaritons of frequency $\omega$ are $q = \pm k_{\mathrm{sp}}(\omega)$,

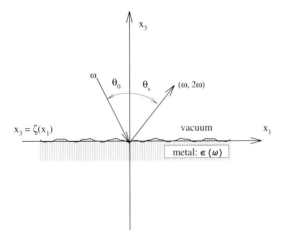

**Fig. 1.** The geometry assumed in scattering from a clean metal surface

where

$$k_{\mathrm{sp}}(\omega) = \frac{\omega}{c} \left[ \frac{\epsilon(\omega)}{\epsilon(\omega)+1} \right]^{\frac{1}{2}} \tag{8}$$

is real when $\epsilon(\omega)$ is real, and is complex when $\epsilon(\omega)$ is complex, with $\mathrm{Re}\, k_{\mathrm{sp}}(\omega) > 0$, $\mathrm{Im}\, k_{\mathrm{sp}}(\omega) > 0$. A plot of the dispersion curve obtained from the dispersion relation (7) is presented in Fig. 2 for the case that $\epsilon(\omega)$ has the simple free-electron form $\epsilon(\omega) = 1 - (\omega_{\mathrm{p}}/\omega)^2$, where $\omega_{\mathrm{p}}$ is the plasma frequency of the conduction electrons in the metal. This curve saturates at the frequency $\omega = \omega_{\mathrm{p}}/\sqrt{2}$ in the limit as $q \to \infty$. In the discussion that follows, we will assume that the frequency $\omega$ of the incident light and the frequency $2\omega$ of the second-harmonic light are both smaller than $\omega_{\mathrm{p}}/\sqrt{2}$, so that a surface plasmon polariton exists at both the fundamental and harmonic frequencies. The existence of these surface waves will be seen as important for the formation of the multiple-scattering effects in the generation of second-harmonic light reflected from a weakly rough random metal surface that are of interest here.

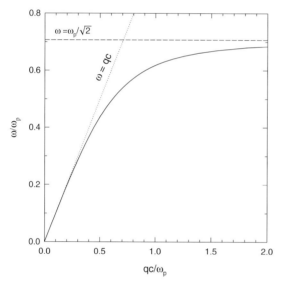

**Fig. 2.** The dispersion curve for surface plasmon polaritons supported by a planar metal surface, characterized by the dielectric function $\epsilon(\omega) = 1 - \omega_{\mathrm{p}}^2/\omega^2$, in contact with vacuum

## 3.1   Linear and Nonlinear Boundary Conditions

The Maxwell equations satisfied by the electromagnetic fields in regions I and II have to be supplemented by boundary conditions at the interface $x_3 =$

$\zeta(x_1)$. At the fundamental frequency $\omega$, these boundary conditions express the continuity of the tangential components of the magnetic and electric fields across the interface and can be written in the forms

$$H_2^{(I)}(x_1, x_3|\omega)\Big|_{x_3=\zeta(x_1)} = H_2^{(II)}(x_1, x_3|\omega)\Big|_{x_3=\zeta(x_1)} \tag{9}$$

$$\frac{\partial}{\partial N} H_2^{(I)}(x_1, x_3|\omega)\Big|_{x_3=\zeta(x_1)} = \frac{1}{\epsilon(\omega)} \frac{\partial}{\partial N} H_2^{(II)}(x_1, x_3|\omega)\Big|_{x_3=\zeta(x_1)}, \tag{10}$$

where

$$\frac{\partial}{\partial N} = -\zeta'(x_1)\frac{\partial}{\partial x_1} + \frac{\partial}{\partial x_3} \tag{11}$$

is the derivative along the normal to the interface $x_3 = \zeta(x_1)$ at each point, directed into the region (I) above the interface.

Turning now to the boundary conditions at the harmonic frequency, we note that homogeneous and isotropic metals possess inversion symmetry. Therefore, the dipolar contribution to the bulk nonlinear polarization, $\chi^{(2)}$, vanishes. The presence of a surface breaks the inversion symmetry, and because both the electromagnetic fields and material parameters vary rapidly at the surface, their gradients give rise to the optical nonlinearity of a metal surface. Therefore the second-harmonic radiation we are interested in is generated in a vacuum–metal interface layer that has finite thickness on a microscopic scale. Consequently, the resulting nonlinear polarization can be taken into account through the boundary conditions for the second-harmonic fields. In what follows, we will neglect the small contribution to the nonlinearity coming from the bulk of the metal and the possible anisotropy of the material constants.

The general form of the nonlinear polarization in a centrosymmetric metal is [22]

$$\mathbf{P}^{NL}(x_1, x_3|2\omega) = \frac{1}{4\pi}[\alpha \mathbf{E}(\nabla \cdot \mathbf{E}) + \beta(\mathbf{E} \cdot \nabla)\mathbf{E}$$

$$+ \gamma \mathbf{E} \times (\nabla \times \mathbf{E}) + \mathbf{E}(\mathbf{E} \cdot \nabla\rho) + \mathbf{E}^2 \nabla\kappa], \tag{12}$$

where $\mathbf{E}$ is the macroscopic electric field and the coefficients in general depend on the distance from the vacuum–metal interface. Their values are to be determined either from experiments or microscopic calculations. Relations among the coefficients can be obtained through the use of the energy conservation law, and they are $\alpha = -\beta = -\gamma$ and $\nabla\rho = \nabla(\alpha + 2k)$ [22]. If both $\alpha$ and $\kappa$ are independent of the distance from the interface, the nonlinear polarization reduces to the form introduced by *Bloembergen* [4]. If we assume that $\rho = \kappa = 0$, and $\alpha = e/(8\pi m\omega^2)$, $\gamma = \beta = e^3 n_0(z)/(4m^2\omega^4)$, where e is the magnitude of the electron charge, $m$ is the electron mass, and $n_0(z)$ is the electron number density as a function of the local coordinate normal to the interface, the nonlinear polarization (12) reduces to that of the free-electron model [23,24,25]. The form of the nonlinear polarization given by (12) together with the relations $\alpha = -\beta = -\gamma$ and $\nabla\rho = \nabla(\alpha + 2\kappa)$ satisfies energy conservation at the vacuum–metal interface; a nonphysical energy source at the interface equal to $\dot{E}^2(\mathbf{E}\cdot\nabla\alpha)$ arises in the model of *Bloembergen* et al. [4], where $\dot{\mathbf{E}}$ is the time derivative of the electric field.

Both the material constants and electromagnetic fields vary strongly across the interface. However, outside the interfacial layer, the fields are finite. This implies that in spite of the singular behavior of the nonlinear polarization, the tangential components of the electric and magnetic fields and the normal component of the displacement vector must be finite. The latter in turn implies singular behavior of the normal component of the electric field across the interfacial layer.

The nonlinear boundary conditions for the harmonic fields are obtained by integrating the Maxwell equations for them across the interfacial layer and then passing to the limit of a vanishing thickness of the layer. In carrying out this calculation, it is convenient to introduce a local coordinate system $(x, y, z)$ with its origin at each point of the surface $x_3 = \zeta(x_1)$ that is defined by the unit vectors $\{\hat{\mathbf{x}}, \hat{\mathbf{y}}, \hat{\mathbf{z}}\}$, where $\hat{\mathbf{x}} = [1, 0, \zeta'(x_1)]/\phi(x_1), \hat{\mathbf{y}} = \hat{\mathbf{x}}_2$, $\hat{\mathbf{z}} = [-\zeta'(x_1), 0, 1]/\phi(x_1)$, and $\phi(x_1) = \{1 + [\zeta'(x_1)]^2\}^{\frac{1}{2}}$. $\hat{\mathbf{x}}$ and $\hat{\mathbf{z}}$ are unit vectors tangent and normal to the interface in the plane perpendicular to its generators. In this coordinate system, all material parameters $\epsilon(\omega)$, $\epsilon(2\omega)$, $\alpha$, $\beta$, $\gamma$, $\rho$, and $\kappa$ depend only on the distance from the interface along the normal to it, $z$. On integrating the tangential component of the equation $\nabla \times \mathbf{H} = -(2i\omega/c)\mathbf{D}$, namely, $\partial\mathbf{H}_t/\partial z = (2i\omega/c)\hat{\mathbf{z}} \times \mathbf{D}_t + \nabla_t H_z$, along a contour around the vacuum–metal interface, with the use of the relations $\mathbf{D} = \epsilon(2\omega, z)\mathbf{E} + 4\pi\mathbf{P}^{\text{NL}}$, we obtain the first nonlinear boundary condition,

$$\mathbf{H}_t^{(I)}(x, 0|2\omega) - \mathbf{H}_t^{(II)}(x, 0|2\omega) = \frac{2i\omega}{c}4\pi \lim_{\eta\to 0} \int_{-\eta}^{\eta} dz\hat{\mathbf{z}} \times \mathbf{P}_t(x, z|2\omega). \tag{13}$$

By using (12) this result can be rewritten as

$$\mathbf{H}_t^{(I)}(x, 0|2\omega) - \mathbf{H}_t^{(II)}(x, 0|2w) = \frac{2i\omega}{c}\mu_3\hat{\mathbf{z}} \times \mathbf{E}_t(x, 0|\omega)D_z(x, 0|\omega), \tag{14}$$

where the nonlinearity constant $\mu_3$ is given by

$$\mu_3 = \lim_{\eta\to 0} \int_{-\eta}^{\eta} dz \left( \frac{d}{dz}\frac{\alpha(z)}{\epsilon(\omega, z)} + \frac{2}{\epsilon(\omega, z)}\frac{d\kappa(z)}{dz} \right). \tag{15}$$

The function $\epsilon(\omega, z > \eta) = 1$ in vacuum, and $\epsilon(\omega, z < -\eta) = \epsilon(\omega)$ in the metal.

Similarly, on integrating the tangential component of the equation $\nabla \times \mathbf{E} = (2i\omega/c)\mathbf{H}$, namely, $\partial\mathbf{E}_t/\partial z = -(2i\omega/c)\mathbf{H}_t + \nabla_t E_z$, along a contour around the vacuum–metal interface, we obtain the second nonlinear boundary condition,

$$\mathbf{E}_t^{(I)}(x,0|2\omega) - \mathbf{E}_t^{(II)}(x,0|2\omega) = -4\pi \lim_{\eta\to 0} \int_{-\eta}^{\eta} dz \frac{\partial}{\partial t} \frac{P_z^{NL}(x,z|2\omega)}{\epsilon(2\omega,z)}, \qquad (16)$$

where $\epsilon(2\omega, z > \eta) = 1$ in vacuum, $\epsilon(2\omega, z < -\eta) = \epsilon(2\omega)$ in the metal, and $\partial/\partial t = (\partial/\partial x, \partial/\partial y, 0)$. The use of (12) enables this result to be rewritten as

$$\mathbf{E}_t^{(I)}(x,0|2\omega) - \mathbf{E}_t^{(II)}(x,0|2\omega) = -\frac{\partial}{\partial t}[\mu_1 D_z^2(x,0|\omega) + \mu_2 \mathbf{E}_t^2(x,0|\omega)], \qquad (17)$$

where the nonlinearity constants $\mu_1$ and $\mu_2$ are given by

$$\mu_1 = \lim_{\eta\to 0} \int_{-\eta}^{\eta} dz \frac{1}{\epsilon^2(\omega,z)\epsilon(2\omega,z)} \frac{d}{dz}[\alpha(z) + 3\kappa(z)] \qquad (18)$$

$$\mu_2 = \lim_{\eta\to 0} \int_{-\eta}^{\eta} dz \frac{1}{\epsilon(2\omega,z)} \frac{d\kappa(z)}{dz}. \qquad (19)$$

In obtaining the boundary conditions (14) and (17), we kept only the most singular terms in the integrands in (13) and (16), respectively, in the limit as $\eta \to 0$.

The fields entering the boundary conditions (14) and (17) are written in the local coordinate system $(x, y, z)$. On returning to the laboratory coordinate system $(x_1, x_2, x_3)$, we note that when the incident electromagnetic field is p-polarized, so is the scattered field at frequency $\omega$, because cross-polarized scattering is forbidden in the scattering geometry assumed here. Consequently, the nonlinear sources on the right-hand sides of (14) and (17) are nonzero only for p-polarized fields of frequency $2\omega$, so that only p-polarized second-harmonic light is generated in reflection from a one-dimensional random surface. In this case, it is convenient to work with the single nonzero component of the magnetic field in the system, $H_2(x_1, x_3|\Omega)$, with $\Omega = \omega, 2\omega$. The nonlinear boundary conditions (14) and (17) then take the forms

$$H_2^{(I)}(x_1|2\omega) - H_2^{(II)}(x_1|2\omega) = \frac{2ic}{\omega}\frac{\mu_3}{\phi^2(x_1)} L^{(I)}(x_1|\omega) \frac{d}{dx_1} H^{(I)}(x_1|\omega)$$

$$\equiv A(x_1|2\omega), \qquad (20)$$

$$L_2^{(I)}(x_1|2\omega) - \frac{1}{\epsilon(2\omega)} L_2^{(II)}(x_1|2\omega) =$$

$$\frac{2ic}{\omega}\frac{d}{dx_1}\left(\frac{1}{\phi^2(x_1)}\left\{\mu_1\left[\frac{d}{dx_1}H^{(I)}(x_1|\omega)\right]^2 + \mu_2\left[L^{(I)}(x_1|\omega)\right]^2\right\}\right)$$

$$\equiv B(x_1|2\omega), \qquad (21)$$

where we have introduced the source functions

$$H^{(\mathrm{I})}(x_1|\Omega) = H_2^{(\mathrm{I})}(x_1, x_3|\Omega)\Big|_{x_3=\zeta(x_1)} , \tag{22}$$

$$L^{(\mathrm{I})}(x_1|\Omega) = \frac{\partial}{\partial N} H_2^{(\mathrm{I})}(x_1, x_3|\Omega)\Big|_{x_3=\zeta(x_1)} . \tag{23}$$

Now, we turn to the application of these boundary conditions to the problem of second-harmonic generation in the scattering of p-polarized light from weakly rough and strongly rough one-dimensional random metal surfaces.

## 3.2   Strongly Rough Surfaces

The single nonzero component of the magnetic field in the system depicted in Fig. 1 satisfies the following Helmholtz equation in regions I and II, respectively:

$$\left(\frac{\partial^2}{\partial x_1^2} + \frac{\partial^2}{\partial x_3^2} + \frac{\Omega^2}{c^2}\right) H_2^{(\mathrm{I})}(x_1, x_3|\Omega) = 0 \tag{24}$$

and

$$\left(\frac{\partial^2}{\partial x_1^2} + \frac{\partial^2}{\partial x_3^2} + \epsilon(\Omega)\frac{\Omega^2}{c^2}\right) H_2^{(\mathrm{II})}(x_1, x_3|\Omega) = 0, \tag{25}$$

where $\Omega$ stands for either $\omega$ or $2\omega$. The application of Green's second integral identity in the plane [26] to regions I and II in turn yields the equations

$$\theta[x_3 - \zeta(x_1)]H_2^{(\mathrm{I})}(x_1, x_3|\Omega) = H_0(x_1, x_3|\Omega)$$

$$+ \frac{1}{4\pi}\int_{-\infty}^{\infty} \mathrm{d}x_1' \left\{ \left[\frac{\partial}{\partial N'}G_0^{(\Omega)}(x_1, x_3|x_1', x_3')\right]_{x_3'=\zeta(x_1')} H^{(\mathrm{I})}(x_1'|\Omega) \right.$$

$$\left. - \left[G_0^{(\Omega)}(x_1, x_3|x_1', x_3')\right]_{x_3'=\zeta(x_1')} L^{(\mathrm{I})}(x_1'|\Omega) \right\}, \tag{26}$$

$$\theta[\zeta(x_1) - x_3]H_2^{(\mathrm{II})}(x_1, x_3|\Omega) =$$

$$- \frac{1}{4\pi}\int_{-\infty}^{\infty} \mathrm{d}x_1' \times \left\{ \left[\frac{\partial}{\partial N'}G_\epsilon^{(\Omega)}(x_1, x_3|x_1', x_3')\right] H_2^{(\mathrm{II})}(x_1', x_3'|\Omega) \right.$$

$$\left. - G_\epsilon^{(\Omega)}(x_1, x_3|x_1', x_3')\frac{\partial}{\partial N'}H_2^{(\mathrm{II})}(x_1', x_3'|\Omega) \right\}_{x_3'=\zeta(x_1')} , \tag{27}$$

where

$$H_0(x_1, x_3|\Omega) = \begin{cases} H_2^{(\mathrm{I})}(x_1, x_2|\omega)_{\mathrm{inc}} & \text{if}\,\Omega = \omega \\ 0 & \text{if}\,\Omega = 2\omega, \end{cases} \tag{28}$$

with $H_2^{(I)}(x_1, x_3|\omega)_{inc}$ the incident field. $G_0^{(\Omega)}(x_1, x_3|x_1', x_3')$ is Green's function in vacuum,

$$G_0^{(\Omega)}(x_1, x_3|x_1', x_3') = \int_{-\infty}^{\infty} \frac{dq}{2\pi} \frac{2\pi i}{\alpha_0(q, \Omega)} \exp[iq(x_1 - x_1') + i\alpha_0(q, \Omega)|x_3 - x_3'|]$$

$$= i\pi H_0^{(1)} \left\{ (\Omega/c) \left[ (x_1 - x_1')^2 + (x_3 - x_3')^2 \right]^{\frac{1}{2}} \right\}, \qquad (29)$$

where

$$\alpha_0(q, \Omega) = [(\Omega/c)^2 - q^2]^{\frac{1}{2}} \quad q^2 < (\Omega/c)^2$$

$$= i[q^2 - (\Omega/c)^2]^{\frac{1}{2}} \quad q^2 > (\Omega/c)^2, \qquad (30)$$

$H_0^{(1)}(z)$ is a Hankel function of the first kind, and $G_\epsilon^{(\Omega)}(x_1, x_3|x_1', x_3')$ is Green's function for the metal,

$$G_\epsilon^{(\Omega)}(x_1, x_3|x_1', x_3') = \int_{-\infty}^{\infty} \frac{dq}{2\pi} \frac{2\pi}{\beta(q, \Omega)} \exp[iq(x_1 - x_1') - \beta(q, \Omega)|x_3 - x_3'|]$$

$$= i\pi H_0^{(1)} \left\{ n_c(\Omega)[(x_1 - x_1')^2 + (x_3 - x_3')^2]^{\frac{1}{2}} \right\} \qquad (31)$$

with

$$\beta(q, \Omega) = [q^2 - \epsilon(\Omega)(\Omega/c)^2]^{\frac{1}{2}}, \quad \mathrm{Re}\,\beta(q, \Omega) > 0, \mathrm{Im}\,\beta(q, \Omega) < 0, \qquad (32)$$

and

$$n_\epsilon(\Omega) = [\epsilon(\Omega)]^{\frac{1}{2}}, \quad \mathrm{Re}\,n_c(\Omega) > 0, \mathrm{Im}\,n_\epsilon(\Omega) > 0. \qquad (33)$$

By setting $x_3 = \zeta(x_1) + \eta$ in (26) and (27), where $\eta$ is a positive infinitesimal, the following two equations are obtained:

$$H^{(I)}(x, |\Omega) = H_0(x_1|\Omega) + \frac{1}{4\pi} \int_{-\infty}^{\infty} dx_1'$$

$$\times \left\{ \left[ \frac{\partial}{\partial N'} G_0^{(\Omega)}(x_1, x_3|x_1', x_3') \right]_{\substack{x_3 = \zeta(x_1) + \eta \\ x_3' = \zeta(x_1')}} H^{(I)}(x_1'|\Omega) \right.$$

$$\left. - \left[ G_0^{(\Omega)}(x_1, x_3|x_1', x_3') \right]_{\substack{x_3 = \zeta(x_1) + \eta \\ x_3' = \zeta(x_1')}} L^{(I)}(x_1'|\Omega) \right\}, \qquad (34)$$

$$0 = \frac{1}{4\pi} \int_{-\infty}^{\infty} dx_1' \left\{ \left[ \frac{\partial}{\partial N'} G_\epsilon^{(\Omega)}(x_1, x_3|x_1', x_3') \right] H_2^{(II)}(x_1', x_3'|\Omega) \right.$$

$$\left. - G_\epsilon^{(\Omega)}(x_1, x_3|x_1', x_3') \frac{\partial}{\partial N'} H_2^{(II)}(x_1', x_3'|\Omega) \right\}_{\substack{x_3 = \zeta(x_1) + \eta \\ x_3' = \zeta(x_1')}}, \qquad (35)$$

where $H_0(x_1|\Omega) = H_0(x_1, x_3|\Omega)|_{x_3 = \zeta(x_1)}$.

Equations (34) and (35) at the fundamental frequency $\omega$ can be simplified if we use a local impedance boundary condition to relate the source functions $L^{(1)}(x_1|\omega)$ and $H^{(1)}(x_1|\omega)$:

$$L^{(1)}(x_1|\omega) = \frac{K^{(0)}(x_1|\omega)}{\epsilon(\omega)} H^{(1)}(x_1|\omega). \tag{36}$$

The function $K^{(0)}(x_1|\omega)$ is given as an expansion in powers of the product of the skin depth of the metal $d(\omega)$ and the second derivative of the surface profile function [27],

$$K^{(0)}(x_1|\omega) = \frac{\phi(x_1)}{d(\omega)} \left\{ 1 + \frac{d(\omega)}{2} \frac{\zeta''(x_1)}{\phi^3(x_1)} - \frac{d^2(\omega)}{8} \frac{[\zeta''(x_1)]^2}{\phi^6(x_1)} + O[d^3(\omega)] \right\} \tag{37}$$

where

$$d(\omega) = \frac{c}{\omega\sqrt{-\epsilon(\omega)}}, \quad \operatorname{Re} d(\omega) > 0, \operatorname{Im} d(\omega) > 0. \tag{38}$$

The impedance boundary condition (36) is known to represent a good approximation for highly reflective metals [28,29] and has been used successfully in recent numerical work [30,31]. Its use makes (35) redundant and simplifies the numerical problem thereby.

With the use of relation (36), Eq. (34) for the source function $H^{(1)}(x_1|\omega)$ reduces to

$$H^{(1)}(x_1|\omega) = H_0(x_1|\omega) + \frac{1}{4\pi} \int_{-\infty}^{\infty} dx_1'$$

$$\times \left\{ \left[ \frac{\partial}{\partial N'} G_0^{(\omega)}(x_1, x_3|x_1', x_3') \right]_{\substack{x_3 = \zeta(x_1) + \eta \\ x_3' = \zeta(x_1')}} \right.$$

$$- \left[ G_0^{(\omega)}(x_1, x_3|x_1', x_3') \right]_{\substack{x_3 = \zeta(x_1) + \eta \\ x_3' = \zeta(x_1')}}$$

$$\left. \times \frac{K^{(0)}(x_1'|\omega)}{\epsilon(\omega)} \right\} H^{(1)}(x_1'|\omega). \tag{39}$$

This equation is solved numerically by replacing the infinite domain of integration by the finite domain $(-L/2, L/2)$, evaluating the resulting integral over $x_1'$ by a numerical quadrature scheme, and setting $x_1$ equal to the abscissas used in that scheme. The resulting matrix equation is then solved by standard linear equation solver algorithms [32]. With $H^{(1)}(x_1|\omega)$ obtained in this way, $L^{(1)}(x_1|\omega)$ is then calculated by (36).

The scattered field in the vacuum is given by the second term on the right-hand side of (26), which in the far field becomes

$$H_2^{(1)}(x_1, x_3|\omega)_{\mathrm{sc}} = \int_{-\infty}^{\infty} \frac{dq}{2\pi} R(q, \omega) \exp[iqx_1 + i\alpha_0(q, \omega)x_3], \tag{40}$$

where

$$R(q,\omega) = \frac{i}{2\alpha_0(q,\omega)}$$

$$\times \int_{-\infty}^{\infty} dx_1 \left\{ i[q\zeta'(x_1) - \alpha_0(q,w)] - \frac{K^{(0)}(x_1|\omega)}{\epsilon(w)} \right\} H^{(I)}(x_1|\omega)$$

$$\times \exp\{-i[qx_1 + \alpha_0(q,w)\zeta(x_1)]\} . \tag{41}$$

If we introduce the scattering angle $\theta_s$ by $q = (\omega/c)\sin\theta_s$, for plane wave illumination, the contribution to the mean differential reflection coefficient (the fraction of the total energy incident on the surface scattered per unit angle) from the incoherent component of the scattered light is then given by [21]

$$\left\langle \frac{\partial R(\theta_s|\omega)}{\partial \theta_s} \right\rangle_{\text{incoh}} = \frac{\langle|r(\theta_s|\omega)|^2\rangle - |\langle r(\theta_s|\omega)\rangle|^2}{8\pi(\omega/c)L\cos\theta_0} , \tag{42}$$

where $L$ is the length of the surface, $\theta_0$ is the angle of incidence defined by $k = (\omega/c)\sin\theta_0$, and $r(\theta_s|\omega) = R[(\omega/c)\sin\theta_s, \omega]$.

The equations for the source functions $H^{(I)}(x_1|2\omega)$ and $L^{(I)}(x_1|2\omega)$ at the harmonic frequency are obtained from (34) and (35) with aid of the boundary conditions (20) and (21). The resulting equations are

$$H^{(I)}(x_1|2\omega) = \frac{1}{4\pi} \int_{-\infty}^{\infty} dx_1' \left\{ \left[ \frac{\partial}{\partial N'} G_0^{(2\omega)}(x_1, x_3|x_1', x_3') \right]_{\substack{x_3 = \zeta(x_1) + \eta \\ x_3' = \zeta(x_1')}} \right.$$

$$\left. \times H^{(I)}(x_1'|2\omega) - \left[ G_0^{(2\omega)}(x_1, x_3|x_1', x_3') \right]_{\substack{x_3 = \zeta(x_1) + \eta \\ x_3' = \zeta(x_1')}} L^{(I)}(x_1'|2\omega) \right\} , \tag{43}$$

$$Q(x_1|2\omega) = \frac{1}{4\pi} \int_{-\infty}^{\infty} dx_1' \left\{ \left[ \frac{\partial}{\partial N'} G_\epsilon^{(2\omega)}(x_1, x_3|x_1', x_3') \right]_{\substack{x_3 = \zeta(x_1) + \eta \\ x_3' = \zeta(x_1')}} \right.$$

$$\left. \times H^{(I)}(x_1'|2\omega) - \epsilon(2\omega) \left[ G_\epsilon^{(2\omega)}(x_1, x_3|x_1', x_3') \right]_{\substack{x_3 = \zeta(x_1) + \eta \\ x_3' = \zeta(x_1')}} L^{(I)}(x_1'|2\omega) \right\} , \tag{44}$$

where

$$
Q(x_1|2\omega) = \frac{1}{4\pi} \int_{-\infty}^{\infty} dx_1' \left\{ \left[ \frac{\partial}{\partial N'} G_\epsilon^{(2\omega)}(x_1, x_3|x_1'x_3') \right]_{\substack{x_3 = \zeta(x_1) + \eta \\ x_3' = \zeta(x_1')}} \right.
$$

$$
\left. \times A(x_1'|2\omega) - \epsilon(2\omega) \left[ G_\epsilon^{(2\omega)}(x_1, x_3|x_1'x_3') \right]_{\substack{x_3 = \zeta(x_1) + \eta \\ x_3' = \zeta(x_1')}} B(x_1'|2\omega) \right\}. \tag{45}
$$

These equations are solved numerically in the same way as the corresponding equations at the fundamental frequency.

The solutions of this pair of equations enable calculating the scattered field at the harmonic frequency, which is given by the second term on the right-hand side of (26). By using (29), we find that in the far field it can be written as

$$
H_2^{(I)}(x_1, x_3|2\omega)_{sc} = \int_{-\infty}^{\infty} \frac{dq}{2\pi} R(q, 2\omega) \exp[iqx_1 + i\alpha_0(q, 2\omega)x_3], \tag{46}
$$

where

$$
R(q, 2\omega) = \frac{i}{2\alpha_0(q, 2\omega)} \int_{-\infty}^{\infty} dx_1 \left\{ i[q\zeta'(x_1) - \alpha_0(q, 2\omega)]H^{(I)}(x_1|2\omega) \right.
$$

$$
\left. - L^{(I)}(x_1|2\omega) \right\} \exp\{-i[qx_1 + \alpha_0(q, 2\omega)\zeta(x_1)]\}. \tag{47}
$$

If we introduce the scattering angle $\theta_s$ by $q = 2(\omega/c)\sin\theta_s$, we can write the total time-averaged scattered flux at the harmonic frequency crossing the plane $x_3 = \text{const} > \zeta(x_1)_{max}$ as

$$
P_{sc}^{(2\omega)} = \int dx_1 \int dx_2 \text{Re} \, [S_3^c]_{sc}
$$

$$
= L_2 \frac{c^2}{128\pi^2\omega} \int_{-\frac{\pi}{2}}^{\frac{\pi}{2}} d\theta_s |r(\theta_s|2\omega)|^2, \tag{48}
$$

where $L_2$ is the length of the surface along the $x_2$ axis, $[S_3^c]_{sc}$ is the 3-component of the complex Poynting vector of the scattered field at $2\omega$, and

$$
r(\theta_s|2\omega) = -4i\frac{\omega}{c}\cos\theta_s R\left(2\frac{\omega}{c}\sin\theta_s, 2\omega\right). \tag{49}
$$

The magnitude of the total time-average flux incident on the surface $x_3 = \zeta(x_1)$ is

$$
P_{inc} = \left| \int dx_1 \int dx_2 \text{Re} \, [S_3^c]_{inc} \right|, \tag{50}
$$

where $[S_3^c]_{\text{inc}}$ is the 3-component of the complex Poynting vector of the incident field. If we choose a plane wave for the incident field,

$$H_2^{(\text{I})}(x_1, x_3|\omega)_{\text{inc}} = H_0 \exp[ikx_1 - i\alpha_0(k, \omega)x_3], \tag{51}$$

where $H_0$ is the amplitude of the incident wave, we find that

$$P_{\text{inc}} = S|H_0|^2 \frac{c^2 \alpha_0(k, \omega)}{8\pi\omega}, \tag{52}$$

where $S$ is the area of the $x_1 x_2$ plane covered by the random surface. If we introduce the angle of incidence $\theta_0$ by

$$k = (\omega/c)\sin\theta_0, \tag{53}$$

the incident flux (42) becomes

$$P_{\text{inc}} = S|H_0|^2 \frac{c \cos\theta_0}{8\pi}. \tag{54}$$

The total power scattered by the sample at the harmonic frequency is proportional to the square of the irradiance on the surface (incident power per unit area) and the size of the effective source (illuminated area). That is

$$P_{\text{sc}}^{(2\omega)} \propto \left(\frac{P_{\text{inc}}}{S}\right)^2 S, \tag{55}$$

where $S$ is the illuminated area. It is convenient to normalize the scattered power so that the results are independent of the incident power and the illuminated area. Therefore, we define the mean normalized intensity of second-harmonic light generated in reflection by

$$I(\theta_s|2\omega) = \frac{P_{\text{sc}}^{(2\omega)}}{P_{\text{inc}}^2} S. \tag{56}$$

The contribution to the mean normalized intensity from the incoherent component of the scattered harmonic light is therefore

$$\langle I(\theta_s|2\omega)\rangle_{\text{inc}} = \frac{L_2}{2S\omega|H_0|^4 \cos^2\theta_0} \left[\langle|r(\theta_s|2\omega)|^2\rangle - |\langle r(\theta_s|2\omega)\rangle|^2\right]. \tag{57}$$

We present results for a silver surface whose profile constitutes a Gaussian random process with a Gaussian correlation function. To compare with the experimental results reported by *O'Donnell* and *Torre* [19], the statistical parameters that characterize the random profile are chosen as $a = 3.4$ μm and $\delta = 1.81$ μm.

First, in Fig. 3, we show the results of linear scattering calculations at the fundamental wavelength, $\lambda = 1.06$ μm, for normal incidence. The curve has the typical shape of the scattering patterns produced by high-sloped,

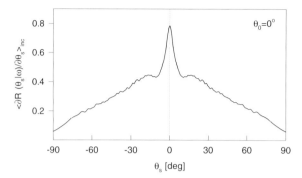

**Fig. 3.** The incoherent component of the mean differential reflection coefficient as a function of the scattering angle $\theta_s$ for the scattering of p-polarized light of wavelength $\lambda = 1.064$ μm from a random silver surface characterized by the Gaussian power spectrum (25), the roughness parameters $a = 3.4$ μm and $\delta = 1.81$ μm, and a dielectric constant $\epsilon(\omega) = -56.25 + i0.60$. The sampling on the surface, of length $L = 40\lambda$, was $\Delta x = \lambda/20$, and the curve shows the result of averaging over $N_p = 2000$ realizations of the surface. The angle of incidence is $\theta_0 = 0°$

randomly rough surfaces. The most noteworthy feature in the curve is a pronounced and well-defined enhanced backscattering peak in the $\theta_s = 0°$ direction. Note that since we are showing only the incoherent component of the mean differential reflection coefficient, the coherent or specular component has been excluded, and the peak shown in the figure is not a specular effect. Moreover, for the parameters employed in the simulation, the coherent component is negligible and would not be visible in the figure. These theoretical results for the linear problem are in good agreement with the experimental results of *O'Donnell* and *Torre* [19].

It is known that for high-sloped surfaces, the backscattering enhancement phenomenon is due to the multiple (mainly double) scattering of waves within the valleys of the surface [33,34,35,36]. The situation is shown schematically in Fig. 4. In the far field, the waves that follow multiple scattering paths such as those shown in Figs. 4a and 4b are phase coherent in the vicinity of the backscattering direction, where they interfere constructively. According to this picture, the angular width of the backscattering peak is of the order of $\lambda/a$.

Computer simulation results for the mean normalized second-harmonic scattering intensity are shown in Fig. 5. It can be verified that they are in good agreement with the experimental results of *O'Donnell* and *Torre* [19]. Given the relative simplicity of the model assumed for nonlinear polarization and the fact that there are no fitting parameters, this could be even a bit surprising. We note also that the scattering curve has some similarities to

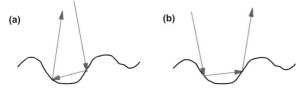

**Fig. 4.** Illustrative diagram of the double scattering processes in the valleys of the surface

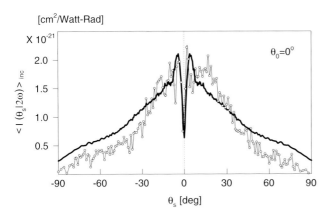

**Fig. 5.** The incoherent component of the mean normalized second-harmonic intensity as a function of the scattering angle $\theta_s$ for the scattering of p-polarized light from a random silver surface characterized by the Gaussian power spectrum (5) and the roughness parameters $a = 3.4$ μm, and $\delta = 1.81$ μm, and dielectric constants $\epsilon(\omega) = -56.25 + i0.60$ and $\epsilon(2\omega) = -11.56 + i0.37$. The thick line represents the results of the numerical simulation, and the open circles represent the experimental results of *O'Donnell* and *Torre* [19]. The incident plane wave has a wavelength $\lambda = 1.064$ μm, and the surface, of length $L = 40\lambda$, was sampled at an interval $\Delta x = \lambda/20$. The curve shows the result of averaging over $N_p = 2000$ realizations of the surface. The angle of incidence is $\theta_0 = 0°$

that shown in Fig. 3 but, significantly, instead of a peak, there is now a dip in the backscattering direction.

It is now known [19,21] that the origin of the dip observed in the backscattering direction lies in the destructive interference between waves multiply scattered within the valleys of the surface. To demonstrate this, we first calculate the angular distribution of the scattered light at the fundamental and harmonic frequencies in a single scattering approximation. Assuming an incident plane wave and a locally flat surface (tangent plane model), we can calculate the linear and nonlinear source functions involved. For the fundamental frequency, we can write (Kirchhoff approximation)

$$H_K^{(I)}(x_1|\omega) = [1 + \varrho_p(x_1)]H(x_1|\omega)_{inc} \,, \tag{58}$$

$$L_K^{(I)}(x_1|\omega) = -i\frac{\omega}{c}[1 - \varrho_p(x_1)]\phi(x_1)\cos\theta_1 H(x_1|\omega)_{inc} \,, \tag{59}$$

where $\varrho_p(x_1)$ represents the Fresnel reflection coefficient for p-polarized light from a planar vacuum–metal interface corresponding to the local angle of incidence $\theta_1(x_1)$, defined as the angle of incidence measured from the normal to the surface at the point $[x_1, \zeta(x_1)]$. The Fresnel reflection coefficient is given by

$$\varrho_p(x_1) = \frac{\epsilon(\omega)\cos\theta_1 - \sqrt{\epsilon(\omega) - \sin^2\theta_1}}{\epsilon(\omega)\cos\theta_1 + \sqrt{\epsilon(\omega) - \sin^2\theta_1}} \,, \tag{60}$$

with

$$\sin\theta_1 = \frac{1}{\phi(x_1)}\left[\sin\theta_0 - \zeta'(x_1)\cos\theta_0\right] \,, \tag{61}$$

$$\cos\theta_1 = \frac{1}{\phi(x_1)}\left[\cos\theta_0 + \zeta'(x_1)\sin\theta_0\right] \,. \tag{62}$$

Once the linear source functions are known, the functions $A(x_1|2\omega)$ and $B(x_1|2\omega)$ can be determined from (20) and (21). We find that

$$A(x_1|2\omega) = \frac{2i\omega}{c}\mu_3(\omega)\cos\theta_1\sin\theta_1\left[1 - \varrho_p^2(x_1)\right][H(x_1|\omega)_{inc}]^2 \,, \tag{63}$$

$$B(x_1|2\omega) = \left(\frac{2\omega}{c}\right)^2\mu_1(\omega)\phi(x_1)\sin^3\theta_1\left[1 + \varrho_p(x_1)\right]^2[H(x_1,|\omega)_{inc}]^2 \,. \tag{64}$$

Now, for a tilted flat surface, the second-harmonic source functions may be found by first writing the second-harmonic fields above and below the interface as plane waves multiplied by some coefficients (of reflection and transmission). These coefficients can be determined by requiring that the fields and their unnormalized normal derivatives satisfy the nonlinear boundary conditions given by (20) and (21). Finally, the nonlinear source functions are found by evaluating the vacuum field and its normal derivative on the surface. In our tangent plane approximation, we find that

$$H_K^{(I)}(x_1|2\omega) = \varrho^{(2\omega)}(x_1)[H(x_1|\omega)_{inc}]^2 \,, \tag{65}$$

$$L_K^{(I)}(x_1|2\omega) = \frac{2i\omega}{c}\phi(x_1)\cos\theta_1\varrho^{(2\omega)}(x_1)[H(x_1|\omega)_{inc}]^2 \,, \tag{66}$$

where $\varrho^{(2\omega)}(x_1)$ represents the amplitude of the second-harmonic plane wave reflected from a flat, tilted surface illuminated by a unit amplitude plane wave, and is given by

$$\varrho^{(2\omega)}(x_1) = \frac{2i\omega}{c} \sin\theta_1 \left[ \epsilon(2\omega)\cos\theta_1 + \sqrt{\epsilon(2\omega) - \sin^2\theta_1} \right]^{-1}$$

$$\times \left\{ \mu_3(\omega)\sqrt{\epsilon(2\omega) - \sin^2\theta_1} \cos\theta_1 \left[1 - \varrho_p^2(x_1)\right] \right.$$

$$\left. - \mu_1(\omega)\epsilon(2\omega)\sin^2\theta_1 \left[1 + \varrho_p(x_1)\right]^2 \right\}. \tag{67}$$

The physical processes taken into account in this approximation are illustrated in Fig. 6, and can be summarized as follows. Light of frequency $\omega$ hits the surface. The nonlinearities of the locally flat surface generate light of frequency $2\omega$ that is then scattered into the local specular direction as shown in Fig. 6a. Further interactions of the second-harmonic radiation with the surface are neglected, which may lead to unphysical situations, such as that illustrated in Fig. 6b. So, the approximation is the equivalent of the Kirchhoff approximation in linear scattering. The mean normalized second-harmonic intensity calculated on the basis of the nonlinear source functions (65) and (66) produces results that are very different from those obtained with the full calculations; we conclude, then, that the observed effects are not due to single scattering.

Now, we turn our attention to the multiple scattering contributions to the mean second-harmonic intensity distribution. For this, we consider first the single and double scattering contributions to the mean differential reflection coefficient at the fundamental frequency. It is known that by solving the scattering equations iteratively, one can calculate separately, according to the order of scattering, the different contributions to the scattered field. The iterative method of solution was studied in detail by *Liszka* and *McCoy* [37] for a case that corresponds to a perfectly conducting one-dimensional surface in s-polarization. The technique has been used to show that the backscattering enhancement phenomenon is due to multiple scattering [35]. An extension of this method, applicable to metallic and dielectric surfaces, has been proposed by *Sentenac* and *Maradudin* [38]. Following these authors and employing a local impedance boundary condition, we can write the following matrix

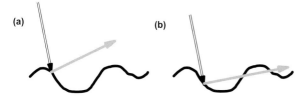

(a)     (b)

**Fig. 6a,b.** Illustrative diagram of the single scattering processes that produce second-harmonic scattered light. The *double line black arrows* represent light of frequency $\omega$, and the *thick gray arrows* represent light of frequency $2\omega$

equation for the source function at the fundamental frequency [21]:

$$H^{(\mathrm{I})}(x_m|\omega) = [1 + \varrho_{\mathrm{p}}(x_m)]H(x_m|\omega)_{\mathrm{inc}}$$

$$+ \frac{1}{2}\sum_{n=1}^{N}\left\{[1 + \varrho_{\mathrm{p}}(x_m)]\left[\mathcal{H}_{mn}^{(0)} - \mathcal{L}_{mn}^{(0)}\frac{K^{(0)}(x_n|\omega)}{\epsilon(\omega)}\right]\right.$$

$$\left. -[1 - \varrho_{\mathrm{p}}(x_m)]\left[\mathcal{H}_{mn}^{(\epsilon)} - \mathcal{L}_{mn}^{(\epsilon)}K^{(0)}(x_n|\omega)\right]\right\}H^{(\mathrm{I})}(x_n|\omega), \quad (68)$$

where, as before, $\varrho_{\mathrm{p}}(x_m)$ represents the Fresnel reflection coefficient corresponding to a planar surface that is tangent to the rough surface at point $x_m$. Explicit expressions for the matrix elements $\mathcal{H}_{mn}^{(0,\epsilon)}$ and $\mathcal{L}_{mn}^{(0,\epsilon)}$ are given in [38]. The first term on the right-hand side of the equation represents the Kirchhoff approximation.

A solution to this equation can be written in the form of a Neumann–Liouville series:

$$H^{(\mathrm{I})}(x_m|\omega) = H^{(1)}(x_m|\omega) + H^{(2)}(x_m|\omega) + H^{(3)}(x_m|\omega) + \ldots, \quad (69)$$

where

$$H^{(1)}(x_m|\omega) = [1 + \varrho_{\mathrm{p}}(x_m)]H(x_m|\omega)_{\mathrm{inc}}, \quad (70)$$

$$H^{(s)}(x_m|\omega) = \frac{1}{2}\sum_{n=1}^{N}\left\{[1 + \varrho_{\mathrm{p}}(x_m)]\left[\mathcal{H}_{mn}^{(0)} - \mathcal{L}_{mn}^{(0)}\frac{K^{(0)}(x_n|\omega)}{\epsilon(\omega)}\right]\right.$$

$$\left. -[1 - \varrho_{\mathrm{p}}(x_m)]\left[\mathcal{H}_{mn}^{(\epsilon)} - \mathcal{L}_{mn}^{(\epsilon)}K^{(0)}(x_n|\omega)\right]\right\}H^{(s-1)}(x_n|\omega).$$

$$(71)$$

If we denote by $r^{(s)}(\theta_s|\omega)$ the contribution to the scattering amplitude $r(\theta_s|\omega)$ calculated from $H^{(s)}(x_m|\omega)$, it represents, in the geometrical optics limit, the contribution of $s$ scattering events to the scattering amplitude.

Now, let us consider the consequences of making this kind of approximation for the source functions at the fundamental frequency on the calculations of the scattering distribution at the second-harmonic frequency. Some of the multiple scattering processes that give rise to the second-harmonic scattering pattern are shown in Fig. 7. As far as the linear part of the scattering is concerned, the Kirchhoff approximation is sufficient to describe the processes shown in Fig. 7a,b. However, in contrast with our previous approximation (Fig. 6a), a multiple scattering solution for the second-harmonic field is required to account for these processes. In Fig. 8a, we present results for the mean normalized second-harmonic intensity based on a single scattering approximation for the linear scattering problem and the complete solution of

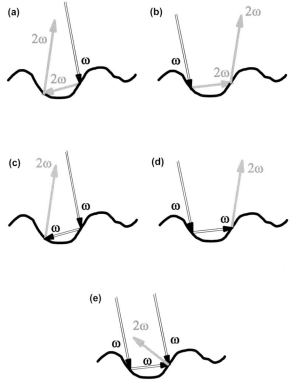

**Fig. 7a-e.** Illustrative diagram of some of the multiple scattering processes that produce second-harmonic scattered light. The *double line black arrows* represent light of frequency $\omega$, and the *thick gray arrows* represent light of frequency $2\omega$

the scattering equations for the second-harmonic field. In other words, for each realization of the profile, we use the Kirchhoff approximation to determine the source functions at the frequency $\omega$ and solve the full system of matrix equations given by (43) and (44) to determine the source functions at $2\omega$. This procedure takes into account the processes illustrated in Fig. 7a,b. It may be seen in Fig. 8a that when these multiple scattering processes are incorporated into the solution, a sharp minimum in the backscattering direction appears in the angular distribution of the mean second-harmonic intensity.

The processes shown in Fig. 7c,d contain double scattering processes at the fundamental frequency. So, to include these processes, we use a double scattering approximation for the source functions at frequency $\omega$ and solve the full system of equations at $2\omega$. Calculations of the mean normalized second-harmonic intensity based on this procedure are shown in Fig. 8b. The contribution of these processes to the mean second-harmonic scattering pattern also leads to a minimum in the backscattering direction. The curve with the thick line shown in Fig. 8c contains single and double scattering processes

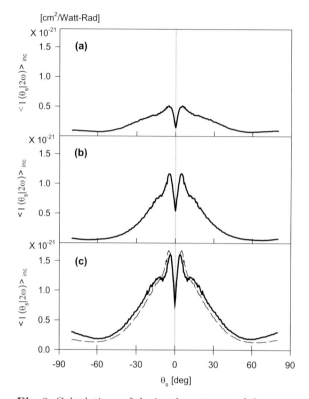

**Fig. 8.** Calculations of the incoherent part of the mean normalized second-harmonic intensity as a function of the scattering angle $\theta_s$ based on the iterative solutions of the linear problem for the scattering of p-polarized light from a random silver surface characterized by the Gaussian power spectrum (5) and the roughness parameters $a = 3.4$ µm and $\delta = 1.81$ µm. Other parameters of the simulation are as in Fig. 5, and the angle of incidence is $\theta_0 = 0°$. The curves have (**a**) the single scattering contributions in the linear scattering and all contributions at the harmonic frequency, (**b**) pure double scattering contributions in the linear scattering and all contributions at the harmonic frequency, and (**c**) the single and double scattering contributions in the linear scattering and all contributions at the harmonic frequency. In (**c**), the curve shown with the dashed line represents the sum of the curves shown in (**a**) and (**b**)

in the linear scattering calculations and a full solution in the second-harmonic calculations.

The results shown in Fig. 8 demonstrate that, as suggested by *O'Donnell* and *Torre* [19], the minimum in the backscattering direction in the angular distribution of the mean second-harmonic intensity is a consequence of multiple scattering (and nonlinear mixing of the light) within the valleys of the surface. As in linear optics, it is also natural to expect that some of these processes are coherent and that the backscattering effects are due to the interference between multiply scattered waves. However, in contrast with the linear case, the interference in second-harmonic light appears to be destructive.

Since the dependence of the second-harmonic wave on the incident one is quadratic, the accumulated phase of the wave of frequency $\omega$ is doubled at the point where the second-harmonic wave is generated. Thus, apart from the phase shifts acquired by the wave on the reflections, the relative phase difference between the processes illustrated in Fig. 7a,b should be zero in the backscattering direction. That interference effects are indeed a plausible explanation for the dip may be recognized from the fact that the amplitude reflection coefficient for second-harmonic generation, (67) is an odd function of the local angle of incidence on the surface. We conclude, then, that the processes depicted in Fig. 7a,b are phase coherent in the backscattering direction and that the waves that follow these two paths are $\pi$ radians out of phase, producing destructive interference in the far field. In the linear problem, the interference is constructive because the pairs of paths illustrated in Fig. 4a,b become reciprocal in the backscattering direction. On the other hand, reciprocity does not apply in second-harmonic generation and, moreover, there seems to be some degree of antireciprocity when source and detector are interchanged.

The arguments given above can be repeated and applied to the processes shown in Fig. 4c,d. Again, we conclude that these processes are coherent with each other and that the corresponding scattered waves interfere destructively. We also remark that there seems to be no coherency between the paths shown in Figs. 4a,c or 7d or between the paths shown in Fig. 7b,c, or d.

However, to the order of scattering considered, the processes shown in Fig. 7a–d are not the only ones that contribute to the second-harmonic scattering distribution. The process depicted in Fig. 7e is also possible, and it is worth noting that its contribution is already included in the results shown with the thick line in Fig. 8c. However, we also notice that this process involves the generation of second-harmonic light in a direction that is near normal to the local tangent plane, making its contribution relatively small. Furthermore, paths such as this (Fig. 7e) do not seem to have a coherent partner.

To substantiate our arguments and statements, we show with the dashed line curve of Fig. 8c, the sum of the scattering intensities shown in Fig. 8a,b. The curve represents, then, the incoherent sum of the contributions from the pairs of processes shown in Fig. 7a,b, and those shown in Fig. 7c,d. It contains no contributions due to the interference between these two pairs of processes. The two curves presented in Fig. 8c are very similar and present only minor differences in the region that surrounds the backscattering dip. The similarity between these two curves implies that, as expected, the processes of Fig. 7e do not contribute significantly. In addition, these curves demonstrate the relative lack of coherence between the contributions from the pairs of processes illustrated in Fig. 7a,b, and those illustrated in Fig. 7c,d.

Our results indicate that, to the order of multiple scattering considered, the main contributions to the second-harmonic scattering pattern come from the processes shown in Fig. 7a–d, and that only the contributions corresponding to the pair of processes shown in Fig. 7a,b, and those corresponding to the pair Fig. 7c,d are coherent with each other.

## 3.3   Weakly Rough Surfaces

When the random metal surface is weakly rough, second-harmonic generation of light reflected from such a surface can be treated perturbatively. Such a treatment is useful when interpreting the results of the numerical simulations as well as experimental data, since it yields analytic expressions for the various contributions to the process of second-harmonic generation whose content can be readily understood.

We assume that the surface roughness satisfies the condition for the validity of the Rayleigh hypothesis, $|\zeta'(x_1)| \ll 1$ [39,40,41]. In this case, the Fourier representation of the fields in the vacuum above the surface,

$$H_2^{(I)}(x_1, x_3|\Omega) = H_2^{(I)}(x_1, x_3|\Omega)_{\text{inc}} + \int_{-\infty}^{\infty} \frac{dq}{2\pi} R(q, \Omega) e^{iqx_1} e^{i\alpha_0(q,\Omega)x_3}, \quad (72)$$

where

$$H_2^{(I)}(x_1, x_3|\omega)_{\text{inc}} = H_0 e^{i(kx_1 - \alpha_0(k,\omega)x_3)} \quad (73)$$

is the incident field and $H_2^{(I)}(x_1, x_3|2\omega)_{\text{inc}} = 0$ can be continued onto the surface $x_3 = \zeta(x_1)$ itself. To derive the equations for the scattering amplitude $R(q, \Omega)$, we substitute in (26) the values of the functions $H^{(II)}(x_1', x_3'|\Omega)$ and $\partial H^{(II)}(x_1', x_3'|\Omega)/\partial N'$ evaluated at $x_3' = \zeta(x_1')$ and expressed in terms of the source functions $H^{(I)}(x_1'|\Omega)$ and $L^{(I)}(x_1'|\Omega)$ with the use of the linear (when $\Omega = \omega$) and nonlinear (when $\Omega = 2\omega$) boundary conditions across the interface $x_3' = \zeta(x_1')$, and set $x_3 > \zeta(x_1)$. Using the explicit expressions for Green's function $G_\epsilon^{(\Omega)}(x_1, x_3|x_1', x_3')$, (31), we obtain

$$0 = \int_{-\infty}^{\infty} \frac{dq}{2\pi} M_\Omega(p|q) R(q, \Omega) + N_\Omega(p|k), \quad (74)$$

where

$$M_\Omega(p|q) = i\,[1 - \epsilon(\Omega)]\, \frac{pq + i\beta(p, \Omega)\alpha_0(q, \Omega)}{i\beta(p, \Omega) - \alpha_0(q, \Omega)}$$

$$\times I[i\beta(p, \Omega) - \alpha_0(q, \Omega)|p - q], \quad (75)$$

and

$$I(\gamma|Q) = \int_{-\infty}^{\infty} dx_1 e^{-iQx_1} e^{-i\gamma\zeta(x_1)}. \quad (76)$$

The driving term $N_\Omega(p|k)$ on the right-hand side of (74) has different representations for the fundamental and harmonic fields. When $\Omega = \omega$, it is related to the incident field

$$N_\omega(p|k) = iH_0 \left[1 - \epsilon(\omega)\right] \frac{pk - i\beta(p,\omega)\alpha_0(k,\omega)}{i\beta(p,\omega) + \alpha_0(k,\omega)}$$

$$\times I[i\beta(p,\omega) + \alpha_0(k,\omega)|p - k], \tag{77}$$

and when $\Omega = 2\omega$, it is the source produced by the surface nonlinear polarization

$$N_{2\omega}(p|k) = -\int_{-\infty}^{\infty} dx_1 e^{-ipx_1} e^{\beta(q)\zeta(x_1)} \left\{[\beta(q,2\omega) + iq\zeta'(x_1)]\right.$$

$$\times A(x_1|2\omega) - \epsilon(2\omega)B(x_1|2\omega)\}, \tag{78}$$

where $A(x_1|2\omega)$ and $B(x_1|2\omega)$ defined in (20) and (21) are the nonlinear current and charge densities, respectively. Note that the nonlinear source $Q(x_1|2\omega)$, which enters the exact equation (44) for the source functions and is given by (45), is related to the nonlinear driving term $N_{2\omega}(q|k)$ through

$$Q(x_1|2\omega) = \int_{-\infty}^{\infty} \frac{dq}{2\pi} e^{iqx_1} e^{-\beta(q,2\omega)\zeta(x_1)} \frac{1}{2\beta(q,2\omega)} N_{2\omega}(q|k). \tag{79}$$

The explicit expression for the nonlinear source term $N_{2\omega}(q|k)$ is

$$N_{2\omega}(q|k) = -\frac{2c}{\omega} \int_{-\infty}^{\infty} \frac{dp}{2\pi} \int_{-\infty}^{\infty} \frac{dr}{2\pi} \int_{-\infty}^{\infty} \frac{dt}{2\pi} I[i\beta(q,2\omega)|q - p - r - t]$$

$$\times \left\{ i\mu_3 \frac{q(q - p - r - t) - \beta^2(q,2\omega)}{i\beta(q,2\omega)} pH(p|\omega)L(r|\omega) \right.$$

$$-\epsilon(2\omega)(p + r + t)\left[\mu_1 prH(p,\omega)H(r|\omega)\right.$$

$$\left. -\mu_2 L(p|\omega)L(r|\omega)\right]\right\}\Phi(t), \tag{80}$$

where

$$\Phi(t) = \int_{-\infty}^{\infty} dx_1 e^{-itx_1} \phi^{-2}(x_1), \tag{81}$$

and $H(p|\omega)$ and $L(p|\omega)$ are the Fourier coefficients of the source functions $H^{(1)}(x_1|\omega)$ and $L^{(1)}(x_1|\omega)$, respectively.

To solve the linear scattering problem, we seek the solution of (74) in the form

$$R(q,\omega) = -2\pi\delta(q - k) - 2iG^{(\omega)}(q|k)\alpha_0(k,\omega), \tag{82}$$

where we have introduced Green's function $G^{(\omega)}(q|k)$ associated with the randomly rough interface between the vacuum and the scattering medium. We define it as the solution of the equation

$$G^{(\omega)}(q|k) = 2\pi\delta(q - k)G_0(k, \omega) + \int_{-\infty}^{\infty} \frac{dp}{2\pi} G^{(\omega)}(q|p)V^{(\omega)}(p|k)G_0(k, \omega)$$

$$= 2\pi\delta(q - k)G_0(k, \omega) + G_0(q, \omega)\int_{-\infty}^{\infty} \frac{dp}{2\pi} V^{(\omega)}(q|p)G^{(\omega)}(p|k), \quad (83)$$

where

$$G_0(q, \omega) = \frac{i\epsilon(\omega)}{\epsilon(\omega)\alpha_0(q, \omega) + i\beta(q, \omega)} \tag{84}$$

is Green's function associated with a planar surface. The scattering potential $V^{(\omega)}(p|k)$ satisfies an integral equation which can be readily obtained if we combine the two equations obtained by substituting (82) in (74), and multiplying (83) by $M_w(p|q)$ and integrating the result over $q$. The resulting equation for the scattering potential has the form [42]

$$\int_{-\infty}^{\infty} \frac{dq}{2\pi} [N_\omega(p|q) - M_\omega(p|q)] \frac{V^{(\omega)}(q|k)}{2\alpha_0(q, \omega)}$$

$$= \frac{N_\omega(p, k) + M_\omega(p, k)[1 + 2i\alpha_0(k, \omega)G_0(k, \omega)]}{2\alpha_0(k, \omega)G_0(k, \omega)}. \tag{85}$$

Now, if we introduce the proper self–energy $M(q|p)$ by the relation

$$\int_{-\infty}^{\infty} \frac{dq}{2\pi} V^{(\omega)}(p|q)G^{(\omega)}(q|k) = \int_{-\infty}^{\infty} \frac{dq}{2\pi} M(p|q)\langle G^{(\omega)}(q|k)\rangle \tag{86}$$

and recall that due to the stationarity of $\zeta(x_1)$ the averaged Green's function $\langle G^{(\omega)}(q|k)\rangle = 2\pi\delta(q - k)G(k, \omega)$, (83) takes the form

$$G^{(\omega)}(q|k) = G_0(q, \omega)2\pi\delta(q - k) + G_0(q, \omega)M(q|k)G(k, \omega). \tag{87}$$

Since, due to the stationarity of $\zeta(x_1)$, the averaged proper self–energy is $\langle M(q|k)\rangle = 2\pi\delta(q - k)M(k, \omega)$, then the averaged Green's function is given by

$$\langle G^{(\omega)}(q|k)\rangle = 2\pi\delta(q - k)\frac{1}{G_0^{-1}(k, \omega) - M(k, \omega)}. \tag{88}$$

If we express $G_0(q, \omega)$ in terms of $G(q, \omega)$ and $M(q, \omega)$ and substitute it in (83), we obtain a more convenient equation for Green's function

$$G^{(\omega)}(p|k) = G(p, \omega)2\pi\delta(p - k)$$

$$+ G(p, \omega)\int_{-\infty}^{\infty} \frac{dq}{2\pi} \left[ V^{(\omega)}(p|q) - \langle M(p|q)\rangle \right] G^{(\omega)}(q|k). \tag{89}$$

From this equation, the equation for the proper self–energy $M(q|k)$ can be readily derived if we multiply it from the left by $V^{(\omega)}(q|p)$, integrate on $q$, and use definition (86). The resulting equation has the form

$$M(p|k) = V^{(\omega)}(p|k) + \int_{-\infty}^{\infty} \frac{dq}{2\pi} M(p|q) G(q, \omega) \left[ V^{(\omega)}(p|k) - \langle M(q|k) \rangle \right]. \tag{90}$$

Now, we introduce a new operator $t_\omega(q|k)$ by the relation

$$t_\omega(q|k) G(q, \omega) = \int_{-\infty}^{\infty} \frac{dq}{2\pi} \left[ V^{(\omega)}(p|q) - \langle M(p|q) \rangle \right] G^{(\omega)}(q|k). \tag{91}$$

The operator $t_\omega(q|k)$ has the very useful property $\langle t_\omega(q|k) \rangle = 0$ and satisfies the integral equation

$$t_\omega(p|k) = V^{(\omega)}(p|k) - \langle M(p|k) \rangle$$

$$+ \int_{-\infty}^{\infty} \frac{dq}{2\pi} t_\omega(p|q) G(q, \omega) \left[ V^{(\omega)}(q|k) - \langle M(q|k) \rangle \right]$$

$$= V^{(\omega)}(p|k) - \langle M(p|k) \rangle$$

$$+ \int_{-\infty}^{\infty} \frac{dq}{2\pi} \left[ V^{(\omega)}(p|q) - \langle M(p|q) \rangle \right] G(q, \omega) t_\omega(q|k), \tag{92}$$

which is obtained by multiplying (89) by $V^{(\omega)}(q|p) - \langle M(q|p) \rangle$, integrating over $p$, and using definition (91).

The scattering amplitude $R(q, \omega)$ is now expressed in terms of $t_\omega(q|k)$ and the averaged Green function, and has the form

$$R(q, \omega) = 2\pi\delta(q - k) R^{(\omega)}(k) - 2iG(q, \omega) t_\omega(q|k) G(k, \omega) \alpha_0(k, \omega) \tag{93}$$

where

$$R^{(\omega)}(k) = -1 - 2i\alpha_0(k, \omega) G(k, \omega) \tag{94}$$

is the specular reflection coefficient for a rough surface.

To deal with the fields of frequency $2\omega$, we also introduce Green's function $G^{(2\omega)}(q|p)$ associated with rough surface by the relation

$$R(p, 2\omega) = - \int_{-\infty}^{\infty} \frac{dq}{2\pi} G^{(2\omega)}(p|q) N_{2\omega}(q|k) \tag{95}$$

and postulate that it satisfies the equation

$$G^{(2\omega)}(p|k) = G_0(p, 2\omega) 2\pi\delta(p - k)$$

$$+ G_0(p, 2\omega) \int_{-\infty}^{\infty} \frac{dq}{2\pi} V^{(2\omega)}(p|q) G^{(2\omega)}(q|k). \tag{96}$$

By comparing this equation with the reduced Rayleigh equation, (74), keeping in mind relation (95), we can show that the scattering potential in this case is a function that has the form

$$V^{(2\omega)}(q|p) = -\frac{1}{\epsilon(2\omega)}M_{2\omega}(q|p) + G_0^{-1}(q, 2\omega)2\pi\delta(q-p)$$

$$\equiv -i\frac{1-\epsilon(2\omega)}{\epsilon(2\omega)}\frac{qp + i\beta(q, 2\omega)\alpha_0(p, 2\omega)}{i\beta(q, 2\omega) - \alpha_0(p, 2\omega)}$$

$$\times J[i\beta(q, 2\omega) - \alpha_0(p, 2\omega)|q-p],\qquad(97)$$

where

$$J(\gamma|Q) = \int_{-\infty}^{\infty} dx_1 e^{-iQx_1}\left[e^{-i\gamma\zeta(x_1)} - 1\right].\qquad(98)$$

Now, if we introduce the operator $t_{2\omega}(q|p)$ and the proper self-energy $M(q|p)$, as in the problem of linear scattering, we obtain a very convenient representation of the scattering amplitude $R(q, 2\omega)$ in terms of $t_{2\omega}(q|p)$ and the averaged Green function $G(q, 2\omega)$,

$$R(p, 2\omega) = G(p, 2\omega)Q(p|2k)$$

$$+G(p, 2\omega)\int_{-\infty}^{\infty}\frac{dq}{2\pi}t_{2\omega}(p|q)G(q, 2\omega)Q(q|2k),\qquad(99)$$

where

$$Q(q|2k) = -\frac{1}{\epsilon(2\omega)}N_{2\omega}(q|k).\qquad(100)$$

To analyze the main features of the angular dependence of the intensity of the generated light, which was defined in Sect. 3.2 and in terms of $R(q, 2\omega)$ has the form

$$\langle I(\theta_s|2\omega)\rangle_{\text{inc}} = \frac{8\omega}{L_1c^2|H_0|^4}\frac{\cos^2\theta_s}{\cos^2\theta_0}\left[\langle|R(q, 2\omega)|^2\rangle - |\langle R(q, 2\omega)\rangle|^2\right],\qquad(101)$$

where $q = (2\omega/c)\sin\theta_s$ and $k = (\omega/c)\sin\theta_0$, we need explicit expressions for the different contributions to it. From the form of the equation for the scattering amplitude $R(q, 2\omega)$, (99), it is clear that the the first term on the right-hand side of it, i.e., the nonlinear driving term $Q(q|2k)$, describes the generation of light at the harmonic frequency due to the nonlinear mixing of scattered fundamental waves. The effects of the multiple scattering of surface plasmon polaritons of frequency $\omega$ are contained in this term. The integral term in (99) describes the multiple scattering of waves of frequency $2\omega$, which are generated due to the presence of the nonlinear source $Q(q|2k)$ and, in particular, contains the effects of the multiple scattering of surface plasmon polaritons of frequency $2\omega$.

To clarify the analysis that we are going to present, we subdivide $Q(q|2k)$ into four contributions with different physical meanings in the following manner. Using (93), we can write the explicit expressions for the Fourier coefficients of the source functions in the forms,

$$H(q,\omega) = -2i\alpha_0(k,\omega)G(k,\omega)H_0\left[2\pi\delta(q-k)\right.$$

$$\left.+G(q,\omega)t_\omega(q|k) + H_{sc}(q,\omega)\right], \qquad (102)$$

where

$$H_{sc}(q,\omega) = \frac{1}{2}\{J[\alpha_0(k,\omega)|q-k] + J[-\alpha_0(k,\omega)|q-k]\}$$

$$+\frac{i\beta(k,\omega)}{\epsilon(\omega)}2\alpha_0(k,\omega)\{J[\alpha_0(k,\omega)|q-k] - J[-\alpha_0(k,\omega)|q-k]\}$$

$$+\int_{-\infty}^{\infty}\frac{dp}{2\pi}J[-\alpha_0(p,\omega)|q-p]G(p,\omega)t_\omega(p|k), \qquad (103)$$

and

$$L(q,\omega) = -2\alpha_0(k,\omega)G(k,\omega)H_0$$

$$\times\left[2\pi\delta(q-k)\frac{i\beta(k,\omega)}{\epsilon(\omega)} - \alpha_0(q,\omega)G(q,\omega)t_\omega(q|k) + L_{sc}(q|k)\right], \quad (104)$$

with

$$L_{sc}(q|k) = \frac{(\omega^2/c^2) - qk}{2\alpha_0(k,\omega)}\{J[\alpha_0(k,\omega)|q-k] - J[-\alpha_0(k,\omega)|q-k]\}$$

$$+\frac{(\omega^2/c^2) - qk}{2\alpha_0^2(k,\omega)}\frac{i\beta(k,\omega)}{\epsilon(\omega)}\{J[\alpha_0(k,\omega)|q-k]$$

$$+J[-\alpha_0(k,\omega)|q-k]\} + -\int_{-\infty}^{\infty}\frac{dp}{2\pi}\frac{(\omega^2/c^2) - qp}{\alpha_0(p,\omega)}$$

$$J[-\alpha_0(p,\omega)|q-p]G(p,\omega)t_\omega(p|k). \qquad (105)$$

Since the function $N_{2\omega}(q|k)$, given by (82), is composed of the products of the source functions at frequency $\omega$, we can separate three important contributions to $Q(q|2k)$. The first contribution contains only the product of $\delta$ functions, $(2\pi)^2\delta(p-k)\delta(p'-k)$. It describes the nonlinear mixing of the fields of frequency $\omega$ which would be specular if the metal–vacuum interface were planar. The second contribution to $Q(q|2k)$ contains the product of a $\delta$ function and terms with Green's function, e. g., $2\pi\delta(p-k)G(p',\omega)t_\omega(p'|k)$, and describes the interaction of the "specular" and scattered fields, including nonlinear mixing of the excited surface plasmon polaritons with the incident light. The third important contribution contains the product of two

Green's functions of surface plasmon polaritons of frequency $\omega$, $G(p,\omega)t_\omega(p|k)$ $G(p',\omega)t_\omega(p'|k)$. It describes nonlinear mixing of the scattered fields and includes the mixing of co- and contrapropagating surface plasmon polaritons. Finally, the fourth term, which we will denote by $Q_{\rm sc}(q|2k)$, contains the roughness-induced corrections to the first three terms in the nonlinear source $Q(q|2k)$ According to the classification we have just given, we separate explicitly the three main contributions to the nonlinear driving source, so that we obtain

$$Q(q|2k) = \frac{8ic}{\omega}\alpha_0^2(k,\omega)H_0^2G^2(k,\omega)\left[\Gamma_0(q)2\pi\delta(q-2k)\right.$$

$$+\Gamma_1(q,k)G(q-k,\omega)t_\omega(q-k|k)$$

$$+\int_{-\infty}^{\infty}\frac{dp}{4\pi}\Gamma_2(q,p)G(q-p,\omega)G(p,\omega)t_\omega(q-p|k)t_\omega(p|k)$$

$$\left.+\,Q_{\rm sc}(q|2k)\right]\,. \tag{106}$$

The explicit expressions for the nonlinear coefficients entering (106) are

$$\Gamma_0(q) = -\mu_3 k\frac{\beta(q,2\omega)}{\epsilon(2\omega)}\frac{\beta(k,\omega)}{\epsilon(\omega)} - q\left\{\mu_1 kq - \mu_2\left[\frac{\beta(k,\omega)}{\epsilon(\omega)}\right]^2\right\}\,, \tag{107}$$

$$\Gamma_1(q,k) = \mu_3\frac{i\beta(q,2\omega)}{\epsilon(2\omega)}\left[(q-k)\frac{i\beta(k,\omega)}{\epsilon(\omega)} - k\alpha_0(q-k,\omega)\right]$$

$$-2q\left[\mu_1(q-k)k - \mu_2\frac{i\beta(k,\omega)}{\epsilon(\omega)}\alpha_0(q-k,\omega)\right]\,, \tag{108}$$

$$\Gamma_2(q,p) = \mu_3\frac{i\beta(q,2\omega)}{\epsilon(2\omega)}(q-p)\alpha_0(q,\omega) + q\left[\mu_1(q-p)p\right.$$

$$\left.+\mu_2\alpha_0(q-p,\omega)\alpha_0(p,\omega)\right]\,. \tag{109}$$

An explicit expression for the function $Q_{\rm sc}(q|2k)$ is presented in [43].

To calculate the mean intensity of the second-harmonic light given by (101), we need to calculate $\langle|R(q,2\omega)|^2\rangle - |\langle R(q,2\omega)\rangle|^2$. In terms of the operator $t_{2\omega}(q|k)$, this quantity takes the form

$$\langle|R(q,2\omega)|^2\rangle - |\langle R(q,2\omega)\rangle|^2 = |G(q,2\omega)|^2\left\{\langle|Q(q|k)|^2\rangle - |\langle Q(q|k)\rangle|^2\right.$$

$$+L_1\tau_{2\omega}(q|k)|G(2k,2\omega)|^2|Q(2k|2k)|^2$$

$$+L_1 \tilde{\tau}_{2\omega}(q|k) + 2\mathrm{Re} \int_{-\infty}^{\infty} \frac{dp}{2\pi} \langle t_{2\omega}(q|p)$$

$$G(p, 2\omega)Q(p|k)Q^*(q|k)\rangle_{\mathrm{c}} \Bigg\} , \tag{110}$$

where $\langle \cdot \rangle_{\mathrm{c}}$ denotes the cumulant average [44]. In writing (110), we have introduced the notation

$$\langle t_\Omega(q|p)t_\Omega^*(q|p') \rangle = 2\pi\delta(p-p')L_1\tau_\Omega(q|p) \tag{111}$$

for the averaged reducible vertex function $\tau_\Omega(q|p)$ in the problem of linear scattering of light from a rough surface, and the specular component of the field of frequency $2\omega$ generated at the rough surface, $Q(2k|2k)$, is determined by

$$\langle Q(p|2k) \rangle \equiv Q(2k|2k)2\pi\delta(p-2k) . \tag{112}$$

The function $\tilde{\tau}_{2\omega}(q|2k)$ in (110) is the analog of the reducible vertex function in the problem of nonlinear scattering and is given by

$$\tilde{\tau}_{2\omega}(q|2k) = \int_{-\infty}^{\infty} \frac{dp}{2\pi} \int_{-\infty}^{\infty} \frac{dp'}{2\pi} \Bigg[ \frac{1}{L_1} \langle t_{2\omega}(q|p)G(p, 2\omega)Q(p|2k)t_{2\omega}^*(q|p')$$

$$G^*(p', 2\omega)Q^*(p'|2k) \rangle_{\mathrm{c}}$$

$$+\tau_{2\omega}(q|p)|G(p, 2\omega)|^2 \langle |Q(p|2k)|^2 \rangle_{\mathrm{c}}$$

$$+\frac{1}{L_1} \langle t_{2\omega}(q|p)G(p, 2\omega)Q^*(p'|2k) \rangle_{\mathrm{c}} \langle t_{2\omega}(q|p')$$

$$G^*(p', 2\omega)Q(p|2k) \rangle_{\mathrm{c}} \Bigg] . \tag{113}$$

As in the experiments of [17] and [18], to separate different mechanisms for the interplay of the nonlinearity and roughness of the surface, we study features of the mean intensity of the generated light under different scattering conditions imposed by the power spectrum of the surface roughness. To do this, we calculate the mean intensity of the second-harmonic light generated in reflection from a one-dimensional, random silver surface characterized by the rectangular power spectrum (6) for the two particular cases where $h_1 = 1$, and $h_2 = 0$, and where $h_1 = 0$, and $h_2 = 1$. In our calculations, the wavelength of the incident light was chosen 1.064 μm, as in the experiments of [17] and [18], so that the wavelength of the generated light is 0.532 μm. The dielectric constants of silver at the fundamental and harmonic frequencies are then $\epsilon(\omega) = -56.25 + i0.60$ and $\epsilon(2\omega) = -11.56 + i0.37$, respectively, which ensures that surface plasmon polaritons exist at both frequencies. The real parts of their wave numbers are $k_{sp}(\omega) = 1.009008(\omega/c)$ and $k_{sp}(2\omega) = 1.0462234(2\omega/c)$, respectively.

To illustrate the specific effects of the rectangular power spectrum, in Fig. 9 we present the results of linear scattering calculations at the fundamental frequency $\omega$ for three angles of incidence $\theta_0 = 0°$ (Fig. 9a), $\theta_0 = 8°$ (Fig. 9b), and $\theta_0 = 10°$ (Fig. 9c) where the power spectrum is centered at the wave number $k_{sp}(\omega)$ of the surface plasmon polaritons at frequency $\omega$ and has a half width equal to $(\omega/c)\sin\theta_{max}$. The rms height of the surface roughness is $\delta = 10.8$ nm, and the characteristic angle $\theta_{max}$ is $\theta_{max} = 15°$. In this case, the light whose angle of incidence is within the range $-\theta_{max} < \theta_0 < \theta_{max}$

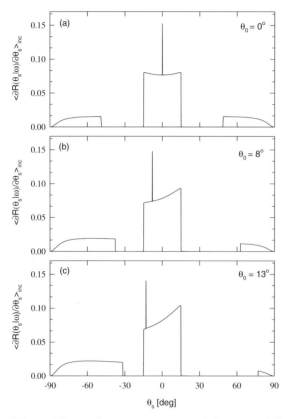

**Fig. 9.** The incoherent component of the mean differential reflection coefficient as a function of the scattering angle $\theta_s$ for the scattering of p-polarized light of wavelength $\lambda = 1.064$ μm from a random silver surface characterized by the rectangular power spectrum (6) centered at the wave number of surface plasmon polaritons of wavelength $\lambda$, with roughness parameters $\delta = 10.8$ nm and $\theta_{max} = 15°$, and a dielectric constant $\epsilon(\omega) = -56.25 + i0.60$. The angles of incidence are (**a**) $\theta_0 = 0°$, (**b**) $\theta_0 = 8°$, and (**c**) $\theta_0 = 13°$

is converted effectively into surface plasmon polaritons of frequency $\omega$, and the surface plasmon polaritons of frequency $\omega$ are converted effectively into the light that propagates into the vacuum within the range of scattering angles $-\theta_{max} < \theta_s < \theta_{max}$. The plots display single–scattering wings at large angles of scattering and an almost rectangular distribution coming from the double–scattering processes mediated by the surface plasmon polaritons of frequency $\omega$. The height of the enhanced backscattering peak in Fig. 9 is exactly twice the background intensity, as expected in linear scattering theory.

In the calculations of the nonlinear scattering problem, we use the numerical values of the nonlinear parameters $\mu_1(\omega)$ and $\mu_3(\omega)$, which are given by (18) and (15), calculated in the framework of the free-electron model. They are $\mu_1 = (0.2384987 \times 10^{-14} + i6.384 \times 10^{-17})$ CGSE, and $\mu_3 = (0.6818093 \times 10^{-14} + i1.316 \times 10^{-18})$ CGSE. We note, that in the free-electron model, the nonlinear coefficient $\mu_2$, given by (19), vanishes. When using the Agranovich and Darmanyan model of surface nonlinear polarization, we estimated the nonlinear coefficients in the following manner. Since the free-electron model describes the experimental results on second-harmonic generation in reflection from a planar metal surface fairly well [24], we assume that the values of the nonlinear coefficients $\mu_1$ and $\mu_3$ in the Agranovich and Darmanyan model coincide with those in the free-electron model. From these coefficients, we can express the phenomenological constants $\alpha(z)$ and $\kappa(z)$ entering the expression for the surface nonlinear polarization in the Agranovich and Darmanyan model in terms of the parameters appearing in the free-electron model. Having $\alpha(z)$ and $\kappa(z)$ in hand, we obtain an expression for $\mu_2$ in the form

$$\mu_2^{(0)} = \frac{2}{3}\beta \ln[\epsilon(\omega)/\epsilon(2\omega)],\tag{114}$$

so that $\mu_2 = (0.7732004 \times 10^{-14} + i9.871 \cdot 10^{-17})$ CGSE.

First we analyze the processes of multiple scattering of surface plasmon polaritons of frequency $2\omega$. As in [17], we suppress the excitation of surface plasmon polaritons of frequency $\omega$, so that the features in the angular distribution of the mean intensity of the scattered light of frequency $2\omega$ are due to the multiple scattering of surface plasmon polaritons of frequency $2\omega$. This is achieved by using the rectangular power spectrum (6), centered at the wave numbers of surface plasmon polaritons of frequency $2\omega$, $k_{sp}(2\omega)$, with a half width equal to $(2\omega/c)\sin\theta_{max}$. In this case, the excitation of surface plasmon polaritons of frequency $\omega$ through single-scattering processes is forbidden, and the strong conversion of surface plasmon polaritons of frequency $2\omega$ into volume electromagnetic waves radiated into vacuum in the given range of the scattering angle is ensured.

In Fig. 10a–c, we present the mean intensity of the second-harmonic light, calculated by numerical simulations (solid lines) and the perturbative approach (dashed lines), when p-polarized light is incident on a one-dimensional, random silver surface, characterized by the power spectrum (6) with $h_1 = 0$

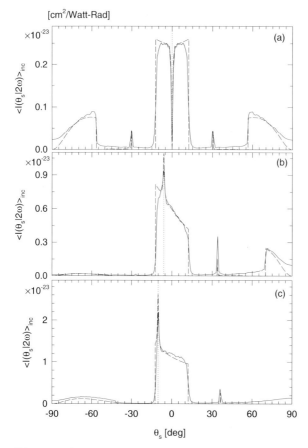

[cm²/Watt-Rad]

**Fig. 10.** The mean normalized intensity of the second-harmonic light as a function of the scattering angle $\theta_s$ for the scattering of p-polarized light from a randomly rough silver surface whose roughness is characterized by the rectangular power spectrum (6), centered at the wave numbers of surface plasmon polaritons of frequency $2\omega$, with roughness parameters $\delta = 11.1$ nm and $\theta_m = 12.2°$. The nonlinear coefficients are given by the free-electron model. The angles of incidence are (**a**) $\theta_0 = 0°$, (**b**) $\theta_0 = 6°$, and (**c**) $\theta_0 = 10°$. The *solid lines* represent the results of the numerical simulations; the *dashed lines* represent the perturbative results

and $h_2 = 1$, an rms height $\delta = 11.1$ nm, and $\theta_{\max} = 12.2°$. The power spectrum is centered at the wave numbers $\pm k_{\mathrm{sp}}(2\omega)$ of the surface plasmon polariton at frequency $2\omega$ and has a half width equal to $(2\omega/c)\sin\theta_{\max}$. The nonlinear parameters used in the calculations, $\mu_1(\omega)$, $\mu_2(\omega)$, $\mu_3(\omega)$, are given by the free-electron model. The results of numerical simulations represent the averages of results obtained from 3000 realizations of the random surface.

The plots in Fig. 10a–c look similar to the plots of the differential reflection coefficient for linear scattering presented in Fig. 9. They display the single-scattering wings at large angles of scattering and an almost rectangular distribution coming from the double-scattering processes mediated by the surface plasmon polaritons of frequency $2\omega$. However, in contrast to the results for linear scattering, the mean intensity of the second-harmonic light displays a dip in the retroreflection direction when the fundamental light is incident normally on the surface, $\theta_0 = 0°$ (Fig. 10a), and a peak in the retroreflection direction for larger angles of incidence $\theta_0 = 6°$ (Fig. 10b) and $\theta_0 = 10°$ (Fig. 10c). Two weak peaks positioned at $\theta_s = \pm 30.3°$ when the angle of incidence is $0°$, at $\theta_s = 33.83°$ and $\theta_s = -26.89°$ when the angle of incidence is $6°$, and at $\theta_s = 36.25°$ and $\theta_s = -24.69°$ when the angle of incidence is $10°$, are also displayed.

Although the results presented in Fig. 10 are in quite good quantitative agreement with the experimental results of [17], they disagree with them qualitatively, since only a dip in the retroreflection direction was observed in the experiments of [17] for all angles of incidence of the fundamental light.

In Fig. 11a–c, we present plots of the mean intensity of second-harmonic light generated in reflection from the same surface used in the calculations of the results presented in Fig. 10, but with the surface nonlinear polarization given by the Agranovich and Darmanyan model, in which we assume that $\mu_1 = 0$. In this case, a dip in the retroreflection direction is displayed for all angles of incidence.

In view of the power spectrum, the central distribution of the intensity of the light of frequency $2\omega$ in Fig. 10–11 is due to the roughness-induced radiation of surface plasmon polaritons of frequency $2\omega$ and is determined by the contribution of the second, $\tau_{2\omega}(q|2k)|G(2k, 2\omega)|^2|Q(2k|2k)|^2$, and third, $\tilde{\tau}_{2\omega}(q|2k)$, terms on the right-hand side of (110). As is well known in linear scattering theory [5], the averaged reducible vertex function $\tau_\Omega(q|p)$ displays a Lorentzian-enhanced backscattering peak whose position is determined by the condition $q + p = 0$ and whose half width at half maximum equals the decay rate of the surface plasmon polaritons propagating along the rough surface. Therefore, the contribution to the mean intensity of the second-harmonic from the second term on the right-hand side of (110), $\tau_{2\omega}(q|2k)|G(2k, 2\omega)|^2|Q(2k|2k)|^2$, can be viewed as a linear scattering one; the only difference is that the "amplitude" of the incident field, that is, $G(2k, 2\omega)Q(2k|2k)$, depends on the angle of incidence (the processes associated with this term are illustrated schematically in Fig. 12a). Due to the presence of $\tau_{2\omega}(q|2k)$, this contribution displays an enhanced second-harmonic generation peak. The position of the peak is determined by the condition $q + 2k = 0$. Since $q = (2\omega/c) \sin\theta_s$, the condition $q = -2k$ is equivalent to $\sin\theta_s = -\sin\theta_0$, i.e., the enhanced second-harmonic generation peak occurs in the retroreflection direction. The height of the peak as a function of the angle of incidence is determined by the effective amplitude of the field being scattered, $G(2k, 2\omega)Q(2k|2k)$, i.e., by the coherent component of the nonlinear source function of frequency $2\omega$. The strongest contribution to it is the

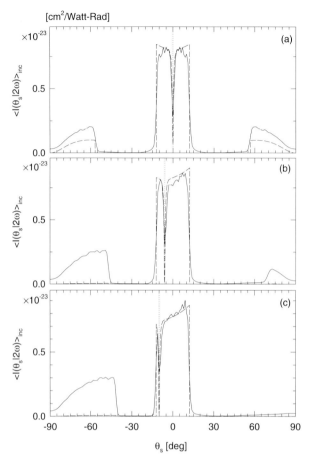

**Fig. 11.** The same as Fig. 10, except that the nonlinear coefficients are given by the Agranovich and Darmanyan model with $\mu_1 = 0$

specular contribution to the term $Q(q|2k)$, (106), which is governed by the effective nonlinear coefficient $\Gamma_0(2k)$, which describes the specular generation of p-polarized light reflected from a planar surface. Since this coefficient is proportional to $k$, in contrast to the situation in linear scattering, the peak of the enhanced second-harmonic generation in this case will be absent when light is incident normally on the surface. We note that this contribution to the mean intensity of the generated light always displays a peak in the retroreflection direction.

The second important, and much stronger, term is $\tilde{\tau}_{2\omega}(q|2k)$. In a sense, this term is analogous to the contributions from the ladder and maximally crossed diagrams in the linear problem of the scattering of light from a randomly rough surface. However, in the nonlinear scattering problem, the amplitude of the field being scattered is the nonlinear source function $Q(q|2k)$.

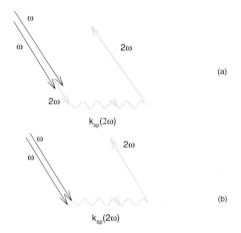

$\omega$

$\omega$

$2\omega$

(a)

$2\omega$

$k_{sp}(2\omega)$

$\omega$

$\omega$

$2\omega$

(b)

$k_{sp}(2\omega)$

**Fig. 12a,b.** Diagrams of multiple scattering processes involving surface plasmon polaritons of frequency $2\omega$. The *solid line arrows* represent light of frequency $\omega$, the *wavy gray line arrows* represent surface plasmon polaritons of frequency $2\omega$, and the *solid gray line arrows* represent light of frequency $2\omega$

Therefore, it describes the scattering of the waves of frequency $2\omega$ generated through the nonlinear mixing of the diffusely scattered fundamental radiation. To second order in the nonlinear coefficients, the function $\tilde{\tau}_{2\omega}(q|2k)$ has the form

$$\tilde{\tau}_{2\omega}(q|2k) = \int_{-\infty}^{\infty} \frac{dp}{2\pi} \left[ \tau_{2\omega}(q|p)|G(p,2\omega)|^2 \langle |Q(p|2k)|^2 \rangle_0 \right.$$

$$+ \langle t_{2\omega}(q,p)Q^*(q+2k-p|2k) \rangle_0 G(p,2\omega)G^*(q+2k-p,2\omega)$$

$$\left. \times \langle t_{2\omega}^*(q,q+2k-p)Q(p|2k) \rangle_0 \right] , \tag{115}$$

where we have introduced the notation $\langle f(q,p)g(p',q') \rangle = 2\pi\delta(q-p+p'-q')$ $\langle f(q,p)g(q'-q+p,q') \rangle_0$.

As in [5], to calculate the most important contributions to $\tilde{\tau}_{2\omega}(q|2k)$, we will use the pole approximation for Green's function $G(q,\Omega)$ [5]. In this approximation, we obtain the following expression for $\tilde{\tau}_{2\omega}(q|2k)$:

$$\tilde{\tau}_{2\omega}(q|2k) = \frac{C^2(2\omega)}{\Delta_t(2\omega)} \left\{ \tau_{2\omega}[q|k_{sp}(2\omega)]\langle |Q[k_{sp}(2\omega)|2k]|^2 \rangle_0 \right.$$

$$\left. + \tau_{2\omega}[q|-k_{sp}(2\omega)]\langle |Q[-k_{sp}(2\omega)|2k]|^2 \rangle_0 \right]$$

$$+ \frac{4\Delta_t(2\omega)C^2(2\omega)}{(q+2k)^2 + 4\Delta_t^2(2\omega)}$$

$$\times \mathrm{Re}\ \{\mathcal{P}_{2\omega}[q, k_{\mathrm{sp}}(2\omega)|q + 2k - k_{\mathrm{sp}}(2\omega), 2k]$$

$$\mathcal{P}_{2\omega}^*[q, q + 2k - k_{\mathrm{sp}}(2\omega)|k_{\mathrm{sp}}(2\omega), 2k]$$

$$+\mathcal{P}_{2\omega}[q, -k_{\mathrm{sp}}(2\omega)|q + 2k + k_{\mathrm{sp}}(2\omega), 2k]$$

$$\mathcal{P}_{2\omega}^*[q, q + 2k + k_{\mathrm{sp}}(2\omega)| - k_{\mathrm{sp}}(2\omega), 2k]\}$$

$$+\frac{1}{2}C^2(2\omega)\tau_{\mathrm{MC}}(q, -q|2\omega)|G(q, 2\omega)|^2 \langle|Q(-q|2k)|^2\rangle_0, \tag{116}$$

where

$$\mathcal{P}_{2\omega}(q, p|p', q') = \langle t_{2\omega}(q, p)Q^*(p'|q')\rangle_0 \tag{117}$$

and

$$C(\Omega) = \frac{(|\epsilon_1(\Omega)|)^{3/2}}{\epsilon_1^2(\Omega) - 1} \tag{118}$$

is the residue of Green's function at the poles $q = \pm k_{\mathrm{sp}}(\Omega)$. The decay rate $\Delta_t(\Omega)$ of the surface plasmon polaritons is

$$\Delta_t(\Omega) = \Delta_\epsilon(\Omega) + \Delta_{\mathrm{sp}}(\Omega), \tag{119}$$

where

$$\Delta_\epsilon(\Omega) = \frac{1}{2}k_{\mathrm{sp}}(\Omega)\frac{\epsilon_2(\Omega)}{[\epsilon_1(\Omega) + 1]\epsilon_1(\Omega)} \tag{120}$$

is the decay rate of the surface plasmon polaritons due to ohmic losses and

$$\Delta_{\mathrm{sp}}(\Omega) = C(\Omega)\mathrm{Im}\,M[k_{\mathrm{sp}}(\Omega)] \tag{121}$$

is the decay rate of surface plasmon polaritons due to their scattering by the surface roughness.

From (116), it is seen that the function $\tilde{\tau}_{2\omega}(q|2k)$ displays a Lorentzian peak centered at $q = -2k$, i.e., in the retroreflection direction. However, the height of the peak is determined by the Fourier component of the nonlinear source through which the excitation of surface plasmon polaritons of frequency $2\omega$ occurs. As a result, the height of the peak depends on the particular trajectory of the scattering path. In our case, it is the "incident" surface plasmon polaritons of frequency $2\omega$ that are excited due to the nonlinear mixing of the fundamental waves, which are multiply scattered. Schematic illustrations of the processes contained in the term under discussion are presented in Fig. 12b. The mean intensities of these "incident" surface plasmon polaritons are determined by the nonlinear source functions and are proportional to $Q[k_{\mathrm{sp}}(2\omega)|2k]$ and $Q[-k_{\mathrm{sp}}(2\omega)|2k]$. Generally speaking, both the nonlinear excitation and the roughness-induced radiation of the surface plasmon polaritons of frequency $2\omega$ are nonreciprocal processes, and, as a result, there is no reason to expect a backscattering enhancement peak in the angular distribution of the intensity of the generated light. Depending on the phases acquired in the processes of excitation and radiation, a peak or a dip can occur.

From the expression for the nonlinear source function, (106), one can see that the strongest contributions to $Q[\pm k_{sp}(2\omega)|2k]$ come from the terms that are governed by the effective nonlinearities $\Gamma_0[\pm k_{sp}(2\omega)]$, (107), and $\Gamma_1[\pm k_{sp}(2\omega)|k]$, (108). We can see that the terms with the nonlinear constants $\mu_2$ and $\mu_3$ are proportional to the wave vector of the intermediate excitations, in our case $\pm k_{sp}(2\omega)$, whereas the term with the nonlinear coefficient $\mu_1$ is a quadratic function of it. Therefore, when the surface plasmon polaritons of frequency $2\omega$ propagating in opposite directions are excited in the processes of nonlinear mixing, they will have a phase difference $\pi$ if the nonlinear constant $\mu_1$ is zero or small compared to $\mu_2$ and $\mu_3$. This is always the case at normal, or almost normal, incidence of the fundamental light, since the term with $\mu_1$ is also linear in the tangential component of the wave vector of the incident light. Therefore, at normal incidence, only a dip can be formed in the retroreflection direction. With an increase of the angle of incidence, the term with $\mu_1$ becomes dominant and, as a result, a peak in the retroreflection direction will be formed. The angle of incidence at which the dip disappears and the peak begins to evolve is determined by the relative magnitudes of the nonlinear coefficients, and also by the dielectric functions of the medium, $\epsilon(\omega)$ and $\epsilon(2\omega)$, as well. When $\mu_1 = 0$ (Fig. 11), the angular distribution of the intensity displays a dip in the retroreflection direction for all angles of incidence. The strong dependence of the intensity of the generated light on the angle of incidence displayed in the plots in Fig. 10 also shows the dominant role of the nonlinear constant $\mu_1$ when the calculations are carried out within the framework of the free-electron model.

It should be noted that peaks at $q = k \pm k_{sp}(\omega)$ are present in the gap between the central distribution and the single-scattering wings in Fig. 10, as well as in the experimental data of [17]. They are due to the resonant nonlinear mixing of the incident light with the surface plasmon polaritons of frequency $\omega$. For the given power spectrum of the surfaces under study, surface plasmon polaritons of frequency $\omega$ can be excited in higher order scattering processes, in fact in the third event of scattering by the surface roughness. Therefore, the resonant peaks due to the nonlinear mixing of surface plasmon polaritons of frequency $\omega$ with the incident light can arise, but only in higher order scattering processes. They can be seen in Fig. 10 but are too weak to be seen in Fig. 11.

Now, we turn to the analysis of the effects of the multiple scattering of surface plasmon polaritons of frequency $\omega$. To do this, we assume that the power spectrum of the surface roughness is now such that the fundamental light incident on the surface at angles of incidence $|\theta_0| < \theta_{max}$ is strongly coupled into surface plasmon polaritons of frequency $\omega$, and the conversion of

surface plasmon polaritons of frequency $2\omega$, excited due to nonlinear scattering processes, into volume waves in the vacuum is suppressed. This is achieved
by using the rectangular power spectrum, (26), centered at the wave numbers
of surface plasmon polaritons of frequency $\omega$, $k_{sp}(\omega)$, with a half width equal
to $(\omega/c)\sin\theta_{max}$.

In Fig. 13a–c, we present the mean intensity of the second-harmonic light,
calculated by numerical simulations when p-polarized light is incident on
a one-dimensional, random silver surface at three angles of incidence, $\theta_0 = 0°$
(Fig. 13a), $\theta_0 = 8°$ (Fig. 13b), and $\theta_0 = 13°$ (Fig. 13c). The nonlinear coefficients are given by the free-electron model. The surface roughness is characterized by the power spectrum (6) with $h_1 = 1$ and $h_2 = 0$, with an rms height
$\delta = 28.3\,\mathrm{nm}$, and $\theta_{max} = 15°$. This power spectrum is centered at the wave
number $k_{sp}(\omega)$ of the surface plasmon polaritons at frequency $\omega$. The angular distributions of the mean intensity of the second-harmonic light presented
in Fig. 13 differ considerably from the angular distributions of the intensity
of the light scattered linearly from the same surface, which are presented
in Fig. 9. In the gaps between the single scattering wings and the central
distribution, two strong resonant peaks are also displayed, at $\theta_s = \pm 30.3°$
when the angle of incidence is $0°$, at $\theta_s = 35.03°$ and $\theta_s = -25.78°$ when the
angle of incidence is $8°$, and at $\theta_s = 38.1°$ and $\theta_s = -23.1°$ when the angle
of incidence is $13°$. These are the peaks whose positions are determined by
the condition $q = k \pm k_{sp}(\omega)$ and are associated with the resonant nonlinear
interaction of the excited surface plasmon polariton of frequency $\omega$ with the
incident light. The most striking feature of the plots in Fig. 13 is a narrow dip
at $\theta_s = 0$ that is present in the plots at small angles of incidence and evolves
into a peak when the angle of incidence is $13°$. Note that the intensity of the
light of frequency $2\omega$ in this case is an order of magnitude greater than in
the case where the excitation of surface plasmon polaritons of frequency $\omega$ is
forbidden (Fig. 10). A weak dip in the retroreflection direction is displayed in
Fig. 13b. Although the main features of the experimental results of [18] are
displayed by the plots presented in Fig. 13, no dips or peaks in the direction
normal to the mean surface and in the retroreflection direction were observed
experimentally.

In Fig. 14, we present plots of the the intensity of second-harmonic light of
light reflected from the same surface used in calculating the results presented
in Fig. 13, but with the surface nonlinear polarization given by the Agranovich
and Darmanyan model where we have set $\mu_1 = 0$. From the plots, we can see
that when the fundamental light is incident normally on the surface, a narrow
dip occurs in the direction normal to the mean surface, independent of the
values of the nonlinear coefficients $\mu_1$, $\mu_2$, and $\mu_3$. With increasing angle
of incidence, the dip evolves into a peak when the nonlinear coefficient $\mu_1$
is nonzero, whereas when only $\mu_1 = 0$, only a dip appears in the angular
dependence of the mean intensity. A weak dip in the retroreflection direction
is displayed when the angle of incidence is $8°$ in both cases.

[cm²/Watt-Rad]

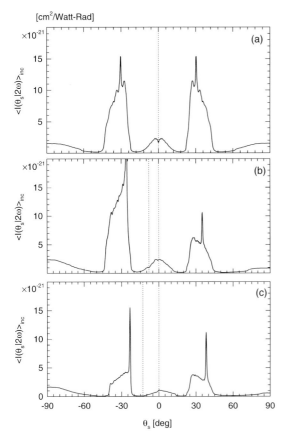

**Fig. 13.** The mean normalized intensity of the second-harmonic light as a function of the scattering angle $\theta_s$ for the scattering of p-polarized light from a randomly rough silver surface whose roughness is characterized by the rectangular power spectrum (6), centered at the wave numbers of surface plasmon polaritons of frequency $\omega$, with roughness parameters $\delta = 28.3$ nm and $\theta_m = 15°$. The nonlinear coefficients are given by the free-electron model. The angles of incidence are (**a**) $\theta_0 = 0°$, (**b**) $\theta_0 = 8°$, and (**c**) $\theta_0 = 13°$. The plots represent the results of the numerical simulations

In Fig. 15, we present plots of the the intensity of the second-harmonic light for the same cases for which the results presented in Fig. 13 were obtained, except that the rms height of the surface roughness is taken as $\delta = 10.8$ nm instead of 28.3 nm. These results were obtained by the numerical and perturbative approaches developed in the preceding sections. The parameters of the surface roughness used in the calculations of the results presented in Fig. 15 are the parameters of surface 1 studied in [18], and for the numerical calculations illustrated in Fig. 13 and Fig. 14, we have used the parameters of the surface 3. For the latter sets of parameters, the per-

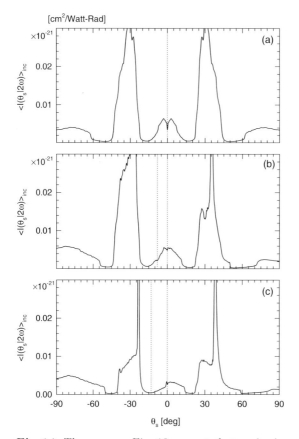

[cm²/Watt-Rad]

**Fig. 14.** The same as Fig. 13, except that $\mu_2$ is given by (114), and $\mu_1 = \mu_3 = 0$

turbation theory approach breaks down since $\sqrt{|\epsilon(\omega)|}(\omega\delta/c) = 1.25575$. The experimental curves for the mean intensity of the harmonic light presented in [18] also support the conclusion that the perturbation theory is applicable only to surface 1, the most weakly rough surface used in the study.

First let us analyze the general features of the angular distribution of the mean intensity of the scattered light The second-harmonic light generated through the nonlinear mixing of the surface plasmon polaritons of the fundamental frequency emerges into vacuum in the angular range determined by the conditions

$$-\frac{1}{2}\left(\sin\theta_{\max} - \sin\theta_0\right) < \sin\theta_s < \frac{1}{2}\left(\sin\theta_{\max} + \sin\theta_0\right), \tag{122}$$

$$-\frac{1}{2}\left[n_{\mathrm{sp}}(\omega) + \sin\theta_{\max}\right] < \sin\theta_s < -\frac{1}{2}\left[n_{\mathrm{sp}}(\omega) - \sin\theta_{\max}\right],$$

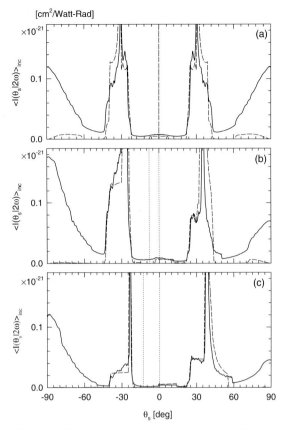

**Fig. 15.** The mean normalized intensity of the second-harmonic light as a function of the scattering angle $\theta_s$ for the scattering of p-polarized light from a randomly rough silver surface whose roughness is characterized by the rectangular power spectrum (6), centered at the wave numbers of surface plasmon polaritons of frequency $\omega$, with roughness parameters $\delta = 10.8\,\mathrm{nm}$ and $\theta_m = 15°$. The nonlinear coefficients are given by the free-electron model. The angles of incidence are (**a**) $\theta_s = 0°$, (**b**) $\theta_s = 8°$, and (**c**) $\theta_s = 13°$. The plots represent the results of the numerical (*solid line*) and perturbative (*dashed line*) calculations

and

$$\frac{1}{2}\left[n_{\mathrm{sp}}(\omega) - \sin\theta_{\max}\right] < \sin\theta_s < \frac{1}{2}\left[n_{\mathrm{sp}}(\omega) + \sin\theta_{\max}\right] . \tag{123}$$

In contrast, the surface plasmon polaritons of frequency $2\omega$ excited through the nonlinear interaction are converted into vacuum light that radiates within the angular range determined by the conditions

$$\frac{1}{2}\left[n_{\mathrm{sp}}(\omega) - \sin\theta_{\max}\right] - n_{\mathrm{sp}}(2\omega) < \sin\theta_{\mathrm{s}}$$

$$< \frac{1}{2}\left[n_{\mathrm{sp}}(\omega) + \sin\theta_{\max}\right] - n_{\mathrm{sp}}(2\omega) \quad (124)$$

and

$$-\frac{1}{2}\left[n_{\mathrm{sp}}(\omega) + \sin\theta_{\max}\right] + n_{\mathrm{sp}}(2\omega) < \sin\theta_{\mathrm{s}}$$

$$< -\frac{1}{2}\left[n_{\mathrm{sp}}(\omega) - \sin\theta_{\max}\right] + n_{\mathrm{sp}}(2\omega). \quad (125)$$

In (124) and (125), $n_{\mathrm{sp}}(\Omega) = [k_{\mathrm{sp}}(\Omega)c]/\Omega$ is the refractive index of the surface plasmon polaritons of frequency $\Omega$. The single scattering processes give a contribution to the intensity of the second-harmonic light only in the angular ranges

$$\frac{1}{2}\left[n_{\mathrm{sp}}(\omega) - \sin\theta_{\max}\right] + \sin\theta_0 < \sin\theta_{\mathrm{s}}$$

$$< \frac{1}{2}\left[n_{\mathrm{sp}}(\omega) + \sin\theta_{\max}\right] + \sin\theta_0 \quad (126)$$

and

$$-\frac{1}{2}\left[n_{\mathrm{sp}}(\omega) + \sin\theta_{\max}\right] + \sin\theta_0 < \sin\theta_{\mathrm{s}}$$

$$< -\frac{1}{2}\left[n_{\mathrm{sp}}(\omega) - \sin\theta_{\max}\right] + \sin\theta_0 . \quad (127)$$

The plots of Figs. 13–15, indeed, display a nonzero intensity of the scattered light of frequency $2\omega$ only in these angular intervals.

The effects of the multiple scattering of surface plasmon polaritons of frequency $\omega$ influence all of the terms contributing to the mean differential intensity of the second-harmonic light. The first term in (110), $\langle|Q(q|2k)|^2\rangle - |\langle Q(q|2k)\rangle|^2$, describes the second-harmonic generation of volume waves of frequency $2\omega$ through the nonlinear mixing of the scattered fundamental light, including the multiply-scattered surface plasmon polariton of frequency $\omega$, and the remaining contributions describe those processes in which the generated waves of frequency $2\omega$ have been scattered by the rough surface. The latter give rise to a structureless background of the intensity of the second-harmonic light because the power spectrum chosen in this case forbids the conversion of surface plasmon polaritons of frequency $2\omega$ into radiative waves in vacuum. The main features of the angular distribution of the intensity of light of the harmonic frequency can be described if we keep only those contributions to the nonlinear driving term $Q(q|2k)$ that contain the operators $t_\omega(p|k)$, i.e., the second and third terms in (106). The second term in (106), which contains the nonlinear coefficient $\Gamma_1(q,k)$, gives the contribution to the mean differential intensity of the harmonic light of the form

$$\chi_1(q|k) = |\Gamma_1(q,k)|^2 |G(q-k,\omega)|^2 \tau_\omega(q-k|k)\,. \tag{128}$$

This term describes the intensity of the light of frequency $2\omega$ generated by the nonlinear interaction of the incident light with the scattered waves of frequency $\omega$. (Schematic illustrations of these processes are presented in Fig. 16a–c.) The function $\chi_1(q|k)$ contains the product of two highly peaked functions, $|G(q-k,\omega)|^2\tau_\omega(q-k|k)$ and, as a result, displays three peaks of the enhanced second-harmonic generation. Two strong resonant peaks at $q = k \pm k_{\mathrm{sp}}(\omega)$ are due to the resonant interaction of the incident light with the excited surface plasmon polaritons of frequency $\omega$ [7,45] (Fig. 16a,b) and appear already in the single scattering processes due to the presence of the factor $|G(q-k,\omega)|^2$ [7,45]. The coherent interference of multiply-scattered surface plasmon polaritons of frequency $\omega$ leads to the appearance of a peak in the direction normal to the surface because the reducible vertex function $\tau_\omega(q-k|k)$ displays a Lorentzian peak centered at $q-k+k = 0$. The function $\tau_\omega(q-k|k)$ describes the second-harmonic generation by the nonlinear mixing of the incident light with volume waves of frequency $\omega$ emerging into the vacuum after being multiply scattered by the surface roughness. It displays a peak when $q = 0$, i.e., in the direction normal to the mean surface. The presence of this peak can be understood easily since it is due to the mixing of the incident light and the light scattered in the retroreflection direction, i.e., in the direction of the enhanced backscattering (Fig. 16c). In this case the contrapropagating beams of volume waves interact nonlinearly giving rise to the waves of frequency $2\omega$ propagating into the vacuum in the direction normal to the surface. However, not only the reducible vertex function $\tau_\omega(q-k|k)$ depends on the angle of scattering. The effective nonlinear coefficient $\Gamma_1(q,k)$ is also a function of $q$ and, at small values of $q$, is proportional to $q$; therefore, the efficiency of the nonlinear mixing in such processes vanishes when $q = 0$, i.e., in the direction normal to the mean surface. Generally speaking, depending on the values of the nonlinear coefficients $\mu_1$, $\mu_2$, and $\mu_3$ and the dielectric functions $\epsilon(2\omega)$ and $\epsilon(\omega)$, the processes we have just discussed can lead to a peak with a dip at the top.

A much stronger contribution to the intensity of the light of frequency $2\omega$ comes from the nonlinear mixing of the multiply-scattered surface plasmon polaritons of frequency $\omega$, propagating in opposite directions. They are illustrated in Fig. 16d and are described by the third term in the nonlinear source (106). Its contribution to the mean differential intensity has the form

$$\chi_3(q|k) = \int_{-\infty}^{\infty} \frac{\mathrm{d}p}{4\pi} |\Gamma_2(q,p) + \Gamma_2(q,q-p)|^2 |G(q-p,\omega)G(p,\omega)|^2$$

$$\times \tau_\omega(q-p|k)\tau_\omega(p|k)\,. \tag{129}$$

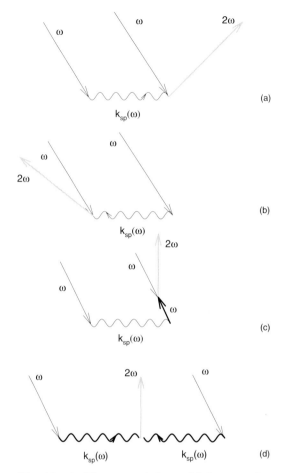

**Fig. 16a-d.** Diagrams of the multiple scattering processes involving surface plasmon polaritons of frequency $\omega$. The *solid line arrows* represent light of frequency $\omega$, the *wavy gray line arrows* represent surface plasmon polaritons of frequency $\omega$, and the *solid gray line arrows* represent light of frequency $2\omega$

In the pole approximation for Green's functions, it has the form

$$\chi_3(q,k) = \frac{C^2(\omega)}{2\Delta_t(\omega)} \frac{C^2(\omega)}{q^2 + 4\Delta_t^2(\omega)} \Big\{ \tau_\omega[k_{\mathrm{sp}}(\omega)|k]\tau_\omega[q - k_{\mathrm{sp}}(\omega)|k]$$

$$\times |\Gamma_2[q, k_{\mathrm{sp}}(\omega)] + \Gamma_2[q, q - k_{\mathrm{sp}}(\omega)]|^2$$

$$+\tau_\omega[-k_{\mathrm{sp}}(\omega)|k]\tau_\omega[q + k_{\mathrm{sp}}(\omega)|k]$$

$$\times |\Gamma_2[q, k_{\mathrm{sp}}(\omega)] + \Gamma_2[q, q - k_{\mathrm{sp}}(\omega)]|^2$$

$$+C^2(\omega)\tau'_\omega(-k,k)\tau_\omega(q+k|k)|G(k,\omega)G(q+k,\omega)|^2|$$

$$\times|\Gamma_2(q,-k)+\Gamma_2(q,q+k)|^2\Big\},\tag{130}$$

where $\tau'_\omega(-k,k)$ is the residue of $\tau_\omega(q,k)$ at the pole $q=-k$. Although this function also has a Lorentzian factor centered at $q=0$, the efficiency of the nonlinear mixing of the contrapropagating surface plasmon polaritons is determined by the effective nonlinear coefficient $\Gamma_2[q,\pm k_{\rm sp}(\omega)]+\Gamma_2[q,q\mp k_{\rm sp}(\omega)]$. This effective nonlinear coefficient is also linear in $q$ for small $q$ due to the symmetry of the surface nonlinear polarization. As is well known, the symmetry of the nonlinear polarization of a metal surface forbids such processes [16]. Therefore, this contribution displays a dip rather than a peak in the direction normal to the mean surface. The depth of this dip depends strongly on the values of the material parameters and the angle of incidence of the fundamental light.

In the nonlinear source $Q(q|2k)$, (106), we have separated two leading terms that contain the scattering of surface plasmon polaritons of frequency $2\omega$. However, the remaining term $Q_{\rm sc}(q|2k)$ contains higher order processes with the participation of surface plasmon polaritons of frequency $\omega$, which can give rise to qualitative features in the angular distribution of the mean differential intensity of the light of frequency $2\omega$. We list here the most important processses left out of our description. The first are the processes of nonlinear mixing in which at least one of the participating waves is the volume wave emerging into the vacuum after the scattering of surface plasmon polaritons of frequency $\omega$ by the surface roughness. In particular, the nonlinear mixing of the enhanced backscattered radiation with the enhanced backscattered surface plasmon polariton beam leads to the appearance of resonant peaks or dips when $q=-k\pm k_{\rm sp}(\omega)$ In a sense, these processes are analogous to the processes of the resonant mixing of the incident light with the surface plasmon polaritons of frequency $\omega$ which lead to the peaks/dips at $q=k\pm k_{\rm sp}(\omega)$ (Figs. 13–15). However, these peaks are weak and are displayed in the plots of Figs. 13–15, as well as in the experimental curves of [18], as a weak structure on the left and right shoulders of the rectangular distributions of the intensity in the angular ranges $-45°<\theta_{\rm s}<-20°$ and $20°<\theta_{\rm s}<45°$.

We have also left out of our discussion the processes of nonlinear mixing of the backscattered waves of frequency $\omega$. They display a peak or a dip in the retroreflection direction, since the waves of frequency $\omega$ scattered into the retroreflection direction are coherent. No surface plasmon polaritons of frequency $2\omega$ participate in the formation of the peak/dip. The dip is clearly seen in the plots in Figs. 13–15.

In addition, the last, and possibly the strongest, processes are the processes of nonlinear mixing of copropagating surface plasmon polaritons of frequency $\omega$. The latter processes are nonresonant since the frequency dispersion of the dielectric function of the metal breaks the phase matching conditions.

It should be pointed out that only the multiply–scattered waves participate in the processes of nonlinear mixing listed, and the resulting angular pattern of the intensity of the generated light does not depend on the angle of incidence. This is why the rectangular distributions of the intensity in the angular ranges $-45° < \theta_s < -20°$ and $20° < \theta_s < 45°$ in Figs. 13–15 do not move with an increase of the angle of incidence.

The nonlinear interaction of the multiply-scattered surface plasmon polaritons of frequency $\omega$ can also lead to the excitation of surface plasmon polaritons of frequency $2\omega$. These types of processes are of a higher order in the surface profile function and usually are weak. What is more, the particular form of the power spectrum used in our calculations makes them even weaker or totally forbids them. However, the nonlinear mixing of copropagating surface plasmon polaritons of frequency $\omega$ can result in the excitation of surface plasmon polaritons of frequency $2\omega$ on the rough surface. These processes can give quite a strong contribution to the intensity of the second-harmonic radiation independent of the power spectrum and display, depending on the effective nonlinearities $\Gamma_2[q, \pm k_{sp}(\omega)]$, Lorentzian peaks or dips at $q = \pm[k_{sp}(2\omega) - 2k_{sp}(\omega)]$. For the particular values of the dielectric functions $\epsilon(\omega)$ and $\epsilon(2\omega)$ assumed in our calculations, these peaks/dips occur at $\theta_s = \pm1.1°$, that is, in the vicinity of the direction normal to the surface. Possibly, the additional weak structure in the vicinity of the normal direction present in Figs. 13–15 can be attributed to this mechanism.

# 4    The Kretschmann Geometry

As we noted in the introduction, the first experimental studies of second-harmonic generation of light reflected from, or transmitted through, randomly rough metal surfaces were carried out in the Kretschmann ATR geometry. In several of these studies [8,9,11], the random, unilluminated surface of the metal film was in contact with a nonlinear quartz crystal. In others [10,12,13], it was in contact with vacuum. In this section, we present theories of second-harmonic generation in the reflection of light from, and in its transmission through, a metal film with a one-dimensional random surface when that surface is in contact with vacuum, so that the nonlinear interactions occur at the silver surfaces. A theory of second-harmonic generation in transmission when the random metal surface is in contact with a nonlinear quartz crystal has been presented in [46], and the interested reader is referred to that work for the details of the analysis.

The system we consider in this section is depicted in Fig. 17. It consists of a dielectric prism characterized by a real, positive, dielectric constant $\epsilon_0$ in the region $x_3 > D$ (region I), a metal film characterized by an isotropic, complex, frequency-dependent, dielectric function $\epsilon(\omega) = \epsilon_1(\omega) + i\epsilon_2(\omega)$ in

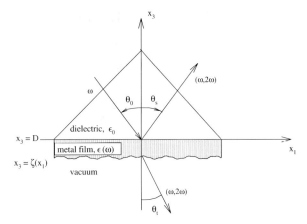

**Fig. 17.** The Kretschmann attenuated total reflection geometry

the region $\zeta(x_1) < x_3 < D$ (region II), and vacuum in the region $x_3 < \zeta(x_1)$ (region III). This system is illuminated through the prism by p-polarized light of frequency $\omega$, whose plane of incidence is the $x_1x_3$ plane. Therefore, it is convenient to work with the single nonzero component of the magnetic field in it, $H_2(x_1, x_3|\omega)$.

At each frequency $\omega$ in the interval $0 < \omega < \omega_{\max}$, where $\omega_{\max}$ is the solution of $\epsilon_0 + \operatorname{Re}\epsilon(\omega) = 0$, the structure depicted in Fig. 17 supports two p-polarized surface plasmon polaritons, whose wave numbers are denoted by $q_1(\omega)$ and $q_2(\omega)$, with $|q_1(\omega)| < |q_2(\omega)|$. In the absence of the surface roughness $[\zeta(x_1) \equiv 0]$, these wave numbers are obtained as the solutions of the dispersion relation

$$[\epsilon_0\beta(q,\omega) + \epsilon(\omega)\beta_{\mathrm{p}}(q,\omega)][\epsilon(\omega)\beta_0(q,\omega) + \beta(q,\omega)]$$

$$= -[\epsilon_0\beta(q,\omega) - \epsilon(\omega)\beta_0(q,\omega)][\epsilon(\omega)\beta_0(q,\omega) - \beta(q,\omega)]\exp[-2\beta(q,\omega)D],$$

$$(131)$$

where $\beta_{\mathrm{p}}(q,\omega) = [q^2 - \epsilon_0(\omega/c)^2]^{\frac{1}{2}}$, with $\operatorname{Re}\beta_{\mathrm{p}}(q,\omega) > 0$, $\operatorname{Im}\beta_{\mathrm{p}}(q,\omega) < 0$; $\beta(q,\omega) = [q^2 - \epsilon(\omega)(\omega/c)^2]^{\frac{1}{2}}$, with $\operatorname{Re}\beta(q,\omega) > 0$, $\operatorname{Im}\beta(q,\omega) < 0$; and $\beta_0(q,\omega) = [q^2 - (\omega/c)^2]^{\frac{1}{2}}$, with $\operatorname{Re}\beta_0(q\omega) > 0$, $\operatorname{Im}\beta_0(q,\omega) < 0$.

If, for the time being, we neglect the imaginary part of $\epsilon(\omega)$, we find that true surface plasmon polaritons exist only in the region $q > \sqrt{\epsilon_0}(\omega/c)$, where both $\beta_{\mathrm{p}}(q,\omega)$ and $\beta_0(q,\omega)$ are real and positive, so that the electromagnetic field in both the prism and the vacuum decays exponentially with increasng distance from the metal film. In the limit $D \to \infty$, the vanishing of the first factor on the left-hand side of (131) is the dispersion relation for a surface plasmon polariton localized to the prism metal interface; the vanishing of the second factor is the dispersion relation for a surface plasmon polariton localized to the metal–vacuum interface.

The solution of (131) for finite values of $D$ in the frequency range of interest to us yields two real values of $q$, $q_1(\omega)$ and $q_2(\omega)$. In this frequency range, $q_2(\omega) > \sqrt{\epsilon_0}(\omega/c)$, so that both $\beta_p(q_2(\omega),\omega)$ and $\beta_0(q_2\omega),\omega)$ are real and positive. The field associated with this mode has its maximum at the prism–metal interface and decays exponentially both into the prism and into the vacuum. This is a true surface plasmon polariton, which in the limit of a thick film becomes the surface plasmon polariton at a prism–metal interface. The solution $q_1(\omega)$ is larger than $\omega/c$ but is smaller than $\sqrt{\epsilon_0}(\omega/c)$, so that $\beta_0(q_1(\omega),\omega)$ is real, whereas $\beta_p(q_1,(\omega),\omega)$ is imaginary. The field associated with this mode has its maximum at the metal–vacuum interface and decays exponentially into the vacuum; however, it radiates into the prism. Thus, this mode is a leaky wave. In the limit of a thick film, this mode becomes the surface plasmon polariton at a metal–vacuum interface. The fact that this mode is a leaky wave for finite values of $D$ makes it possible to excite it in the Kretschmann ATR configuration.

The dispersion curves resulting from the solution of (131) are plotted in Fig. 18 for the case that $\epsilon_0 = 2.25$ and $\epsilon(\omega)$ has the simple free-electron form $\epsilon(\omega) = 1 - (\omega_p/\omega)^2$, where $\omega_p$ is the plasma frequency of the conduction electrons in the metal. The portion of one branch of the dispersion curve that corresponds to the leaky wave is depicted by the dashed curve and represents a plot of $\omega$ as a function of $\mathrm{Re}\, q$.

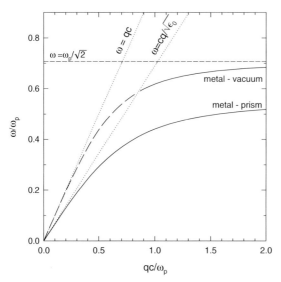

**Fig. 18.** The dispersion curves for the surface plasmon polaritons supported by the Kretschmann attenuated total reflection geometry in the absence of surface roughness. The dielectric constant of the prism is $\epsilon_0 = 2.25$; the dielectric function of the metal is $\epsilon(\omega) = 1 - \omega_p^2/\omega^2$; the medium below the film is vacuum. The leaky mode is depicted by the *dashed curve*

When the imaginary part of $\epsilon(\omega)$ is restored, both wave numbers $q_1(\omega)$ and $q_2(\omega)$ become complex. Their imaginary parts, however, are generally small enough compared with their real parts that the dispersion curves plotted in Fig. 18 are almost unchanged, except that the entire horizontal axis, it must now be understood, is labeled R$\epsilon q$.

The fact that one of the branches of the dispersion curves obtained from the solution of (131) is that of a leaky wave for a certain range of frequencies (Fig. 18) gives rise to the possibility of additional peaks in the angular dependence of the incoherent component of the scattered light. These leaky wave peaks are single-scattering phenomena and hence, are generally more intense than multiple-scattering phenomena such as the enhanced backscattering peaks. The scattering angles at which these peaks occur can be determined simply by equating the $x_1$ component of the wave vector of the scattered light to the wave number of the leaky surface plasmon polariton,

$$\mathrm{Re}\,[q_1(\omega)] = \pm\sqrt{\epsilon_0}(\omega/c)\sin\theta_{\mathrm{leaky}}. \tag{132}$$

Therefore, the angles at which the leaky wave peaks occur are independent of the angle of incidence. Because $\mathrm{Re}\,[q_1(\omega)]$ is smaller than $\sqrt{\epsilon_0}(\omega/c)$ but larger than $(\omega/c)$, there are no leaky wave peaks in transmission, because the dielectric constant is unity in the vacuum region and the corresponding peaks lie in the nonradiative region of the optical spectrum. Because the leaky wave peaks significantly contribute to the angular dependence of the intensity of the scattered light, we may expect that multiple-scattering effects will manifest themselves more clearly under conditions where the leaky wave peaks are absent.

In the rest of this section, both a weakly rough and a strongly rough metal–vacuum interface will be considered. Underlying the theoretical treatment of both types of random surfaces are the boundary conditions at the planar prism–metal interface and at the random metal–vacuum interface at the harmonic frequency.

## 4.1   Linear and Nonlinear Boundary Conditions

The boundary conditions satisfied by the field components $H_2^{(i)}(x_1, x_3|\omega)(i = \mathrm{I, II, III})$ at the fundamental frequency $\omega$ express the continuity of the tangential components of the magnetic and electric fields across the interfaces $x_3 = D$ and $x_3 = \zeta(x_1)$. They can be written in the forms

$$H_2^{(\mathrm{I})}(x_1, x_3|\omega)\Big|_{x_3=D} = H_2^{(\mathrm{II})}(x_1, x_3|\omega)\Big|_{x_3=D}$$

$$\frac{1}{\epsilon_0}\frac{\partial}{\partial x_3}\,H_2^{(\mathrm{I})}(x_1, x_3|\omega)\Big|_{x_3=D} = \frac{1}{\epsilon(\omega)}\frac{\partial}{\partial x_3}\,H_2^{(\mathrm{II})}(x_1, x_3|\omega)\Big|_{x_3=D} \tag{133}$$

and

$$H_2^{(II)}(x_1, x_3|\omega)\Big|_{x_3=\zeta(x_1)} = H_2^{(III)}(x_1, x_3|\omega)\Big|_{x_3=\zeta(x_1)}$$

$$\frac{1}{\epsilon(\omega)} \frac{\partial}{\partial N} H_2^{(II)}(x_1, x_3|\omega)\Big|_{x_3\zeta(x_1)} = \frac{\partial}{\partial N} H_2^{(III)}(x_1, x_3|\omega)\Big|_{x_3=\zeta(x_1)}, \quad (134)$$

respectively.

As in the derivation of the boundary conditions at the harmonic frequency at a vacuum–metal interface, a central role in the derivation of the boundary conditions at the harmonic frequency at the prism–metal film interface and at the metal film–vacuum interface in the Kretschmann ATR geometry is played by the nonlinear polarization in a centrosymmetric metal film. In this case, we take it to have the form [22]

$$\mathbf{P}_{I,III}^{NL} = \frac{1}{4\pi}[\alpha_{I,III}\mathbf{E}(\nabla \cdot \mathbf{E}) + \beta_{I,III}(\mathbf{E} \cdot \nabla)\mathbf{E}$$

$$+\gamma_{I,III}\mathbf{E} \times (\nabla \times \mathbf{E}) + \mathbf{E}(\mathbf{E} \cdot \nabla \rho_{I,III}) + (\mathbf{E} \cdot \mathbf{E})\nabla \kappa_{I,III}], \quad (135)$$

where the subscripts I and III denote the prism–metal and metal–vacuum interfaces, respectively, $\mathbf{E}$ is the macroscopic electric field, and the coefficients in general depend on the distance from the corresponding interface.

The nonlinear boundary conditions for the harmonic fields are obtained by integrating the Maxwell equations for them across the interfacial layers and then passing to the limit of vanishing thicknesses of the layers. Without going into the details of their derivation, which mimics their derivation at a rough vacuum–metal interface presented in Sect. 3.1, we note that when the incident field is p-polarized, the fields at the harmonic frequency $2\omega$ are also p-polarized. The nonlinear boundary conditions at the prism–metal interface $x_3 = D$ can then be written in the forms

$$H_2^{(I)}(x_1, x_3|\omega)\Big|_{x_3=D} - H_2^{(II)}(x_1, x_3|\omega)\Big|_{x_3=D}$$

$$= \frac{2ic}{\omega}\mu_3^{(I)}\frac{1}{\epsilon_0}L^{(I)}(x_1|\omega)\frac{d}{dx_1}H^{(I)}(x_1|\omega)$$

$$\equiv A^{(I)}(x_1|2\omega) \qquad (136)$$

$$\frac{1}{\epsilon_0}\frac{\partial}{\partial x_3} H_2^{(I)}(x_1, x_3|\omega)\Big|_{x_3=D} - \frac{1}{\epsilon(2\omega)}\frac{\partial}{\partial x_3} H_2^{(II)}(x_1, x_3|\omega)\Big|_{x_3=D}$$

$$= \frac{2ic}{\omega}\frac{d}{dx_1}\left[\mu_1^{(I)}\left(\frac{d}{dx_1}H^{(I)}(x_1|\omega)\right)^2 + \mu_2^{(I)}\frac{1}{\epsilon_0^2}L^{(I)}(x_1|\omega)^2\right]$$

$$\equiv B^{(I)}(x_1|2\omega), \qquad (137)$$

where the nonlinearity constants are given by

$$\mu_1^{(I)} = \lim_{\eta \to 0} \int_{D-\eta}^{D+\eta} dx_3 \frac{1}{\epsilon(2\omega, x_3)} \left\{ [\alpha_I(x_3) + \beta_I(x_3)] \frac{d}{dx_3} \frac{1}{\epsilon^2(\omega, x_3)} \right.$$

$$\left. + \frac{1}{\epsilon^2(\omega, x_3)} \frac{d}{dx_3} [\rho_I(x_3) + \kappa_I(x_3)] \right\} , \tag{138}$$

$$\mu_2^{(I)} = \lim_{\eta \to 0} \int_{D-\eta}^{D+\eta} dx_3 \frac{1}{\epsilon(2\omega, x_3)} \frac{d\kappa_I(x_3)}{dx_3} , \tag{139}$$

$$\mu_3^{(I)} = \lim_{\eta \to 0} \int_{D-\eta}^{D+\eta} dx_3 \left[ \alpha_I(x_3) \frac{d}{dx_3} \frac{1}{\epsilon(\omega, x_3)} + \frac{1}{\epsilon(\omega, x_3)} \frac{d\rho_I(x_3)}{dx_3} \right] . \tag{140}$$

At the same time, the nonlinear boundary conditions at the metal film–vacuum interface $x_3 = \zeta(x_1)$ become

$$H_2^{(II)}(x_1, x_3|\omega)\Big|_{x_3=\zeta(x_1)} - H_2^{(III)}(x_1, x_3|\omega)\Big|_{x_3=\zeta(x_1)}$$

$$= \frac{2ic}{\omega} \frac{\mu_3^{(III)}}{\phi^2(x_1)} L^{(III)}(x_1|\omega) \frac{d}{dx_1} H^{(III)}(x_1|\omega)$$

$$\equiv A^{(III)}(x_1|2\omega) \tag{141}$$

$$\frac{1}{\epsilon(2\omega)} \frac{\partial}{\partial N} H_2^{(II)}(x_1, x_3|\omega)\Big|_{x_3=\zeta(x_1)} - \frac{\partial}{\partial N} H_2^{(III)}(x_1, x_3|2\omega)\Big|_{x_3=\zeta(x_1)}$$

$$= \frac{2ic}{\omega} \frac{d}{dx_1} \left( \frac{1}{\phi^2(x_1)} \left\{ \mu_1^{(III)} \left[ \frac{d}{dx_1} H^{(III)}(x_1|\omega) \right]^2 + \mu_2^{(III)} L^{(III)}(x_1|\omega)^2 \right\} \right) . \tag{142}$$

The nonlinearity coefficients in the case are

$$\mu_1^{(III)} = \lim_{\eta \to 0} \int_{-\eta}^{\eta} \frac{1}{\epsilon(2\omega, z)} \left\{ \frac{1}{2} [\alpha_{III}(z) + \beta_{III}(z)] \frac{d}{dz} \frac{1}{\epsilon^2(\omega, z)} \right.$$

$$\left. + \frac{1}{\epsilon^2(\omega, z)} \frac{d}{dz} [\rho_{III}(z) + \kappa_{III}(z)] \right\} , \tag{143}$$

$$\mu_2^{(\mathrm{III})} = \lim_{\eta \to 0} \int_{-\eta}^{\eta} dz \frac{1}{\epsilon(2\omega, z)} \frac{d\kappa_{\mathrm{III}}(z)}{dz} , \tag{144}$$

$$\mu_3^{(\mathrm{III})} = \lim_{\eta \to 0} \int_{-\eta}^{\eta} dz \left[ \alpha_{\mathrm{III}}(z) \frac{d}{dz} \frac{1}{\epsilon(\omega, z)} + \frac{1}{\epsilon(\omega, z)} \frac{d\rho_{\mathrm{III}}(z)}{dz} \right] , \tag{145}$$

where $z$ is a local coordinate normal to the interface $x_3 = \zeta(x_1)$ at each point. The boundary conditions (136)–(137) and (141)–(142) will be used in both the perturbative calculations of second-harmonic generation in a weakly rough metal film and in the computer simulations for a more strongly rough metal film.

## 4.2   Strongly Rough Surfaces

A monochromatic p-polarized beam of light of frequency $\omega$, whose plane of incidence is the $x_1 x_3$ plane illuminates the prism–metal interface $x_3 = D$ from the side of the prism. In this case, it is convenient to work with the single nonzero component of the magnetic vector at the fundamental frequency $\omega$.

The $x_2$ components of the magnetic fields at $\omega$ and $2\omega$ satisfy Helmholtz equations in each of the regions: $x_3 > D$ (region I); $\zeta(x_1) < x_3 < D$ (region II); and $x_3 < \zeta(x_1)$ (region III). By applying Green's second integral identity in the plane [26] to each of these regions in turn, we obtain the following equations:

$$\theta(x_3 - D)H_2^{(\mathrm{I})}(x_1, x_3 | \Omega) = H_0(x_1, x_3 | \Omega) + \frac{1}{4\pi} \int_{-\infty}^{\infty} dx_1'$$

$$\times \left\{ \left[ \frac{\partial}{\partial x_1'} G_{\epsilon_0}^{(\Omega)}(x_1, x_3 | x_1', x_3') \right]_{x_3' = D} H^{(\mathrm{I})}(x_1' | \Omega) \right.$$

$$\left. - \left[ G_{\epsilon_0}^{(\Omega)}(x_1, x_3 | x_1', x_3') \right]_{x_3' = D} L^{(\mathrm{I})}(x_1' | \Omega) \right\} \tag{146}$$

$$\theta[x_3 - \zeta(x_1)]\theta(D - x_3)H_2^{(\mathrm{II})}(x_1, x_3 | \Omega) = -\frac{1}{4\pi} \int_{-\infty}^{\infty} dx_1'$$

$$\times \left\{ \left[ \frac{\partial}{\partial x_3'} G_{\epsilon}^{(\Omega)}(x_1, x_3 | x_1', x_3') \right] H_2^{(\mathrm{II})}(x_1', x_3' | \Omega) \right.$$

$$- G_\epsilon^{(\Omega)}(x_1, x_3 | x_1', x_3') \frac{\partial}{\partial x_3'} H_2^{(\mathrm{II})}(x_1', x_3' | \Omega) \Bigg\}_{x_3' = D}$$

$$+ \frac{1}{4\pi} \int_{-\infty}^{\infty} dx_1' \left\{ \left[ \frac{\partial}{\partial N'} G_\epsilon^{(\Omega)}(x_1, x_3 | x_1', x_3') \right] H_2^{(\mathrm{II})}(x_1', x_3' | \Omega) \right.$$

$$\left. - G_\epsilon^{(\Omega)}(x_1, x_3 | x_1', x_3') \frac{\partial}{\partial N'} H_2^{(\mathrm{II})}(x_1', x_3' | \Omega) \right\}_{x_3' = \zeta(x_1')} ; \quad (147)$$

$$\theta[\zeta(x_1) - x_3] H_2^{(\mathrm{III})}(x_1, x_3 | \Omega) = -\frac{1}{4\pi} \int_{-\infty}^{\infty} dx_1'$$

$$\times \left\{ \left[ \frac{\partial}{\partial N'} G_0^{(\Omega)}(x_1, x_3 | x_1', x_3') \right]_{x_3' = \zeta(x_1')} H^{(\mathrm{III})}(x_1' | \Omega) \right.$$

$$\left. - \left[ G_0^{(\Omega)}(x_1, x_3 | x_1', x_3') \right]_{x_3' = \zeta(x_1')} L^{(\mathrm{III})}(x_1' | \Omega) \right\}, \quad (148)$$

respectively. The source functions $H^{(\mathrm{I,III})}(x_1 | \Omega)$ and $L^{(\mathrm{I,III})}(x_1 | \Omega)$ are defined by

$$H^{(\mathrm{I})}(x_1 | \Omega) = H_2^{(\mathrm{I})}(x_1, x_3 | \Omega) \big|_{x_3 = D} , \qquad (149)$$

$$L^{(\mathrm{I})}(x_1 | \Omega) = \frac{\partial}{\partial x_3} H_2^{(\mathrm{I})}(x_1, x_3 | \Omega) \big|_{x_3 = D} , \qquad (150)$$

$$H^{(\mathrm{III})}(x_1 | \Omega) = H_2^{(\mathrm{III})}(x_1, x_3 | \Omega) \big|_{x_3 = \zeta(x_1)} , \qquad (151)$$

$$L^{(\mathrm{III})}(x_1 | \Omega) = \frac{\partial}{\partial N} H_2^{(\mathrm{III})}(x_1, x_3 | \Omega) \big|_{x_3 = \zeta(x_1)} , \qquad (152)$$

and

$$H_0(x_1, x_3 | \Omega) = H_2^{(\mathrm{I})}(x_1, x_3 | \omega)_{\mathrm{inc}} , \quad \Omega = \omega ,$$

$$= 0 , \quad \Omega = 2\omega . \qquad (153)$$

The functions $G_{\epsilon_0}^{(\Omega)}(x_1, x_3 | x_1', x_3')$, $G_{\epsilon}^{(\Omega)}(x_1, x_3 | x_1', x_3')$, and $G_0^{(\Omega)}(x_1, x_3 | x_1', x_3')$ are Green's functions defined by the following expressions:

$$G_{\epsilon_0}^{(\Omega)}(x_1, x_3 | x_1', x_3') = \int_{-\infty}^{\infty} \frac{dq}{2\pi} \frac{2\pi i}{\alpha_p(q, \Omega)} \exp[iq(x_1 - x_1')$$

$$+ i\alpha_p(q, \Omega)|x_3 - x_3'|]$$

$$= i\pi H_0^{(1)} \left\{ \sqrt{\epsilon_0}(\Omega/c)[(x_1 - x_1')^2 + (x_3 - x_3')^2]^{\frac{1}{2}} \right\},$$

$$G_{\epsilon}^{(\Omega)}(x_1, x_3 | x_1', x_3') = \int_{-\infty}^{\infty} \frac{dq}{2\pi} \frac{2\pi}{\beta(q, \Omega)} \exp[iq(x_1 - x_1') - \beta(q, \Omega)|x_3 - x_3'|]$$

$$= i\pi H_0^{(1)} \left\{ n_\epsilon(\Omega)(\Omega/c)[(x_1 - x_1')^2 + (x_3 - x_3')^2]^{\frac{1}{2}} \right\},$$

$$G_0^{(\Omega)}(x_1, x_3 | x_1', x_3') = \int_{-\infty}^{\infty} \frac{dq}{2\pi} \frac{2\pi i}{\alpha_0(q, \Omega)} \exp[iq(x_1 - x_1')$$

$$+ i\alpha_0(q, \Omega)|x_3 - x_3'|]$$

$$= i\pi H_0^{(0)} \{(\Omega/c)[(x_1 - x_1')^2 + (x_3 - x_3')^2]^{\frac{1}{2}} \}, \tag{154}$$

where

$$\alpha_p(q, \Omega) = \left[ \epsilon_0(\Omega/c)^2 - q^2 \right]^{\frac{1}{2}}, \quad q^2 < \epsilon_0(\Omega/c)^2,$$

$$= i \left[ q^2 - \epsilon_0(\Omega/c)^2 \right]^{\frac{1}{2}}, \quad q^2 > \epsilon_0(\Omega/c)^2,$$

$$\beta(q, \Omega) = \left[ q^2 - \epsilon(\Omega)(\Omega/c)^2 \right]^{\frac{1}{2}}, \quad \text{Re}\,\beta(q, \Omega) > 0, \text{Im}\,\beta(q, \Omega) < 0,$$

$$\alpha_0(q, \Omega) = \left[ (\Omega/c)^2 - q^2 \right]^{\frac{1}{2}}, \quad q^2 < (\Omega/c)^2,$$

$$= i \left[ q^2 - (\Omega/c)^2 \right]^{\frac{1}{2}}, \quad q^2 > (\Omega/c)^2, \tag{155}$$

and

$$n_\epsilon(\Omega) = [\epsilon(\Omega)]^{\frac{1}{2}}, \quad \text{Re}\,n_\epsilon(\Omega) > 0, \text{Im}\,n_\epsilon(\Omega) > 0, \tag{156}$$

and $H^{(0)}(z)$ is a Hankel function of the first kind.

The scattered field in the prism is given by the second term on the right-hand side of (146), which in the far field becomes

$$H_2^{(I)}(x_1, x_3 | \Omega)_{\text{sc}} = \int_{-\infty}^{\infty} \frac{dq}{2\pi} R(q, \Omega) \exp[iqx_1 + i\alpha_p(q, \Omega)x_3], \qquad (157)$$

where

$$R(q, \Omega) = \frac{1}{2i\alpha_p(q, \Omega)} \int_{-\infty}^{\infty} dx_1 \exp[-iqx_1 - i\alpha_p(q, \Omega)D]$$

$$[\alpha_p(q, \Omega)H^{(I)}(x_1 | \Omega) + L^{(I)}(x_1 | \Omega)]. \qquad (158)$$

Similarly, the transmitted field in the vacuum region is given by the right-hand side of (148), which in the far field becomes

$$H_2^{(III)}(x_1, x_3 | \Omega)_{\text{tr}} = \int_{-\infty}^{\infty} \frac{dq}{2\pi} T(q, \Omega) \exp[iqx_1 - i\alpha_0(q, \Omega)x_3], \qquad (159)$$

where

$$T(q, \Omega) = \frac{1}{2i\alpha_0(q, \Omega)} \int_{-\infty}^{\infty} dx_1 \exp[-iqx_1 + i\alpha(q, \Omega)\zeta(x_1)]$$

$$\times \left\{ i[q\zeta'(x_1) + \alpha_0(q, \Omega)]H^{(III)}(x_1 | \Omega) - L^{(III)}(x_1 | \Omega) \right\}. \qquad (160)$$

The equations for the source functions $H^{(I,III)}(x_1 | \omega)$ and $L^{(I,III)}(x_1 | \omega)$ at the fundamental frequency $\omega$ that enter the nonlinear boundary conditions (136)–(137) and (141)–(142) are obtained by setting $x_3 = D + \eta$, where $\eta$ is a positive infinitesimal, in (146) and (147), setting $x_3 = \zeta(x_1) + \eta$ in (147) and (148), and using the linear boundary conditions (133) and (134) at $x_3 = D$ and $x_3 = \zeta(x_1)$, respectively. The results are

$$H^{(I)}(x_1 | \omega) = H(x_1 | \omega)_{\text{inc}} + \frac{1}{4\pi} \int_{-\infty}^{\infty} dx_1'$$

$$\left\{ \left[ \frac{\partial}{\partial x_3'} G_{\epsilon_0}^{(\omega)}(x_1, x_3 | x_1', x_3') \right]_{\substack{x_3 = D + \eta \\ x_3' = D}} \right.$$

$$\times H^{(I)}(x_1' | \omega) - \left[ G_{\epsilon_0}^{(\omega)}(x_1, x_3 | x_1', x_3') \right]_{\substack{x_3 = D + \eta \\ x_3' = D}} L^{(I)}(x_1' | \omega) \right\} \qquad (161)$$

$$0 = \frac{1}{4\pi} \int_{-\infty}^{\infty} dx_1'$$

$$\left\{ -\left[ \frac{\partial}{\partial x_3'} G_\epsilon^{(\omega)}(x_1, x_3 | x_1', x_3') \right]_{\substack{x_3 = D + \eta \\ x_3' = D}} H^{(I)}(x_1' | \omega) \right.$$

$$+ \frac{\epsilon(\omega)}{\epsilon_0} \left[ G_\epsilon^{(\omega)}(x_1, x_3 | x_1', x_3') \right]_{\substack{x_3 = D + \eta \\ x_3' = D}} L^{(I)}(x_1' | \omega)$$

$$+ \left[ \frac{\partial}{\partial N'} G_\epsilon^{(\omega)}(x_1, x_3 | x_1', x_3') \right]_{\substack{x_3 = D \\ x_3' = \zeta(x_1')}} H^{(III)}(x_1' | \omega)$$

$$\left. - \epsilon(\omega) \left[ G_\epsilon^{(\omega)}(x_1, x_3 | x_1', x_3') \right]_{\substack{x_3 = D \\ x_3' = \zeta(x_1')}} L^{(III)}(x_1' | \omega) \right\} , \tag{162}$$

$$H^{(III)}(x_1 | \omega) = \frac{1}{4\pi} \int_{-\infty}^{\infty} dx_1'$$

$$\left\{ -\left[ \frac{\partial}{\partial x_3'} G_\epsilon^{(\omega)}(x_1, x_3 | x_1', x_3') \right]_{\substack{x_3 = \zeta(x_1) \\ x_3' = D}} H^{(I)}(x_1' | \omega) \right.$$

$$+ \frac{\epsilon(\omega)}{\epsilon_0} \left[ G_\epsilon^{(\omega)}(x_1, x_3 | x_1', x_3') \right]_{\substack{x_3 = \zeta(x_1) \\ x_3' = D}} L^{(I)}(x_1' | \omega)$$

$$+ \left[ \frac{\partial}{\partial N'} G_\epsilon^{(\omega)}(x_1, x_3 | x_1', x_3') \right]_{\substack{x_3 = \zeta(x_1) + \eta \\ x_3' = \zeta(x_1')}} H^{(III)}(x_1' | \omega)$$

$$- \epsilon(\omega) \left[ G_\epsilon^{(\omega)}(x_1, x_3 | x_1', x_3') \right]_{\substack{x_3 = \zeta(x_1) + \eta \\ x_3' = \zeta(x_1')}}$$

$$\left. L^{(III)}(x_1' | \omega) \right\} , \tag{163}$$

$$0 = \frac{1}{4\pi} \int_{-\infty}^{\infty} dx_1'$$

$$\left\{ -\left[ \frac{\partial}{\partial N'} G_0^{(\omega)}(x_1, x_3 | x_1', x_3') \right]_{\substack{x_3 = \zeta(x_1) + \eta \\ x_3' = \zeta(x_1')}} H^{(III)}(x_1' | \omega) \right.$$

$$\left. + \left[ G_0^{(\omega)}(x_1, x_3 | x_1', x_3') \right]_{\substack{x_3 = \zeta(x_1) + \eta \\ x_3' = \zeta(x_1')}} L^{(III)}(x_1' | \omega) \right\}. \tag{164}$$

These integral equations are solved by converting them into matrix equations. This is done by replacing the infinite range of integration by the finite range $(-L/2, L/2)$, evaluating the resulting integrals over $x_1'$ by a numerical quadrature scheme, and finally setting $x_1$ equal to the values of the abscissas used in the numerical quadrature scheme.

To obtain the source functions at the harmonic frequency $2\omega$, we set $x_3 = D + \eta$, where $\eta$ is a positive infinitesimal, in (146) and (147), set $x_3 = \zeta(x_1) + \eta$ in (147) and (148), and use the nonlinear boundary conditions (136), (137), and (141), (142) at $x_3 = D$ and $x_3 = \zeta(x_1)$, respectively. The resulting equations can be written as

$$H^{(I)}(x_1 | 2\omega) = \frac{1}{4\pi} \int_{-\infty}^{\infty} dx_1' \left\{ \left[ \frac{\partial}{\partial x_3'} G_{\epsilon_0}^{(2\omega)}(x_1, x_3 | x_1', x_3') \right]_{\substack{x_3 = D + \eta \\ x_3' = D}} \right.$$

$$\left. \times H^{(I)}(x_1' | 2\omega) - \left[ G_{\epsilon_0}^{(2\omega)}(x_1, x_3 | x_1', x_3') \right]_{\substack{x_3 = D + \eta \\ x_3' = D}} L^{(I)}(x_1' | 2\omega) \right\}, \tag{165}$$

$$Q^{(I)}(x_1 | 2\omega) = \frac{1}{4\pi} \int_{-\infty}^{\infty} dx_1' \left\{ -\left[ \frac{\partial}{\partial x_3'} G_{\epsilon}^{(2\omega)}(x_1, x_3 | x_1', x_3') \right]_{\substack{x_3 = D + \eta \\ x_3' = D}} \right.$$

$$\times H^{(I)}(x_1' | 2\omega) + \frac{\epsilon(2\omega)}{\epsilon_0} \left[ G_{\epsilon}^{(2\omega)}(x_1, x_3 | x_1', x_3') \right]_{\substack{x_3 = D + \eta \\ x_3' = D}} L^{(I)}(x_1' | 2\omega)$$

$$+ \left[ \frac{\partial}{\partial N'} G_{\epsilon}^{(2\omega)}(x_1, x_3 | x_1', x_3') \right]_{\substack{x_3 = D \\ x_3' = \zeta(x_1')}} H^{(\mathrm{III})}(x_1' | 2\omega)$$

$$- \epsilon(2\omega) \left[ G_{\epsilon}^{(2\omega)}(x_1, x_3 | x_1', x_3') \right]_{\substack{x_3 = D \\ x_3' = \zeta(x_1')}} L^{(\mathrm{III})}(x_1' | 2\omega) \Bigg\} \qquad (166)$$

$$H^{(\mathrm{III})}(x_1 | 2\omega) = Q^{(\mathrm{III})}(x_1 | 2\omega) + \frac{1}{4\pi} \int_{-\infty}^{\infty} dx_1'$$

$$\Bigg\{ - \left[ \frac{\partial}{\partial x_3'} G_{\epsilon}^{(2\omega)}(x_1, x_3 | x_1', x_3') \right]_{\substack{x_3 = \zeta(x_1) \\ x_3' = D}}$$

$$\times H^{(\mathrm{I})}(x_1' | 2\omega) + \frac{\epsilon(2\omega)}{\epsilon_0} \left[ G_{\epsilon}^{(2\omega)}(x_1, x_3 | x_1', x_3') \right]_{\substack{x_3 = \zeta(x_1) \\ x_3' = D}} L^{(\mathrm{I})}(x_1' | 2\omega)$$

$$+ \left[ \frac{\partial}{\partial N'} G_{\epsilon}^{(2\omega)}(x_1, x_3 | x_1', x_3') \right]_{\substack{x_3 = \zeta(x_1) + \eta \\ x_3' = \zeta(x_1')}} H^{(\mathrm{III})}(x_1' | 2\omega)$$

$$- \epsilon(2\omega) \left[ G_{\epsilon}^{(2\omega)}(x_1, x_3 | x_1', x_3') \right]_{\substack{x_3 = \zeta(x_1) + \eta \\ x_3' = \zeta(x_1')}} L^{(\mathrm{III})}(x_1' | 2\omega) \Bigg\}, \qquad (167)$$

$$0 = \frac{1}{4\pi} \int_{-\infty}^{\infty} dx_1'$$

$$\Bigg\{ - \left[ \frac{\partial}{\partial N'} G_0^{(2\omega)}(x_1, x_3 | x_1', x_3') \right]_{\substack{x_3 = \zeta(x_1) + \eta \\ x_3' = \zeta(x_1')}} H^{(\mathrm{III})}(x_1' | 2\omega)$$

$$+ \left[ G_0^{(2\omega)}(x_1, x_3 | x_1', x_3') \right]_{\substack{x_3 = \zeta(x_1) + \eta \\ x_3' = \zeta(x_1')}} L^{(\mathrm{III})}(x_1' | 2\omega) \Bigg\}, \qquad (168)$$

where

$$Q^{(I)}(x_1|2\omega) = \frac{1}{4\pi} \int_{-\infty}^{\infty} dx_1'$$

$$
\left\{ -\left[ \frac{\partial}{\partial x_3'} G_\epsilon^{(2\omega)}(x_1, x_3|x_1', x_3') \right]_{\substack{x_3 = D + \eta \\ x_3' + D}} A^{(I)}(x_1'|2\omega) \right.
$$

$$
+\epsilon(2\omega) \left[ G_\epsilon^{(2\omega)}(x_1, x_3|x_1', x_3') \right]_{\substack{x_3 = D + \eta \\ x_3' = D}} B^{(I)}(x_1'|2\omega)
$$

$$
-\left[ \frac{\partial}{\partial N'} G_\epsilon^{(2\omega)}(x_1, x_3|x_1', x_3') \right]_{\substack{x_3 = D \\ x_3' = \zeta(x_1')}} A^{(III)}(x_1'|2\omega)
$$

$$
\left. +\epsilon(2\omega) \left[ G_\epsilon^{(2\omega)}(x_1, x_3|x_1', x_3') \right]_{\substack{x_3 = D \\ x_3 = \zeta(x_1')}} B^{(III)}(x_1'|2\omega) \right\}, \quad (169)
$$

$$Q^{(III)}(x_1|2\omega) = A^{(III)}(x_1|2\omega) + \frac{1}{4\pi} \int_{-\infty}^{\infty} dx_1'$$

$$
\left\{ \left[ \frac{\partial}{\partial x_3'} G_\epsilon^{(2\omega)}(x_1, x_3|x_1', x_3') \right]_{\substack{x_3 = \zeta(x_1) \\ x_3' = D}} \right.
$$

$$
\times A^{(I)}(x_1'|2\omega) - \epsilon(2\omega) \left[ G_\epsilon^{(2\omega)}(x_1, x_3|x_1', x_3') \right]_{\substack{x_3 = \zeta(x_1) \\ x_3' = D}} B^{(I)}(x_1'|2\omega)
$$

$$
+\left[ \frac{\partial}{\partial N'} G_\epsilon^{(2\omega)}(x_1, x_3|x_1', x_3') \right]_{\substack{x_3 = \zeta(x_1) + \eta \\ x_3' = \zeta(x_1')}} A^{(III)}(x_1'|2\omega)
$$

$$
\left. -\epsilon(2\omega) \left[ G_\epsilon^{(2\omega)}(x_1, x_3|x_1', x_3') \right]_{\substack{x_3 = \zeta(x_1) + \eta \\ x_3' = \zeta(x_1')}} B^{(III)}(x_1'|2\omega) \right\}. \quad (170)
$$

We note that the functions $Q^{(I)}(x_1|2\omega)$ and $Q^{(III)}(x_1|2\omega)$ play the role of sources in (166) and (167). Equations (165)–(170) are solved numerically by the approach described earlier.

When the source functions $H^{(\mathrm{I,III})}(x_1|2\omega)$ and $L^{(\mathrm{I,III})}(x_1|2\omega)$ have been obtained in this fashion, the amplitudes of the scattered and transmitted fields in the far field can be calculated from (146) and (148), with the results

$$H_2^{(\mathrm{I})}(x_1,x_3|2\omega)_{\mathrm{sc}} = \int_{-2\sqrt{\epsilon_0}(\omega/c)}^{2\sqrt{\epsilon_0}(\omega/c)} \frac{\mathrm{d}q}{2\pi} R(q,2\omega)$$

$$\times \exp[\mathrm{i}qx_1 + \mathrm{i}\alpha_{\mathrm{p}}(q,2\omega)x_3]\,, \tag{171}$$

with

$$R(q,2\omega) = \frac{1}{2\mathrm{i}\alpha_{\mathrm{p}}(q,2\omega)} \int_{-\infty}^{\infty} \mathrm{d}x_1 \exp[-\mathrm{i}qx_1 - \mathrm{i}\alpha_{\mathrm{p}}(q,2\omega)D]$$

$$\times [\mathrm{i}\alpha_{\mathrm{p}}(q,2\omega)H^{(\mathrm{I})}(x_1|2\omega) + L^{(\mathrm{I})}(x_1|2\omega)], \tag{172}$$

and

$$H_2^{(\mathrm{III})}(x_1,x_3|2\omega)_{\mathrm{tr}} = \int_{-2(\omega/c)}^{2(\omega/c)} \frac{\mathrm{d}q}{2\pi} T(q,2\omega) \exp[\mathrm{i}qx_1 - \mathrm{i}\alpha_0(q,2\omega)x_3], \tag{173}$$

with

$$T(q,2\omega) = \frac{1}{2\mathrm{i}\alpha_0(q,2\omega)} \int_{-\infty}^{\infty} \mathrm{d}x_1 \exp[-\mathrm{i}qx_1 + \mathrm{i}\alpha_0(q,2\omega)\zeta(x_1)]$$

$$\times \{\mathrm{i}[q\zeta'(x_1) + \alpha_0(q,2\omega)]H^{(\mathrm{III})}(x_1|2\omega) - L^{(\mathrm{III})}(x_1|2\omega)\}\,. \tag{174}$$

We introduce the angles of scattering and transmission by

$$q = 2\sqrt{\epsilon_0}(\omega/c)\sin\theta_{\mathrm{s}}\,, \qquad q = 2(\omega/c)\sin\theta_{\mathrm{t}}\,, \tag{175}$$

respectively. Then the total, time-averaged, scattered flux crossing the plane $x_3 = \mathrm{const} > D$ can be written as

$$P_{\mathrm{sc}}^{(2\omega)} = L_2 \frac{c^2}{128\pi^2\omega\epsilon_0} \int_{-\frac{\pi}{2}}^{\frac{\pi}{2}} \mathrm{d}\theta_{\mathrm{s}} |r(\theta_{\mathrm{s}}|2\omega)|^2, \tag{176}$$

where

$$r(\theta_{\mathrm{s}}|2\omega) = 4\mathrm{i}\sqrt{\epsilon_0}(\omega/c)\cos\theta_{\mathrm{s}} R[2\sqrt{\epsilon_0}(\omega/c)\sin\theta_{\mathrm{s}}, 2\omega] \tag{177}$$

and the total, time-averaged, transmitted flux crossing the plane $x_3 = \mathrm{const} < \zeta(x_1)_{\mathrm{min}}$ is given by

$$P_{\mathrm{tr}}^{(2\omega)} = L_2 \frac{c^2}{128\pi^2\omega} \int_{-\frac{\pi}{2}}^{\frac{\pi}{2}} \mathrm{d}\theta_{\mathrm{t}} |t(\theta_{\mathrm{t}}|2\omega)|^2\,, \tag{178}$$

where

$$t(\theta_t|2\omega) = 4i(\omega/c)\cos\theta_t T[2(\omega/c)\sin\theta_t, 2\omega].  \tag{179}$$

The efficiency of the second-harmonic generation in reflection is defined as the total power of the scattered harmonic light normalized by the square of the power of the incident field and multiplied by the illuminated area $S$:

$$I(\theta_s|2\omega) = \frac{P_{sc}^{(2\omega)}}{P_{inc}^2}S.  \tag{180}$$

Similarly, the efficiency of second-harmonic generation in transmission is given by

$$I(\theta_t|2\omega) = \frac{P_{tr}^{(2\omega)}}{P_{inc}^2}S.  \tag{181}$$

The efficiencies defined in this way do not depend on the incident power, and therefore are convenient to use in experimental measurements of second-harmonic generation.

For the incident field, we use a superposition of an infinite number of incoming plane waves,

$$H_2^{(I)}(x_1,x_3|\omega)_{inc} = H_0\sqrt{\epsilon_0}\frac{\omega}{c}\frac{w}{2\sqrt{\pi}}\int_{-\frac{\pi}{2}}^{\frac{\pi}{2}} d\theta \exp[-\epsilon_0\frac{\omega^2 w^2}{4c^2}(\theta-\theta_0)^2]$$

$$\times \exp\{i\sqrt{\epsilon_0}(\omega/c)(x_1\sin\theta - (x_3-D)\cos\theta\},  \tag{182}$$

where $\theta_0$ is the angle of incidence. The magnitude of the total time-averaged flux incident on the interface $x_3 = D$ is then

$$P_{inc} = \left| \int_{-\infty}^{\infty} dx_1 \int_{-\frac{1}{2}L_2}^{\frac{1}{2}L_2} dx_2 \text{Re}\left[S_3^c\right]_{inc} \right|$$

$$= L_2|H_0|^2 \frac{cw}{16\sqrt{2\pi\epsilon_0}}\left\{\text{erf}\left[\sqrt{\epsilon_0}\frac{\omega w}{\sqrt{2}c}\left(\frac{\pi}{2}-\theta_0\right)\right]\right.$$

$$\left.+\text{erf}\left[\sqrt{\epsilon_0}\frac{\omega w}{\sqrt{2}c}\left(\frac{\pi}{2}+\theta_0\right)\right]\right\},  \tag{183}$$

where $S_3^c$ is the three-component of the complex Poynting vector, $L_2$ is the length of the surface along the $x_2$ axis, and $\text{erf}(z)$ is the error function.

The contributions to the mean normalized scattered and transmitted intensities rom the incoherent components of the scattered and transmitted harmonic light are then given by

$$\langle I(\theta_s|2\omega)\rangle_{\text{inc}} = SL_2 \frac{c^2}{128\pi^2\omega\epsilon_0} \frac{[\langle|r(\theta_s|2\omega)|^2\rangle - |\langle r(\theta_s|2\omega)\rangle|^2]}{P_{\text{inc}}^2},$$

$$\langle I(\theta_t|2\omega)\rangle_{\text{inc}} = SL_2 \frac{c^2}{128\pi^2\omega} \frac{[\langle|t(\theta_t|2\omega)|^2\rangle - |\langle t(\theta_t|2\omega)\rangle|^2]}{P_{\text{inc}}^2}, \qquad (184)$$

respectively.

In the numerical calculations based on the preceding results, we used the phenomenological nonlinearity coefficients $\mu_\alpha^{(\text{I,III})}(\alpha = 1, 2, 3)$ calculated on the basis of the free-electron model [23,24,25]. Although this model does not satisfy the energy conservation law, as noted above, it yields analytic expressions for these coefficients which provide a good fit to experimental data on second-harmonic generation reflected from a planar silver surface [24]. The expressions for these coefficients are [47]

$$\mu_1^{(\text{I})} = -2\beta\left\{A\ln\left[\frac{\epsilon(\omega)}{\epsilon(2\omega)}\right] - B\left[\frac{1}{\epsilon_0} - \frac{1}{\epsilon(\omega)}\right] - \frac{C}{2}\left[\frac{1}{\epsilon_0^2} - \frac{1}{\epsilon^2(\omega)}\right]\right\},$$

$$\mu_2^{(\text{I})} = 0,$$

$$\mu_3^{(\text{I})} = -\beta\left[\frac{1}{\epsilon(\omega)} - \frac{1}{\epsilon_0}\right], \qquad (185)$$

and

$$\mu_1^{(\text{III})} = \frac{1}{3}\beta\left(\frac{\{\epsilon(\omega) - 1\}\{\epsilon(\omega) - 3\}}{\epsilon^2(\omega)} - \frac{4}{3}\ln\left[\frac{\epsilon(\omega)}{\epsilon(2\omega)}\right]\right),$$

$$\mu_2^{(\text{III})} = 0,$$

$$\mu_3^{(\text{III})} = \beta\left[\frac{1 - \epsilon(\omega)}{\epsilon(\omega)}\right], \qquad (186)$$

where

$$\beta = e/(8\pi m\omega^2), \qquad (187)$$

with e the magnitude of the electronic charge and $m$ the electron mass,

$$A = -\frac{4 - 3\alpha}{3\epsilon_0\alpha}B,$$

$$B = -\frac{2}{3\epsilon_0} + \frac{4 - 3\alpha}{3\epsilon_0\alpha}C,$$

$$C = \frac{2}{3}\frac{2 + \epsilon_0\alpha}{\epsilon_0\alpha}, \qquad (188)$$

and

$$\alpha = \frac{1 - \epsilon(\omega)}{\epsilon_0 - \epsilon(\omega)} .$$ (189)

In Fig. 19, we present the contributions to the mean normalized scattered and transmitted intensities from the incoherent component of the second-harmonic light scattered from and transmitted through a rough silver film of mean thickness $D = 50$ nm when it is illuminated through the prism ($\epsilon_0 = 1.5$) at an angle of incidence $\theta_0 = 0°$ by p-polarized light of wavelength $\lambda = 1.063$ μm [$\epsilon(\omega) = -56.9 + i0.6, \epsilon(2\omega) = -11.56 + i0.27$]. The roughness of the back, random, surface is characterized by the Gaussian power spectrum (5) and the roughness parameters $\delta - 7$ nm and $a = 600$ nm. The numerical solution of the dispersion equation (41) for the film with planar surfaces yields two surface plasmon polaritons at the frequency of the incident light, the real parts of whose wave numbers are $q_1(\omega) = 1.009(\omega/c)$, $q_1(\omega) = 1.242(\omega/c)$. The real parts of the wave numbers of the surface plasmon polaritons at the harmonic frequency are $q_1(2\omega) = 1.046(2\omega/c)$, $q_2(2\omega) = 1.323(2\omega/c)$. $q_1(\omega)$ and $q_1(2\omega)$ are the wave numbers of the leaky waves. The imaginary parts of these wave numbers are three orders of magnitude smaller than their real parts, so they will be neglected in what follows.

Four prominent peaks are observed in both reflection and transmission in the results plotted in Fig. 19. These peaks occur due to the resonant nonlinear mixing of the surface plasmon polaritons of frequency $\omega$, excited through the roughness of the metal film–vacuum interface by the incident light, with the incident light. Through this interaction, the surface plasmon polaritons are converted into volume electromagnetic waves at the harmonic frequency $2\omega$. This process gives rise to four peaks in the angular dependence of the intensity of the incoherent component of the second-harmonic light, in reflection and transmission. The positions of these peaks can be obtained from the equations

$$k \pm q_{1,2}(\omega) = q_{sc}^{2\omega} ,$$ (190)

$$k \pm q_{1,2}(\omega) = q_{tr}^{2\omega} ,$$ (191)

where $k = \sqrt{\epsilon_0}(\omega/c)\sin\theta_0$ is the $x_1$ component of the wave vector of the incident light, $q_{sc}^{2\omega} = 2\sqrt{\epsilon_0}(\omega/c)\sin\theta_s$ is the $x_1$ component of the wave vector of the scattered second-harmonic light, and $q_{tr}^{2\omega} = (2\omega/c)\sin\theta_t$ is the $x_1$ component of the wave vector of the transmitted second-harmonic light.

We note that the process leading to (191) is not mediated by surface plasmon polaritons at the second-harmonic frequency since the Gaussian power spectrum we have assumed ensures the excitation of the surface plasmon polaritons of frequency $\omega$, whereas the excitation of surface plasmon polaritons of frequency $2\omega$ is effectively suppressed. This results because in the latter case, the wave number of the surface plasmon polariton lies in the wings of the power spectrum (especially at normal incidence). However, it can be seen

[cm²/Watt-Rad]

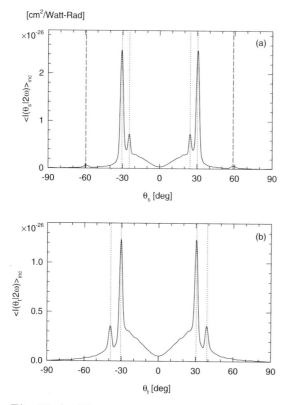

**Fig. 19a,b.** The mean normalized scattered and transmitted intensities of the second-harmonic light generated in the Kretschmann geometry with a silver film whose roughness is characterized by the Gaussian power spectrum (5). $\theta_0 = 0°, \lambda = 1.063\ \mu m,\ D = 50\ nm, \epsilon_0 = 1.5,\ \epsilon(\omega) = -56.9 + i0.6,\ \epsilon(2\omega) = -11.56 + i0.27, \delta = 7\ nm, a = 600\ nm;\ L = 25\ \mu m,\ g = L/4, N = 400, N_p = 1000$

from Fig. 19 that even the weak excitation of surface plasmon polaritons of frequency $2\omega$ gives rise to small leaky peaks in reflection marked by the dashed lines. The positions of the leaky peaks are given by

$$q_1(2\omega) = \pm 2\sqrt{\epsilon}(\omega/c)\sin\theta_{\text{leaky}}, \tag{192}$$

where $q_1(2\omega)$ is the wave number of the leaky surface plasmon polariton at the harmonic frequency and $\theta_{\text{leaky}}$ is the angle that defines the positions of the leaky peaks.

   In Fig. 20, we plot the contributions to the mean normalized scattered and transmitted intensities from the incoherent components of the light scattered and transmitted at frequency $2\omega$ for the same parameters used in obtaining Fig. 19 but with $\theta_0 = 5°$. The positions of the resonant peaks are shifted according to (190), and the right-hand peaks are more intense than the left-

[cm²/Watt-Rad]

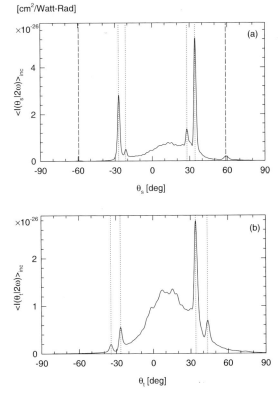

**Fig. 20a,b.** The same as Fig. 19 but with $\theta_0 = 5°$

hand peaks. The leaky peaks retain the same positions, and the right leaky peak is more intense than the left one.

Thus far, we have considered only small angles of incidence. However, it is worthwhile to study the case where the angle of incidence is optimal for exciting a surface plasmon polariton. If the angle of incidence is optimal, the incident light couples directly into a surface plasmon polariton, even if the film has planar surfaces. The optimal angle is obtained by matching the $x_1$ component of the wave vector of the incident light with the wave number of the surface plasmon polariton,

$$\sqrt{\epsilon_0}(\omega/c)\sin\theta_{\text{opt}} = q_{1,2}(\omega).\tag{193}$$

We assume the following parameters characterizing the scattering system: $\epsilon_0 = 16$, $D = 50\,\text{nm}$, $\lambda = 1.063\,\mu\text{m}$, $\epsilon(\omega) = -56.9 + i0.6$. For these parameters, we find from (193) that $q_1(\omega) = 1.009\omega/c$ and $q_2(\omega) = 4.73\omega/c$. There is therefore only one optimal angle, namely, the one corresponding to $q_1(\omega)$ in (193), which has the value $\theta_{\text{opt}} = 14.611°$. It can be seen from (191) by using the condition (193) that at the optimal angle of incidence, there must be a resonance peak in the direction normal to the mean surface in both reflection and transmission. We have pointed out earlier that for small angles of incidence, the positions of the resonance peaks shift according to (191), so that if the condition (193) for the optimal excitation of surface plasmon polaritons is satisfied, then one of the right-hand peaks in reflection or transmission occurs at a scattering angle equal to zero.

These qualitative arguments are illustrated by the results presented in Fig. 21 for the contributions to the mean normalized scattered and transmitted intensities of the second-harmonic light when the angle of incidence is optimal. The following values were used in obtaining these results: $\theta_0 = 14.61°$, $\lambda = 1.063$ μm, $\epsilon_0 = 16$, $D = 50$ nm, $\epsilon(\omega) = -56.9 + i0.6$, $\epsilon(2\omega) = -11.56 + i0.37$, $\delta = 8$ nm, $a = 300$ nm. The results plotted in Fig. 21 display distinct peaks in the direction normal to the mean surface both in reflection and transmission. A well-defined peak is also present in the retroreflection

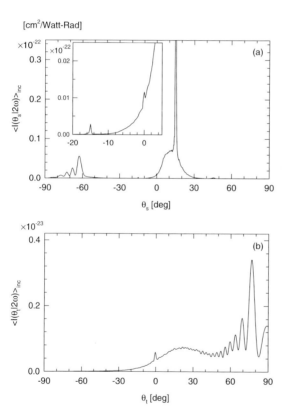

**Fig. 21a,b.** The mean normalized scattered and transmitted intensities of the second-harmonic light generated in the Kretschmann geometry with a silver film whose roughness is characterized by the Gaussian power spectrum (5). The angle of incidence is optimal for exciting a surface plasmon polariton. $\theta_0 = 14.61°$, $\lambda = 1.063$ μm, $D = 50$ nm, $\epsilon_0 = 16$, $\epsilon(\omega) = -56.9 + i0.6$, $\epsilon(2\omega) = -11.56 + i0.27$, $\delta = 8$ nm, $a = 300$ nm, $L = 30$ nm, $g = L/5$, $N = 400$, $N_p = 1000$

direction, but there is no enhanced transmission peak, because for the parameters chosen, it should occur in the nonradiative region of transmission angles. These results, therefore, confirm the existence of peaks in the direction normal to the mean surface. However, the amplitudes of these peaks can be very small, especially in reflection.

## 4.3 Weakly Rough Surfaces

As already mentioned in the preceding section, the angular distribution of the intensity of transmitted light of frequency $2\omega$ displays a peak in the direction normal to the mean surface. The origin of this peak, however, needs some further clarification. So, in the perturbative treatment of second-harmonic generation in transmission, we will be interested only in the origin of a peak of the enhanced second-harmonic generation in the direction normal to the mean surface. Therefore, in this subsection, we will derive an integral equation only for the transmitted amplitude. We assume again that the surface roughness satisfies the condition for the validity of the Rayleigh hypothesis, $|\zeta'(x_1)| \ll 1$ [39,40,41]. In this case, the Fourier integral representation of the transmitted field given by (159) can be continued onto the interface $x_3 = \zeta(x_1)$. The equation for the transmitted amplitude $T(q|k)$ can be obtained in the following manner. First, we set $x_3 < D$ in (146) and use the linear (when $\Omega = \omega$) or nonlinear (when $\Omega = 2\omega$) boundary conditions to replace the source functions $H^{(1)}(x_1'|\Omega)$ and $L^{(1)}(x_1'|\Omega)$ through $H^{(II)}(x_1', x_3'|\Omega)$ and $\partial H^{(II)}(x_1', x_3'|\Omega)/\partial x_3'$ at $x_3' = D$. Using the explicit expression for Green's function $G_{\epsilon_0}^{(\Omega)}(x_1, x_3|x_1', x_3')$, (154), we obtain the relation between the Fourier components of the field and its normal derivative evaluated at the interface $x_3 = D - \eta$, $\eta \to 0$, inside the film, which has the form

$$L^{(II)}(q, D) = \frac{i\alpha_p(q, \Omega)}{\epsilon_0} H^{(II)}(q, D) + J_0(q|\Omega) \,, \tag{194}$$

where we have denoted

$$H^{(II)}(q, D) = \int_{-\infty}^{\infty} dx_1 e^{-iqx_1} H^{(II)}(x_1, x_3|\Omega)_{x_3=D} \,, \tag{195}$$

$$L^{(II)}(q, D) = \int_{-\infty}^{\infty} dx_1 e^{-iqx_1} \frac{\partial}{\partial x_3} H^{(II)}(x_1, x_3|\Omega)_{x_3=D} \,, \tag{196}$$

and the function $J_0(q|\Omega)$ has different forms for the fundamental and second-harmonic fields:

$$J_0(q|\omega) = -\frac{2i\alpha_p(q, \omega)}{\epsilon_p} H_0 2\pi\delta(q - k) \,, \tag{197}$$

and

$$J_0(q|2\omega) = \int_{-\infty}^{\infty} dx_1 e^{-iqx_1} \left[ \frac{i\alpha_p(q, 2\omega)}{\epsilon_0} A^{(1)}(x_1|2\omega) - B^{(1)}(x_1|2\omega) \right]. \tag{198}$$

If in (147) we replace $H^{(II)}(x_1', x_3'|\Omega)$ and $\partial H^{(II)}(x_1', x_3'|\Omega)/\partial N'$ evaluated at the rough interface $x_3' = \zeta(x_1')$ by the source functions $H^{(III)}(x_1'|\Omega)$ and $L^{(III)}(x_1'|\Omega)$ through the linear (when $\Omega = \omega$) or nonlinear (when $\Omega = 2\omega$) boundary conditions, set first $x_3 > D$ and then $x_3 < \zeta(x_1)$, and use the explicit expression for Green's function $G_\epsilon^{(\Omega)}(x_1, x_3|x_1', x_3')$, (154), we obtain a pair of integral equations

$$0 = \left[ -\beta(q, \Omega) H^{(II)}(q, D) + L^{(II)}(q, D) \right] e^{-\beta(q,\Omega)D}$$

$$+ \int_{-\infty}^{\infty} dx_1 e^{-iqx_1} e^{\beta(q,\Omega)\zeta(x_1)} \left\{ [\beta(q, \Omega) + q\zeta'(x_1)] H^{(III)}(x_1|\Omega) \right.$$

$$\left. - \epsilon(2\omega) L^{(III)}(x_1|\Omega) \right\} + J_1(q, \Omega), \tag{199}$$

$$0 = \left[ \beta(q, \Omega) H^{(II)}(q, D) + L^{(II)}(q, D) \right] e^{\beta(q,\Omega)D}$$

$$+ \int_{-\infty}^{\infty} dx_1 e^{-iqx_1} e^{-\beta(q,\Omega)\zeta(x_1)} \left\{ [-\beta(q, \Omega) + q\zeta'(x_1)] H^{(III)}(x_1|\Omega) \right.$$

$$\left. + \epsilon(2\omega) L^{(III)}(x_1|\Omega) \right\} + J_2(q, \Omega), \tag{200}$$

where for the fundamental fields, $J_1(q|\omega) = J_2(q|\omega) = 0$, and for the harmonic fields,

$$J_1(q, 2\omega) = \int_{-\infty}^{\infty} dx_1 e^{-iqx_1} e^{\beta(q,2\omega)\zeta(x_1)} \left\{ [\beta(q, 2\omega) + q\zeta'(x_1)] A^{(III)}(x_1|2\omega) \right.$$

$$\left. - \epsilon(2\omega) B^{(III)}(x_1|2\omega) \right\}, \tag{201}$$

$$J_2(q, 2\omega) = \int_{-\infty}^{\infty} dx_1 e^{-iqx_1} e^{-\beta(q,2\omega)\zeta(x_1)} \left\{ [-\beta(q, 2\omega) \right.$$

$$\left. + q\zeta'(x_1)] A^{(III)}(x_1|2\omega) + \epsilon(2\omega) B^{(III)}(x_1|2\omega) \right\}. \tag{202}$$

Now, if we use the expression for the transmitted field given by (159) and eliminate the source functions $H^{(II)}(q, D)$ and $L^{(II)}(q, D)$ from (194), (199), and (200), we obtain a single integral equation for the transmitted amplitude $T(q, \Omega)$ in the form

$$\int_{-\infty}^{\infty} \frac{dp}{2\pi} M_\Omega(q|p) T(p, \Omega) = -N_\Omega(q|k), \tag{203}$$

where

$$M_\Omega(q|p) = i(1 - \epsilon(\Omega)) \left\{ \left[ \beta(q,\Omega) + \frac{i\alpha_p(q,\Omega)}{\epsilon_0} \right] e^{-\beta(q,\Omega)D} \right.$$

$$\times \frac{qp - i\beta(q,\Omega)\alpha_0(p\Omega)}{i\beta(q,\Omega) + \alpha_0(p\Omega)} I[i\beta(q,\Omega) + \alpha_0(p\Omega)|q - p]$$

$$+ \left[ \beta(q,\Omega) - \frac{i\alpha_p(q,\Omega)}{\epsilon_0} \right] e^{\beta(q,\Omega)D}$$

$$\left. \times \frac{qp + i\beta(q,\Omega)\alpha_0(p\Omega)}{-i\beta(q,\Omega) + \alpha_0(p,\Omega)} I[-i\beta(q,\Omega) + \alpha_0(p,\Omega)|q - p] \right\}, \quad (204)$$

and the function $N_\Omega(q|k)$ is

$$N_\omega(q|k) = -4\beta(q,\omega) \frac{i\alpha_p(q,\omega)}{\epsilon_0} H_0 2\pi \delta(q - k), \quad (205)$$

when $\Omega = \omega$ and is the nonlinear source function when $\Omega = 2\omega$:

$$N_{2\omega}(q|k) = 2\beta(q,2\omega) \left[ \frac{i\alpha_p(q,2\omega)}{\epsilon(2\omega)} A^{(I)}(q|2\omega) - B^{(I)}(q|2\omega) \right]$$

$$+ \int_{-\infty}^{\infty} dx_1 e^{-iqx_1} \left( \left[ \beta(q,2\omega) + \frac{i\alpha_p(q,2\omega)}{\epsilon_0} \right] e^{\beta(q,2\omega)[\zeta(x_1)-D]} \right.$$

$$\times \left\{ [\beta(q,2\omega) + iq\zeta'(x_1)] A^{(III)}(x_1|2\omega) - \epsilon(2\omega) B^{(III)}(x_1|2\omega) \right\}$$

$$+ \left[ \beta(q,2\omega) - \frac{i\alpha_p(q,2\omega)}{\epsilon_0} \right] e^{-\beta(q,2\omega)[\zeta(x_1)-D]}$$

$$\times \left\{ [-\beta(q,2\omega) + iq\zeta'(x_1)] \right.$$

$$\left. \left. \times A^{(III)}(x_1|2\omega) - \epsilon(2\omega) B^{(III)}(x_1|2\omega) \right\} \right), \quad (206)$$

where $A^{(I)}(q|2\omega)$ and $B^{(I)}(q|2\omega)$ are the Fourier coefficients of the nonlinear current and charge density $A^{(I)}(x_1|2\omega)$ and $B^{(I)}(x_1|2\omega)$. The first term in the nonlinear source, (206), is associated with the prism–metal film interface, and the second is associated with the metal film–vacuum interface.

To solve the linear scattering problem, we seek the solution of the reduced Rayleigh equation, (203), in the form

$$T(q,\omega) = 2H_0 G^\omega(q|k) \frac{i\beta(k,\omega)}{\epsilon(\omega)} t_{01}(k,\omega) e^{-\beta(k,\omega)D}, \quad (207)$$

where $G^\omega(q|k)$ is Green's function associated with the three-layer system with a randomly rough interface between the vacuum and the metal and

$$t_{01}(k,\omega) = \frac{2\alpha_p(k,\omega)\epsilon(\omega)}{\alpha_p(k,\omega)\epsilon(\omega) + i\beta(k,\omega)\epsilon_0}. \tag{208}$$

As for the free metal surface, we define $G^\omega(q|k)$ as the solution of (83) with Green's function $G_0(q,\omega)$ associated with the planar three-layer structure, which has the form

$$G_0(q,\omega) = \frac{ie^{\beta(q,\omega)D}[\alpha_p(q,\omega)\epsilon(\omega) + i\beta(q,\omega)\epsilon_0]}{D(q,\omega)}, \tag{209}$$

where

$$D(q,\omega) = [\alpha_p(q,\omega)\epsilon(\omega) + i\beta(q,\omega)\epsilon_0][\alpha_0(q,\omega)\epsilon(\omega) + i\beta(q,\omega)]e^{\beta(q,\omega)D}$$

$$- [\alpha_p(q,\omega)\epsilon(\omega) - i\beta(q,\omega)\epsilon_0][\alpha_0(q,\omega)\epsilon(\omega) - i\beta(q,\omega)]e^{-\beta(q,\omega)D}. \tag{210}$$

By following every step of the solution of the problem of linear scattering described in Sect. 3.3, we arrive at the expression for the transmission amplitude

$$T(q,\omega) = H_0 T_0(k,\omega)[2\pi\delta(q-k) + G(q,\omega)t_\omega(q|k)], \tag{211}$$

where

$$T_0(k,\omega) = -2i\frac{\alpha_p(k,\omega)}{\epsilon_0(\omega)}t_{01}(k,\omega)e^{-\beta(k,\omega)D}G(k,\omega) \tag{212}$$

is the specular transmission coefficient and the operator $t_\omega(q|k)$ is the solution of (91). The only difference from the solution described in Sect. 3.3 is that the scattering potential $V^{(\omega)}(q|k)$ is given by the function $M_\omega(q|k)$, (204), in which one should replace the functions $I(\gamma|Q)$ by the functions $J(\gamma|Q)$.

The equation for the transmission amplitude at frequency $2\omega$ coincides in its form with the reduced Rayleigh equation for $R(q,2\omega)$ of Sect. 3.3. Therefore, we can immediately write the formal solution,

$$T(q,2\omega) = G(q,2\omega)Q(q|2k) + G(q,2\omega)$$

$$\int_{-\infty}^{\infty}\frac{dp}{2\pi}t_{2\omega}(q|p)G(p,2\omega)Q(p|2k), \tag{213}$$

where

$$Q(q|2k) = -\frac{iN_{2\omega}(q|k)}{\epsilon(2\omega)}. \tag{214}$$

In Fig. 22, we present the mean intensity of the second-harmonic light, calculated by using the small–amplitude perturbation approach, when

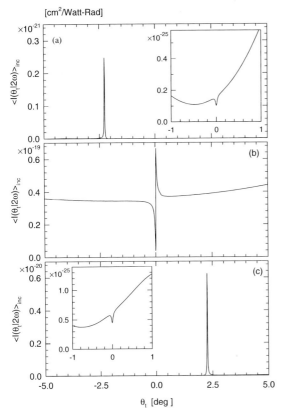

**Fig. 22.** The mean normalized transmitted intensity of the second-harmonic light as a function of the angle of transmission $\theta_t$ for the scattering of p-polarized light through a randomly rough silver film–vacuum interface. The interface roughness is characterized by the Gaussian power spectrum (5) with roughness parameters $\delta = 5\,\text{nm}$ and $a = 113.5\,\text{nm}$. The nonlinear coefficients are given by the free-electron model. The angles of incidence are (**a**) $\theta_0 = 22°$, (**b**) $\theta_0 = 24°$, and (**c**) $\theta_0 = 26°$. The optimal angle for excitation of surface plasmon polaritons at the film–vacuum interface is $24°$

p-polarized light is incident through a prism on a one-dimensional, random silver film–vacuum interface, characterized by the Gaussian power spectrum (5) with an rms height $\delta = 5\,\text{nm}$ and a transverse correlation length $a = 113.5\,\text{nm}$. The mean thickness of the film is $D = 55\,\text{nm}$, and the angles of incidence are $\theta_0 = 22°$ (a), $\theta_0 = 24°$ (b), and $\theta_0 = 26°$ (c). The refractive index of the prism is $n_0 = 2.479$, so that the optimal angle for the excitation of surface plasmon polaritons is $\theta_0 = 24°$. The nonlinear coefficients $\mu_{1,2,3}^{(\mathrm{I,III})}$ were calculated on the basis of the free-electron model [47].

To analyze the main features of the angular dependence of the intensity of the generated light, we need explicit expressions for the different contributions to the mean differential intensity of the harmonic light. The effects of the multiple scattering of surface plasmon polaritons of frequency $\omega$, which are of interest to us here, are contained in the nonlinear driving term $Q(q|2k)$ on the right-hand side of (213). The integral term in (213) describes the multiple scattering of the waves of frequency $2\omega$ and, in particular, contains the effects of the multiple scattering of surface plasmon polaritons of frequency $2\omega$. Since these effects do not include a peak in the direction normal to the mean surface, which is of primary interest to us here, we omit their contribution to the mean differential intensity. Since both contributions to $N_{2\omega}(q|k)$, and consequently to $Q(q|2k)$ contain the product of the transmission amplitudes $T(p,\omega)T(p',\omega)$, by using the expression for the transmission amplitude $T(q,\omega) = \mathcal{T}_0(k,\omega)H_0[2\pi\delta(q-k) + G(q,\omega)t_\omega(q|k)]$ (211) we can subdivide each term in $Q(q|2k)$ into three contributions with different physical meanings. The first contributions, which contain only the product of $\delta$ functions, $(2\pi)^2\delta(p-k)\delta(p'-k)$, describe the nonlinear mixing of the fields of frequency $\omega$ which would be specular if the film–vacuum interface were planar. The part of $Q(q|2k)$ that contains the product of a $\delta$ function and a term with Green's function, e. g., $2\pi\delta(p-k)G(p',2\omega)t_\omega(p'|k)$, describes the interaction of the "specular" and scattered fields, including the nonlinear mixing of the excited surface plasmon polaritons with the incident light. Finally, the part of $Q(q|2k)$ that contains the product of two Green's functions, $G(p,\omega)t_\omega(p|k)G(p',\omega)t_\omega(p'|k)$, describes the nonlinear mixing of the scattered fields and includes the mixing of co- and contrapropagating surface plasmon polaritons. According to the classification we have just given, we separate these three contributions to the nonlinear driving source explicitly, and include all of the roughness-induced corrections to them in the additional source function $\mathcal{Q}_{\mathrm{sc}}(q|2k)$, so that we obtain

$$Q(q,2\omega) = \frac{2ic}{\omega}\mathcal{T}_0^2(k,\omega)H_0^2\left[\Gamma_0(k)2\pi\delta(q-2k)\right.$$

$$+\Gamma_1(q,k)G(q-k,\omega)t_\omega(q-k|k)$$

$$+\int_{-\infty}^{\infty}\frac{dp}{4\pi}\Gamma_2(q,p)G(q-p,\omega)G(p,\omega)t_\omega(q-p|k)t_\omega(p|k)$$

$$\left.+\mathcal{Q}_{\mathrm{sc}}(q|2k)\right]\,,\tag{215}$$

where the nonlinear coefficients $\Gamma_i$ are given by $\Gamma_i = \gamma_i^{(\mathrm{I})} + \gamma_i^{(\mathrm{III})}$, and explicit expressions for $\gamma_i^{(\mathrm{I},\mathrm{III})}$ are presented in the Appendix. The most resonant contributions to the intensity of the harmonic light have the form

$$\langle I(\theta_{\mathrm{t}}|2\omega)\rangle_{\mathrm{res}} = \frac{512\omega^3}{\epsilon_0 c^4} \left| t_{10}(k,\omega)e^{-\beta(k,\omega)D} \right|^2$$

$$\times \cos^2\theta_{\mathrm{s}}\cos^2\theta_0 |G(q,2\omega)|^2 |G(k,\omega)|^4$$

$$\times \left[ |\Gamma_1(q,k)|^2 |G(q-k,\omega)|^2 \tau_\omega(q-k|k) \right.$$

$$+ \int_{-\infty}^{\infty} \frac{\mathrm{d}p}{2\pi} |\Gamma_2(q,p)|^2 |G(q-p,\omega)G(p,\omega)|^2$$

$$\left. \times \tau_\omega(q-p|k)\tau_\omega(p|k) \right], \tag{216}$$

where we have defined $\langle t_\omega(q|k)t_\omega^*(p|k)\rangle = 2\pi\delta(q-p)\tau_\omega(q|k)$, and $\tau_\omega(q|k)$ is the averaged reducible vertex function that can be calculated by using, for example, the pole approximation for Green's function [5].

The presence of Green's functions $|G(q,2\omega)|^2$ and $|G(k,\omega)|^4$ in (216) leads to the amplification of nonlinear processes at the metal film–vacuum interface in the Kretschmann geometry, because Green's function $G(q,2\omega)$ has poles at the wave numbers of the surface plasmon polaritons, $k_{\mathrm{sp}}(2\omega)$ and $G(k,\omega)$ has poles at the wave numbers of the surface plasmon polaritons, $k_{\mathrm{sp}}(\omega)$. Thus, when the angle of incidence is optimal for the excitation of surface plasmon polaritons of frequency $\omega$, the enhancement of the intensity of the second-harmonic light can reach several orders of magnitude compared with that for scattering from a single surface. This is true for both the coherent and incoherent parts of the intensity. When the angle of scattering is optimal for the detection of surface plasmon polaritons of frequency $2\omega$, resonance enhancement of the intensity can also occur. However, this enhancement can be observed only in the intensity of the reflected light.

The second interesting feature of (216) is the term that contains the factor $|G(q-k,\omega)|^2$. This term describes the second-harmonic generation due to nonlinear mixing of the incident light with the surface plasmon polaritons that are excited due to the surface roughness. Its presence leads to the appearance of strong resonant peaks at $q = k \pm k_{\mathrm{sp}}(\omega)$ and $q = k \pm k_{\mathrm{sp}}^{(p)}(\omega)$. This is the origin of the peak of the strong enhancement of second-harmonic generation in reflection from a randomly rough metal surface that was first observed in [48] and analyzed by *Deck* and *Grygier* [45] in the framework of first-order perturbation theory in the small roughness. These peaks of enhanced second-harmonic generation are due to the resonant interaction of the incident light and forward/backward propagating surface plasmon polaritons excited in a single scattering event. No interference effects are involved in the formation of the peaks. The main reason for using the Kretschmann geometry in the experiments [8,9,10,11,12,13] was to excite surface plasmon polaritons associated with the metal–vacuum (metal–nonlinear crystal) interface through the ATR phenomenon. Thus, the angle of incidence was tuned so that $k = k_{\rm sp}(\omega)$. In this case, the peaks due to the poles of $G(q - k, \omega)$ move so that the peak due to the mixing of the incident light with the forward propagating surface plasmon polariton moves into the nonradiative region, and that due to the mixing of the incident light with the backward propagating surface plasmon polariton moves in the direction of the normal to the mean surface, $q = k - k_{\rm sp}(\omega) = 0$. The strength and the shape of the peak is determined by the effective nonlinear coefficient $\Gamma_1(q, k)$. This effective nonlinear coefficient is linear in $q$ for small $q$. This is the result of the symmetry of the surface nonlinear polarization that forbids second-harmonic generation by contrapropagating surface plasmon polaritons [16]. Since the incident light is the surface plasmon polariton propagating in the forward direction, when the angle of incidence is optimal for exciting surface plasmon polaritons of frequency $\omega$, its nonlinear mixing with the scattered surface plasmon polaritons propagating in the backward direction is forbidden by this symmetry. As a result, the resonant peak has an antiresonant shape. The width of the peak is determined by the decay rate of the surface plasmon polaritons on the rough interface $\mathrm{Im}\,[k_{\rm sp}(\omega)] = \Delta_{\rm tot}(\omega) = \Delta_\epsilon(\omega) + \Delta_{\rm sc}(\omega)$, where $\Delta_\epsilon(\omega)$ is the decay rate of the surface plasmon polaritons of frequency $\omega$ due to ohmic losses and $\Delta_{\rm sc}(\omega)$ is their decay rate due to their roughness-induced scattering into other surface plasmon polaritons.

The effects of the multiple scattering of surface plasmon polaritons are contained in the function $\tau(q|k)$. As shown in [5,49], because our scattering system supports two surface plasmon polaritons at frequency $\omega$, the function $\tau(q|k)$ contains a superposition of two Lorentzian peaks centered at $q + k = 0$ (enhanced backscattering) with half widths $\Delta_{\rm tot}(\omega)$ and $\Delta_{\rm tot}^{(p)}(\omega)$. It also contains Lorentzian peaks centered at $q + k \pm [k_{\rm sp}(\omega) - k_{\rm sp}^{(p)}(\omega)] = 0$ (satellite peaks) with a half width $\Delta_{\rm tot}(\omega) + \Delta_{\rm tot}^{(p)}(\omega)$. In these results, $\Delta_{\rm tot}^{(p)}(\omega)$ is the decay rate of the surface plasmon polariton associated with the prism–metal interface. Therefore, $\tau(q - k|k)$ entering the second term in (216) has a peak at $q = 0$. Since this contribution arises due to the nonlinear mixing of the incident and scattered waves, it describes the coherent generation of the second-harmonic light by the incident and backscattered radiation, enhanced due to the multiple scattering of surface plasmon polaritons by the roughness.

The strongest contribution associated with the multiple scattering of surface plasmon polaritons of frequency $\omega$ comes, however, from the second term in (216), which describes the nonlinear mixing of the multiply-scattered surface plasmon polaritons of frequency $\omega$ propagating in opposite directions. In the pole approximation for Green's function [5], this contribution has the form

$$\langle I(\theta_t|2\omega)\rangle_{\text{res}} = \frac{512\omega^3}{\epsilon_0 c^4} \left| t_{10}(k,\omega) e^{-2\beta(k,\omega)D} \right|^2 |G(q,2\omega)|^2 |G(k,\omega)|^2$$

$$\times \left( \frac{C^4(\omega)}{\Delta_{\text{tot}}(\omega)} \frac{1}{q^2 + 4\Delta_{\text{tot}}^2(\omega)} \left\{ |\Gamma_2[q, k_{\text{sp}}(\omega)]|^2 \tau[k_{\text{sp}}(\omega)|k] \tau[q - k_{\text{sp}}(\omega)|k] \right. \right.$$

$$\left. + |\Gamma_2[q, -k_{\text{sp}}(\omega)]|^2 \tau[-k_{\text{sp}}(\omega)|k] \tau[q + k_{\text{sp}}(\omega)|k] \right\}$$

$$+ C_{\text{p}}^4(\omega) \Delta_{\text{tot}}^{(p)}(\omega) \frac{1}{q^2 + 4[\Delta_{\text{tot}}^{(p)}(\omega)]^2}$$

$$\left\{ |\Gamma_2[q, k_{\text{sp}}^{(p)}(\omega)]|^2 \tau[k_{\text{sp}}^{(p)}(\omega)|k] \tau[q - k_{\text{sp}}^{(p)}(\omega)|k] \right.$$

$$\left. \left. + |\Gamma_2[q, -k_{\text{sp}}^{(p)}(\omega)]|^2 \tau[-k_{\text{sp}}^{(p)}(\omega)|k] \tau[q + k_{\text{sp}}^{(p)}(\omega)|k] \right\} \right), \tag{217}$$

where $C(\omega)$ and $C_{\text{p}}(\omega)$ are the residues of Green's function at the poles $\pm k_{\text{sp}}(\omega)$ and $\pm k_{\text{sp}}^{(p)}(\omega)$, respectively. We have not included in (217) the contribution from the possible satellite peaks [49], since their positions are far from the direction normal to the mean surface. As expected, the result given by (217) contains Lorentzian factors centered at $q = 0$. However, the efficiency of the nonlinear mixing of the contrapropagating surface plasmon polaritons is determined by the effective nonlinear coefficients $\Gamma_2[q, \pm k_{\text{sp}}(\omega)]$ and $\Gamma_2[q, \pm k_{\text{sp}}^{(p)}(\omega)]$. Because the nonlinearity of the system is due only to the film interfaces, these effective nonlinear coefficients are linear in $q$ for small $q$. This is the manifestation of the well-known fact that the symmetry of the nonlinear polarization of a metal surface forbids such processes [16]. As a result, the contribution given by (217) displays a dip rather than a peak in the direction normal to the mean surface. The depth of this dip depends strongly on the values of the material parameters and the angle of incidence of the fundamental light.

## 5   Conclusions

We have reviewed the theoretical studies of the generation and scattering of second-harmonic light at randomly rough metallic interfaces. For strongly rough clean metal surfaces, the study was based on a Monte Carlo type computer simulation of the problem, in which the nonlinearities of the surface were modeled using a free-electron model for the polarizability of the metal. We have found that the angular distribution of the scattered light at the harmonic frequency displays well-defined dips in the backscattering direction, and the results are in quantitative agreement with the available experimental data [19]. By solving the scattering equations iteratively, we have shown that the observed features are due to destructive interference between waves that have been multiply scattered in the valleys of the surface.

The perturbative analysis of the linear and nonlinear processes that occur on a rough clean metal surface, presented in this contribution, allowed us to separate different processes of scattering and nonlinear mixing leading to the experimentally observed features and provided a complete explanation of all of the features present in the angular dependence of the second-harmonic generation intensity measured in the experiments of [17] and [18]. The comparison with the experimental results of [18], where no distinct dip or peak was observed, except for a shallow minimum in the direction normal to the mean surface, suggests that an important role in the surface nonlinear polarization is played by the nonlinear constant $\mu_2$. Thus, on the basis of the results of our analysis, we can conclude that the Agranovich and Darmanyan model of the surface nonlinear polarization, rather than the free-electron model, should be used to describe the experimental results. From our results, it also follows that the nonlinear coefficient $\mu_1$, which is the largest nonlinear coefficient in the free-electron model, should be considerably smaller in value, and $\mu_2$ should have a nonzero value. In principle, a careful comparison of theoretical and experimental results gives the possibility of estimating the nonlinear parameters entering the expression for surface nonlinear polarization.

Turning now to second-harmonic generation in the Kretschmann ATR geometry, the theoretical study of this effect was prompted by the desire to resolve the difference between recent experimental studies in which a peak in the angular dependence of the intensity of the transmitted second-harmonic light was observed in the direction normal to the mean scattering surface [10,13] and a study in which no peak in this direction was observed [12] and to provide an explanation for the origin of this peak that refined that given in [10] and [13]. The perturbative calculations described in Sect. 4.3 have shown that for a general angle of incidence, the interference effects in the multiple scattering of surface plasmon polaritons lead, due to the symmetry of the nonlinear polarization, to the appearance of a dip rather than a peak in the direction normal to the mean scattering surface in transmission at the harmonic frequency. However, when the angle of incidence is chosen optimally for the excitation of surface plasmon polaritons at the metal film–vacuum interface, a much stronger mechanism leads to the appearance of a peak in the direction normal to the mean interface in transmission. It is nonlinear mixing of the incident light, which in this case is the resonantly excited surface plasmon polariton, and the backward propagating surface plasmon polariton excited through the surface roughness that form the peak. No interference effects are involved in the formation of the peak. Since this peak already appears in the contribution from single-scattering processes, it is considerably stronger than the weak feature associated with multiple scattering. When the angle of incidence is shifted from the optimal, the peak moves away from the direction normal to the mean interface, so that the weak dip that is due to multiple scattering effects can be observed. However, the efficiency of second-harmonic generation decreases significantly as the angle of incidence departs from optimal.

The computer simulation results for second-harmonic generation in the reflection and transmission of light through a randomly rough metal film in the Kretschmann geometry show that for small angles of incidence, four resonance peaks occur in the angular dependence of the intensity of the second-harmonic light, both in reflection and transmission. These peaks arise from resonance processes in which the incident light, through the surface roughness, excites two surface plasmon polaritons of frequency $\omega$ at the metal–vacuum interface, which propagate in the forward and backward directions and then interact nonlinearly with the incident light to produce peaks at angles of scattering and transmission given by (199) and (200). Not all of these peaks were observed in the results of the small-amplitude perturbation theory calculations of second-harmonic generation on a weakly rough surface of a metal film because the focus of these calculations was the peak in the direction normal to the mean scattering surface. When the angle of incidence is optimal for exciting the surface plasmon polaritons, as in the experimental studies reported in [10,12,13], one of these resonance peaks is in the direction normal to the mean scattering surface, both in reflection and in transmission. These peaks are well-defined. However, their amplitudes are small. A peak is also present in the retroreflection direction, whose amplitude, however, is also small. This is due mainly to the fact that for constructive interference of multiply-scattered optical paths to occur in reflection, each interfering optical path must cross a strongly attenuating film at least twice. In transmission, it must do so only once. No enhanced transmission peak is observed because it would occur in the nonradiative region.

Thus, the results of both the perturbative and computer simulation studies of the angular dependence of the intensity of second-harmonic light generated in the Kretschmann geometry confirm the existence of a peak in the direction normal to the mean surface in reflection and transmission. The existence of a peak in the retroreflection direction is also confirmed.

## Acknowledgments

The work of T. A. L. and A. A. M. was supported in part by Army Research Office Grant DAAD 19-99-1-0321. The work of E.R.M. was supported in part by CONACYT Grant 32254–E.

## Appendix

In this Appendix, we present the explicit expressions for the effective nonlinear coefficients $\gamma_i^{(\mathrm{I,III})}$:

$$\gamma_0^{(\mathrm{I})}(k) = t_{10}(2k, 2\omega)\mathrm{e}^{-\beta(2k,2\omega)D}\left\{\mu_3^{(\mathrm{I})}ik\frac{\alpha_{\mathrm{p}}(2k, 2\omega)}{\epsilon_0}\frac{\beta(k,\omega)}{\epsilon(\omega)}f_+(k)f_-(k)\right.$$

$$\left.-2k\left[\mu_1^{(\mathrm{I})}k^2f_+^2(k) + \mu_2^{(\mathrm{I})}\left(\frac{i\beta(k,\omega)}{\epsilon(\omega)}\right)^2 f_-^2(k)\right]\right\}, \tag{218}$$

$$\gamma_1^{(\mathrm{I})}(q, k) = t_{10}(q, 2\omega)\mathrm{e}^{-\beta(q,2\omega)D}$$

$$\left\{\mu_3^{(\mathrm{I})}i\frac{\alpha_{\mathrm{p}}(q, 2\omega)}{\epsilon_0}\left[k\frac{\beta(q-k,\omega)}{\epsilon(\omega)}f_+(k)f_-(q-k)\right.\right.$$

$$\left.+ (q-k)\frac{\beta(k,\omega)}{\epsilon(\omega)}f_+(q-k)f_-(k)\right]$$

$$-2q\left[\mu_1^{(\mathrm{I})}(q-k)kf_+(q-k)f_+(k)\right.$$

$$\left.\left.-\mu_2^{(\mathrm{I})}\frac{\beta(k,\omega)}{\epsilon(\omega)}\frac{\beta(q-k,\omega)}{\epsilon(\omega)}f_-(q-k)f_+(k)\right]\right\}, \tag{219}$$

$$\gamma_2^{(\mathrm{I})}(q, p) = t_{10}(q, 2\omega)\mathrm{e}^{-\beta(q,2\omega)D}$$

$$\left\{\mu_3^{(\mathrm{I})}\frac{\alpha_{\mathrm{p}}(q, 2\omega)}{\epsilon_0}\left[p\frac{i\beta(q-p,\omega)}{\epsilon(\omega)}f_+(p)f_-(q-p)\right.\right.$$

$$\left.+(q-p)\frac{i\beta(p,\omega)}{\epsilon(\omega)}f_+(q-p)f_-(p)\right]$$

$$-2q\left[\mu_1^{(\mathrm{I})}(q-p)pf_+(q-p)f_+(p)\right.$$

$$\left.\left.-\mu_2^{(p)}\frac{\beta(q-p,\omega)}{\epsilon(\omega)}\frac{\beta(p,\omega)}{\epsilon(\omega)}f_-(q-p)f_+(p)\right]\right\}, \tag{220}$$

with

$$f_\pm(q) =$$

$$\frac{[\alpha_0(q,\omega)\epsilon(\omega) + \mathrm{i}\beta(q,\omega)]\mathrm{e}^{\beta(q,\omega)D} \mp [\alpha_0(q,\omega)\epsilon(\omega) - \mathrm{i}\beta(q,\omega)\mathrm{e}^{\beta(q,\omega)D}]}{2\mathrm{i}\beta(q,\omega)}, \quad (221)$$

and

$$\gamma_0^{(\mathrm{III})}(k) = \mu_3^{(\mathrm{III})}\mathrm{i}k\frac{\beta(2k,2\omega)}{\epsilon(2\omega)}\alpha_0(k,\omega)g_+(2k)$$

$$-2kg_-(2k)\left[\mu_1^{(\mathrm{III})}k^2 + \mu_2^{(\mathrm{III})}\alpha_0^2(k,\omega)\right], \quad (222)$$

$$\gamma_1^{(\mathrm{III})}(q,k) = \mu_3^{(\mathrm{III})}\mathrm{i}\frac{\beta(q,2\omega)}{\epsilon(2\omega)}g_+(q)\left[k\alpha_0(q-k,\omega) + (q-k)\alpha_0(k,\omega)\right]$$

$$-2qg_-(q)\left[\mu_1^{(\mathrm{III})}(q-k)k + \mu_2^{(\mathrm{III})}\alpha_0(k,2\omega)\alpha_2(q-k,\omega)\right], \quad (223)$$

$$\gamma_2^{(\mathrm{III})}(q,p) = \mu_3^{(\mathrm{III})}\mathrm{i}\frac{\beta(q,2\omega)}{\epsilon(2\omega)}g_+(q)\left[p\alpha_0(q-p,\omega) + (q-p)\alpha_0(p,\omega)\right]$$

$$-2qg_-(q)\left[\mu_1^{(\mathrm{III})}(q-p)p + \mu_2^{(\mathrm{III})}\alpha_0(q-p,\omega)\alpha_0(p,\omega)\right], \quad (224)$$

with

$$g_\pm(q) = 1 \pm \frac{\alpha_\mathrm{p}(q,2\omega)\epsilon(2\omega) - \mathrm{i}\beta(q,2\omega)\epsilon_0}{\alpha_\mathrm{p}(q,2\omega)\epsilon(2\omega) + \mathrm{i}\beta(q,2\omega)\epsilon_0}\mathrm{e}^{-2\beta(q,2\omega)D}. \quad (225)$$

# References

1. A. R. Bishop, D. K. Campbell, St. Pnevmatikos (Eds.), *Disorder and Nonlinearity* (Springer, Berlin Heidelberg 1992); A. R. Bishop, S. Jimenez, L. Vazquez (Eds): *Fluctuation Phenomena: Disorder and Nonlinearity* (World Scientific, Singapore 1995)
2. F. Brown, R. E. Parks, A. M. Sleeper, Nonlinear optical reflection from a metallic boundary, Phys. Rev. Lett. **14**, 1029–1031 (1965)
3. S. S. Jha, Nonlinear optical reflection from a metal surface, Phys. Rev. Lett. **15**, 412–414 (1965)
4. N. Bloembergen, R. K. Chang, S. S. Jha, C. H. Lee, Optical second-harmonic generation in reflection from media with inversion symmetry, Phys. Rev. **174**, 813–822 (1968)
5. A. R. McGurn, A. A. Maradudin, V. Celli, Localization effects in the scattering of light from a randomly rough grating, Phys. Rev. B **31**, 4866–4871 (1985)

6. A. R. McGurn, A. A. Maradudin, An analogue of enhanced backscattering in the transmission of light through a thin film with a randomly rough surface, Opt. Commun. **72**, 279–285 (1989)

7. A. R. McGurn, T. A. Leskova, V. M. Agranovich, Weak-localization effects in the generation of second harmonics of light at a randomly rough vacuum-metal grating, Phys. Rev. B **44**, 11441–11456 (1991)

8. X. Wang, J. H. Simon, Directionally scattered optical second-harmonic generation with surface plasmons, Opt. Lett. **16**, 1475–1477 (1991)

9. H. J. Simon, Y. Wang, L. B. Zhou, Z. Chan, Coherent backscattering of optical second-harmonic generation with long-range surface plasmons, Opt. Lett. **17**, 1268–1270 (1992)

10. O. A. Aktsipetrov, V. N. Golovkina, O. I. Kapusta, T. A. Leskova, N. N. Novikova, Anderson localization effects in the second harmonic generation at a weakly rough metal surface, Phys. Lett. A **170**, 231–234 (1992)

11. Y. Wang, H. J. Simon, Coherent backscattering of optical second-harmonic generation in silver films, Phys. Rev. B **47**, 13695–13699 (1993)

12. L. Kuang, H. J. Simon, Diffusely scattered second harmonic generation from a silver film due to surface plasmons, Phys. Lett. A **197**, 257–261 (1995)

13. S. I. Bozhevolnyi, K. Pedersen, Second harmonic generation due to surface plasmon localization, Surf. Sci. **377–379**, 384–387 (1997)

14. E. Kretschmann, The determination of the optical constants of metals by excitation of surface plasmons, Z. Phys. A **241**, 313–324 (1971)

15. D. Sarid, Long-range surface-plasma waves on very thin metal films, Phys. Rev. Lett. **47**, 1927–1930 (1981)

16. M. Fukui, J. E. Sipe, V. C. Y. So, G. I. Stegeman, Nonlinear mixing of oppositely travelling surface plasmons, Solid State Commun. **27**, 1265–1267 (1978)

17. K. A. O'Donnell, R. Torre, C. S. West, Observations of backscattering effects in second-harmonic generation from a weakly rough metal surface, Opt. Lett. **21**, 1738–1740 (1996)

18. K. A. O'Donnell, R. Torre, C. S. West, Observations of second harmonic generation from randomly rough metal surfaces, Phys. Rev. B **55**, 7985–7992 (1997)

19. K. A. O'Donnell, R. Torre, Second-harmonic generation from a strongly rough metal surface, Opt. Commun. **138**, 341–344 (1997)

20. M. Leyva-Lucero, E. R. Méndez, T. A. Leskova, A. A. Maradudin, J. Q. Lu, Multiple-scattering effects in the second-harmonic generation of light in reflection from a randomly rough surface, Opt. Lett. **21**, 1809–1811 (1996)

21. M. Leyva-Lucero, E. R. Méndez, T. A. Leskova, A. A. Maradudin, Destructive interference effects in the second harmonic light generated at randomly rough surfaces, Opt. Commun. **161**, 79–94 (1999)

22. V. M. Agranovich, S. A. Darmanyan, Theory of second-harmonic generation upon reflection of light from a medium with a center of inversion, JETP Lett. **35**, 80–82 (1982)

23. D. Maystre, M. Nevière, R. Reinisch, Nonlinear polarization inside metals: a mathematical study of the free-electron model, Appl. Phys. A **39**, 115–121 (1986)

24. J. L. Coutaz, D. L. Maystre, M. Nevieére, R. Reinisch, Optical second harmonic generation from silver at 1.064 μm pump wavelength, J. Appl. Phys. **62**, 1529–1531 (1987)

25. M. Neviére, P. Vincent, D. Maystre, R. Reinisch, J. L. Coutaz, Differential theory for metallic gratings in nonlinear optics: second harmonic generation, J. Opt. Soc. Am. B **5**, 330–336 (1988)
26. A. Danese, *Advanced Calculus*, (Allyn Bacon, Boston 1965) Vol. I p. 123
27. R. Garcia-Molina, A. A. Maradudin, T. A. Leskova, The impedance boundary condition for a curved surface, Phys. Rep. **194**, 351–359 (1990)
28. R. A. Depine, Improvement of a differential method for metallic diffraction gratings by means of an integral method, J. Mod. Opt. **34**, 1135–1139 (1987)
29. M. E. Knotts, K. A. O'Donnell, Backscattering enhancement from a conducting surface with isotropic roughness, Opt. Commun. **99**, 1–6 (1993)
30. M. E. Knotts, T. R. Michel, K. A. O'Donnell, Comparisons of theory and experiment in light scattering from a random rough surface, J. Opt. Soc. A **10**, 928–941 (1993)
31. E. R. Méndez, A. G. Navarette, R. E. Luna, Statistics of the polarization properties of one-dimensional randomly rough surfaces, J. Opt. Soc. Am. A **12**, 2507–2516 (1995)
32. W. H. Press, S. A. Teukolsky, W. T. Vetterling, B. P. Flannery, *Numerical Recipes in Fortran*, 2nd. ed. (Cambridge University Press, Cambridge 1992) Chap. 2
33. E. R. Méndez, K. A. O'Donnell, Observation of depolarization and backscattering enhancement in light scattering from Gaussian random surfaces, Opt. Commun. **61** 91–95 (1987)
34. K. A. O'Donnell, E. R. Méndez, An experimental study of scattering from characterized random surfaces, J. Opt. Soc. Am. A **4**, 1194–1205 (1987)
35. A. A. Maradudin, T. Michel, A. R. McGurn, E. R. Méndez, Enhanced backscattering of light from a random grating, Ann. Phys. **203**, 255–307 (1990)
36. A. A. Maradudin, E. R. Méndez, T. Michel, Backscattering effects in the elastic scattering of p-polarized light from a large-amplitude random metallic grating, in *Scattering in Volumes and Surfaces*, M. Nieto-Vesperinas, J. C. Dainty (Eds.) (North-Holland, Amsterdam 1990), pp. 157–174
37. E. G. Liszka, J. J. McCoy, Scattering at a rough boundary – extensions of the Kirchhoff approximation, J. Acoust. Soc. Am. **71**, 1093–1100 (1982)
38. A. Sentenac, A. A. Maradudin, A reformulation of the one-dimensional surface field integral equations, Waves Random Media **3**, 343–354 (1993)
39. Lord Rayleigh, *The Theory of Sound* (MacMillan, London, 1895) Vol. II, pp. 89, 297–311
40. R. Petit, M. Cadilhac, Sur la diffraction d'une onde plane par un réseau infiniment conducteur, C. R. Acad. Sci. Paris B **262**, 468–471 (1966)
41. N. R. Hill, V. Celli, Limits of convergence of the Rayleigh method for surface scattering, Phys. Rev. B **17**, 2478–2481 (1978)
42. G. C. Brown, V. Celli, M. Coopersmith, M. Haller, Unitary and reciprocal expansions in the theory of light scattering from a grating, Surf. Sci. **129**, 507–515 (1983)
43. T. A. Leskova, A. A. Maradudin, M. Leyva-Lucero, E. R. Méndez, Multiple-scattering effects in the second-harmonic generation of light in reflection from a randomly rough surface, to be published
44. R. Kubo, Generalized cumulant expansion method, J. Phys. Soc. Jpn. **17**, 1100–1120 (1962)
45. R. T. Deck, R. K. Grygier, Surface–plasmon enhanced harmonic generation at a rough metal surface, Appl. Opt. **23**, 3202–3213 (1984)

46. T. A. Leskova, M. Leyva-Lucero, E. R. Méndez, A. A. Maradudin, I. V. Novikov, The surface enhanced second harmonic generation of light from a randomly rough metal surface in the Kretschmann geometry, Opt. Commun. **183**, 529–545 (2000)
47. I. V. Novikov, Optical interactions at rough surfaces, Ph.D. Thesis, University of California, Irvine (1998)
48. C. K. Chen, A. R. B. de Castro, Y. R. Shen, Surface–enhanced second-harmonic generation, Phys. Rev. Lett. **46**, 145–148 (1981)
49. See, for example, V. Freilikher, E. Kanzieper, A. A. Maradudin, Coherent scattering enhancement in systems bounded by rough surfaces, Phys. Rep. **288**, 127–204 (1997)

# Index

# Topics in Applied Physics

Printing (Computer to Film): Saladruck Berlin
Binding: Stürtz AG, Würzburg